Department of the Environment

Building Research Establishment
Princes Risborough Laboratory

Handbook of Hardwoods
2nd Edition
Revised by R. H. Farmer B.A., D.Sc.(Tech), F.R.I.C., F.I.W.Sc.

London: Her Majesty's Stationery Office

© Crown copyright 1972

First Published 1972
Second Impression 1975

HER MAJESTY'S STATIONERY OFFICE

Government Bookshops

49 High Holborn, London WC1V 6HB
13a Castle Street, Edinburgh EA2 3AR
41 The Hayes, Cardiff CF1 1JW
Brazennose Street, Manchester M60 8AS
Southey House, Wine Street, Bristol BS1 2BQ
258 Broad Street, Birmingham B1 2HE
80 Chichester Street, Belfast BT1 4JY

Government publications are also available
through booksellers

ISBN 0 11 470541 0

Printed in England for Her Majesty's Stationery Office
by Ebenezer Baylis & Son Ltd., The Trinity Press, Worcester, and London

Preface

The *Handbook of Hardwoods* was first published as a single volume comprising both home-grown and imported timbers in 1956, and since that time has only been subjected to minor revisions. A full revision of the book has now been undertaken and the opportunity has been taken in the present edition to include descriptions of a number of timbers that are comparatively new to the United Kingdom market. At the same time some timbers that were described in the earlier book and appear now to be of little interest have been omitted. In all, the book provides full descriptions of 117 hardwoods and a further 103 are described more briefly.

In the descriptions of the timbers, which have been completely re-written, a uniform arrangement of headings and sub-headings has been adopted for easy reference. The information on the timbers has been brought up to date and additional data, e.g. on strength properties, working properties and plywood manufacture, have been included where they are available. All numerical data are given in both metric and imperial units. Although much of the information is derived from tests carried out at the Princes Risborough Laboratory, data from other reliable sources have also been freely used.

J. B. Dick
Director Building Research Establishment
1972

Building Research Establishment
Princes Risborough Laboratory
Princes Risborough
Aylesbury, Buckinghamshire

Contents

	Page
INTRODUCTION	1
The Weights of Timbers	1
Strength Properties	1
Shrinkage and Movement	2
Working Properties	3
Wood Bending Properties	5
Veneer and Plywood	5
Defects Caused by Wood-Boring Insects	6
Resistance to Marine Borers	7
Natural Durability	8
Amenability to Preservative Treatment	9
Uses	10
THE TIMBERS	11
Appendix I Properties of Hardwoods	214
Appendix II Types of Saws	226
Appendix III Kiln Schedules	229
INDEX OF BOTANICAL NAMES	233
INDEX OF TRADE AND LOCAL NAMES	237

Introduction

This book aims to provide information that will assist users of hardwoods to select the timbers best suited to their purposes and to process them in the most satisfactory manner, having regard to the individual features of each timber. The first edition of the book has been completely re-written and additional information on some properties of the timbers has been included. For easy reference a uniform arrangement of headings and sub-headings has been employed in the descriptions of all the more important timbers. Those of lesser importance are described more briefly, with a smaller number of headings.

Information on the trees, their size and distribution, is based largely on published literature. Much of the technical information on the timbers is derived from tests carried out at the Princes Risborough Laboratory, or at similar research institutions. Tests on timber are necessarily carried out on a limited amount of material. Although care is taken to obtain as nearly as possible representative material for test, it should be understood that the properties of any timber are liable to considerable variation. This must be taken into consideration when putting to practical use information on the properties and processing of the timbers, particularly where numerical data are concerned.

Information on uses has been obtained from various sources and from general experience. No attempt has been made to list in detail all the uses to which a timber may be put, but an indication is given of the features of the timber affecting its utilisation and of the types of use to which it is suited.

Attention is drawn to the Table in Appendix I, which summarises in general terms the more important properties of most of the timbers described. This is intended only as a general guide: for more detailed information the full descriptions of the timbers should be consulted.

Throughout the book numerical data are quoted in metric (SI) units, followed by the British units in parentheses. The following factors may be used to convert British units into SI units:

1 in = 25·4 mm
1 ft = 0·305 m
1 lb = 0·454 kg
1 lb/ft³ = 16·0 kg/m³
1 lbf/in² = 0·00690 N/mm²

The Weights of Timbers

The weight of a piece of wood clearly varies with the amount of water that it contains. For this reason it is important that, when the weight, or more strictly the density, of a timber is quoted, the moisture content at which the weight was determined should be stated. In the descriptions of the timbers the weight when dried refers to a moisture content of 12 per cent. The weight of a timber at any other moisture content within the range of, say, 5 to 25 per cent can be estimated with fair accuracy by adding or subtracting 0·5 per cent of the given weight for each 1 per cent moisture content above or below 12 per cent.

In many cases information on average weights of timbers is taken from published literature and the moisture content of the wood to which the weight relates is not stated. In such cases the term 'seasoned' is used in the present book and indicates that the precise moisture content at which the weight was determined is not known. In general, however, the use of this term should be avoided where possible.

In all species a considerable variation in weight is found to occur, apart from that arising from differences in moisture content. The average weights given in the text are only approximate and where sufficient information is available to justify quoting a range, this has been done.

Strength Properties

Timber, like all other materials of construction, has the ability to resist applied or external forces. This resistance or strength involves a number of specific mechanical properties, and it is largely these that determine the suitability of different species of timber for the various purposes for which they are used. Many other factors have to be considered as well in the selection of a species for a particular purpose, but in general there are few instances where the choice does not depend to some degree upon one or more of its mechanical properties. A basic knowledge of the strength properties of a timber is therefore essential if it is to be used efficiently.

Various factors, such as moisture content, density and temperature affect the values obtained for the strength of timber, and in order to achieve comparable results a standard test procedure must be adopted. The strength properties quoted for the individual species have been derived from tests on small clear specimens – i.e. specimens free from all defects, 20 × 20 mm square in cross section. The specimens were tested in accordance with the British Standard No 373:1957.

Average values are quoted for selected strength properties and provide the basis for a general comparison of one timber with another; however, it must be remembered that individual pieces of a timber can differ appreciably from the average figures quoted.

The wood density is a major factor determining the strength of a species; in general, species with a high density have high strength properties and vice versa. Even so, it is not practical to select a species for a particular purpose on density alone, because there may be marked differences in certain specific properties between timbers of the same density. Thus European ash, having the same density and bending strength as European beech, has an energy absorbing capacity approximately 50 per cent higher than that of beech, by virtue of which it has acquired its special reputation in the manufacture of sports goods, particularly hockey sticks, and also in the manufacture of tool handles. Both uses require a high degree of 'toughness'.

When comparing the strength of various species of timber it is essential that the moisture content be known. Above the fibre saturation point (25 to 30 per cent) changes in moisture content have no apparent effect on the strength of wood. Below the fibre saturation point, most strength properties increase with decrease in moisture content, although not always to the same extent; the properties of resistance to suddenly applied loads and toughness of some species may show slight decreases as a result of drying.

Average values are given in the text for bending strength (modulus of rupture), stiffness (modulus of elasticity) and compression parallel to the grain (maximum compression strength). In Appendix I, these three properties, with the addition of impact bending (resistance to suddenly applied loads), are divided into four broad groups, which in many instances provide all that is required in assessing the suitability of a timber for a particular purpose or for comparing one timber with another. Brief notes on the strength properties are given below as a guide to their significance.

1 Maximum Bending Strength (Modulus of Rupture)

This property is a measure of the ultimate bending strength of timber and the values given apply only to the size of specimen and loading conditions employed in the test.

2 Stiffness

This property is of importance in determining the deflection of a beam under load – the greater the stiffness, the less the deflection. It is usually considered in conjunction with bending strength, as for many uses stiffness is the controlling factor in the design.

3 Compression Parallel to Grain (Maximum Compression Strength)

This property measures the ability of a timber to withstand loads when applied on the end grain. It is of importance where use as short columns or props is contemplated. When determining the strength of a long column or strut, as distinct from a short column, the critical property is the stiffness (above) of the material.

4 Impact Bending (Resistance to Suddenly Applied Loads)

This is one measure of toughness of timber. The test procedure is to drop a constant weight from increasing heights on to a beam supported near the ends; the resistance of a timber to a suddenly applied load is indicated by the height of the drop causing fracture of the beam.

Shrinkage and Movement

Shrinkage measurements were obtained from the changes in width of plain-sawn (tangential) and quarter-sawn (radial) boards when kiln dried from the green state to a moisture content of 12 per cent, and are expressed as percentages of the green dimension. They are also given as inches per foot to the nearest one-sixteenth inch.

The term 'movement' is used in referring to the dimensional changes that take place when timber which has been dried is subjected to changes in atmospheric conditions. To determine the movement values quoted, test samples, after being kiln dried to 12 per cent moisture content, were conditioned first in air at 90 per cent relative humidity, and then in air at 60 per cent relative humidity, the temperature being 25°C (77°F) in both cases. The moisture content values of the samples when in equilibrium at the two humidities are given, together with the movements corresponding to the particular moisture content range.

It is necessary to stress here that shrinkage and movement are not directly related one to the other. For example, it is possible that a wood may shrink quite appreciably in drying from the green to 12 per cent moisture content, yet it may undergo comparatively small dimensional changes when subjected to a given range of atmospheric conditions in service. The reason is that the so-called 'fibre saturation point', or the moisture content value at which appreciable shrinkage begins to take place, varies between different species. Furthermore, the moisture content change of one timber corresponding to any given range of atmospheric conditions often differs considerably from that of another (see afzelia and gedu nohor).

Shrinkage values therefore are useful only in estimating roughly the dimensional allowances necessary in converting green material. It must be pointed out that a further allowance must also be added for possible losses owing to distortion.

Movement values, on the other hand, give some indication of how the dried timber will tend to behave when subjected to atmospheric changes in service. A so-called stable timber is one that exhibits comparatively small dimensional changes in passing from the 90 per cent to the 60 per cent relative humidity conditions, together with small distortional propensities.

Details of the kiln schedules recommended for drying the various timbers are given in Appendix III.

Working Properties

The notes on woodworking properties of timbers are based on standard tests carried out at the Princes Risborough Laboratory, on other published information and, where possible, on experience with the timber in industry. Standard tests have not been carried out on some species and for these the assessment is made using the other two sources of information.

The standard tests are carried out on material kiln dried to between 10 and 14 per cent moisture content. The increased brittleness of material of low moisture content assists the planing of timber with wavy or interlocked grain, as the chips break more easily and such picking up as occurs is shallower than it is at higher moisture contents. In the case of denser timbers, drier material will have a greater resistance to cutting, increased dulling effect on tools, and a greater tendency to produce tool vibration. In most cases green or partially dried timber saws more readily than dry material and factors such as excessive blunting will be less evident.

All tests are carried out on commercial machines, maintained free from excessive wear and using standard tools operating at the settings and speeds recommended by the machine manufacturer. The standard test procedure includes an examination of the following factors:

Blunting
The blunting of cutting edges on saws and cutters is classified as slight, moderate or severe. The rate of wear is a function of timber density and in some species is affected by the presence of silica or other abrasive substances. Blunting can vary considerably within a species. Severe blunting may occasionally occur with a timber classed as moderate and some parcels of timber classed as severe are found to machine more readily than the classification implies.

Sawing
Three types of sawing operation are recognised, rip-sawing, cross-cutting and band-sawing.

The rip-sawing tests have been carried out using the spring set circular saws listed in Appendix II, Table 1, and the most suitable type in terms of the number of teeth and the hook angle determined from an examination of the sawn surface, the behaviour of the blade and the feeding force required. It is recognised that in the United Kingdom users will be changing to the rip-saws specified in B.S. 411:1969 (see Appendix II, Table 2) and recommendations are therefore made from this list. The types of saw listed in Appendix II, Table 3, are used for cross-cutting. A satisfactory classification is given when all three saws are suitable.

For both rip-sawing and cross-cutting tungsten carbide tipped saws are being used more extensively, particularly for cutting the denser or more abrasive timbers, and for other timbers when long life between sharpenings is a requirement.

The band-sawing tests have been confined to the use of the narrow blade specified in Appendix II, Table 4. No tests have been made using a wide band-saw, and the type of blade suggested for each species is based on the classification of B.S. 4411:1969 (see Appendix II, Table 5). It should be noted that the density classification is applicable to most species but for timber containing silica or other abrasive substances special saws and tooth shapes may be required.

Machining

The two most important factors in planing are feed speed and cutting angle (rake). Feed speed is independent of the species and is chosen in relation to the number of cutters on and the rotational speed of the block, so that the pitch of the cutter marks on the planed surface is suitable for the end use of the material. When producing a fine pitch it is important to ensure that the bite is sufficient for the cutter to cut rather than scrape. The latter condition causes rapid tool wear.

The cutting angle is determined by the species being planed and three angles are used: 30° (covering the range 30° to 35°), 20° and 15°. The largest angle is suitable for most species, but when interlocked grain is present a reduced angle of 20° or even 15° may be necessary to avoid picking up of the surface. Most planers are supplied with blocks giving a cutting angle of between 30° and 35° but special blocks with a reduced cutting angle can be supplied, or when using a standard block a smaller angle can be obtained by grinding or honing a face bevel on the cutter. A satisfactory planing classification indicates that an acceptable surface is obtained with the 30° to 35° cutting angle.

Some low density timbers give a woolly surface when planed with the normal 30° cutting angle, and this is noted against the relevant species. The occurrence of this defect can only be minimised by ensuring that the cutting edges are maintained in a sharp condition and this can be achieved more easily if a reduced sharpness angle is used. The sharpness angle should not be reduced below 30°, as otherwise rapid wear of the cutting edges will occur. In addition, a reduced angle should not be used with those species which contain hard knots, as this will result in the cutting edge quickly becoming damaged.

High-speed steel cutters are suitable for most species, but for those timbers whose blunting is classed as 'severe', and also when long runs between sharpening are required on 'moderate' timbers, cutters tipped with tungsten carbide are preferred.

In addition to planing, the standard tests include moulding with a French head, collars, and square block, recessing with a single wing tool and mortising with a chain and hollow square chisels. Boring tests are also made using two- and three-wing, straight and twisted fluted bits. If these operations are carried out without difficulty, then the species is classed as being satisfactory. When, on the other hand, difficulty is experienced, for example arrises may tend to break out, drills and recessing bits may char the wood or excessive force may be required during mortising with the hollow chisel, then these are noted.

Nailing

The nailing characteristics of the species are determined by hand driving 12 gauge wire nails into the face of 13 mm ($\frac{1}{2}$ inch) thick boards, 13 mm from the cross-cut end and from each edge. If the wood splits to the cross-cut end and beyond the nail in the opposite direction then this is noted.

Gluing

The notes refer to timber in the solid form. Where difficulties are encountered with veneers they are noted under plywood manufacture. It is assumed that good gluing practice will be used, i.e. that the wood will be at the correct moisture content, the surfaces will be good and clean, the glue spread will be uniform and adequate and the pressure and temperature will be controlled. The gluing characteristics of a timber are stated as: good, variable, difficult.

Wood Bending Properties

In assessing the bending properties of a timber, the most important factor is the minimum radius of curvature at which a reasonable percentage of faultless bends can be made for a given thickness of clear material. This radius varies according to whether the timber is bent with or without a supporting strap after a suitable softening treatment, or is bent cold in the form of thin laminations at a moisture content of about 12 per cent. The ratio radius/thickness at which breakages during bending do not exceed 5 per cent is determined by tests.

Other assessments are obtained from observations of the end pressures, bending moments and general behaviour of selected pieces bent to a standard radius of curvature.

Classification of timbers according to their steam bending properties is, however, based mainly on the minimum bending radius of sound, clear specimens, 25 mm (1 inch) thick, at a moisture content of about 25 per cent. The specimens are subjected to saturated steam at atmospheric pressure for a period of not less than 45 minutes before bending. The following classification of the steam bending properties of woods has been adopted by the Laboratory and, for purposes of comparison, it may be noted that with a very good bending species such as home-grown beech (*Fagus sylvatica*), specimens 25 mm (1 inch) thick may safely be bent to a radius as small as 38 mm (1·5 inch), or a specimen 50 mm (2 inch) thick to a radius of about 75 mm (3·0 inch). It should be noted that with some species sufficient material was not available for comprehensive tests.

Radius of curvature at which breakages during bending should not exceed 5%		Classification of bending properties (material supported, and 25 mm thick)
mm	in	
Less than 150	Less than 6	Very good
150–250	6–10	Good
260–500	11–20	Moderate
510–750	21–30	Poor
Exceeding 750	Exceeding 30	Very poor

Bends of smaller radii than those given may be obtained and utilised if, for example, certain bending defects that may then occur can be removed in the final machining and finishing operations. On the other hand it must be stressed that faultless bends of the radii listed for the various species can be produced only by using selected material and efficient bending methods.

Veneer and Plywood

Figured veneers for decorative purposes are usually cut from selected flitches in a slicer, a process that is feasible with almost all species. For plywood manufacture, however, veneer must be produced by rotary cutting in a lathe and a continuous supply of suitable logs is essential. Plywood is made for purposes ranging from the technical to the aesthetic. For the former, strength and durability are of more importance than appearance. For the latter, veneers of high quality and uniformity of colour are used for faces and, if defects are present, may also be used for the cores. Some species, although unpopular for faces, may yield admirable core veneers.

Some timbers are unsuitable even for the core of plywood because, for example, of their difficult drying characteristics or the tendency of the finished board to warp. Some of these, however, may prove acceptable for plywood intended for sale in a local market. Technically, there are few timbers which, because of weight alone, are unsuitable for plywood manufacture but in practice those outside the range 400 to 720 kg/m³ (25 to 45 lb/ft³) are seldom employed.

The use of a timber for plywood manufacture has been considered and described under the following basic categories:

1 It is known to be used commercially; its availability in the United Kingdom is also noted.
2 Its use is confined to local markets.
3 It has been found by test to be suitable, though commercial evaluation may be lacking.
4 It has been found by test, or is generally considered, to be unsuitable because of, e.g. weight, irregular colour, difficulty in gluing, liability to distort or split badly, etc.

5 If no information is available this fact is stated.

The propensity of plywood to split after two years' exposure to the weather has been determined for a number of species on panels with 1·5 mm face plies and is recorded as one of two grades, defined as follows:

Grade I   Splitting not readily visible to the naked eye by reason of its fineness or of the masking effect of the structure of the wood.
Grade II   Splitting distinctly visible to the naked eye.

Defects Caused by Wood-Boring Insects

All timbers are at some stage in their life susceptible to attack by one species or other of wood-boring insects, but in practice only a small proportion become infested. Infestation may take place in standing trees, in woodwork which has been in service for many years, or at any intermediate stage depending on the species of insect and the species and condition of the wood. Infestation in the tree or log usually dies out as the timber dries after felling but evidence of attack remains permanently in the dry wood as wormholes. Some species of wood can be attacked by a greater variety of insects than others, and with few exceptions the sapwood is more susceptible than the heartwood which can be immune to most wood-boring insects.

The Princes Risborough Laboratory issues a series of leaflets, papers and technical notes on the various classes of wood-boring insects. They are dealt with comprehensively in *Insect and Marine Borer Damage to Timber and Woodwork* by J. D. Bletchly (HMSO).

In hardwoods the following are the most common types of wood-boring insects causing damage to the wood.

Ambrosia beetles (Platypodidae and Scolytidae)
The defects caused by these insects may occur in heartwood as well as in sapwood. They take the form of numerous circular holes and short tunnels of 1 to 3 mm diameter according to the species of insect responsible, and are often associated with a dark discoloration of the surrounding wood. The full extent of the damage is not usually evident until the logs are converted. The infestation becomes extinct when the timber is converted and dried and cannot recur or spread to other converted timber. Since the holes are seldom sufficiently large or numerous to cause any significant weakness the timber can safely be used where its appearance is of no consequence. This defect may occur in the standing tree, but much more commonly in the freshly felled log. The frequency of occurrence depends upon the time that elapses between felling and conversion. It is more common in timbers from tropical countries but since almost every species of wood is liable to attack, mention is made in the text only of those timbers where damage has been found to be particularly troublesome.

Longhorn beetles (Cerambycidae) and jewel beetles (Buprestidae)
There are many species in these families widely distributed in tropical and temperate forests. Most of them attack only trees and logs and on conversion and drying the infestation dies out, leaving large oval-shaped tunnels packed with bore dust. The damage is seldom of structural significance and can easily be eliminated during manufacture.

Powder-post beetles (Lyctidae and Bostrychidae)
These insects affect only the sapwood, and attack most of the larger-pored hardwood species. Infestation can take place at any time during the first few years after drying, but is usually initiated in the sawmill or factory, and the wood may be

reduced to powder. Bostrychid damage occurs in the tropics and dies out when the timber over-winters in this country. Such wood will, however, be susceptible to indigenous *Lyctus* beetles. The susceptibility of timber to this type of insect attack is governed primarily by the starch content, which may be affected by methods of storing logs or drying after conversion.

Furniture beetles (Anobiidae)

These insects, which include the common furniture beetle and the death watch beetle, are widespread in most temperate countries and are troublesome in furniture, panelling and structural timbers. Wickerwork, and the older types of plywood, are highly susceptible to attack by the common furniture beetle. Sapwood is preferred but heartwood is not immune, especially if slight fungal decay is present. With few exceptions, tropical hardwoods appear to be immune from attack by the common furniture beetle.

The furniture beetle *Ptilinus pectinicornis* causes damage very similar to that of the common furniture beetle in a small number of fine-grained temperate hardwoods.

Termites (Order Isoptera)

Information on the resistance of hardwoods to attack by termites in tropical and sub-tropical countries is incomplete and data are often conflicting. Some timbers achieve a local reputation for immunity but the resistance of any particular species varies according to local conditions and the kind of termites involved, e.g. sub-terranean termites nesting in the ground or dry wood termites which attack timber direct and maintain no ground contact. Only a few species of timber have established any reputation for world-wide resistance. In the sections on individual timbers mention is made of the results of the tests that are at present available.

Resistance to Marine Borers

Timber used in sea or brackish waters is subject to attack by marine-boring animals such as *Teredo* spp. (shipworm), *Limnoria* (gribble), etc. Marine borers are widely distributed, but they are particularly destructive in tropical waters. Around the coast of Great Britain *Limnoria* appears to be generally active and although *Teredo* attack is spasmodic it is always liable to occur.

Most timbers have not sufficient resistance to marine borers to be used untreated in waters where they are liable to be attacked; and as conditions vary from place to place, before a timber is selected for use untreated for marine work in any part of the world, advice should be sought from the nearest Forest Products Laboratory about the degree of hazard that may exist in the particular waters concerned. Because of the variation in conditions no mention is made in the descriptions of the timbers of this property of resistance to marine borer attack.

The timbers listed below, some of which are described fully in the text, are generally recognised as being resistant. None of them, however, is immune from attack.

Afrormosia	(*Pericopsis elata*)
African padauk	(*Pterocarpus soyauxii*)
Andaman padauk	(*Pterocarpus dalbergioidies*)
Basralocus	(*Dicorynia guianensis*)*
Belian	(*Eusideroxylon zwageri*)*
Brush box	(*Tristania conferta*)*
Ekki	(*Lophira alata*)*
Greenheart	(*Ocotea rodiaei*)*
Iroko	(*Chlorophora excelsa*)
Ironbark	(*Eucalyptus* spp.)
Jarrah	(*Eucalyptus marginata*)
Kapur	(*Dryobalanops* spp.)
Manbarklak	(*Eschweilera subglandulosa*)*
Muninga	(*Pterocarpus angolensis*)
Okan	(*Cylicodiscus gabunensis*)*
Opepe	(*Nauclea diderrichii*)*
Pyinkado	(*Xylia dolabriformis*)*
Red louro	(*Ocotea rubra*)*
Southern blue gum	(*Eucalyptus globulus*)
Teak	(*Tectona grandis*)
Turpentine	(*Syncarpia laurifolia*)*

Those timbers marked with an asterisk* are in general believed to be the best for marine work. All sapwood should be excluded from timbers that are to be used untreated, and with some timbers, notably greenheart, it may be necessary to use only large logs and to reject those of smaller diameter which contain a large proportion of sapwood.

For marine piling and wharf construction it is advisable, where possible, to use timber that has been impregnated under pressure with preservatives. Suitable preservatives are coal tar creosote, creosote–coal tar solutions, and copper/chrome/arsenic water-borne mixtures. Long experience has shown that when timber is thoroughly impregnated with a high loading of one of these preservatives it will last for a very long time. For this reason it is best to choose a timber which is easy to treat with preservatives.

Natural Durability

In this country the term natural durability generally refers to the resistance of a timber to fungal decay in service and it is used in this sense in this note. The natural durability of a timber is of importance in situations where the timber is liable to become damp*, for example, where it is used outdoors or in certain indoor situations where there is a risk of moisture penetration or condensation. When timber is used in situations where it can always be kept dry natural durability is unimportant for there is then no risk of decay.

The decay resistance of most timbers varies a great deal and even pieces cut from the same tree will often show wide differences, so it is only possible to speak of durability in approximate terms. For this reason, timbers have been classified into five broad grades based on the performance of their heartwood in ground contact. The five grades are as follows:

Grade of durability	Approximate life in contact with the ground (years)
Very durable	More than 25
Durable	15–25
Moderately durable	10–15
Non-durable	5–10
Perishable	Less than 5

All available information has been used in classifying the timbers, including laboratory tests and field tests made both in this country and abroad and also records of performance in actual use. Where there are insufficient data to enable a firmer assessment to be made, a timber has been given a classification marked 'Provisional'.

The classification is primarily a relative one but, from the results of field tests carried out by the Princes Risborough Laboratory since 1932, it has been possible to give some quantitative meaning to each grade in the Table. The average life stated for each grade relates to material of 50 × 50 mm (2 × 2 inch) section. Larger sizes will, of course, last longer, and in general the increase will be in direct proportion to the thickness (least dimension) and not the cross-sectional area. This applies irrespective of whether the timber is hardwood or softwood, perishable or durable. For example, a 100 × 100 mm (4 × 4 inch) stake will last only about twice as long as a 50 × 50 mm (2 × 2 inch) one and so on.

Timber used externally, but not in contact with the ground, will generally have a much longer life than that indicated by its classification.

The durability grade given refers to heartwood, because the sapwood of almost all timbers is either perishable or non-durable. It is essential to remember this when dealing with timbers which sometimes contain a high proportion of sapwood, as this may give them a shorter life than their classification suggests.

---

* Above about 20 per cent moisture content.

When choosing a timber for a structure which is to be exposed to conditions conducive to decay, it is of the utmost importance to decide, before the timber is acquired, how the required durability is to be obtained. Usually there is a choice between using a naturally durable wood and a less durable one treated with a preservative. If a less durable wood is chosen it is desirable to select one which can be easily treated. This is particularly important where a very long life is required. Where a timber has to be selected for other properties or where the choice is limited by price or availability, the question of durability is often neglected and there is then a risk that the timber chosen may not be either durable enough in its natural state nor permeable enough to permit adequate preservative treatment.

Amenability to Preservative Treatment

The ease with which a timber can be impregnated with preservatives is important when it is to be used under conditions favourable to decay or to attack by insects or marine borers. Examples of such uses in which this property is particularly important are fence posts, marine piling, poles and railway sleepers. Only a few timbers are durable enough in their natural state to give long service when used in situations where they are liable to become damp, and it is now common practice to employ preserved timber, which is not only often cheaper but, if the preservative treatment is correctly carried out, will usually outlast a durable wood.

It is not possible to preserve all timbers equally well. Some are virtually impenetrable and cannot be given a satisfactory treatment, whereas others are permeable and can be heavily impregnated with preservatives. Where a long life is required under exposed conditions it is essential therefore to choose a timber which can be well impregnated. In this connection it is important to remember that the sapwood of a timber, although nearly always perishable, is usually more permeable than the heartwood. Consequently, round timbers containing an outer band of sapwood can generally be impregnated much more satisfactorily, and can be made to last longer than sawn material of the same species.

As heartwood and sapwood usually differ greatly in permeability, their amenability to treatment must be considered separately. Many hardwoods contain only a small amount of sapwood, but there are some which have a relatively wide band of sapwood. It is necessary to have information on the proportion of sapwood present when advising on the best treatment of these timbers.

The terms used in this Handbook to describe the extent to which a timber can be impregnated under pressure with preservatives have been arbitrarily defined from the results of standardised tests using a standard grade of creosote as the preservative material. Other liquids, for example water-borne and organic solvent type preservatives, penetrate slightly differently, but the resistance of a timber to impregnation under pressure will generally be of the same order whatever type of preservative is used.

The terms used are as follows:

Permeable

> These timbers can be penetrated completely under pressure without difficulty, and can usually be heavily impregnated by the hot and cold open tank process.

Moderately resistant

> These timbers are fairly easy to treat, and it is usually possible to obtain a lateral penetration of the order of 6 to 18 mm ($\frac{1}{4}$ to $\frac{3}{4}$ inch) in about 2 to 3 hours under pressure, or the penetration of a large proportion of the vessels.

Resistant

> These timbers are difficult to impregnate under pressure and require a long period of treatment. It is often very difficult to penetrate them laterally more than about 3 to 6 mm ($\frac{1}{8}$ to $\frac{1}{4}$ inch).

Extremely resistant

> These timbers absorb only a small amount of preservative even under long pressure treatments. They cannot be penetrated to an appreciable depth laterally and only to a very small extent longitudinally.

Uses Under this heading an attempt is made to relate the technical and aesthetic properties of the timber to the types of use to which it is suited. Typical uses for each timber are mentioned, with special reference to the wood-using industries of the United Kingdom. The list of uses is in no way exhaustive; it is intended to indicate the kind of work for which each timber is suited.

The Timbers

ABURA
Mitragyna ciliata

other names: bahia (France and Ivory Coast); subaha (Ghana).

THE TREE   Reaches a height of over 30 m (100 ft). Clear bole 18 m (60 ft) or more. Diameter 1·0–1·5 m (3–5 ft). Bole free from buttresses. Mainly West Africa.

THE TIMBER
properties

colour: Uniformly light yellowish-brown or pinkish-brown. Plain appearance. Occasional streaks or veins of darker colour.

sapwood: Not usually differentiated from heartwood, but presence of small, irregular, greyish-brown heart can, in square-edged stock, give appearance of stained sapwood.

grain: Moderately straight, sometimes interlocked. Texture moderately fine and very even.

weight: Variable, 460–690 kg/m³ (29–43 lb/ft³), averaging about 560 kg/m³ (35 lb/ft³) at 12 per cent moisture content.

strength: Somewhat lower than European beech.

Moisture Content	Bending Strength		Modulus of Elasticity		Compression parallel to grain	
	N/mm²	lbf/in²	N/mm²	1000 lbf/in²	N/mm²	lbf/in²
Green	54	7900	8100	1180	27·3	3960
12 per cent	83	12 100	9300	1350	46·5	6740

movement: Small.
Moisture content in 90 per cent relative humidity 18 per cent
Moisture content in 60 per cent relative humidity 12·5 per cent
Corresponding tangential movement 1·7 per cent ($\frac{13}{64}$ in/ft)
Corresponding radial movement   1·0 per cent ($\frac{1}{8}$ in/ft)

processing

drying: Dries rapidly and very well. No degrade likely to occur during kiln treatment provided that all shakes are cut out at conversion stage. Kiln Schedule K.
Shrinkage: Green to 12 per cent moisture content:
          Tangential   about 6·5 per cent ($\frac{13}{16}$ in/ft)
          Radial   about 3·5 per cent ($\frac{7}{16}$ in/ft)

working properties:
Blunting: Variable from slight to severe. Abrasive material cannot be detected visually and use of tungsten-carbide tipped tools is therefore an advantage.
Sawing: Rip sawing – Saw type HR54 (HR40 or TC for abrasive material).
Cross cutting – Satisfactory.
Narrow bandsawing – Satisfactory.
Wide bandsawing – Satisfactory, saw type B.
Machining: Satisfactory, but cutting edges must be kept **sharp** to prevent woolly surface.
Nailing: Satisfactory only with thin gauge nails unless pre-bored.
Gluing: Good.

wood bending:
Cannot be steam-bent appreciably without buckling.
Classification – Very poor.
Limiting radius for 3·2 mm ($\frac{1}{8}$ in) laminae (unsteamed): 152 mm (6 in).

plywood manufacture:
Employed for plywood but seldom seen in the United Kingdom.
Movement: 4·5 mm plywood from 30 per cent to 90 per cent relative humidity – 0·11 per cent.
Surface splitting on exposure to weather – Grade I.
staining and polishing: Satisfactory.

durability and preservation

insect attack: Sapwood liable to attack by powder-post beetles and reported from New Zealand to be liable to attack by the common furniture beetle. Not resistant to termites in West Africa.
durability of heartwood: Perishable.
preservative treatment:
Moderately resistant. The sapwood, which constitutes a large part of the tree, is permeable.

uses

A useful general-purpose timber of medium weight and small dimensional movement, which works easily giving a good finish, but lacks character in appearance. Used in the furniture industry as drawer sides, legs and framing, and for shopfitting. Takes stain well and can be matched with decorative veneers, and is therefore used extensively for lipping and mouldings. Used in the building industry for joinery and decorative mouldings and other interior work.

AFARA or LIMBA
Terminalia superba

other names: ofram (Ghana); akom (Cameroon); limbo, chêne-limbo, fraké, noyer du Mayombe (France and parts of West Africa); korina (United States).

Light afara, light limba, limba clair, limba blanc – plain, light coloured wood.
Dark afara, dark limba, limba noir, limba bariolé – figured heartwood.
'White afara' is a name used in Nigeria for the tree; as a timber name it is confusing and should not be used.

THE TREE

Reaches a height of 45 m (150 ft) or more. Clear bole 27 m (90 ft); buttresses up to 2·5 m (8 ft). Diameter above buttresses 1·0–1·5 m (3–5 ft). Tropical Africa.

THE TIMBER properties

colour: Usually pale straw to yellow-brown throughout, somewhat resembling light oak; but some logs contain irregular dark heart with greyish-brown or nearly black markings, giving an attractive appearance sometimes suggesting figured walnut.
sapwood: Not very clearly distinguishable from heartwood; may be up to 130–150 mm (5 or 6 in) wide.
grain: Varies from straight to somewhat irregular or slightly interlocked. Texture moderately coarse, but even.
brittleheart: Present in some logs.
weight: Very variable, partly owing to presence of lightweight brittleheart. Recorded range 400 to 790 kg/m³ (25–49 lb/ft³), but sound timber usually 480 to 640 kg/m³ (30–40 lb/ft³), averaging 550 kg/m³ (34 lb/ft³) at 12 per cent moisture content.
strength: Very variable; on average, probably about half-way between European beech and obeche. For work where strength is important care must be taken to exclude material containing brittleheart.

Moisture Content	Bending Strength		Modulus of Elasticity		Compression parallel to grain	
	N/mm²	lbf/in²	N/mm²	1000 lbf/in²	N/mm²	lbf/in²
Green	–	–	–	–	–	–
12 per cent	83	12 100	10 600	1530	37·9	5490

	movement:	Small.
		Moisture content in 90 per cent relative humidity 18 per cent
		Moisture content in 60 per cent relative humidity 12 per cent
		Corresponding tangential movement 1·3 per cent ($\frac{5}{32}$ in/ft)
		Corresponding radial movement 1·0 per cent ($\frac{1}{8}$ in/ft)

processing   **drying:**   Dries rapidly and well with little or no checking. Distortion very small. Kiln Schedule J.

Shrinkage: Green to 12 per cent moisture content – no data.

working properties:

The presence of interlocked grain in some material affects many woodworking operations.

Blunting: Slight.

Sawing: Rip-sawing – Saw type HR54.
Cross-cutting – Satisfactory.
Narrow bandsawing – Satisfactory.
Wide bandsawing – Saw type B, satisfactory.

Machining: Satisfactory. In planing, a 20° cutting angle is an advantage where irregular grain is present.

Nailing: Satisfactory.

Gluing: Good.

wood bending: No precise data, but bending properties probably very poor.

plywood manufacture:

Employed for plywood and usually available in the United Kingdom.

Movement: 4·5 mm plywood from 30 per cent to 90 per cent relative humidity – 0·14 per cent.

Surface splitting on exposure to weather – Grade I.

staining and polishing: Satisfactory. The grain requires filling.

durability and preservation

insect attack: Logs liable to severe attack by ambrosia (pinhole borer) beetles. Sapwood liable to attack by powder-post beetles. Has been reported non-resistant to termites in West Africa.

durability of heartwood: Non-durable.

preservative treatment: Reported to be moderately resistant.

uses   A medium-weight timber of good dimensional stability, sometimes having an attractive appearance. Widely used in veneer and plywood industry; selected material is sliced for decorative veneers. Sawn timber used in furniture production (framing, table legs, chair seats, drawer sides, etc.), interior joinery, coffins, and general utility purposes, but timber containing brittleheart must be avoided where good strength properties are required.

'AFRICAN WALNUT'
Lovoa trichilioides

other names: 'Benin walnut', 'Nigerian golden walnut', 'Nigerian walnut', 'Ghana walnut' (Great Britain); apopo, sida (Nigeria); bibolo (Cameroon); dibetou, noyer d'Afrique, noyer de Gabon (France and parts of West Africa); alona wood, congowood, lovoa wood (United States).

THE TREE   Reaches a height of 45 m (150 ft). Bole cylindrical, 18–27 m (60–90 ft); short buttresses. Diameter about 1·2 m (4 ft). West Africa.

THE TIMBER properties   **colour:**   Bronze shade of yellowish-brown, sometimes marked with dark streaks or veins, which probably first suggested the name African walnut. Apart from colour, the timber is more like African mahogany and should not be confused with the true walnuts (*Juglans* spp.).

sapwood: Clearly distinguishable from heartwood. Buff-coloured, up to 75 mm (3 in) wide.

grain: Usually interlocked, producing well marked ribbon grain or stripe figure when cut on quarter. Texture moderately fine.

weight: 480–650 kg/m³ (30–40 lb/ft³), averaging about 550 kg/m³ (34 lb/ft³) at 12 per cent moisture content.

strength: About half-way between obeche and European beech.

Moisture Content	Bending Strength		Modulus of Elasticity		Compression parallel to grain	
	N/mm²	lbf/in²	N/mm²	1000 lbf/in²	N/mm²	lbf/in²
Green	57	8200	7300	1060	29·8	4320
12 per cent	82	11 900	9200	1340	48·2	6990

movement: Small.
Moisture content in 90 per cent relative humidity   18 per cent
Moisture content in 60 per cent relative humidity   13 per cent
Corresponding tangential movement   1·3 per cent ($\frac{5}{32}$ in/ft)
Corresponding radial movement   0·9 per cent ($\frac{7}{64}$ in/ft)

processing

drying: Dries fairly rapidly without much degrade. Existing shakes inclined to extend slightly and some distortion may occur.
Kiln Schedule E.
Shrinkage: Green to 12 per cent moisture content:
Tangential   about 5·0 per cent ($\frac{5}{8}$ in/ft)
Radial   about 2·0 per cent ($\frac{1}{4}$ in/ft)

working properties: Interlocked grain affects many machining operations.
Blunting: Slight.
Sawing: Rip-sawing – Saw type HR 54.
Cross-cutting – Satisfactory.
Narrow bandsawing – Satisfactory.
Wide bandsawing – Saw type B.
Machining: Satisfactory, but cutting angle of 15° required to minimise tearing of interlocked grain in planing and sharp cutting edges necessary to maintain good finish on end-grain during boring and recessing.
Nailing: Satisfactory.
Gluing: Good.

wood bending:
Classification – Moderate.
Ratio radius/thickness for solid bends (steamed):
Supported: 18   Unsupported: 32
Limiting radius for 3·2 mm ($\frac{1}{8}$ in) laminae (unsteamed): 150 mm (6 in)

plywood manufacture: No information available.

staining and polishing: Satisfactory.

durability and preservation

insect attack: Sapwood liable to attack by powder-post beetles. Has been reported moderately resistant to termites in West Africa. Attacked by dry wood termites in West Africa.

durability of heartwood: Moderately durable.

preservative treatment: Extremely resistant. Sapwood moderately resistant.

uses A medium-weight timber of good appearance and moderate durability. Primarily a decorative timber, used fairly widely in the furniture industry, both as veneer and solid, for cabinet work and lipping. In the building and joinery trade used for flush doors, shop-fitting, decorative interior joinery and panelling. Employed in the motor industry for window and door cappings.

AFRORMOSIA
Pericopsis elata
(Formerly *Afrormosia elata*)

other name: kokrodua (Ghana).

THE TREE

Reaches a height of 45 m (150 ft). Clear but somewhat irregularly shaped bole, length 30 m (100 ft), with buttresses 1·8–2·5 m (6–8 ft) high. Diameter about 1·0 m (3 ft), sometimes more. West Africa, mainly Ghana and Ivory Coast.

THE TIMBER
properties

colour: Yellowish-brown, somewhat resembling teak but on exposure to light, progressively darkening to dark brown.

sapwood: Clearly distinguishable from heartwood; light-coloured, rarely more than 25 mm (1 in) wide.

grain: Varies from straight to interlocked. Texture moderately fine.

weight: From 620–780 kg/m³ (39–49 lb/ft³), averaging about 690 kg/m³ (43 lb/ft³) at 12 per cent moisture content.

chemical staining:
Dark stains liable to appear on the wood if it comes into contact with iron or iron compounds under damp conditions.

strength: Slightly higher than European beech.

Moisture Content	Bending Strength		Modulus of Elasticity		Compression parallel to grain	
	N/mm^2	lbf/in^2	N/mm^2	$1000\ lbf/in^2$	N/mm^2	lbf/in^2
Green	108	15 600	11 400	1650	53·8	7800
12 per cent	134	19 400	12 500	1810	71·5	10 350

movement: Small.
Moisture content in 90 per cent relative humidity   15 per cent
Moisture content in 60 per cent relative humidity   11 per cent
Corresponding tangential movement   1·3 per cent ($\frac{5}{32}$ in/ft)
Corresponding radial movement   0·7 per cent ($\frac{5}{64}$ in/ft)

processing

drying: Dries rather slowly but very well with little degrade apart from slight distortion.
Kiln Schedule J.
Shrinkage: Green to 12 per cent moisture content:
Tangential   about 2·5 per cent ($\frac{5}{16}$ in/ft)
Radial   about 1·5 per cent ($\frac{3}{16}$ in/ft)

working properties: Interlocked grain affects many machining operations.
Blunting: Moderate.
Sawing: Rip-sawing – Saw type HR 54. Tungsten-carbide tipping an advantage for long runs.
Cross-cutting – Satisfactory apart from burning with saw type 3.
Narrow bandsawing – Satisfactory.
Wide bandsawing – Saw type B.
Machining: Satisfactory provided cutting edges are kept sharp and timber is well supported during drilling and mortising. A cutting angle of 20° required to minimise tearing of interlocked grain in planing.
Nailing: Not satisfactory unless pre-bored.
Gluing: Good.

wood bending:

Inclined to distort during steaming and radii quoted cannot be achieved if small knots are present.

Classification – Moderate.

Ratio radius/thickness for solid bends (steamed):

Supported: 14   Unsupported: 29

Limiting radius for 3·2 mm ($\frac{1}{8}$ in) laminae (unsteamed): 165 mm ($6\frac{1}{2}$ in).

plywood manufacture:

Employed for plywood and usually available in the United Kingdom.

Movement: 4·5 mm plywood from 30 per cent to 90 per cent relative humidity: 0·14 per cent.

Surface splitting on exposure to weather – Grade II.

staining and polishing: Satisfactory.

durability and preservation

insect attack: Reported highly resistant to termites in West Africa.

durability of heartwood: Very durable.

preservative treatment: Extremely resistant.

uses

Outstanding for small dimensional movement, high durability rating and good appearance combined with good strength properties and satisfactory working and drying properties. Suitable for a wide range of exterior and interior work such as boat building, joinery, furniture and flooring. Widely used as decorative veneer as well as in solid timber form.

AFZELIA
Afzelia spp

Mainly *A. bipindensis* and *A. pachyloba* with some *A. africana* (all West and Central Africa), and *A. quanzensis* (East Africa).

other names: apa, aligna (Nigeria); doussié (Cameroon); mkora, mbembakofi (Tanzania); chanfuta, mussacossa (Mozambique).

THE TREE

A. bipindensis and *A. pachyloba* reach a height of over 30 m (100 ft) and diameter of 1·0–1·2 m (3–4 ft). Buttresses irregular 1·0–1·2 m (3–4 ft) high. Bole fairly cylindrical. West Africa.

A. africana is smaller and sometimes of poorer stem form. West Africa.

A. quanzensis reaches a height of 21–25 m (70–80 ft). Clear bole 3·5–6·0 m (12–20 ft). Diameter about 1·2 m (4 ft), occasionally more. East Africa.

THE TIMBER

No distinction is made commercially between the different species of *Afzelia*. The following data are based mainly on tests of *A. bipindensis*, except where other species are mentioned.

properties

colour: Reddish-brown.

sapwood: 25–50 mm (1–2 in) wide, light-coloured, clearly distinguishable from heartwood.

grain: Varies from straight to moderately interlocked. Texture rather coarse. Logs of *A. quanzensis* examined contained very distorted grain which may have been related to growth conditions.

weight: Variable; 620–950 kg/m³ (39–59 lb/ft³), averaging 820 kg/m³ (51 lb/ft³) at 12 per cent moisture content.

chemical staining:

Contains afzelin, a yellow dye-stuff which, under moist conditions, can discolour textiles.

strength: Comparable to European beech, but somewhat stronger in compression parallel to grain.

Data for afzelia from West Africa:

Moisture Content	Bending Strength		Modulus of Elasticity		Compression parallel to grain	
	N/mm^2	lbf/in^2	N/mm^2	$1000\,lbf/in^2$	N/mm^2	lbf/in^2
Green	–	–	–	–	–	–
12 per cent	125	18 100	13 100	1900	79·2	11 490

Limited tests on *A. quanzensis* which had very distorted grain showed somewhat lower values.

movement: Small.

	A. bipindensis	*A. quanzensis*
Moisture content in 90 per cent relative humidity	14 per cent	15 per cent
Moisture content in 60 per cent relative humidity	9·5 per cent	11 per cent
Corresponding tangential movement	1·0 per cent ($\frac{1}{8}$ in/ft)	0·5 per cent ($\frac{1}{16}$ in/ft)
Corresponding radial movement	0·5 per cent ($\frac{1}{16}$ in/ft)	0·5 per cent ($\frac{1}{16}$ in/ft)

processing

drying: Dries satisfactorily but very slowly. Degrade not likely to be severe, but slight distortion, extension of existing shakes and fine checking may occur.
Kiln Schedule E.
Shrinkage: Green to 12 per cent moisture content:
Tangential about 1·5 per cent ($\frac{3}{16}$ in/ft)
Radial about 1·0 per cent ($\frac{1}{8}$ in/ft)

working properties:
Blunting: Moderate.
Sawing: Rip-sawing – Saw type HR 54 or HR 60.
Cross-cutting – Satisfactory with saw type 2; burning likely to occur with type 3 and tooth vibration with type 1.
Narrow bandsawing – Satisfactory.
Wide bandsawing – Saw type B or C.
Machining: Planing – cutting angle of 15° required to minimise tearing of interlocked grain.
Moulding – Difficult, owing to breaking away of arrises.
Other operations – Satisfactory.
Nailing: Pre-boring necessary.
Gluing: Difficult.

wood bending: Considerable variation in bending properties.
A. quanzensis can be bent to smaller radii than *A. bipindensis*, but limited tests on material containing wild grain and reaction wood produced bends unsuitable for most purposes because of distortion during steaming and setting. *A. bipindensis* does not distort appreciably, but some resin exudation occurs during steaming.
Classification – Moderate.
Ratio radius/thickness for solid bends (steamed):

	A. bipindensis	*A. quanzensis*
Supported	18	9
Unsupported	34	14
Limiting radius for $\frac{1}{8}$ in (3·2 mm) laminae (unsteamed)	240 mm (9$\frac{1}{2}$ in)	180 mm (7 in)

plywood manufacture:
Unsuitable for plywood manufacture because of its high density and difficulty in gluing.

staining and polishing:
Difficult to stain where pores contain yellow or white deposits. Polishes satisfactorily provided the grain is filled.

durability and preservation

insect attack: Sapwood liable to attack by powder-post beetles. Has been reported moderately resistant to termites in West, East and South Africa.

durability of heartwood: Very durable.

preservative treatment:
Extremely resistant. Sapwood reported to be moderately resistant.

uses

A heavy timber having good strength properties together with outstanding durability and stability. Used for high-class joinery, both indoors and out-of-doors, e.g. staircases, counter tops, fittings in banks and public buildings, door and window frames and sills. In transport, used for road and railway wagon underframes, framing of tank wagons, and wooden floors. Makes an attractive floor suitable for normal pedestrian traffic. Possesses good resistance to acids and has been used successfully for construction of vats and filter presses for use with acids and other chemicals. Contains a yellow dye-stuff which may stain fabrics under moist conditions (see above) and is not therefore recommended for kitchen and laundry use.

AGBA
Gossweilerodendron balsamiferum

other names: tola branca, white tola (Angola); tola (Zaire and France). The name tola is also applied to *Oxystigma oxyphyllum* (formerly *Pterygopodium oxyphyllum*), sometimes known as tola mafuta, etc.

THE TREE

Reaches a height of 60 m (200 ft). Clear bole 25–30 m (80–100 ft), diameter 1·5–2·0 m (5–7 ft). Exported mainly from Nigeria and Portuguese Cabinda.

THE TIMBER properties

colour: Uniform yellow-brown or straw-brown.

sapwood: About 100 mm (4 in) wide, usually paler than heartwood.

grain: Straight to moderately interlocked. Texture fine.

weight: Averages about 510 kg/m³ (32 lb/ft³) at 12 per cent moisture content.

brittleheart: Fairly common, especially in large logs and extending as high as 15 m (50 ft) up the tree.

resin: Occurs in large quantities in some logs, in pockets or in shakes near the heart. Other logs practically free from resin. Not usually troublesome after drying, though the timber may retain a slightly resinous odour.

strength: About half-way between obeche and European beech, but careful selection required, since timber containing brittleheart is appreciably weaker.

Moisture Content	Bending Strength		Modulus of Elasticity		Compression parallel to grain	
	N/mm²	lbf/in²	N/mm²	1000 lbf/in²	N/mm²	lbf/in²
Green	52	7500	6000	870	24·3	3520
12 per cent	81	11 800	7600	1100	43·0	6270

movement: Small.
Moisture content in 90 per cent relative humidity   18 per cent
Moisture content in 60 per cent relative humidity   12 per cent
Corresponding tangential movement   1·3 per cent ($\frac{5}{32}$ in/ft)
Corresponding radial movement   0·7 per cent ($\frac{5}{64}$ in/ft)

processing	drying:	Dries fairly rapidly with very little tendency to distort or split. Some gum exudation likely to occur, but not excessively except in pieces containing pith.

processing

drying: Dries fairly rapidly with very little tendency to distort or split. Some gum exudation likely to occur, but not excessively except in pieces containing pith.

Kiln Schedule J.

Shrinkage: Green to 12 per cent moisture content:

Tangential about 3·0 per cent ($\frac{3}{8}$ in/ft)
Radial about 1·5 per cent ($\frac{3}{16}$ in/ft)

working properties:

Interlocked grain and occasional gum deposits affect many machining operations.

Blunting: Slight.

Sawing: Rip-sawing — Saw type HR 54.

Cross-cutting — Satisfactory.

Narrow bandsawing — Satisfactory.

Wide bandsawing — Saw type B.

Machining: Satisfactory, but cutting angle of 20° required to minimise tearing of interlocked grain. Slight breaking away of arrises may occur during moulding.

Nailing: Satisfactory with care.

Gluing: Good.

wood bending: Moderately good. Some resin exudation accompanies steaming.

Classification — Moderate.

Ratio radius/thickness for solid bends (steamed):

Supported: 20 Unsupported: 16

Limiting radius for 3·2 mm ($\frac{1}{8}$ in) laminae (unsteamed): 110 mm ($4\frac{2}{8}$ in).

plywood manufacture:

Employed for plywood and usually available in the United Kingdom.

Movement of plywood — No information.

Surface splitting on exposure to weather — Grade I.

staining and polishing: Satisfactory when filled.

durability and preservation

insect attack: Sapwood liable to attack by the common furniture beetle.

durability of heartwood: Durable.

preservative treatment: Resistant. Sapwood permeable.

uses

A useful general-purpose timber, particularly where light weight, satisfactory working properties and natural durability are required. Where strength is important careful selection is necessary. The slight resinous odour makes it objectionable for items which may come into contact with foodstuffs. Often used as an alternative to oak in furniture, including school and church furniture, and for coffins. An excellent joinery wood, both for interior and exterior use. In boat building, suitable for planking and laminated frames and for interior joinery. Also used for lorry and trailer flooring.

ALBIZIA, WEST AFRICAN
Albizia ferruginea and spp.

other names: okuro (Ghana); ayinre (Nigeria).

Of the species of *Albizia* occurring in West Africa *A. ferruginea* is likely to be of greatest interest to the United Kingdom market. The other species (principally *A. adianthifolia* and *A. zygia*) are lighter in weight and paler in colour than *A. ferruginea*. Species of *Albizia* also occur in East Africa. The following information refers to *A. ferruginea*.

THE TREE

Grows to a height of 37 m (120 ft) or more, diameter about 1·0 m (3 ft), with a clear, straight bole of length 9–12 m (30–40 ft). West Africa.

19

THE TIMBER
properties

colour: Variable from mid-brown to dark red-brown, with an attractive appearance.

sapwood: Distinct from heartwood, pale yellow or straw-coloured, about 50 mm (2 in) wide.

grain: Interlocked and sometimes variable in direction. Texture coarse.

weight: Ranges from 580 to 820 kg/m³ (36–51 lb/ft³), average about 700 kg/m³ (44 lb/ft³), seasoned.

strength: Generally somewhat lower than European beech, but varies with density.

Moisture Content	Bending Strength		Modulus of Elasticity		Compression parallel to grain	
	N/mm^2	lbf/in^2	N/mm^2	$1000\,lbf/in^2$	N/mm^2	lbf/in^2
Green	–	–	–	–	–	–
12 per cent	105	15 200	10 700	1550	64·7	9380

movement: Small.
Moisture content in 90 per cent relative humidity   16 per cent
Moisture content in 60 per cent relative humidity   11·5 per cent
Corresponding tangential movement   1·2 per cent ($\frac{9}{64}$ in/ft)

processing

drying: Limited tests indicate that it dries with little degrade, but very slowly in thick sizes.
Kiln Schedule F.

working properties:
Care required in machining owing to interlocked and irregular grain. Fine dust may cause irritation of the nose.
Blunting: Moderate.
Sawing: Rip-sawing – Saw type HR 54.
           Cross-cutting – Satisfactory.
           Narrow bandsawing – Satisfactory.
           Wide bandsawing – Saw type B.
Machining: Satisfactory, but with a tendency to breaking out when machined across the grain, as in recessing and on arrises.
           Planing – A cutting angle of 15° required to prevent tearing out on quarter-sawn surfaces.
Nailing: Satisfactory when pre-bored.
Gluing: No information.

wood bending: Considerable variation in bending properties is probable.
Classification – Moderate.
Ratio radius/thickness for solid bends (steamed):
        Supported: 20   Unsupported: 40
Limiting radius for 3·2 mm ($\frac{1}{8}$ in) laminae (unsteamed): 178 mm (7 in)

plywood manufacture: No information.

staining and polishing: Satisfactory when filled.

durability and preservation

insect attack: Sapwood liable to attack by powder-post beetles. Heartwood reported to be highly resistant to termites in West Africa.

durability of heartwood: Very durable.

preservative treatment: Extremely resistant. The sapwood is permeable.

uses

The timber has small movement and its woodworking qualities are satisfactory. It should be suitable for joinery and general carpentry.

ALDER
Alnus glutinosa

other names: Black alder (Great Britain).
Grey alder (*Alnus incana*) grows in northern Europe and western Siberia; its timber appears to be indistinguishable from that of *A. glutinosa*.

THE TREE
Reaches a height of 15–27 m (50–90 ft). Clear bole 6–12 m (20–40 ft) long, diameter 0·3–1·2 m (1–4 ft). Europe, Western Asia and North Africa.

THE TIMBER
properties

colour: Pale when first cut, darkening to light reddish-brown. Lustreless surface, with dark lines or streaks formed by broad rays.
sapwood: Not visually distinguishable from heartwood.
grain: Straight, except occasionally in butts. Texture fine.
weight: Averages 530 kg/m^3 (33 lb/ft^3) at 12 per cent moisture content.
strength: About half-way between obeche and European beech.

Moisture Content	Bending Strength		Modulus of Elasticity		Compression parallel to grain	
	N/mm^2	lbf/in^2	N/mm^2	1000 lbf/in^2	N/mm^2	lbf/in^2
Green	49	7100	7600	1100	21·7	3140
12 per cent	80	11 600	8800	1270	41·1	5960

movement: No information available.

processing

drying: Dries well and fairly rapidly.
Kiln Schedule J.
Shrinkage: Green to 12 per cent moisture content:
Tangential about 6·5 per cent ($\frac{13}{16}$ in/ft)
Radial about 4·0 per cent ($\frac{1}{2}$ in/ft)
working properties:
Blunting: Slight.
Sawing: Rip-sawing – Saw type HR 54.
Cross-cutting – Satisfactory.
Narrow bandsawing – Satisfactory.
Wide bandsawing – Saw type A.
Machining: Satisfactory provided cutting edges are thin and kept sharp.
Nailing: Satisfactory.
Gluing: Good.
wood bending:
Inferior to many temperate-grown hardwoods such as beech. Pronounced tendency for drying checks to occur on ends of bends during setting.
Classification – Moderate.
Ratio radius/thickness for solid bends (steamed):
Supported: 14   Unsupported: 18
Limiting radius for 3·2 mm ($\frac{1}{8}$ in) laminae (unsteamed): 188 mm ($7\frac{2}{8}$ in).
plywood manufacture:
Employed for plywood and usually available in the United Kingdom.
Movement: 1·5 mm plywood from 30 per cent to 90 per cent relative humidity – 0·28 per cent.
Surface splitting on exposure to weather: No information.
staining and polishing: Satisfactory.

durability and preservation

insect attack: Sapwood liable to attack by the common furniture beetle, but immune from attack by powder-post beetles.
durability of heartwood: Perishable.
preservative treatment: Permeable.

| uses | Its comparative lightness, softness and ease of working make the timber suitable for hat blocks, broom and brush backs, toys and general turnery. Sawn material and plywood are mainly imported from the Continent. Home-grown timber is generally felled at the pole stage. |

ALSTONIA
Alstonia congensis
and *A. boonei*

other names: patternwood, stoolwood (East and West Africa); mujwa (Uganda); ahun, awun, duku (Nigeria); tsongutti (Zaire); sindru (Ghana); emien (parts of West Africa).

THE TREE Average height about 36 m (120 ft). Length of clear bole about 25 m (80 ft), diameter up to 1·0 m (3 ft), usually fluted. West and Central Africa.

**THE TIMBER
properties** Timber nearly white. Sapwood wide, not distinguishable from heartwood. Grain usually straight, texture moderately fine and even. Appearance of timber marred by presence of slit-like passages (latex traces) running radially through the wood. A soft and light timber, weight about 400 kg/m³ (25 lb/ft³) at 12 per cent moisture content. Strength low, comparable to obeche. Movement small.

processing Dries rapidly and well (Kiln Schedule H). Not troublesome to work, but sharp cutting edges necessary and reduced sharpness angle an advantage. Glues and nails well. Bending properties very poor.

**durability
and preservation** Heartwood perishable. Permeable to preservatives.

uses A soft, light timber, useful only where appearance and strength are not important. Appearance spoilt by presence of latex traces, which also render it unsuitable for plywood manufacture. Liable to discoloration by staining fungi.

ANINGERIA
Aningeria spp.

other names: anegré (Ivory Coast); landosan (Nigeria); mukali, kali (Angola); mukangu, muna (Kenya); osan (Uganda).
Aningeria comprises four tropical African species: *A. robusta* (West Africa), *A. altissima* (West and East Africa), *A. adolfi-friederici* (East Africa), and *A. pseudo-racemosa* (Tanzania).

THE TREE A tall tree, commonly 30–36 m (100–120 ft) in height, sometimes up to 45 m (150 ft). Straight, cylindrical bole, length up to 25 m (80 ft), diameter above buttresses generally 1·0–1·2 m (3–4 ft). Tropical Africa, particularly common in parts of East Africa.

**THE TIMBER
properties** The various species of *Aningeria* have rather similar woods, somewhat like birch in appearance. Heartwood yellowish-white to pale brown, sometimes with a pinkish tint. Lustrous but fairly plain in appearance, although quartered surfaces sometimes show a growth ring figure. Grain usually straight, but sometimes wavy producing a mottle figure. Weight variable, generally in the range 480–580 kg/m³ (30–36 lb/ft³), seasoned. Strength somewhat lower than European beech.

processing Reported to dry without degrade in air or kiln drying. Some tendency for blue-stain to develop in early stages of air drying. Kiln Schedule E. Some species of *Aningeria* are siliceous and have a blunting effect on tools. Reported from East Africa that *A. altissima* saws easily, but *A. adolfi-friederici* is difficult to saw. Requires support when cross-cutting, boring and mortising to avoid breaking out. Glues well and has been shown by tests to be suitable for plywood manufacture.

**durability
and preservation** Heartwood perishable (provisional), and probably permeable to preservatives.

uses Seen mainly as sliced veneer of continental European production. *Aningeria* is usually a plain, pale, fine textured wood and as veneer has been used for surface printed finishes.

ANTIARIS
Antiaris africana
and *A. welwitschii*

other names: Oro, ogiovu (Nigeria); chenchen, kyenkyen (Ghana); kirundu (Uganda).

THE TREE Reaches a height of 36–45 m (120–150 ft). Clear bole up to 21 m (70 ft), straight and cylindrical, diameter 0·6–1·5 m (2–5 ft). West, Central and East Africa.

THE TIMBER
properties

colour: White or light yellow-brown, somewhat resembling obeche.
sapwood: Not visually distinguishable from heartwood, but may be up to 150 mm (6 in) wide.
grain: Interlocked. Texture medium to rather coarse.
weight: From 370 to 530 kg/m³ (23–33 lb/ft³), averaging 430 kg/m³ (27 lb/ft³) at 12 per cent moisture content.
strength: Comparable to obeche.

Moisture Content	Bending Strength		Modulus of Elasticity		Compression parallel to grain	
	N/mm²	lbf/in²	N/mm²	1000 lbf/in²	N/mm²	lbf/in²
Green	–	–	–	–	–	–
12 per cent	59	8600	7200	1050	37·4	5430

movement: Small.
Moisture content in 90 per cent relative humidity 22 per cent
Moisture content in 60 per cent relative humidity 13 per cent
Corresponding tangential movement 1·8 per cent ($\frac{7}{32}$ in/ft)
Corresponding radial movement 0·8 per cent ($\frac{3}{32}$ in/ft)

processing

drying: Dries fairly rapidly, but tends to distort. Twist may be a serious defect and thick material tends to end-split.
Kiln Schedule A.
Shrinkage: Green to 12 per cent moisture content:
Tangential about 3·0 per cent ($\frac{3}{8}$ in/ft)
Radial about 1·5 per cent ($\frac{3}{16}$ in/ft)

working properties:
Interlocked grain and soft, fibrous texture affect many machining operations.
Blunting: Slight.
Sawing: Rip-sawing – Saw type HR 54.
Cross-cutting – Considerable break-out at back of cut, particularly with saw type 2.
Narrow bandsawing – Satisfactory, though with considerable break-out at bottom of cut.
Wide bandsawing – Saw type A.
Machining: Cutting edges must be kept sharp to prevent tearing during machining operations across the end-grain.
Planing – Cutting angle of 20° necessary to minimise tearing of interlocked grain.
Boring and mortising – Adequate support necessary to prevent break-out at exit.
Nailing: Satisfactory.
Gluing: Good.

wood bending: Limited data indicate poor bending properties.
Classification – Very poor.
plywood manufacture:
Employed for plywood and usually available in the United Kingdom.
Movement: 1·5 mm plywood from 30 per cent to 90 per cent relative humidity – 0·15 per cent.
Surface splitting on exposure to weather – Grade II.
staining and polishing: Satisfactory.

durability and preservation

Sapwood very susceptible to sapstain when green.
insect attack: Logs liable to severe attack by ambrosia (pinhole borer) beetles in Africa, and this may extend into the heartwood. Logs also liable to attack by forest longhorn or Buprestid beetles. Sapwood susceptible to attack by powder-post beetles.
durability of heartwood: Perishable.
preservative treatment: Permeable.

uses

Similar in character to obeche, though generally regarded as inferior. Successful utilisation depends on rapid extraction from the forest and protection from degrade by insect and fungal attack, and from blue-stain.
For furniture, a good utility timber which takes stain and polishes well. Suitable for carcassing, shelves, etc. Used as plywood for tea chests in East Africa and might find wider use as plywood. As veneer it is somewhat brittle but sliced veneer has an attractive stripe figure.

APPLE
Malus sylvestris

Tree rarely exceeds 9 m (30 ft) in height. Small, crooked bole, 0·2–0·3 m (8 in–1 ft) in diameter. Europe and South-western Asia.

A heavy timber, weight about 700 kg/m³ (44 lb/ft³) at 12 per cent moisture content. Texture fine and even.

Dries slowly with a marked tendency to distort; Kiln Schedule A. Moderate blunting effect in sawing. Finishes cleanly in most cutting operations, but liable to tear in planing unless cutting angle is reduced to 20°.

Used mainly for small fancy turnery and inlay work.

ASH, AMERICAN
Fraxinus spp.

The timber is derived mainly from three species, *Fraxinus americana* (white ash), *F. pennsylvanica* (green ash) and *F. nigra* (black ash or brown ash). They are all sold as ash, but black ash is commonly distinguished from the other species.

THE TREE

Fraxinus americana may reach a height of 30–36 m (100–120 ft), usually less. Bole well formed, diameter commonly 0·6–0·9 m (2–3 ft). Trees of the other species are smaller. Eastern half of United States and Canada.

THE TIMBER properties

White ash: Sapwood nearly white, rather narrow in old forest trees, wider and less clearly demarcated from heartwood in second growth trees. Heartwood greyish-brown, sometimes with a reddish tinge. Grain usually straight, texture coarse. Weight variable, average about 660 kg/m³ (41 lb/ft³), seasoned.
It is common practice to separate the heavier and stronger material (tough ash) from the less dense material (soft ash). The better grades of tough ash have good strength properties, especially toughness.
Black ash: Similar in general appearance to white ash but darker in colour. Average weight about 560 kg/m³ (35 lb/ft³), seasoned. Less strong and tough than white ash.

processing | Reported to dry without distortion. Working properties satisfactory; moderate blunting effect on tools. Pre-boring probably necessary in nailing the harder material. Gluing good. Wood bending properties variable but generally very good; not tolerant of pin knots. Employed for plywood manufacture but seldom seen in the United Kingdom.

durability and preservation | Sapwood liable to attack by powder-post beetles and by the common furniture beetle. The heartwood is non-durable, but reported to be easy to treat.

uses | Tough ash is suitable for many of the same purposes as European ash and is used extensively for handles of striking tools (axes, hammers, etc.) and for spades, etc., where toughness is important. Used also for railway coach and vehicle construction and framing of caravans. Soft ash is used for cabinet work and interior joinery, for which purposes it is preferred because of its milder nature.

ASH, EUROPEAN
Fraxinus excelsior

other names: English, French, Belgian ash etc., according to origin.

THE TREE | Reaches a height of 25–30 m (80–100 ft). Clear bole 10–15 m (30–50 ft) according to growth conditions. Diameter 0·6–1·5 m (2–5 ft). Europe, south of about 60°N, and Asia minor.

THE TIMBER properties

colour: White to light brown, temporarily turning pink when freshly cut. Trees occasionally contain irregular dark brown or black heart, not necessarily associated with decay.

sapwood: Not usually visually distinguishable from heartwood.

grain: Straight. Contrast between porous early wood and dense late wood produces decorative figure in plain-sawn timber or rotary cut veneer. Texture coarse, owing to bands of porous early wood.

weight: Variable, according to growth conditions. Range 510–830 kg/m³ (32–52 lb/ft³), averaging 690 kg/m³ (43 lb/ft³), at 12 per cent moisture content.

growth rate: As a general rule, timber having growth rings 1·5–6 mm wide is to be preferred, especially for the more exacting purposes. Very slow-grown ash having narrower rings contains a high proportion of porous early wood, with consequent reduction in density and strength.*

strength: Comparable to European beech, but outstandingly high in toughness. Ash has greater toughness than any other home-grown hardwood.

Moisture Content	Bending Strength		Modulus of Elasticity		Compression parallel to grain		Impact (toughness) Max. drop of hammer	
	N/mm²	lbf/in²	N/mm²	1000 lbf/in²	N/mm²	lbf/in²	m	in
Green	66	9600	9500	1380	27·2	3940	1·17	46
12%	116	16 800	11 900	1720	53·3	7730	1·07	42

*See PRL Technical Note No. 54 — "Selecting Ash by Inspection".

movement: Medium.
                Moisture content in 90 per cent relative humidity 22 per cent
                Moisture content in 60 per cent relative humidity 13 per cent
                Corresponding tangential movement 2·5 per cent ($\frac{5}{16}$ in/ft)
                Corresponding radial movement   1·5 per cent ($\frac{3}{16}$ in/ft)

processing

drying: Dries fairly rapidly with little splitting or checking. Tendency to distort unless the kiln temperature is kept low. Severe end-splitting sometimes occurs. Material which has shrunk excessively or become distorted during kiln drying responds well to reconditioning treatment.
Kiln Schedule D.

Shrinkage: Green to 12 per cent moisture content:
            Tangential   about 7·0 per cent ($\frac{7}{8}$ in/ft)
            Radial   about 4·5 per cent ($\frac{9}{16}$ in/ft)

working properties:
Blunting: Moderate.
Sawing: Rip-sawing – Saw type HR54, satisfactory.
             Cross-cutting – Saw type 2 most satisfactory.
             Narrow bandsawing – Satisfactory.
             Wide bandsawing – Saw type B.
Machining: Satisfactory.
Nailing: Pre-boring advisable except with less dense material.
Gluing: Satisfactory.

wood bending:
Has excellent steam-bending properties except when irregular grain or knots are present. Bending qualities may also be influenced by the conditions of growth of the tree.
Classification – Very good.
Ratio radius/thickness for solid bends (steamed):
        Supported: 2·5   Unsupported: 12
Limiting radius for 3·2 mm ($\frac{1}{8}$ in) laminae (unsteamed) – 120 mm ($4\frac{3}{4}$ in).

plywood manufacture:
Employed in plywood manufacture but seldom seen in the United Kingdom.

staining and polishing: Good.

durability and preservation

insect attack: Trees and logs liable to attack by forest longhorn or Buprestid beetles. Sapwood liable to attack by powder-post beetles and by the common furniture beetle. The ash bark beetle, which frequently tunnels in the bark of ash logs, causes no damage to the timber.

durability of heartwood: Perishable.

preservative treatment:
Moderately resistant. Material containing black heart is resistant.

uses

Ash varies considerably in quality. Good selected material is outstanding for its toughness and its good bending properties. Used widely for sports goods (tennis and other racquets, hockey sticks, baseball bats, cricket stumps, billiard cues, gymnasium appliances, etc.) and for handles of tools, such as picks, shovels, axes and hammers, and handles of fishing landing nets. Suitable for furniture parts, both in cabinet work and chairs. Employed extensively in road vehicles and agricultural implements for parts where toughness and weight are important, but should be used with care on account of its lack of natural durability. Used in boat-building for bent parts and tillers, oars, boat hooks, deck beams and frames for canoes and canvas boats. Other uses include fancy turnery and bent handles for walking sticks and umbrellas.

ASH, JAPANESE
Fraxinus mandschurica

other name: tamo (Japan).

Usually pale brown, darker in colour than European ash. Weight about 580 kg/m³ (36 lb/ft³), seasoned.

Generally used for purposes where toughness is not of prime importance. Suitable for cabinet work, and widely employed in Japan and Manchuria for plywood. Has been imported in the form of figured veneers.

ASPEN, CANADIAN
Populus tremuloides

other names: aspen, trembling aspen (Canada).

THE TREE Reaches an average height of 12–18 m (40–60 ft), but sometimes much larger (up to 27 m, 90 ft). Average diameter 0·2–0·3 m (8–12 in), maximum 0·6 m (2 ft). Canada and USA (New England and Lake States, and Rocky Mountain States).

THE TIMBER
properties Like other poplars, aspen tends to be affected by tension wood.
colour: Greyish-white or pale brown.
sapwood: No clear distinction from heartwood.
grain: Straight, but inclined to be woolly. Texture fine and even.
weight: Average about 450 kg/m³ (28 lb/ft³), seasoned.
strength: Somewhat higher than obeche, particularly in stiffness.

Moisture Content	Bending Strength		Modulus of Elasticity		Compression parallel to grain	
	N/mm²	lbf/in²	N/mm²	1000 lbf/in²	N/mm²	lbf/in²
Green	40	5800	8500	1230	16·8	2440
12 per cent	71	10 300	10 500	1530	37·7	5470

(Data from Canadian Department of Forestry)

movement: No data.

processing Tension wood, where present, tends to produce a woolly surface and is inclined to distort.

drying: Inclined to distort unless care is taken in piling.
Kiln Schedule E.
Shrinkage: No information.

working properties:
Blunting: Slight.
Sawing: Rip-sawing – Saw type HR 54.
Cross-cutting – Satisfactory.
Narrow bandsawing – Satisfactory.
Wide bandsawing – Saw type A.
Machining: Thin, sharp cutting edges, produced by means of a reduced grinding angle, required to prevent a woolly finish during planing and to avoid crumbling during cutting operations across the end-grain. A cutting angle of 30° is satisfactory.
Nailing: Satisfactory.
Gluing: Good.

c

wood bending: No data available. Unlikely to be suitable for solid bending.
plywood manufacture:
Employed for plywood and usually available in the United Kingdom.
staining and polishing: Satisfactory.

durability and preservation

insect attack: No information available.
durability of heartwood: Non-durable.
preservative treatment:
Probably extremely resistant. Sapwood reported to be moderately resistant.

uses Used extensively for veneer, particularly for matches and chip and other fruit baskets, and for wood wool. A useful utility timber suitable for boxes, crates, interior parts of furniture, etc. Widely used in North America for manufacture of pulp and paper, and for plywood and chipboard.

AVODIRÉ
Turraeanthus africanus

THE TREE
Reaches a height of 34 m (110 ft). Clear bole 8–15 m (25–50 ft) long, diameter 0·5–0·6 m (1½–2 ft), occasionally more. Stem irregular and twisted. West Africa, chiefly Ghana and Ivory Coast.

THE TIMBER
properties
colour: Cream or pale yellow with natural lustre, darkening to golden yellow.
sapwood: Not visually distinguishable from heartwood.
grain: Sometimes straight, but often wavy or irregularly interlocked, producing an unusual, attractive mottled figure when cut on the quarter. Texture moderately fine.
weight: Average 550 kg/m³ (34 lb/ft³) at 12 per cent moisture content.
strength: Somewhat lower than European beech.

Moisture Content	Bending Strength		Modulus of Elasticity		Compression parallel to grain	
	N/mm^2	lbf/in^2	N/mm^2	$1000\,lbf/in^2$	N/mm^2	lbf/in^2
Green	–	–	–	–	–	–
12 per cent	92	13 400	9600	1390	51·4	7450

movement: Small.
Moisture content in 90 per cent relative humidity   18 per cent
Moisture content in 60 per cent relative humidity   12 per cent
Corresponding tangential movement   1·8 per cent ($\frac{7}{32}$ in/ft)
Corresponding radial movement   1·0 per cent ($\frac{1}{8}$ in/ft)

processing
drying: Dries fairly rapidly, with some tendency to cup and twist. Existing shakes liable to extend, and some splitting may occur in and around knots.
Kiln Schedule E.
Shrinkage: Green to 12 per cent moisture content:
Tangential   about 3·5 per cent ($\frac{7}{16}$ in/ft)
Radial   about 2·0 per cent ($\frac{1}{4}$ in/ft)
working properties:
Grain often wavy or irregularly interlocked, affecting machining operations.
Blunting: Slight.
Sawing: Rip-sawing – Saw type HR 54, satisfactory.
Cross-cutting – Satisfactory.
Narrow bandsawing – Satisfactory.
Wide bandsawing – Saw type A.

Machining: A cutting angle of 15° is required to give a satisfactory planed finish owing to interlocked grain. French head most suitable for moulding.

Nailing: Satisfactory when pre-bored.

Gluing: Good.

wood bending:
Generally poor, with considerable variation from piece to piece.
Classification – Very poor.

plywood manufacture: No information.

staining and polishing:
Satisfactory, apart from a tendency for quarter-sawn surfaces to stain unevenly.

durability and preservation

insect attack: Reported to be non-resistant to termites in West Africa.

durability of heartwood: Non-durable.

preservative treatment: Extremely resistant. Sapwood permeable.

uses

Suitable for high-class interior joinery and for use in furniture as an alternative to sycamore. Well known as a decorative veneer.

AYAN
Distemonanthus benthamianus

other names: movingui (France); ayanran (Nigeria); bonsamdua (Ghana); distemonanthus (Great Britain).

THE TREE

Average height 27 m (90 ft) and diameter 0·8 m (2½ ft), but under favourable conditions reaches a height of 38 m (125 ft) and diameter 1·4 m (4½ ft). Bole clean, reasonably straight and cylindrical, but somewhat undulating, with weakly developed buttresses. West Africa, mainly Cameroon, Ghana and Nigeria.

THE TIMBER properties

colour: Varies from lemon-yellow to yellow-brown. A dark streak is sometimes present in the darker material.

sapwood: Narrow, straw-coloured, and fairly distinct from heartwood.

grain: Often interlocked and sometimes wavy. Fine texture. Some logs yield handsomely figured timber of considerable decorative value.

weight: From 600 to 770 kg/m³ (37–48 lb/ft³), averaging about 670 kg/m³ (42 lb/ft³) at 12 per cent moisture content. The heavier material tends to be darker in colour.

chemical staining:
Contains a yellow extractive which, under moist conditions, is liable to stain fabrics.

strength: Comparable to European beech.

Moisture Content	Bending Strength		Modulus of Elasticity		Compression parallel to grain	
	N/mm^2	lbf/in^2	N/mm^2	$1000\ lbf/in^2$	N/mm^2	lbf/in^2
Green	–	–	–	–	–	–
12 per cent	108	15 700	11 400	1650	57·3	8310

movement: Small.

Moisture content in 90 per cent relative humidity   15 per cent
Moisture content in 60 per cent relative humidity   11 per cent
Corresponding tangential movement   1·3 per cent ($\frac{5}{32}$ in/ft)
Corresponding radial movement   0·8 per cent ($\frac{3}{32}$ in/ft)

processing	**drying:**	Dries with little degrade.
		Kiln Schedule F.
		Shrinkage: No data.
	working properties:	
		Grain often irregular and interlocked. May contain variable amounts of silica.
	Blunting:	Moderate to severe.
	Sawing:	Rip-sawing – Saw type HR 54, or HR 40 or TC when severe blunting is encountered or when long production runs are planned. Gum building up on saw necessitates use of increased set.
		Cross-cutting – Considerable breaking out at bottom and back of cut. Saw type 2 most suitable.
		Narrow bandsawing – A saw with increased tooth pitch and set recommended.
		Wide bandsawing – Saw type B (hard tipped).
	Machining:	Planing – A 20° cutting angle required for satisfactory planing of material having interlocked grain.
		Moulding – Satisfactory apart from a tendency for square block to cause tearing and breaking away of arrises.
		Boring – Gum build-up causes charring, and blunting is rapid. Timber requires supporting. Blunting may be severe on recessing tools.
		Mortising – Satisfactory, but blunting on hollow square chisel may be rapid.
	Nailing:	Some splitting occurs unless wood is pre-bored.
	Gluing:	Good.
	wood bending:	
		Classification – Moderate.
		Ratio radius/thickness for solid bends (steamed):
		Supported: 20 Unsupported: 32
		Limiting radius for 3·2 mm ($\frac{1}{8}$ in) laminae (unsteamed): 180 mm ($7\frac{1}{8}$ in.)
	plywood manufacture:	Believed to be usable for plywood manufacture.
	Movement:	1·5 mm plywood from 30 per cent to 90 per cent relative humidity – 0·22 per cent.
		Surface splitting on exposure to weather – Grade I.
	staining and polishing:	Satisfactory when filled.

durability and preservation

Insect attack: Has been reported moderately resistant to termites in West Africa.

durability of heartwood: Moderately durable.

preservative treatment: Resistant.

uses

A moderately heavy timber having good dimensional stability and suitable for many of the purposes for which oak is used. A useful joinery timber for door frames, windows and sills, suitable also for cabinet work. Has been used in road and railway vehicle building for frames and cab bodies. Suitable for domestic flooring and for gymnasium floors, being fairly resilient.

Contains a yellow dye-stuff which is liable to stain fabrics under moist conditions (see above), and the timber is therefore not recommended for kitchen and laundry equipment.

BALSA
Ochroma lagopus

THE TREE

Grows rapidly, reaching a height of 20 m (70 ft) and diameter of about 0·6 m (2 ft) in 6 to 7 years. Tropical America, mainly Ecuador.

THE TIMBER
properties

Most of the commercial timber is sapwood.

colour: White to oatmeal, sometimes with a pinkish or yellowish tinge and often a silky lustre. Central core of large logs sometimes pale brown.

	grain:	Straight. Texture even.
	weight:	Varies according to growth conditions and position in tree, from 40 kg/m³ (2½ lb/ft³) to over 320 kg/m³ (20 lb/ft³) at 12 per cent moisture content. In commercial material weight ranges from about 80 kg/m³ (5 lb/ft³) to 250 kg/m³ (16 lb/ft³) at 12 per cent moisture content.
	strength:	In keeping with its low density, balsa is the weakest of the commercial timbers. Considerably lower in strength than obeche. Data for material of density 176 kg/m³:

Moisture Content	Bending Strength		Modulus of Elasticity		Compression parallel to grain	
	N/mm²	lbf/in²	N/mm²	1000 lbf/in²	N/mm²	lbf/in²
Green	–	–	–	–	–	–
12 per cent	23	3300	3200	460	15·5	2250

movement: Small.
Moisture content in 90 per cent relative humidity   21·0 per cent
Moisture content in 60 per cent relative humidity   11·0 per cent
Corresponding tangential movement   2·0 per cent ($\frac{1}{4}$ in/ft)
Corresponding radial movement   0·6 per cent ($\frac{5}{64}$ in/ft)

processing

drying: Serious risk of extensive splitting in the log unless converted very soon after felling. Kiln drying of converted stock preferable to air drying to minimise splitting and warping.
Kiln Schedule H, except for timber of the heavier weight class.
Shrinkage: No data.

working properties:
Blunting: Slight.
Sawing: Thin gauge saws should be used.
Rip-sawing – Saw type HR 60.
Cross-cutting – Satisfactory.
Narrow bandsawing – Satisfactory.
Wide bandsawing – Saw type A.
Machining: Cutting edges should be very sharp and thin to prevent crumbling when machining across the end-grain and to avoid a stringy or woolly finish when planing.
Nailing: Takes nails easily, but is too soft to hold them firmly.
Gluing: Good.

wood bending: Cannot be bent appreciably without buckling.
Classification – Very poor.

plywood manufacture: No information.

staining and polishing: Satisfactory, although the timber is very absorbent.

durability and preservation

insect attack: Logs liable to attack by forest longhorn or Buprestid beetles. Sapwood liable to attack by powder-post beetles, and reported to be attacked by the common furniture beetle in Australia. Said to be attacked by dry-wood termites in West Indies.

durability of heartwood: Perishable.

preservative treatment: Resistant. The sapwood is permeable.

uses

A timber of very low density and correspondingly high buoyancy and insulating value. Largely used as insulating material in cold stores and refrigerated ships. Used as core material in lightweight sandwich constructions: end-grain balsa is particularly suitable for this purpose. Its lightness renders it suitable for rafts, life belts, etc., and for use in aircraft construction. Suitable for packaging fragile articles and also widely used in toys, model aircraft and novelties.

BANAK
Virola koschnyi

other names: sangre palo, palo de sangre (British Honduras); tapsava (United States).

THE TREE Reaches a height of over 40 m (140 ft) and diameter of about 1 m (3 ft). Heavily buttressed. Straight, cylindrical bole, length 18 m (60 ft) above buttresses. Central America. Important species in British Honduras.

THE TIMBER
properties

colour: Pinkish-brown to brownish-grey, lustrous. Veins containing red gum sometimes present.

sapwood: Not clearly distinguishable from heartwood.

grain: Straight. Texture medium, uniform.

weight: Averages about 530 kg/m³ (33 lb/ft³) at 12 per cent moisture content.

strength: About half-way between obeche and European beech, but relatively high in stiffness.

Moisture Content	Bending Strength		Modulus of Elasticity		Compression parallel to grain	
	N/mm^2	lbf/in^2	N/mm^2	$1000\,lbf/in^2$	N/mm^2	lbf/in^2
Green	45	6500	9500	1380	21·9	3170
12 per cent	79	11 400	11 100	1610	41·0	5940

movement: No information available.

processing

drying: Dries rather slowly with a marked tendency to check and split. Distortion may be appreciable, sometimes accompanied by collapse.
Kiln Schedule C.
Shrinkage: Green to 12 per cent moisture content:
Tangential   about 5·5 per cent ($\frac{11}{16}$ in/ft)
Radial   about 3·0 per cent ($\frac{3}{8}$ in/ft)

working properties:
Blunting: Slight.
Sawing: Rip-sawing – Saw type HR 54.
Cross-cutting – Satisfactory.
Narrow bandsawing – Satisfactory.
Wide bandsawing – Saw type A.
Machining: Satisfactory provided that cutting edges are sharp.
Nailing: Satisfactory.
Gluing: Good.

wood bending: Unsuitable owing to severe buckling when compressed.
Classification – Poor.

plywood manufacture:
Employed for plywood but seldom seen in the United Kingdom.
Movement of plywood and surface splitting during weathering:
No information.

staining and polishing: Good.

durability and preservation

insect attack: Logs liable to severe attack by ambrosia (pinhole borer) beetles, and trees and logs liable to attack by forest longhorn or Buprestid beetles. Sapwood can be attacked by powder-post and common furniture beetles. Reported to be attacked by dry-wood termites in Honduras.

durability of heartwood: Perishable.
preservative treatment: Permeable.

uses A plain, easily worked wood of medium density. Suitable for joinery, box-making, and other general interior work.

BANGA WANGA
*Amblygonocarpus
obtusangulus*

A much branched tree reaching a height of 20 m (60 ft) or more. East Africa.

Heartwood rich red-brown, sometimes with narrow streaks of darker colour. Bears some resemblance to Rhodesian teak. Sapwood pale brown or pinkish-brown, sharply defined from heartwood. Grain interlocked, texture moderately fine. A very hard and heavy timber, weight about 960 kg/m³ (60 lb/ft³), seasoned. Small movement.

Kiln Schedule B suggested for drying the timber. Reported to be difficult to saw and work, picking up in planing. Will take a high polish if a smooth surface is obtained. Heartwood very durable and reported to be resistant to termites. Sapwood liable to attack by powder-post beetles.

A hard and heavy timber of good appearance and having high resistance to abrasion. Suitable for heavy constructional work and makes a hard-wearing and decorative flooring.

BASRALOCUS
Dicorynia guianensis
(*D. paraensis*)

other name: angélique.

THE TREE Reaches a height of 30 m (100 ft) or more. Long clear bole up to 1·5 m (5 ft) in diameter above buttresses. Surinam, French Guiana and Brazil.

**THE TIMBER
properties** Variable in colour and sometimes separated into two types having darker and lighter coloured wood. The darker and redder timber (angélique rouge) is the type available commercially. Heartwood reddish when freshly cut, darkening to dull brown or purple-brown. Sapwood narrow, paler in colour and fairly sharply defined. Grain generally straight or slightly interlocked. Texture medium. Weight varies from about 710 to 900 kg/m³ (44–56 lb/ft³), seasoned. Strength slightly higher than European beech.

processing Should be dried slowly. Stated to have a tendency to check and split in drying and to be liable to collapse. Working properties variable depending on the density and presence of silica. Tungsten carbide tipped tools required when working the dry timber. Glues well and finishes smoothly.

**durability
and preservation** Heartwood very durable and extremely resistant to marine borer attack. Probably extremely resistant to preservative treatment.

uses A very good timber for marine construction work, dock and harbour work, lock gates, etc., on account of its good strength properties and high resistance to decay and marine borer attack. Also suitable for heavy construction work, bridge flooring, boat framing, railway sleepers, etc. Reported to have good resistance to acids and has been used in Holland for barrel and vat staves. Has also been used for flooring.

BASSWOOD
Tilia americana

other name: American lime (Great Britain).

THE TREE

Usually grows to a height of 18–21 m (60–70 ft), sometimes up to 30 m (100 ft). Bole straight, frequently clear of branches for much of its length, diameter 0·6–0·8 m (2–2½ ft) or sometimes more. Eastern United States and Canada.

THE TIMBER
properties

colour: Sapwood nearly white, wide, merging into the slightly darker heartwood.
grain: Usually straight. Texture fine and even.
weight: Average about 420 kg/m³ (26 lb/ft³), seasoned.
strength: Slightly higher than obeche, especially in stiffness.

Moisture Content	Bending Strength		Modulus of Elasticity		Compression parallel to grain	
	N/mm²	lbf/in²	N/mm²	1000 lbf/in²	N/mm²	lbf/in²
Green	36	5200	6800	980	15·9	2300
12 per cent	64	9300	8800	1280	34·9	5060

(Data from Canadian Department of Forestry)

movement: No information.

processing

drying: Dries readily with little degrade.
Kiln Schedule K.
Shrinkage: Green to 12 per cent moisture content:
Tangential   about 5·0 per cent ($\frac{5}{8}$ in/ft)
Radial   about 3·5 per cent ($\frac{7}{16}$ in/ft)
working properties:
Blunting: Slight.
Sawing: Rip-sawing – Saw type HR 54.
Cross-cutting – Satisfactory.
Narrow bandsawing – Satisfactory.
Wide bandsawing – Saw type A.
Machining: Satisfactory; finishes smoothly and cleanly.
Nailing: Good.
Gluing: Good.
wood bending: Reported to have poor bending properties.
plywood manufacture:
Employed for plywood and usually available in the United Kingdom.
No information on movement of plywood or surface splitting during weathering.
staining and polishing: Satisfactory.

durability and preservation

insect attack: Trees and logs liable to be damaged by forest longhorn beetles. Sapwood liable to attack by the common furniture beetle and by *Ptilinus pectinicornis*.
durability of heartwood: Non-durable.
preservative treatment: Permeable.

uses

A soft, pale-coloured wood, similar in general characteristics to European lime, but lighter in weight. Used in the woodware trade and for turnery and hand carving. Employed for certain parts of musical instruments including piano keys, and for mallet heads and small tools. Used in United States for plywood, picture frames, toys, laundry and dairy appliances and other purposes. Free from odour and therefore useful for food containers.

BEECH, EUROPEAN
Fagus sylvatica

other names: English, Carpathian, Danish, French, Rumanian, Yugoslavian beech, etc., according to origin (Great Britain).

THE TREE Reaches a height of 30 m (100 ft), occasionally 45 m (150 ft). Diameter about 1·2 m (4 ft), sometimes greater. Clear bole 10–15 m (30–50 ft), according to growth conditions. Europe, approximately between latitude 40°N and 60°N, and western Asia.

THE TIMBER
properties

colour: Whitish to very pale brown, darkening on exposure to slightly reddish-brown. Some trees show a darker coloured core or 'red heart', often with dark veining.
The practice of steaming, common in south-east Europe, changes the colour to pink or light red.

sapwood: Not normally distinguishable from heartwood.

grain: Usually straight. Broad rays distinctly visible, especially on longitudinal surfaces. Texture fine and even.

weight: Variable according to growth conditions and climate. Home-grown and northern European beech, typically hard and dense, averages 720 kg/m³ (45 lb/ft³) at 12 per cent moisture content; central European beech, milder and less dense, averages 670 kg/m³ (42 lb/ft³) at 12 per cent moisture content.

strength: One of the strongest of home-grown timbers. Variation in density is reflected in strength properties.
Used as a standard of comparison for other timbers in this book.

Moisture Content	Bending Strength		Modulus of Elasticity		Compression parallel to grain	
	N/mm²	lbf/in²	N/mm²	1000 lbf/in²	N/mm²	lbf/in²
Green	65	9400	9800	1420	27·6	4010
12 per cent	118	17 100	12 600	1830	56·3	8170

movement: Large.
Moisture content in 90 per cent relative humidity   20 per cent
Moisture content in 60 per cent relative humidity   12 per cent
Corresponding tangential movement   3·1 per cent ($\frac{3}{8}$ in/ft)
Corresponding radial movement   1·7 per cent ($\frac{13}{64}$ in/ft)

processing

drying: Dries fairly well and fairly rapidly but is classed as a moderately refractory timber. Tendency to check, split and distort, and shrinkage in drying is very considerable.
Kiln Schedule D.
Shrinkage: Green to 12 per cent moisture content:
Tangential   about 9·5 per cent ($1\frac{1}{8}$ in/ft)
Radial   about 4·5 per cent ($\frac{9}{16}$ in/ft)

working properties:
Density and working properties variable according to conditions and locality of growth.
Blunting: Moderate (variable).
Sawing: Tendency for the saw to bind when green timber is converted.
Rip-sawing – Saw type HR 54, or HR 60 for dense material.
Cross-cutting – Burning and tooth vibration possible with the tougher material. Saw type 2 most satisfactory.
Narrow bandsawing – Satisfactory.
Wide bandsawing – Saw type B.

Machining: Tendency to burn during drilling. Otherwise all operations satisfactory. Cutting angle of 30° satisfactory in planing. Very good turning properties.

Nailing: Pre-boring necessary.

Gluing: Good.

wood bending:

Steam bending properties exceptionally good; pieces containing knots or irregular grain may be bent successfully. Beech of Continental origin is more variable than home-grown material.

Classification – Very good.

Ratio radius/thickness for solid bends (steamed):

	Supported	Unsupported
Home-grown	1·5	13
Danish	1·7	14·5
Rumanian	1·6	16

Limiting radius for 3·2 mm ($\frac{1}{8}$ in) laminae (unsteamed):

Home-grown	112 mm ($4\frac{2}{5}$ in)
Danish	135 mm ($5\frac{1}{4}$ in)
Rumanian	114 mm ($4\frac{1}{2}$ in)

plywood manufacture:

Employed for plywood and usually available in the United Kingdom.

Movement: 4·5 mm plywood from 30 per cent to 90 per cent relative humidity – 0·23 per cent.

Surface splitting on exposure to weather – Grade II.

staining and polishing: Satisfactory.

durability and preservation

insect attack: Bark and sapwood of logs and converted timber liable to attack by the longhorn beetle. Immune from attack by the powder-post beetles but liable to attack by furniture beetles. Timber in old buildings in England and Wales liable to attack by the death-watch beetle. Reported to be attacked by dry-wood termites in West Africa.

durability of heartwood: Perishable.

preservative treatment: Permeable.

uses

In the United Kingdom beech is used in larger quantities than any other hardwood. The largest consumer is the furniture industry which uses both home-grown and imported timber. The timber is heavy and strong, usually straight-grained, of plain appearance and has good turning and bending properties. Beech from Central Europe is often steamed; this gives it a pinkish tint, preferred for some purposes. In furniture manufacture it is used for solid parts in cabinet work, and its strength makes it particularly suitable for chairs, school desks, etc. Also used for joinery, general turnery, tool handles, brush backs and handles, bobbins, domestic woodware, sports goods and parts of musical instruments. Satisfactory for domestic flooring. Beech plywood is manufactured and imported from the Continent. Beech wood flour is used in manufacture of carbon products.

BEECH, JAPANESE
Fagus crenata and spp.

other name: buna (Japan).

The timber is derived from two or possibly three species of *Fagus*. The trees are similar in size to European beech. The timber resembles European beech in appearance and general character. Weight rather variable, average about 620 kg/m³ (39 lb/ft³), seasoned. The timber is milder and lighter in weight than home-grown beech and is more comparable to central European beech. It is reported to be more stable than European beech. Used for the same purposes as European beech.

BERLINIA
Berlinia spp.

other names: ekpogoi (Nigeria); abem (Cameroon); ebiara (Gaboon).
Several species of *Berlinia* including *B. confusa*, *B. grandiflora* and *B. occidentalis*, occur in West Africa and there is some confusion in nomenclature. Information given below is based on tests on *B. confusa* from Nigeria.

THE TREE

Reaches a height of 37 m (120 ft). Bole generally clear and free from buttresses, but not very straight and may be fluted at the base. Diameter up to 0·8 m (2½ ft). West Africa.

THE TIMBER
properties

colour: Heartwood pinkish-brown to deep red brown, with purple or brown irregular streaks.

sapwood: Variable in width ranging from 25 mm (1 in) to 300 mm (12 in) but commonly 100–150 mm (4–6 in). Dull white, often with a pink tint.

grain: Interlocked, sometimes very irregular. Texture rather coarse and open. Dark-coloured gum streaks containing a hard, black kino sometimes present.

weight: Variable, 550–820 kg/m^3 (34–51 lb/ft^3), averaging about 700 kg/m^3 (44 lb/ft^3), seasoned.

brittleheart: May be present in large logs.

strength: Comparable to European beech.

Moisture Content	Bending Strength		Modulus of Elasticity		Compression parallel to grain	
	N/mm^2	lbf/in^2	N/mm^2	1000 lbf/in^2	N/mm^2	lbf/in^2
Green	72	10 500	9100	1320	34·0	4930
12 per cent	105	15 300	10 800	1570	53·0	7690

movement: Medium.
Moisture content in 90 per cent relative humidity   19 per cent
Moisture content in 60 per cent relative humidity   12 per cent
Corresponding tangential movement   2·0 per cent ($\frac{1}{4}$ in/ft)
Corresponding radial movement   1·2 per cent ($\frac{9}{64}$ in/ft)

processing

drying: Dries rather slowly but well, apart from isolated cases of moderate distortion. A little extension of original shakes may occur. Tendency for mould growths, with associated discoloration, to develop during drying. Kiln Schedule E.
Shrinkage: Green to 12 per cent moisture content:
Tangential   about 5·0 per cent ($\frac{5}{8}$ in/ft)
Radial   about 3·0 per cent ($\frac{3}{8}$ in/ft)

working properties:
Grain usually interlocked, affecting machining properties. Wide sapwood sometimes has irregular grain.
Blunting: Moderate.
Sawing: Rip-sawing – Saw type HR 54, or type HR 60 for denser heartwood material to avoid tooth vibration.
Cross-cutting – Satisfactory apart from saw type 3 which causes burning. Other cross-cut saws tend to produce a fibrous finish.
Narrow bandsawing – Considerable breaking out at bottom of cut, otherwise satisfactory.
Wide bandsawing – Saw type B.

Machining:	Planing – A cutting angle of 20° necessary to produce a good finish. Moulding – French head produced best finish but blunted rather rapidly.
Nailing:	Satisfactory with care. Pre-boring necessary when nailing close to edges.
Gluing:	Good.

wood bending:
Cannot be bent when pin knots are present. Steaming is accompanied by slight resin exudation.
Classification – Moderate.
Ratio radius/thickness for solid bends (steamed):
Supported: 17·5   Unsupported: 19·5
Limiting radius for 3·2 mm ($\frac{1}{8}$ in) laminae (unsteamed): 137 mm ($5\frac{2}{5}$ in).

plywood manufacture: Unsuitable because of weight and coarse texture.

staining and polishing:
Satisfactory, but care is necessary with the wilder-grained sapwood.

durability and preservation

insect attack: Logs liable to severe attack by ambrosia (pinhole-borer) beetles. Sapwood liable to attack by powder-post beetles. Reported moderately resistant to termite attack in West Africa.

durability of heartwood: Non-durable.

preservative treatment: Resistant. Sapwood permeable.

uses
Primarily a structural timber, but selection to exclude brittleheart and very irregular grain is advisable. Suitable for many of the purposes for which oak and keruing are used, e.g. construction of lorry and bus bodies and frames. The colour and inter-locked and sometimes wavy grain produce a distinctive figure, and selected material can yield an attractive decorative veneer for furniture or panelling.

BINUANG
Octomeles sumatrana

other names: benuang (Sarawak, Indonesia); erima, ilimo (New Guinea).

THE TREE
Reaches a height of 55 m (180 ft), diameter 1·5 m (5 ft). Some trees are heavily buttressed. Bole regular, straight up to 21 m (70 ft). Indonesia, Borneo, Philippines, New Guinea and Western Pacific Islands.

THE TIMBER
properties
Heartwood pale brown or pinkish-brown. Wide sapwood, almost white but drab appearance, fairly clearly distinguishable from heartwood. Grain interlocked, producing broad stripe figure on quarter-sawn surfaces. Texture coarse. Weight ranges from 270 to 470 kg/m³ (17–29 lb/ft³), varying with position in tree, average about 400 kg/m³ (25 lb/ft³), seasoned. Brittleheart is frequently present and is lower in density and strength properties than normal wood. Wood free from brittle-heart is comparable in strength to obeche. Small movement.

processing
Dries slowly with severe degrade, especially in the heartwood-sapwood zone, and knots split moderately badly. Kiln Schedule C, but drying may not be satisfactory even with this mild schedule. Has only a slight blunting effect on cutting edges. Sawing generally satisfactory, but in cross-cutting considerable breaking out may occur at bottom of cut. Machining properties affected by interlocked grain and care required because of the softness of the wood. Blunt cutters produce a woolly finish. Satisfactory in nailing. Employed for plywood manufacture but seldom seen in the United Kingdom. Stains and polishes satisfactorily.

durability and preservation
Sapwood liable to attack by powder-post beetles. Heartwood perishable and moderately resistant to preservative treatment. The wide band of sapwood is permeable.

A timber of the obeche type, but difficult to season without deterioration, liable to fungal staining, and often affected by brittleheart. Suitable for rough carpentry work. Not likely to find a ready market in the United Kingdom.

BIRCH, EUROPEAN
Betula pubescens
and *B. verrucosa*

other names: English, Finnish, Swedish birch, etc., according to origin; silver birch, white birch (Great Britain).

THE TREE

Reaches an average height of 18–21 m (60–70 ft) and diameter of 0·6–1·0 m (2–3 ft) but is occasionally considerably larger. In Scandinavia boles are often straight and clean for 9 m (30 ft), especially when grown in pure stands. In British Isles grows in mixed stands and often has a more irregular crooked bole. Europe, from central Spain northwards, but mainly northern and eastern Europe.

THE TIMBER
properties

colour: White to light brown, bright appearance.

sapwood: Not visually distinguishable from heartwood.

grain: Straight. Texture fine. No conspicuous structural features.

weight: Average 660 kg/m³ (41 lb/ft³) at 12 per cent moisture content.

strength: Comparable to European beech. Toughness comparable to ash when dry though not when green.

Moisture Content	Bending Strength		Modulus of Elasticity		Compression parallel to grain		Impact (toughness) Max. drop of hammer	
	N/mm²	lbf/in²	N/mm²	1000 lbf/in²	N/mm²	lbf/in²	m	in
Green	63	9100	9900	1440	26·3	3820	0·76	30
12 per cent	123	17 800	13 300	1930	59·9	8690	1·04	41

movement: No information available.

processing

drying: Air drying – Dries relatively quickly with some tendency to distort. As the timber is very susceptible to fungal attack, conditions should be arranged to promote rapid drying, both for converted timber and for poles. Kiln drying – Dries fairly rapidly and well, but with a tendency to distort.
Kiln Schedule F.
Shrinkage: Green to 12 per cent moisture content:
  Tangential about 8·0 per cent (1 in/ft)
  Radial   about 5·0 per cent (⅝ in/ft)

working properties:
Blunting: Moderate.
Sawing: Rip-sawing – Saw type HR54.
     Cross-cutting – Satisfactory.
     Narrow bandsawing – Satisfactory.
     Wide bandsawing – Saw type B.
Machining: Satisfactory, but a cutting angle of 15° is advantageous to prevent tearing of cross-grained material and irregular grain around knots.
Nailing: Satisfactory. Pre-boring may be advisable when nailing near the edges of material with irregular grain.
Gluing: Good.

wood bending:
Good bending properties if free from knots and irregular grain, but these features are commonly present and detract from its suitability for bending purposes.
No data available on limiting radii of curvature.

plywood manufacture:
Employed for plywood manufacture and usually available in the United Kingdom.
Movement: 4·5 mm plywood from 30 per cent to 90 per cent relative humidity — 0·24 per cent.
Surface splitting on exposure to weather — Grade II.

staining and polishing: Satisfactory.

durability and preservation

insect attack: Not attacked by powder-post beetles, but liable to damage by the common furniture beetle.

durability of heartwood: Perishable.

preservative treatment: Permeable.

uses

Widely used as plywood which is manufactured in Finland and USSR. Solid birch, mainly from the Continent, is used in furniture for upholstery framing, and for domestic brushes and brooms, and turnery. When pressure treated with preservatives birch is suitable for posts where driving is not difficult. Used in Scandinavia for pulping.

BIRCH, PAPER
Betula papyrifera

other names: American birch, American or Canadian white birch (Great Britain). There are a number of varieties of *Betula papyrifera*, with some overlapping in their ranges of growth.

THE TREE

Grows to a height of 18–21 m (60–70 ft). Bole long, clean and cylindrical, average diameter 0·5 m (1½ ft). Canada and northern United States.

THE TIMBER
properties

Resembles European birch more closely than does yellow birch. Wide sapwood, nearly white in colour. Heartwood pale brown. Texture fine and uniform. Weight about 620 kg/m³ (39 lb/ft³), seasoned. Somewhat lower in density and strength properties than yellow birch.

processing

Dries satisfactorily; shrinkage somewhat lower than that of yellow birch. Kiln Schedule H. Working properties very similar to those of yellow birch. Glues well. Reported to have moderately good bending properties. Employed for plywood manufacture but no information on properties of plywood. Stains and finishes well.

durability and preservation

Growth defects known as pith flecks caused by insects are sometimes present in heartwood and sapwood. Heartwood non-durable. Probably moderately resistant to preservative treatment.

uses

The best butts are used for plywood and veneer. The timber is a good turnery wood, used for spools, bobbins, dowels and woodware. Also used for toys and parts of agricultural machinery.

BIRCH, YELLOW
Betula alleghaniensis
(formerly *B. lutea*)

other names: Canadian yellow birch, Quebec birch, American birch (Great Britain); betula wood, hard birch (Canada).

Curly grained and strongly figured varieties of this species have been marketed in Great Britain as Canadian silky wood. Sapwood and heartwood sometimes marketed separately as white birch and red birch, respectively. The former is liable to be confused with paper birch.

Shipments may occasionally contain a proportion of sweet birch (*B. lenta*); this species is somewhat denser and darker in colour than yellow birch.

THE TREE　　May reach a height of 30 m (100 ft) and diameter 1·0–1·2 m (3–4 ft), but averages 18–23 m (60–75 ft) in height and 0·6 m (2 ft) in diameter. Straight, cylindrical bole with only slight taper. North America.

THE TIMBER
properties

colour: Varies from light to dark reddish-brown. Since sapwood is normally marketed together with heartwood, a large proportion of commercial timber is fairly light in colour. On rotary cut material growth rings show as darker reddish-brown lines.

sapwood: Noticeably paler in colour than heartwood.

grain: Generally straight. Texture fine and even.

weight: Averages 690 kg/m³ (43 lb/ft³) at 12 per cent moisture content.

strength: Comparable to European beech. Similar to European ash in toughness.

Moisture Content	Bending Strength		Modulus of Elasticity		Compression parallel to grain		Impact (toughness) Max. drop of hammer	
	N/mm²	lbf/in²	N/mm²	1000 lbf/in²	N/mm²	lbf/in²	m	in
Green	60	8700	9700	1400	24·2	3510	1·40	55
12 per cent	121	17 500	13 000	1880	58·5	8480	1·52	60

(Data from US Department of Agriculture)

movement: Large.
Moisture content in 90 per cent relative humidity　21·5 per cent
Moisture content in 60 per cent relative humidity　12 per cent
Corresponding tangential movement　2·5 per cent ($\frac{19}{64}$ in/ft)
Corresponding radial movement　2·2 per cent ($\frac{17}{64}$ in/ft)

processing

drying: Dries rather slowly, with little degrade.
Kiln Schedule G.
Shrinkage:　Green to 12 per cent moisture content:
Tangential　about 4·5 per cent ($\frac{9}{16}$ in/ft)
Radial　　　about 3·5 per cent ($\frac{7}{16}$ in/ft)

working properties:
Blunting:　Moderate.
Sawing:　Rip-sawing – Saw type HR 54.
Cross-cutting – Satisfactory.
Narrow bandsawing – Satisfactory.
Wide bandsawing – Saw type B.
Machining:　Satisfactory, but cutting angle of 15° required to prevent tearing of cross-grained material and irregular grain around knots.
Nailing:　Satisfactory. Pre-boring advisable when nailing near the edges of material with irregular grain.
Gluing:　Good.

wood bending:
Classification – Very good.
Ratio radius/thickness for solid bends (steamed), based on small scale tests only:　Supported: 3　Unsupported: 17
Thin laminae (unsteamed):
No exact data available, but known from its use in manufacture of laminated tennis racquet frames to be very suitable for laminated bent work.

Employed for plywood and available in the United Kingdom.

Movement: 4·5 mm plywood from 30 per cent to 90 per cent relative humidity — 0·27 per cent.

Surface splitting on exposure to weather — Grade II.

staining and polishing: Good.

durability and preservation

insect attack: Growth defects caused by insects and known as pith flecks sometimes present in sapwood and heartwood. Not attacked by powder-post beetles but liable to attack by the common furniture beetle.

durability of heartwood: Perishable.

preservative treatment: Moderately resistant. Sapwood permeable.

uses Best quality logs are used for manufacture of high-grade plywood. The timber is also used for furniture, especially upholstery frames, for turnery, and parts of agricultural implements. As flooring it has high resistance to wear and is suitable for schools, dance halls, gymnasia and light-duty factory floors.

BLACK BEAN
Castanospermum australe

THE TREE A tree of height about 40 m (130 ft), diameter of bole up to 1·2 m (4 ft). Eastern Australia.

THE TIMBER
properties

A timber of attractive appearance, the chocolate-brown background being relieved by narrow, greyish-brown streaks of tissue surrounding the rather large pores. Bears some resemblance to European walnut or Queensland walnut. Grain usually straight but may be slightly interlocked. Texture rather coarse. Weight about 700 kg/m³ (44 lb/ft³), seasoned. A relatively hard and strong timber; medium movement.

processing Dries slowly with a marked tendency to collapse, honeycomb and split. Air drying before kiln drying is recommended. Kiln Schedule C. Working properties generally satisfactory, but may be affected by alternating soft and hard patches. Has a moderate blunting effect on cutting edges. Can be nailed satisfactorily, but gluing properties are variable. Finishes well with stains and polishes.

durability and preservation

Sapwood liable to attack by powder-post beetles. Reported to be moderately resistant to termites in Australia. Heartwood durable and reported to be extremely resistant to preservative treatment. The sapwood is permeable.

uses A very fine decorative wood employed in Australia both as solid wood and as veneer. Suitable for joinery, furniture and interior fittings, after satisfactory drying.

BLACKBUTT
Eucalyptus pilularis

THE TREE Grows to a height of about 45 m (150 ft), with a long, clear, straight bole, diameter about 1 m (3 ft). Eastern Australia.

THE TIMBER
properties

A pale brown to brown timber with a pinkish tinge, and narrow sapwood. Grain straight or occasionally interlocked or wavy, texture moderately fine. Small scattered gum veins are a typical feature. A hard and heavy timber, average weight about 880 kg/m³ (55 lb/ft³), seasoned. Strength properties similar to those of karri, but rather less hard and more easily split.

processing Very prone to check during drying, and collapse may occur. Air drying before kiln drying is suggested. Kiln Schedule C. Has a moderate blunting effect on cutting edges, and in cross-cutting saw type 3 is probably not suitable. In machining a cutting angle of 15° is required to plane material with irregular grain. For nailing, pre-boring is necessary. Can be glued satisfactorily and has been shown by tests to be usable for plywood. Finishes well with stains and polishes.

durability and preservation	Sapwood not liable to attack by powder-post beetles. Heartwood very durable and extremely resistant to preservative treatment.
uses	Too hard to work and too heavy for joinery but has been used for measuring instruments. Used in Australia for heavy construction work and for mallet heads, and sometimes as decorative veneer.

BLACKWOOD, AFRICAN
Dalbergia melanoxylon

THE TREE	A small, often misshapen tree, 4·5–7·5 m (15–25 ft) in height, yielding short logs, 1·0–1·5 m (3–5 ft) in length, which are commonly fluted and often defective. Mainly East African.
THE TIMBER properties	Heartwood dark purplish-brown with black streaks which usually predominate so that the general effect is nearly black. Sharply demarcated from a very narrow, almost white sapwood. Grain direction commonly somewhat variable according to log form. Has a fine and even texture and a slightly oily nature. Exceptionally hard and heavy, weight about 1200 kg/m³ (75 lb/ft³), seasoned. Small movement.
processing	Very mild drying conditions essential to avoid serious degrade; end coating of logs and sawn stock is necessary. Kiln Schedule B. Presents considerable difficulty in working. Causes rapid blunting of saws; stellite or tungsten carbide tipped saw teeth are recommended. In machining, tends to ride on cutters. Excellent for turnery and finishes exceptionally well.
durability and preservation	Sapwood liable to attack by powder-post beetles. Has been reported moderately resistant to termites in East Africa. Heartwood very durable (provisional).
uses	Used primarily in the manufacture of musical instruments, especially wood wind instruments such as clarinets, oboes, chanters of bagpipes, and some types of flutes and recorders. Particularly suitable for these purposes because of a combination of desirable properties, including small dimensional movement and impermeability to movement of air through the wood, and because it works to a fine finish. Used also for turnery for ornamental purposes and for such goods as brush backs, knife handles, chess-men, truncheons, and bearings and slides.

BLACKWOOD, AUSTRALIAN
Acacia melanoxylon

THE TREE		Grows to a height of 25–30 m (80–100 ft), diameter of bole about 1·0 m (3 ft). South-eastern Australia and Tasmania.
THE TIMBER properties	**colour:**	Varies from golden brown to dark brown, sometimes with a reddish tint, and with fairly regular dark brown zones marking the growth rings.
	sapwood:	Straw-coloured.
	grain:	Usually straight, but sometimes slightly interlocked or wavy, giving rise to an attractive fiddleback figure, which is generally associated with a natural lustre, producing a highly decorative appearance. Texture medium and even.
	weight:	Average about 660 kg/m³ (41 lb/ft³), seasoned.
	strength:	Comparable to European beech. Has good resistance to impact.

D

Moisture Content	Bending Strength		Modulus of Elasticity		Compression parallel to grain	
	N/mm^2	lbf/in^2	N/mm^2	$1000\ lbf/in^2$	N/mm^2	lbf/in^2
Green	75	10 900	11 000	1600	35·0	5070
12 per cent	115	16 700	13 200	1910	60·3	8740

(Data from Division of Forest Products, Melbourne)

movement: No information.

processing

drying: Reported to dry easily without degrade in boards up to 25 mm (1 in) thick. Cupping of wide, flat-sawn boards can generally be avoided by weighting the stack; otherwise it can be removed by a final steaming treatment.
Kiln Schedule E.
Shrinkage: Green to 12 per cent moisture content:
Tangential about 5·0 per cent ($\frac{5}{8}$ in/ft)

working properties:
Blunting: Moderate.
Sawing: Rip-sawing – Saw type HR 54 or HR 60.
Cross-cutting – Satisfactory.
Narrow bandsawing – Satisfactory.
Wide bandsawing – Saw type B.
Machining: A cutting angle of 20° required to plane wavy and inter-locked grain satisfactorily.
Other operations: Satisfactory. Turns well.
Nailing: Satisfactory.
Gluing: No information.

wood bending: Reported to have good steam bending properties.
No precise data available.

plywood manufacture: No information available.

staining and polishing: Good.

durability and preservation

insect attack: Sapwood liable to attack by powder-post beetles and by the common furniture beetle. Reported to be non-resistant to termites in South Africa.

durability of heartwood: Durable (provisional).

preservative treatment:
Extremely resistant. The sapwood is moderately resistant.

uses

A fine decorative timber used in Australia for high-quality furniture, panelling, etc., and for shop and bank fittings, interior joinery, handles and ornamental turnery. Also used for bent work in coach building, boat-building, etc.

BOMBAX
Bombax spp.,
principally
B. buonopozense

THE TREE

A large West African tree, producing timber very similar to that of ceiba; the two species are not always distinguished commercially.

THE TIMBER
properties

The wood is light yellowish-brown or pinkish-brown, very liable to fungal discolora-tion. No clear distinction between sapwood and heartwood. Texture coarse. Light in weight, about 350 kg/m³ (22 lb/ft³), seasoned. Appreciably softer than obeche and generally low in strength properties.

processing

Generally similar to ceiba in drying properties and working properties. Tests for plywood manufacture have shown promising results but require confirmation on freshly-extracted logs. Stains and polishes satisfactorily.

durability and preservation	Sapwood liable to attack by powder-post beetles. Reported to be non-resistant to termites in East and West Africa. Heartwood perishable; permeable to preservatives.
uses	Not usually separated from ceiba. Used mainly for blockboard and for the core of plywood. Also used in West Africa for food utensils such as baking trays, for insulation purposes, and for crates and boxes.

BOMBWAY, WHITE
Terminalia procera

other name: badam (India).

A large tree growing in the Andaman Islands. Timber varies in colour from light brown to moderately dark greyish-brown, mostly straight grained, weight about 640 kg/m³ (40 lb/ft³), seasoned. Somewhat lower in strength than beech. Small movement.

Dries fairly well (Kiln Schedule E). Sawing satisfactory, with moderate blunting effect. In planing, a reduction in cutting angle to 15° is required to plane material with interlocked grain satisfactorily. Nails and glues well. Shown by tests to be usable for plywood but the variable colour may be objectionable. Stains and polishes satisfactorily.

Sapwood liable to attack by powder-post beetles. Reported to be non-resistant to termites in India. Heartwood non-durable and moderately resistant to preservative treatment.

Used in India for core stock for plywood and tea chests. Suitable for furniture manufacture, joinery and general carpentry.

BOXWOODS

The trade name boxwood was originally applied to the wood of *Buxus sempervirens* from Europe and eastern Asia. It has been extended to include the South African boxwood, *Buxus macowani* (East London boxwood), and a number of botanically unrelated species with wood resembling true boxwood in general character. The more important of these are 'Maracaibo boxwood' (*Gossypiospermum praecox*), 'San Domingo boxwood' (*Phyllostylon brasiliensis*) and 'Knysna boxwood' (*Gonioma kamassi*).

BOXWOOD, EUROPEAN
Buxus sempervirens

other names: box; Abassian, Iranian or Persian, Turkey boxwood, etc., according to origin (Great Britain).

THE TREE — A small tree, growing to a height of 6–9 m (20–30 ft). Logs are generally 0·9–1·2 m (3–4 ft) long and 0·1–0·2 m (4–8 in) or occasionally up to 0·3 m (1 ft) in diameter. Europe, including parts of south-east England, North Africa and western Asia.

THE TIMBER
properties

colour:	Pale yellow.
grain:	Sometimes straight but often irregular, especially in wood from the small trees familiar in Great Britain. Texture very fine and even.
weight:	Varies from 830–1140 kg/m³ (52–71 lb/ft³), average about 910 kg/m³ (57 lb/ft³), seasoned.
strength:	A hard and strong timber; no precise data available.
movement:	No information.

processing

drying:	Dries very slowly with pronounced tendency to surface check. Liable to split badly if dried in the round, but splitting can be minimised by soaking the bolts before drying in a solution of common salt or urea, provided these chemicals are not objectionable in the subsequent use of the wood. Kiln Schedule B.

45

working properties:

Blunting: Moderate.

Sawing: Rip-sawing – Saw type HR 60.

Cross-cutting – Saw type 2 most suitable.

Narrow bandsawing – Satisfactory.

Wide bandsawing – Saw type C.

Machining: Planing – Pressure bar and shoe pressures should be increased to hold the wood against cutters. A reduction in cutting angle to 20° may be necessary to plane material with irregular grain satisfactorily.

Other operations:

Dull cutting edges may cause burning, particularly in boring and recessing. Turns very well.

Nailing: Pre-boring necessary.

Gluing: Believed to be good.

wood bending: Reported to have good bending properties.

plywood manufacture: No information.

staining and polishing: Good.

durability and preservation

insect attack: Sapwood not liable to attack by powder-post beetles but may be attacked by the common furniture beetle.

durability of heartwood: Durable (provisional).

preservative treatment: No information.

uses

Outstanding for its fine, smooth texture and good turning properties. Used for tool handles, skittles, croquet mallets and other small turnery, such as chess-men and fancy turnery goods. Also used for rollers in certain branches of the textile industry and for shuttles in the silk industry. Rulers, small pulley blocks, and some parts of musical instruments are also made from boxwood.

BOXWOOD, EAST LONDON
Buxus macowani

other name: Cape box (South Africa).

The tree has a clean bole 4·5–6·0 m (15–20 ft) long, average diameter 0·15 m (6 in). Occurs along the south-east coast of Cape Province.

The timber is very similar in appearance and properties to European boxwood (*Buxus sempervirens*), being pale yellow in colour, of fine, even texture, and weighing about 910 kg/m³ (57 lb/ft³), seasoned.

It is stated that small logs may be dried in the round without splitting, but larger sized logs should be halved or cut into dimension stock before drying. When air-dried the timber should be dried very slowly under cover. Very fine, deep surface checks are liable to develop and open up as drying proceeds. Kiln Schedule B.

Working properties very similar to those of European boxwood. Sharp cutters and increased pressure bar and shoe pressures required in planing. No information on wood bending, gluing properties, or plywood manufacture.

Durable (provisional). No information on preservative treatment.

The timber is suitable for many of the same purposes as European boxwood, e.g. for turnery and engraving.

'BOXWOOD, KNYSNA'
Gonioma kamassi

other names: kamassi (South Africa); 'kamassi boxwood' (Great Britain).

Grows to a height of 6 m (20 ft), or sometimes more. Has a clean, straight bole, about 0·3 m (1 ft) in diameter. Found in the coastal strip in the south of Cape Province.

The timber is similar in general character to European and East London boxwood (*Buxus* species). It is a close-grained, fine-textured timber, weight about 930 kg/m³ (58 lb/ft³), seasoned. No strength data available. It is reported that the fine dust has ill-effects (headaches, giddiness, etc.) on some individuals and the use of dust-extraction equipment is advised.

Dries well under slow drying conditions. If the drying is forced the wood splits very badly and numerous surface checks, which may be small but very deep, may appear on the surface of boards. Kiln Schedule C.

Working properties similar to those of European boxwood. Sawing satisfactory, with a moderate blunting effect. Care required in planing and in boring and recessing because of the hardness of the wood. No information on gluing.

Has been used for many of the same purposes as European boxwood and is considered equal to this timber for engravers' work and fancy turnery.

'BOXWOOD, MARACAIBO'
Gossypiospermum praecox

other names: 'Venezuelan boxwood' (UK); 'West Indian boxwood' (UK); zapatero (Venezuela).

A small- to medium-sized tree, growing in northern South America. The timber is derived mainly from Venezuela.

Logs well formed, about 2·5–3·5 m (8–12 ft) long, diameter 0·15–0·3 m (6–12 in), or occasionally more.

Wood very similar to true boxwoods (*Buxus* species). Lemon-yellow or nearly white in colour with a high lustre. Little difference between heartwood and sapwood. Blue-stain common in logs stored in humid conditions. Grain generally straight, texture very fine and uniform. Weight about 800–900 kg/m³ (50–56 lb/ft³), seasoned.

Processing generally similar to that of other boxwoods.

Used for manufacture of precision rules, engravers' blocks, carving and turnery, and as veneers for cabinet work.

'BOXWOOD, SAN DOMINGO'
Phyllostylon brasiliensis

other name: baitoa (Dominican Republic).

A small- to medium-sized tree, height 15–21 m (50–70 ft). Diameter of bole up to 0·6 m (2 ft). Boles of large trees may be irregular or fluted.

Timber generally similar to that of other boxwoods. Heartwood lemon-yellow, sometimes with a tinge of brown and occasionally with dark streaks. Grain fairly straight but sometimes irregular. Texture fine and uniform. Weight about 950 kg/m³ (59 lb/ft³), seasoned.

Processing and uses generally similar to other boxwoods.

BRUSH BOX
Tristania conferta

A large tree, height about 36 m (120 ft), with a long, clean bole, diameter up to 2 m (6 ft). Eastern Australia.

A brown or reddish-brown timber. Grain inclined to be interlocked, texture fine and even. Weight about 900 kg/m³ (56 lb/ft³). Has high-strength properties with good toughness and resistance to wear. Siliceous.

Has a marked tendency to distort and shrink in drying and some collapse may occur in kiln drying from the green. Partial air drying before kiln drying and a final reconditioning treatment are suggested. Kiln Schedule C. Has a severe blunting effect on tools and is fairly hard to work. Finishes cleanly in most operations when straight-grained, but interlocked material requires a cutting angle of 20°. Can be nailed with care. Unsuitable for plywood manufacture because of its weight and processing difficulties.

Heartwood moderately durable. Extremely resistant to preservative treatment.

Suitable for bridge and wharf decking and for heavy flooring. Can also be used for mauls and mallets and for tool handles.

BUBINGA

Guibourtia demeusii,
G. pellegriniana
and *G. tessmannii*

other name: kevazingo (Gaboon).

THE TREE

Large trees, reaching a height of 30 m (100 ft) or more. Clear bole 9–18 m (30–60 ft) long, diameter about 1·0–1·5 m (3–5 ft). Distributions of the three species overlap, but *G. tessmannii* is most common in the Cameroon, *G. pellegriniana* in Gaboon, and *G. demeusii* is more widely distributed.

THE TIMBER
properties

Timbers of the three species are similar in general character. Heartwood red to reddish-brown with purple veining when fresh. On exposure it becomes yellow or medium brown with a reddish tint and veining becomes less conspicuous. Sapwood rather wide and whitish. Texture moderately coarse. Grain sometimes straight or more usually interlocked, but may be very irregular in some logs and these give highly figured veneers. Highly figured wood is considered to be more common in timber from Gabon (shipped as kevazingo) than in material from the Cameroons. The timber is moderately hard and heavy. Weight generally between 800 and 960 kg/m³ (50–60 lb/ft³), seasoned.

processing

Reported to saw and machine without difficulty but care is needed in material having irregular grain. Takes a fine finish.

durability
and preservation

Reported to be highly resistant to termites in West Africa. Heartwood moderately durable (provisional).

uses

Has an attractive appearance and is used in the United Kingdom mainly as veneer for decorative panelling and inlay work. Other uses include high-class furniture and fancy turnery work, e.g. knife handles, brush backs, etc.

CAMPHORWOOD, EAST AFRICAN

Ocotea usambarensis

THE TREE

Reaches a height of 45 m (150 ft). Clear, straight bole generally 9–15 m (30–50 ft), but may be less. Diameter 1·2–1·8 m (4–6 ft), occasionally up to 3 m (10 ft). Large logs frequently have rotten cores and must be selected to produce wide boards. East Africa, mainly Kenya and Tanzania.

THE TIMBER
properties

colour:	Light yellowish-brown, darkening to deep brown on exposure.
sapwood:	Not sharply defined; paler in colour than heartwood.
grain:	Commonly interlocked, producing a stripe figure on quartered material. Texture moderately fine.
weight:	From 510 to 640 kg/m³ (32–40 lb/ft³), averaging 590 kg/m³ (37 lb/ft³) at 12 per cent moisture content.
odour:	The timber has a distinct scent of camphor.
strength:	Somewhat lower than European beech.

Moisture Content	Bending Strength		Modulus of Elasticity		Compression parallel to grain	
	N/mm²	lbf/in²	N/mm²	1000 lbf/in²	N/mm²	lbf/in²
Green	59	8500	8100	1180	30·6	4440
12 per cent	92	13 300	9900	1440	52·3	7590

		movement:	Small.

movement: Small.
Moisture content in 90 per cent relative humidity   14 per cent
Moisture content in 60 per cent relative humidity   11 per cent
Corresponding tangential movement   0·9 per cent ($\frac{7}{64}$ in/ft)
Corresponding radial movement   0·5 per cent ($\frac{1}{16}$ in/ft)

processing

drying: Dries slowly with little degrade. It is particularly difficult to remove the moisture from the centre of thick, quartered planks.
Kiln Schedule G.
Shrinkage: Green to 12 per cent moisture content:
Tangential   about 4·0 per cent ($\frac{1}{2}$ in/ft)
Radial   about 2·5 per cent ($\frac{5}{16}$ in/ft)

working properties: Interlocked grain affects machining properties.
Blunting: Slight.
Sawing: Rip-sawing – Saw type HR 54 satisfactory.
Cross-cutting – Satisfactory.
Narrow bandsawing – Satisfactory.
Wide bandsawing – Saw type B.
Machining: In planing a 20° cutting angle is required to plane inter-locked grain satisfactorily.
Other operations – Satisfactory.
Nailing: Satisfactory.
Gluing: Good.

wood bending:
Has moderately good bending qualities, though inferior to the well-known bending timbers such as beech, ash or oak. Tendency for some distortion to occur during the bending operation and during the setting process. Does not bend satisfactorily if small knots are present.
Classification – Moderate.
Ratio radius/thickness for solid bends (steamed):
Supported: 14   Unsupported: 27
Limiting radius for 3·2 mm ($\frac{1}{8}$ in) laminae (unsteamed): 175 mm (7 in).

plywood manufacture: No information.

staining and polishing: Good.

durability and preservation

insect attack: Sapwood rarely attacked by powder-post beetles.

durability of heartwood: Very durable.

preservative treatment: Extremely resistant. Sapwood permeable.

uses

A medium-weight timber of good appearance and high durability. It is important to select straight grained timber: many trees are ill-shaped and have twisted grain. Suitable for cabinet work, interior and exterior joinery, and vehicle building. Can be used for domestic flooring. Should not be used for draining boards or kitchen work on account of the camphor-like odour.

CANARIUM, AFRICAN
Canarium schweinfurthii

other names: abel (Cameroon); aiélé (France and Ivory Coast); elemi (Nigeria); mwafu (Uganda).

THE TREE

Reaches a height of 37 m (120 ft). Straight, cylindrical bole, length 27 m (90 ft), diameter about 1·2 m (4 ft). Very slight buttresses. Widely distributed in East, Central and West Africa.

THE TIMBER
properties

colour: Pale pinkish-brown.

sapwood: White or straw-coloured, up to 100 mm (4 in) wide.

grain: Interlocked, sometimes producing a very attractive stripe or roe figure when cut on the quarter. Texture coarse.

weight: About 530 kg/m³ (33 lb/ft³) at 12 per cent moisture content.

brittleheart: Sometimes found in large logs.
scent: The freshly cut timber has a pleasant scent.
strength: About half-way between obeche and European beech.

Moisture Content	Bending Strength		Modulus of Elasticity		Compression parallel to grain	
	N/mm²	lbf/in²	N/mm²	1000 lbf/in²	N/mm²	lbf/in²
Green	41	5900	6200	900	21·6	3130
12 per cent	70	10 100	8100	1180	42·5	6160

movement: Medium.
Moisture content in 90 per cent relative humidity   21·5 per cent
Moisture content in 60 per cent relative humidity   13 per cent
Corresponding tangential movement   2·3 per cent ($\frac{9}{32}$ in/ft)
Corresponding radial movement   1·0 per cent ($\frac{1}{8}$ in/ft)

processing

drying: Dries rather slowly and fairly well. Cross-sectional distortion and collapse may be troublesome; tendency to end-splitting, and for original shakes to extend.
Kiln Schedule H.
Shrinkage: Green to 12 per cent moisture content:
Tangential   about 4·5 per cent ($\frac{9}{16}$ in/ft)
Radial   about 2·5 per cent ($\frac{5}{16}$ in/ft)

working properties: Interlocked grain affects machining properties.
Blunting: Severe, due to presence of silica.
Sawing: Rip-sawing – Saw type HR 40, or TC for long runs.
Cross-cutting – Saw type 2 most satisfactory.
Narrow bandsawing – Satisfactory.
Wide bandsawing – Saw type A (tipped teeth).
Machining: Planing – Satisfactory. A reduction in cutting angle to 20° gives improved finish on interlocked faces, provided the cutters are kept sharp. Otherwise a woolly finish is obtained.
Moulding – French head not satisfactory. Collars most satisfactory.
Drilling – Three-wing drills most satisfactory.
General – High-speed steel cutters are satisfactory if kept sharp but blunt cutting edges give a woolly finish.
Nailing: Satisfactory.
Gluing: Good.

wood bending:
Unsatisfactory for steam bending. Severe buckling and fibre rupture occur when the wood is bent even to a large radius of curvature.
Classification – Very poor.
Limiting radius for 3·2 mm ($\frac{1}{8}$ in) laminae (unsteamed): 188 mm ($7\frac{2}{5}$ in).

plywood manufacture:
Employed for plywood and usually available in the United Kingdom.
Movement of plywood and surface splitting – No information.

staining and polishing: Satisfactory.

durability and preservation

insect attack: Sapwood liable to attack by powder-post beetles. Reported to be non-resistant to termites in West Africa.
durability of heartwood: Non-durable.
preservative treatment: Heartwood extremely resistant. Sapwood permeable.

uses

Uses of the timber are restricted by its severe blunting effect on cutting edges. Suitable for the cores of plywood, and can be sliced to produce decorative panelling, as it has an attractive figure and may readily be stained.

50

CANARIUM, INDIAN
Canarium euphyllum

other names: dhup, white dhup (India).

THE TREE

A tree of height about 25–30 m (80–100 ft), diameter of bole 0·6–1·0 m (2–3½ ft). Found mainly in Andaman Islands.

THE TIMBER
properties

Timber nearly white to pinkish- or yellowish-grey, having an attractive stripe or roe figure when quarter-sawn, and a natural lustre. Resembles gaboon in grain and texture. Weight about 400 kg/m³ (25 lb/ft³), seasoned. Strength somewhat higher than obeche.

processing

In air drying, rapid drying conditions are necessary to avoid mould and discoloration. Vertical stacking should give good results. Kiln dries well (Kiln Schedule H). Works easily, with little blunting effect on tools. Saw type SR or HR54 recommended for rip-sawing. In planing quarter-sawn material a cutting angle of 20° is recommended owing to interlocked grain. Grain raising may occur if blunt cutters are used. Nails and glues well. Used for plywood manufacture but seldom seen in the United Kingdom. Stains and polishes well.

durability and preservation

Sapwood liable to attack by powder-post beetles. Reported to be non-resistant to termites in India. Heartwood perishable and extremely resistant to preservative treatment.

uses

In the United Kingdom has been used mainly for shopfitting and other purposes where it is usual to stain the timber. Suitable also for planking, packing cases and interior fittings. Used in India in the match industry.

CEIBA
Ceiba pentandra

other names: fromager (France); fuma (Congo Republic).

The timber of the closely-allied bombax is similar to that of ceiba and is commonly marketed with it.

THE TREE

A large tree, up to 60 m (200 ft) high, with a straight, cylindrical bole 12–15 m (40–50 ft) long and 2 m (6 ft or more) in diameter. Has large buttresses which may extend 8 m (25 ft) up the bole. Tropical West Africa and America.

THE TIMBER
properties

colour: Varies from pale yellowish-brown to pinkish-brown, but is very liable to fungal discoloration.

sapwood: Not clearly demarcated from heartwood.

grain: Interlocked and sometimes irregular in direction. Texture coarse. The wood lacks the high natural lustre and smooth feel of some other lightweight woods, such as obeche and balsa.

weight: Very light but variable, from about 210 to 450 kg/m³ (13–28 lb/ft³), average about 320 kg/m³ (20 lb/ft³), seasoned.

strength: Low in relation to its weight. Limited data indicate that it has about two-thirds the strength of obeche.

movement: No information.

processing

drying: Dries rapidly without marked distortion.
Kiln Schedule J.
No information on shrinkage.

working properties:

A difficult wood to saw cleanly and finish smoothly because of its light weight.

Blunting: Slight.

Sawing: Rip-sawing – Saw type HR 54.

Cross-cutting and Narrow bandsawing – Satisfactory, but sawn surfaces tend to be woolly.

Wide bandsawing – Saw type A.

Machining: Sharp cutting edges necessary to sever fibres cleanly. A reduced sharpness angle is an advantage.

Care is needed in boring, end-grain working and turning.

Nailing: Satisfactory, but nails have poor holding properties.

Gluing: No information.

plywood manufacture:

Peels to give a good veneer provided that logs are fresh and free from insect and fungal attack. Most suitable for core stock.

staining and polishing: Satisfactory.

durability and preservation

insect attack: Sapwood liable to attack by powder-post beetles. Reported to be non-resistant to termites in East and West Africa.

durability of heartwood: Perishable.

preservative treatment: Permeable.

uses

A lightweight wood, rather coarser in texture than obeche. Its successful utilisation requires rapid extraction, conversion and drying to avoid deterioration and staining. Used mainly for plywood cores and blockboard; has also been used for food utensils, crates and boxes and very lightweight joinery.

CELTIS, AFRICAN

Celtis spp., principally *C. adolfi-friderici*, *C. mildbraedii* and *C. zenkeri*

other names: esa (Ghana); ita, ohia (Nigeria).

There are a number of species of *Celtis* in tropical Africa. The three mentioned above are among the most important as a potential source of timber. No information is available for the individual species; it is believed that the following description is applicable to all three species.

THE TREE

Reaches a height of 27–36 m (90–120 ft). Clear bole, diameter 0·8–1·1 m ($2\frac{1}{2}$–$3\frac{1}{2}$ ft), with long buttresses. Tropical Africa.

THE TIMBER
properties

Heartwood and sapwood not easily distinguishable. Whitish or clear light yellow when freshly cut, becoming greyish-white on exposure. Frequently discoloured by fungal stain. Grain sometimes straight but frequently irregular. Texture fairly fine and uniform. Average weight 780 kg/m³ (49 lb/ft³), seasoned. Strength properties somewhat higher than European beech. Medium movement.

processing

Dries fairly rapidly with little degrade. Slight end-splitting and distortion may occur. Kiln Schedule H. Has a moderate blunting effect on cutting edges. Considerable resistance to hand feeding in rip-sawing and bandsawing. In cross-cutting breaking out occurs at bottom and back of cut. In planing a reduction of cutting angle to 15° is necessary to prevent tearing of interlocked and irregular grain. Bending properties moderate. Has been shown by tests to be usable for plywood though the quality of veneer is variable; might be important in local markets. Glues well, and stains and polishes satisfactorily.

durability and preservation   Very susceptible to fungal staining unless preventive measures are taken. Damage by longhorn beetles sometimes present. Sapwood liable to attack by powder-post beetles. Heartwood perishable and moderately resistant to preservative treatment. The sapwood is permeable.

uses   A heavy timber having good strength properties; should be a useful substitute for ash except for wood bending purposes. A good flooring timber which wears smoothly and withstands heavy traffic without serious breakdown. Forms a good substitute for maple for dance floors. The susceptibility of the timber to staining and to attack by ambrosia (pinhole borer) beetles while in the log has restricted its utilisation.

'CENTRAL AMERICAN CEDAR'
Cedrela spp., principally *C. odorata*
and

'SOUTH AMERICAN CEDAR'
Cedrela spp., principally *C. fissilis*

other names: 'British Honduras', 'Honduras', 'Mexican', Nicaraguan', 'Tabasco', 'Trinidad', 'West Indian', 'Brazilian', 'Guyana', 'Peruvian cedar', etc., according to origin; 'cedar', 'cigar box cedar' (United Kingdom); cedro (South America). 'Central American cedar' is sometimes called 'Spanish cedar' in reference to the former Spanish colonies.

THE TREE   Varies in size according to species and locality of growth. Height from 21–30 m (70–100 ft) or sometimes more. Diameter usually 1–2 m (3–6 ft) above buttresses, which may be 2–3 m (6–10 ft) high. Bole straight and cylindrical, clear of branches for 12–18 m (40–60 ft). Central America, West Indies, and South America except Chile.

THE TIMBER
properties   Bears a general resemblance to the softer grades of mahogany, but exhibits a wide variation in its general character, due partly to differences between species, but mainly to the age and conditions of growth of individual trees.

colour: Varies from pale pinkish-brown to dark reddish-brown. Timber from young or fast-grown trees is commonly paler in colour and less resinous than that from mature or more slowly-grown trees.

sapwood: Paler in colour and generally sharply demarcated from heartwood.

grain: Straight or shallowly interlocked. Texture moderately coarse.

odour: The timber is characterised by a distinctive fragrant scent, due to the presence of an oil which may exude and appear on the surface of the timber as a sticky resin.

weight: Very variable from about 370 to 750 kg/m³ (23–47 lb/ft³), average about 480 kg/m³ (30 lb/ft³), seasoned.

strength: No precise data available. Strength properties variable but generally rather high in proportion to its weight.

movement: Small.
Moisture content in 90 per cent relative humidity   21·5 per cent
Moisture content in 60 per cent relative humidity   14·5 per cent
Corresponding tangential movement   1·5 per cent ($\frac{3}{16}$ in/ft)
Corresponding radial movement   1·0 per cent ($\frac{1}{8}$ in/ft)

processing	**drying:** Kiln dries fairly rapidly and satisfactorily, though with some tendency to distort and collapse. Knots tend to split badly but surface checking is not likely to be serious. Individual pieces may distort or collapse appreciably. Kiln Schedule H.

Shrinkage: Green to 12 per cent moisture content:
Tangential   about 4·0 per cent ($\frac{1}{2}$ in/ft)
Radial   about 3·0 per cent ($\frac{3}{8}$ in/ft)

working properties:
Finish may be affected by interlocked grain and occasional gum pockets.
Blunting: Slight.
Sawing: Rip-sawing – Saw type HR 54.
             Cross-cutting – Satisfactory.
             Narrow bandsawing – Satisfactory.
             Wide bandsawing – Saw type A.
Machining: Planing – A good finish produced by reducing the cutting angle to 20°.
             Moulding – French head most satisfactory.
             Other operations – Satisfactory. Sharp cutting edges required to avoid a woolly finish.
Nailing: Satisfactory.
Gluing: Good.

wood bending:
No precise data available. Central American cedar is used in producing hulls for light racing craft in the boat-building industry and is therefore suitable for bends of moderate radius of curvature. South American cedar is reported to have moderately good steam bending properties.

plywood manufacture:
Employed for plywood. Probably included in cedar plywood obtainable in the United Kingdom.
No information on movement of plywood or surface splitting on exposure to weather.

staining and polishing: Good.

durability and preservation

insect attack: Sapwood liable to attack by powder-post beetles. Reported to be highly resistant to termites in the West Indies and moderately resistant in West Africa.

durability of heartwood: Durable.

preservative treatment:
Heartwood of South American cedar extremely resistant. Sapwood reported to be permeable.

uses

Suitable for high-quality cabinet work, interior joinery and panelling. In boat-building it is employed for planking and for skins of racing boats and decks of canoes as it combines durability and light weight. Also used for cigar boxes and sometimes for sound boards for organs.

CHERRY, EUROPEAN
Prunus avium

other names: gean, mazzard, wild cherry (Great Britain).

THE TREE   Reaches a height of 18–25 m (60–80 ft), diameter about 0·6 m (2 ft). Europe, south of about latitude 60° north, and Asia Minor.

THE TIMBER properties

colour: Pale pinkish-brown, darkening somewhat on exposure to light.
sapwood: Lighter in colour than heartwood and moderately well defined.
grain: Generally straight. Texture fine and even.
weight: About 600 kg/m³ (38 lb/ft³) at 12 per cent moisture content.
strength: Slightly weaker than European beech.

Moisture Content	Bending Strength		Modulus of Elasticity		Compression parallel to grain	
	N/mm²	lbf/in²	N/mm²	1000 lbf/in²	N/mm²	lbf/in²
Green	64	9300	8300	1200	27·8	4030
12 per cent	110	15 900	10 200	1480	54·5	7910

movement: Medium.
Moisture content in 90 per cent relative humidity   19 per cent
Moisture content in 60 per cent relative humidity   12·5 per cent
Corresponding tangential movement   2·0 per cent ($\frac{1}{4}$ in/ft)
Corresponding radial movement   1·2 per cent ($\frac{9}{64}$ in/ft)

processing

drying: Dries fairly readily but with a pronounced tendency to warp.
Kiln Schedule A.
Shrinkage: Green to 12 per cent moisture content:
Tangential   about 6·5 per cent ($\frac{13}{16}$ in/ft)
Radial   about 3·5 per cent ($\frac{7}{16}$ in/ft)

working properties:
Blunting: Moderate.
Sawing: Rip-sawing – Saw type HR 54.
Cross-cutting – Satisfactory.
Narrow bandsawing – Satisfactory.
Wide bandsawing – Saw type B.
Machining: Satisfactory provided the material is reasonably straight grained. A reduction in cutting angle to 20° would improve the quality of finish for irregular grained material.
Nailing: No information.
Gluing: Good.

wood bending:
Classification – Very good.
Ratio radius/thickness for solid bends (steamed):
Supported: 2·0   Unsupported: 17
Limiting radius for 3·2 mm ($\frac{1}{8}$ in) laminae (unsteamed): 150 mm (6 in).

plywood manufacture:
Employed in plywood but seldom seen in the United Kingdom.
Movement of plywood and surface splitting: No information.

staining and polishing: Good.

durability and preservation

insect attack: Sapwood liable to attack by the common furniture beetle, but almost immune from attack by powder-post beetles.
durability of heartwood: Moderately durable (provisional).
preservative treatment: No information available.

uses

A decorative wood, generally used in small sections as it is inclined to warp. Very suitable for cabinet and furniture making and for panelling and decorative joinery. Turns well and is used for domestic ware, shuttle pins, toys and parts of musical instruments.

CHESTNUT, SWEET
Castanea sativa

other names: Spanish chestnut, European chestnut (Great Britain).

THE TREE

Reaches a height of 30 m (100 ft) or more and diameter about 1·5 m (5 ft). Clear, straight bole, 6 m (20 ft) or more in length. Also frequently coppice grown to pole size on 15–20 year rotation.

THE TIMBER	colour:	Yellowish-brown, closely resembling oak.
properties	sapwood:	Pale in colour, about 13 mm ($\frac{1}{2}$ in) wide, clearly distinguishable from heartwood.
	grain:	Straight, except in timber from old trees which may be spiral grained. Owing to absence of broad rays, does not have the characteristic silver grain of quartered oak. Texture coarse.
	ring shake:	Liable to be present in timber from old trees.
	weight:	Averages about 540 kg/m³ (34 lb/ft³) at 12 per cent moisture content.
	corrosion:	Owing to its acidic character it tends to accelerate corrosion of metals, especially iron and steel, in contact with it under damp conditions.
	chemical staining:	
		Blue-black iron stains liable to appear if wood is in contact with iron or iron compounds in presence of moisture.
	strength:	About half-way between obeche and European beech.

Moisture Content	Bending Strength		Modulus of Elasticity		Compression parallel to grain	
	N/mm²	lbf/in²	N/mm²	1000 lbf/in²	N/mm²	lbf/in²
Green	52	7600	7200	1040	24·2	3510
12 per cent	79	11 500	8200	1190	44·4	6440

movement: Small.

Moisture content in 90 per cent relative humidity   17·5 per cent
Moisture content in 60 per cent relative humidity   12·5 per cent
Corresponding tangential movement   1·3 per cent ($\frac{5}{32}$ in/ft)
Corresponding radial movement   0·7 per cent ($\frac{5}{64}$ in/ft)

processing

drying: Dries slowly with marked tendency to collapse and honeycomb and to retain patches of moisture. Collapsed timber does not generally recondition satisfactorily.
Kiln Schedule D.
Shrinkage: Green to 12 per cent moisture content:
Tangential   about 5·5 per cent ($\frac{11}{16}$ in/ft)
Radial   about 3·0 per cent ($\frac{3}{8}$ in/ft)

working properties:
Blunting: Slight.
Sawing: Rip-sawing – Saw type HR 54.
Cross-cutting – Satisfactory.
Narrow bandsawing – Satisfactory.
Wide bandsawing – Saw type A or B.
Machining: Satisfactory.
Nailing: Satisfactory.
Gluing: Good.

wood bending:
When bent in the green state very liable to rupture on the inner face, particularly when knots, even of small size, are present. Air-dry wood suitable for bending if free from knots and other defects, but slight wrinkling on the edges may occur.
Classification – Good.
Ratio radius/thickness for solid bends (steamed):
Supported: 6   Unsupported: 15
Limiting radius for 3·2 mm ($\frac{1}{8}$ in) laminae (unsteamed): 190 mm ($7\frac{1}{2}$ in).

plywood manufacture:
Shown by very limited tests to be unsuitable for plywood manufacture because of excessive splitting of veneer during drying.

staining and polishing: Satisfactory.

**durability
and preservation**

insect attack: Sapwood liable to attack by powder-post beetles, the common furniture beetle, and the furniture beetle *Ptilinus pectinicornis*. Sapwood and heartwood of timbers in old buildings in England and Wales liable to be attacked by the death-watch beetle.

durability of heartwood: Durable.

preservative treatment: Extremely resistant.

uses

Resembles oak in appearance, but is lighter in weight and more easily worked. Used for furniture, coffin boards, fencing and gates, and for domestic ware, ornamental bowls and kitchen utensils. Grown as coppice and used, on account of its durability, for cleft fencing, stakes and hop poles. Casks of chestnut staves are used for oils and fats, fruit juices, cheap wines, etc. Also used for barrel hoops and for walking sticks and umbrella handles.

CHICKRASSY
Chukrasia tabularis

other names: chittagong wood (United Kingdom); yinma (Burma).

A slender tree, height about 21–25 m (70–80 ft), with a clean cylindrical bole, length 9 m (30 ft), diameter 0·6 m (2 ft). Parts of India and Burma.

An attractive wood, fragrant when fresh, yellowish-red to red, toning down to yellowish- or reddish-brown, with a high lustre. Sapwood pale, yellowish or brownish, not clearly demarcated from heartwood. Plain sawn timber has an attractive growth ring figure and selected logs show fiddleback, roe figure and mottle. Average weight about 620 kg/m³ (39 lb/ft³), seasoned. Fairly strong and moderately hard. Reported to have small movement.

Dries well, even in large sections. Liable to develop very fine hair checks on the surface which are noticeable after polishing. For air drying, logs should be converted green and the sawn stock carefully piled in open stacks under cover. Kiln Schedule E. Works fairly easily and finishes cleanly in most operations, but a cutting angle of 20° is advisable when planing material with interlocked grain. Nails well and glues well. Used for plywood manufacture but seldom seen in the United Kingdom. Stains and polishes well.

Reported to be moderately resistant to termites in Ceylon. Heartwood non-durable (provisional) and extremely resistant to preservative treatment.

An attractive wood which can be used in the solid for furniture or as decorative veneer for panelling. Used in India for building work provided the timber is not in contact with the ground.

'CHILEAN LAUREL'
Laurelia aromatica

THE TREE

Grows to a height of about 14–15 m (45–50 ft). Bole straight, diameter generally not more than 0·6 m (2 ft), but sometimes up to 0·9 m (3 ft). Chile.

**THE TIMBER
properties**

colour: Variable, yellowish-brown with greenish, grey and purplish streaks having some resemblance to American whitewood.

sapwood: Uniform greyish-brown.

grain: Generally straight. Texture moderately fine.

weight: About 510 kg/m³ (32 lb/ft³) at 12 per cent moisture content.

strength: About half-way between obeche and European beech.

Moisture Content	Bending Strength		Modulus of Elasticity		Compression parallel to grain	
	N/mm²	lbf/in²	N/mm²	1000 lbf/in²	N/mm²	lbf/in²
Green	–	–	–	–	–	–
12 per cent	90	13 100	9800	1420	47·1	6830

movement: Large.
Moisture content in 90 per cent relative humidity   21·5 per cent
Moisture content in 60 per cent relative humidity   12·5 per cent
Corresponding tangential movement   3·0 per cent ($\frac{3}{8}$ in/ft)
Corresponding radial movement   No data

processing

drying: Dries fairly readily, but with a definite tendency to collapse; wood in which collapse has occurred responds well to a reconditioning treatment. Kiln Schedule C.
No data on shrinkage during drying.

working properties:
Blunting: Slight.
Sawing: Rip-sawing – Saw type HR 54.
          Cross-cutting – Satisfactory.
          Narrow bandsawing – Satisfactory.
          Wide bandsawing – Saw type A.
Machining: Satisfactory provided cutting edges are kept sharp.
Nailing: Satisfactory.
Gluing: Good.

wood bending:
Limited tests have shown that the timber is suitable for bends of moderate radius of curvature, but distortion and fractures are liable to occur. Cannot be bent satisfactorily if knots, even of very small size, are present. Classification – Moderate.
Ratio radius/thickness for solid bends (steamed):
          Supported: 17   Unsupported: 19·5
Limiting radius for 3·2 mm ($\frac{1}{8}$ in) laminae (unsteamed): 173 mm ($6\frac{4}{8}$ in).

plywood manufacture: No information.

staining and polishing: Satisfactory.

durability and preservation

insect attack: No information.

durability of heartwood: Non-durable.

preservative treatment:
          Probably moderately resistant. The sapwood is probably permeable.

uses

A lightweight hardwood, similar in its working properties to American whitewood and can be used for similar purposes, e.g. interior parts of furniture and interior joinery.

CHUGLAM, WHITE, and INDIAN SILVER-GREY WOOD
Terminalia bialata

THE TREE

Reaches a height of 30–50 m (100–160 ft) and diameter 0·8–1·5 m (2$\frac{1}{2}$–5 ft). Mainly Andaman Islands.

THE TIMBER

Terminalia bialata produces two commercially recognised timbers: a pale wood known as white chuglam, and a darker, usually figured wood known as Indian silver-grey wood.

properties	colour:	White chuglam – Uniform greyish-yellow. Indian silver-grey wood – Grey or smoky yellow-brown with irregular dark markings producing marbled figure.

properties

colour: White chuglam – Uniform greyish-yellow.
Indian silver-grey wood – Grey or smoky yellow-brown with irregular dark markings producing marbled figure.

sapwood: May comprise the whole tree or only a proportion of it.

grain: Usually straight. Texture medium.

weight: Averages 670 kg/m³ (42 lb/ft³) at 12 per cent moisture content.

strength: Comparable to European beech when green; when dry, slightly lower than European beech.

Moisture Content	Bending Strength		Modulus of Elasticity		Compression parallel to grain	
	N/mm²	lbf/in²	N/mm²	1000 lbf/in²	N/mm²	lbf/in²
Green	80	11 600	11 400	1650	39·9	5790
12 per cent	100	14 500	13 100	1900	49·6	7200

(Data from Indian Forest Records)

movement: No information available.

processing

drying: Air dries very well with no appreciable degrade. Logs before conversion should be well protected against rapid end drying to avoid end splitting and extension of shakes. The timber should be converted green, then stacked under cover for air drying with minimum delay.
Kiln drying proceeds without trouble.
Kiln Schedule E.
No data on shrinkage.

working properties:
Blunting: Moderate.
Sawing: Rip-sawing – Saw type HR 54.
Cross-cutting – Satisfactory.
Narrow bandsawing – Satisfactory.
Wide bandsawing – Saw type B.
Machining: Generally finishes cleanly, but a reduction in cutting angle to 20° may be beneficial when planing wavy-grained material.
Nailing: Satisfactory.
Gluing: Good.

wood bending:
No data available, but experience with other species of *Terminalia* suggests that it is unlikely to possess good bending properties.

plywood manufacture:
Unsuitable because of excessive splitting of veneers during drying.

staining and polishing: Satisfactory.

durability and preservation

insect attack: Sapwood liable to attack by powder-post beetles.

durability of heartwood: Moderately durable (provisional).

preservative treatment:
Indian silver-grey wood – Extremely resistant.
White chuglam – Probably moderately resistant.

uses

The highly figured Indian silver-grey wood is suitable, as solid wood or veneer, for all types of decorative work – furniture, panelling, cabinet work and high-class joinery. The plain wood, white chuglam, is used in India for flooring, plain furniture, tea chests and ships' fittings.

E

CHUMPRAK
Tarrietia cochinchinensis

A timber originating from Thailand and having a general resemblance to the closely-allied Malayan mengkulang. Probably similar to mengkulang in technical properties, though it appears to be more uniform in quality and somewhat darker in colour. Grain usually interlocked, texture moderately coarse. Average weight about 700 kg/m³ (44 lb/ft³), seasoned. Kiln Schedule C is suitable for drying. Suitable for interior construction and fittings, and for light-duty flooring.

COCUS WOOD
Brya ebenus

other name: cocus (United Kingdom).

A small West Indian tree, generally not more than 8 m (25 ft) high and 0·2 m (8 in) in diameter. Logs are generally 75–150 mm (3–6 in) in diameter and 1·2–2·5 m (4–8 ft) long.

Heartwood rich brown, sometimes with an olive hue when fresh, deepening on exposure to dark chocolate-brown or nearly black, usually beautifully veined. Sapwood yellowish, 13–25 mm ($\frac{1}{2}$–1 in) wide. Grain usually straight, texture very fine and uniform. Average weight about 1180 kg/m³ (74 lb/ft³), seasoned. A very hard, heavy, tough and strong wood. Reported to have good stability.

Requires care in drying. Kiln Schedule A. Not unduly difficult to work in relation to its hardness. Finishes very smoothly and turns excellently. Takes a fine polish.

Used for musical instruments, particularly flutes and clarinets, and for turnery, cutlery handles and fancy articles.

COIGUE
Nothofagus dombeyi

other name: coihue (Chile).

THE TREE　　Usually grows to a height of 21–30 m (70–100 ft), diameter about 0·6–1·0 m (2–3 ft), but may grow to a larger size under favourable conditions. Chile.

THE TIMBER
properties　　Has a general resemblance to European beech. Heartwood varies from pale pinkish-brown to reddish- or yellowish-brown. Logs may contain a large proportion of paler coloured sapwood. Pith flecks are reported to occur in the wood and may spoil its appearance. Rather variable in density and general quality; average density about 620 kg/m³ (39 lb/ft³), seasoned. Texture variable, fine to medium. The timber is usually harder and heavier than that of the allied Chilean species rauli (*Nothofagus procera*). Strength somewhat lower than European beech.

processing　　Variable in its drying properties, but generally dries badly with a pronounced tendency to distort. Collapse is liable to occur. Kiln Schedule B. Working properties generally satisfactory, but cutting edges should be kept sharp. Has a moderate blunting effect on tools. Has good bending qualities though inferior to those of well-known bending timbers such as beech, ash or oak. Severe staining occurs when in contact with metal steamers or supporting straps. Stains and polishes satisfactorily.

durability
and preservation　　Heartwood moderately durable (provisional), and probably resistant to preservative treatment. The sapwood is permeable.

uses　　Although resembling beech in appearance coigue is lower in strength and subject to considerable degrade in drying. It has been used for furniture and interior fittings but is generally more suitable for rough work such as packaging, lorry bottoms, etc.

CORDIA, WEST AFRICAN
Cordia millenii
and *C. platythyrsa*

other name: omo (Nigeria).

THE TREE The two species are believed to be similar, but *C. millenii* may be the more important. Grows to a height of 18–30 m (60–100 ft). Bole cylindrical but rarely straight, up to 9–12 m (30–40 ft) long, diameter about 1·0 m (3 ft), above buttresses. Tropical Africa.

THE TIMBER
properties

colour: Varies from pale golden brown to medium brown, occasionally with a pinkish tint.

grain: Somewhat irregular in direction; typically interlocked to give a stripe which is further enhanced on accurately quarter-cut surfaces by a ray figure. Texture coarse.

weight: About 430 kg/m³ (27 lb/ft³), seasoned.

brittleheart: Fairly common.

strength: Rather variable. On average somewhat higher than obeche.

Moisture Content	Bending Strength		Modulus of Elasticity		Compression parallel to grain	
	N/mm²	lbf/in²	N/mm²	1000 lbf/in²	N/mm²	lbf/in²
Green	54	7800	6100	880	26·3	3810
12 per cent	67	9700	6900	1000	35·9	5200

movement: Small.
Moisture content in 90 per cent relative humidity   15 per cent
Moisture content in 60 per cent relative humidity   10·5 per cent
Corresponding tangential movement   1·0 per cent ($\frac{1}{8}$ in/ft)
Corresponding radial movement   0·7 per cent ($\frac{5}{64}$ in/ft)

processing

drying: Dries well and rapidly, with only a slight tendency to bow and twist. A high temperature schedule is necessary to remove pockets of moisture which are liable to remain in the timber.
Kiln Schedule K.
Shrinkage: Green to 12 per cent moisture content:
Tangential   about 2·5 per cent ($\frac{5}{16}$ in/ft)
Radial   about 1·5 per cent ($\frac{3}{16}$ in/ft)

working properties:
Blunting: Slight.
Sawing: Rip-sawing – Saw type HR 54.
Cross-cutting – Satisfactory but some breaking out at bottom of cut.
Narrow bandsawing – Satisfactory.
Wide bandsawing – Saw type A.
Machining: Planing – A reduction of cutting angle to 20° required to avoid tearing of interlocked grain on quarter-sawn faces.
Other operations – Generally satisfactory but some tearing when cutting against the grain, and breaking out at tool exits.
Nailing: Satisfactory.

wood bending: No information.

plywood manufacture: No information.

staining and polishing: Good when filled.

<table>
<tr>
<td>durability
and preservation</td>
<td>insect attack: No information.
durability of heartwood:</td>
</tr>
</table>

**durability
and preservation**

insect attack: No information.

durability of heartwood:
Variable. Outer heartwood very durable but the less dense inner heartwood is probably durable to non-durable.

preservative treatment: Probably resistant.

uses

A lightweight timber, somewhat soft and low in strength properties, but easy to work and stable. Useful for decorative or other parts of furniture and joinery where strength is not important. Might find use in boat-building.

CRABWOOD
Carapa guianensis

other names: andiroba (Brazil); krappa (Surinam); figueroa, tangare (Ecuador).

THE TREE

A large tree usually growing to a height of 30–40 m (100–130 ft), diameter 0·9–1·2 m (3–4 ft). Bole straight and cylindrical, clear for about 15 m (50 ft). Tropical South America and West Indies.

**THE TIMBER
properties**

Crabwood is closely related to the mahoganies (*Swietenia* and *Khaya*). In Guyana three varieties are recognised, varying in colour, density and quality. Colour varies from pale pink to rich red-brown when freshly sawn, darkening to a fairly uniform dull reddish-brown. Resembles a plain mahogany in appearance, but lacks its natural lustre. Sapwood pale brown or oatmeal coloured, 25–50 mm (1–2 in) wide, not always sharply defined. Grain generally straight but sometimes interlocked. Texture medium to coarse. Weight varies from 580 to 740 kg/m³ (36–46 lb/ft³), average about 610 kg/m³ (39 lb/ft³), seasoned. Comparable in strength to European beech. Small movement.

processing

Dries fairly well but rather slowly with a tendency to split in the initial stages. Kiln Schedule C. Has a moderate blunting effect on cutting edges. Sawing satisfactory. In planing material with interlocked grain a cutting angle of 15° is necessary, and care is required in other operations. Glues well and might be used locally for plywood manufacture if splitting of logs could be reduced. Takes stains and polishes satisfactorily.

**durability
and preservation**

Logs liable to severe attack by ambrosia (pinhole-borer) beetles. Heartwood is moderately durable.

uses

Resembles mahogany but is less attractive in appearance. Could be used in the furniture industry and for interior joinery. Used in Guyana for general construction and interior work.

CRAMANTEE
Guarea excelsa

Cramantee from British Honduras is closely related to the two West African species of *Guarea*. A fairly large tree, height up to 30 m (100 ft), diameter sometimes approaching 1·5 m (5 ft). Similar in general appearance to West African guarea, but the pores are larger and bands of soft tissue are broader, less numerous and more distinct. These slight differences are unlikely to affect appreciably the useful properties of the timber.

properties

Heartwood pinkish to deep reddish-brown, distinct but not sharply demarcated from the thick whitish or brownish sapwood. Grain straight or slightly wavy and interlocked. Texture medium. Weight about 570–700 kg/m³ (36–44 lb/ft³), seasoned. Strength somewhat lower than European beech. Small movement.

processing

Dries well but rather slowly with a slight tendency to split, check and distort. Kiln Schedule E. Has a moderate blunting effect on cutting edges. Sawing and machining satisfactory; a cutting angle of 20° recommended in planing. Bending properties moderate. No information on plywood manufacture, but probably suitable on account of its resemblance to West African guarea.

<table>
<tr><td>durability
and preservation</td><td>Heartwood moderately durable. Extremely resistant to preservative treatment, sapwood permeable.</td></tr>
<tr><td>uses</td><td>Generally similar to those of guarea.</td></tr>
</table>

CURUPAY
Anadenanthera macrocarpa
(formerly *Piptadenia macrocarpa*)

other names: cebil, cebil colorado (Argentina); angico preto (Brazil).

A medium-sized tree with a reasonably straight and clear bole, yielding logs of length about 4–7 m (13–23 ft), diameter 0·6–0·9 m (2–3 ft). South America.

properties
Heartwood pale brown, darkening on exposure to reddish-brown with darker coloured, almost black, streaks, giving an attractive appearance. Sapwood yellow-brown or pale pink. Grain usually irregular and sometimes markedly interlocked. Texture fine. The wood is very hard and heavy, weight about 1000–1140 kg/m³ (62–71 lb/ft³), average about 1050 kg/m³ (66 lb/ft³), seasoned. Strength comparable to greenheart. Medium movement.

processing
Dries slowly with little tendency to distort, but is somewhat liable to split and check in kiln drying, especially in thicker dimensions. Kiln Schedule G. Difficult to work on account of its hardness, and has a severe blunting effect on cutting edges. Considerable tooth vibration and overheating occur in sawing. In planing, a reduction in cutting angle to 15° or 10° is necessary when irregular grain is present; resistance to cutting is then very high. Other machining operations rather difficult. No information on gluing properties.

**durability
and preservation**
Heartwood very durable and extremely resistant to preservative treatment.

uses
The timber is too heavy and too difficult to work to be recommended for furniture manufacture or as a general-purpose timber. It turns well and could be utilised for small pieces of decorative turnery. Used in the countries of origin for agricultural purposes requiring high-strength properties and outstanding durability, such as posts, gates, troughs and dips; also for naval construction work, bridging, wagon stock, piling and window and door frames.

DAHOMA
Piptadeniastrum africanum
(formerly *Piptadenia africana*)

other names: agboin, ekhimi (Nigeria); dabema (France and Ivory Coast).

THE TREE
Reaches a height of 45 m (150 ft), diameter 0·9–1·2 m (3–4 ft) above large buttresses. Straight, cylindrical bole 9–15 m (30–50 ft) long above buttresses. West, Central and parts of East Africa.

**THE TIMBER
properties**

colour:	Yellowish-brown.
sapwood:	Light coloured, 50 mm (2 in) or more in width.
grain:	Broadly interlocked, producing prominent stripe on quarter-sawn material. Texture coarse.

weight: 620–780 kg/m³ (39–49 lb/ft³), averaging 690 kg/m³ (43 lb/ft³) at 12 per cent moisture content.

odour: Freshly cut timber has an unpleasant smell which disappears on drying.

chemical staining: Liable to stain in contact with iron under moist conditions.

irritant properties: The fine dust produced in machining or sanding can cause irritation of the skin or nose and throat.

strength: Comparable to European beech, but owing to interlocked grain should not be used in small sections for purposes where strength is important, e.g. ladder rungs.

Moisture Content	Bending Strength		Modulus of Elasticity		Compression parallel to grain	
	N/mm²	lbf/in²	N/mm²	1000 lbf/in²	N/mm²	lbf/in²
Green	76	11 000	9900	1430	36·7	5320
12 per cent	109	15 800	11 200	1620	58·7	8520

movement: Medium.
Moisture content in 90 per cent relative humidity   20 per cent
Moisture content in 60 per cent relative humidity   12 per cent
Corresponding tangential movement   2·9 per cent ($\frac{11}{32}$ in/ft)
Corresponding radial movement   1·6 per cent ($\frac{3}{16}$ in/ft)

processing

drying: Dries slowly, but variable in behaviour. Some material has a marked tendency to collapse and distort; collapse cannot be removed by a reconditioning treatment.
Kiln Schedule A.
Shrinkage: Green to 12 per cent moisture content:
Tangential   about 5·0 per cent ($\frac{5}{8}$ in/ft)
Radial   about 2·5 per cent ($\frac{5}{16}$ in/ft)

working properties:
Interlocked grain and fibrous texture affect machining operations.
Blunting: Moderate, most pronounced in sawing.
Sawing: Rip-sawing – Saw type HR40. Tendency to spring and bind on saw.
Cross-cutting – Considerable breaking out at back and base of cut. Saw type 2 most satisfactory.
Narrow bandsawing – Satisfactory.
Wide bandsawing – Saw type B.
Machining: Planing – A cutting angle of 15° recommended for machining of material with interlocked grain.
Moulding – French head most satisfactory, but finish is fibrous.
Boring – Moderate charring and breaking out at bottom of cut.
Recessing – Considerable tearing and dulling of cutter.
Mortising – Hollow square chisel overheats and chars.
Nailing: Satisfactory.
Gluing: Good.

wood bending:
Inclined to distort severely during steaming and in the bending operation.
Classification – Moderate.
Ratio radius/thickness for solid bends (steamed):
Supported: 11–15   Unsupported: 27–29
Limiting radius for 3·2 mm ($\frac{1}{8}$ in) laminae (unsteamed):
203–216 mm (8–8½ in)

plywood manufacture:
Usable for plywood and might be important in local markets.
Movement: 1·5 mm plywood from 30 per cent to 90 per cent relative humidity – 0·19 per cent.
Surface splitting on exposure to weather – Grade I.

staining and polishing: Satisfactory when filled.

durability and preservation

insect attack: Sapwood liable to attack by powder-post beetles. Reported to be highly resistant to termites in West Africa, but only moderately resistant in South Africa.

durability of heartwood: Durable.

preservative treatment: Resistant. Sapwood moderately resistant.

uses

A timber of moderate weight and good strength properties, but sometimes having irregular grain, and less attractive in appearance than some other woods having similar technical properties. Chiefly useful in large structural sizes. In vehicle building used as bearers and bottoms for lorries and trailers. Suitable as an alternative to oak for building construction, heavy work benches and similar purposes.

DANTA
Nesogordonia papaverifera
(*Cistanthera papaverifera*)

other names: otutu (Nigeria); kotibé (Ivory Coast).

THE TREE

Reaches a height of 27–30 m (90–100 ft) and diameter 0·6–0·8 m (2–2½ ft). Clean, cylindrical bole, length 14–15 m (45–50 ft) above short buttresses. West Africa.

THE TIMBER properties

colour: Reddish-brown, somewhat similar to a dark mahogany.

sapwood: Light-coloured, 50 mm (2 in) or more wide.

grain: Narrowly interlocked, producing stripe figure when cut on quarter. Texture fine.

weight: Averages 740 kg/m³ (46 lb/ft³) at 12 per cent moisture content.

knots and scars:
Appearance sometimes marred by presence of small, sound pin knots and dark streaks of scar tissue.

strength: Comparable to European beech.

Moisture Content	Bending Strength		Modulus of Elasticity		Compression parallel to grain	
	N/mm²	lbf/in²	N/mm²	1000 lbf/in²	N/mm²	lbf/in²
Green	–	–	–	–	–	–
12 per cent	137	19 800	11 700	1690	69·3	10 050

movement: Medium.
Moisture content in 90 per cent relative humidity   21·5 per cent
Moisture content in 60 per cent relative humidity   13·5 per cent
Corresponding tangential movement   2·0 per cent (¼ in/ft)
Corresponding radial movement   1·5 per cent (3/16 in/ft)

processing	**drying:**	Dries rather slowly, but well, with little degrade. Knots tend to split and some 'ribbing' may occur on kiln drying.

processing

drying: Dries rather slowly, but well, with little degrade. Knots tend to split and some 'ribbing' may occur on kiln drying.
Kiln Schedule E.
Shrinkage: Green to 12 per cent moisture content:
Tangential   about 5·0 per cent ($\frac{5}{8}$ in/ft)
Radial about 3·5 per cent ($\frac{7}{16}$ in/ft)

working properties: Interlocked grain affects finishing operations.
Blunting: Moderate.
Sawing: Rip-sawing – Saw type HR 60.
Cross-cutting – Satisfactory.
Narrow bandsawing – Satisfactory.
Wide bandsawing – Saw type B.
Machining: Planing – Material having interlocked grain requires a 15° cutting angle for satisfactory planing.
Moulding – French head most suitable.
Other operations – Satisfactory.
Nailing: Pre-boring advisable.
Gluing: Good.

wood bending:
Considerable variation in bending properties; some material, especially sapwood, inclined to be brittle. Generally suitable for bends of moderate curvature. Pieces containing pin knots may be bent successfully with adequate control of end pressures.
Classification – Moderate.
Ratio radius/thickness for solid bends (steamed):
Supported: 14   Unsupported: 30
Limiting radius for 3·2 mm ($\frac{1}{8}$ in) laminae (unsteamed): 135 mm ($5\frac{1}{4}$ in).

plywood manufacture:
Shown by tests to be usable and might be important in local markets.
Movement: 4·5 mm plywood from 30 per cent to 90 per cent relative humidity – 0·16 per cent.

staining and polishing: Good.

**durability
and preservation**

insect attack: Liable to attack by powder-post beetles. Reported moderately resistant to termite attack in West Africa.
durability of heartwood: Durable.
preservative treatment: Resistant. Sapwood moderately resistant.

uses

A heavy timber having good strength properties, resistance to wear and natural durability, with reasonable stability and ease of working. Suitable for general construction, particularly lorry bodies and floors, and for joinery. Its smooth wearing properties and high resistance to abrasion render it suitable for most types of flooring, particularly where a decorative effect is desired. Also suitable for bench tops, turnery and rifle furniture. Used in West Africa in boat-building for bent timbers in place of rock elm, and for planking.

DEGAME
*Calycophyllum
candidissimum*

other name: lemonwood (United States).

THE TREE

A rather small tree, height about 12–20 m (40–65 ft), diameter up to 0·5 m (20 in). Cuba, Central America, and tropical South America.

**THE TIMBER
properties**

Heartwood brownish, more or less variegated. Sapwood wide, white to brownish-white, not sharply demarcated from heartwood. Grain variable from straight to very irregular. Texture fine and uniform. Average weight about 820 kg/m³ (51 lb/ft³) seasoned. A hard, heavy and tough timber, strength about half-way between European beech and greenheart.

processing

No information on drying. Not unduly difficult to work. Takes polishes well.

durability and preservation	Heartwood reported to be not very resistant to decay.
uses	Regarded as a good alternative to lancewood. Used for archery bows, top joints of fishing rods, small tool handles of exceptional hardness, and for turnery.

DOGWOOD
Cornus florida

other name: cornel.

A small tree with a merchantable bole about 1·2–2·4 m (4–8 ft) long, 150 mm (6 in) or more in diameter. Eastern United States. The commercial timber, which is entirely sapwood, is flesh coloured to pale pinkish-brown. Heartwood, when present, is confined to a very small central core of dark brown wood. The timber is hard and heavy, average weight about 820 kg/m³ (51 lb/ft³), seasoned. Grain straight, texture fine and even. Strength generally similar to European beech, but higher in hardness and shock resistance and slightly lower in stiffness. Large movement.

Dries rather slowly without much distortion or splitting. Kiln Schedule E. Rather hard to work but gives a smooth finish. Turns well. Not resistant to decay because the commercial timber consists of sapwood. Its uses depend upon its hardness and close texture which cause it to remain smooth under continuous wear. Its principal use in this country is for shuttles.

EBONY, AFRICAN
Diospyros spp.
chiefly *D. crassiflora*

other names: Cameroon ebony, Kribi ebony, Gaboon ebony, Nigerian ebony, etc., according to origin (Great Britain).

THE TREE Reaches a height of 15–18 m (50–60 ft), with average diameter of 0·6 m (2 ft). Nigeria, Cameroon and Gaboon.

THE TIMBER Ebony is usually marketed in Britain as short billets of heartwood only.

properties

colour:	Selected timber has an even jet-black colour, but some trees produce heartwood of uneven grey to black colour or with black stripes.
sapwood:	Pale in colour and may comprise a large proportion of the bole.
grain:	Straight to slightly interlocked. Texture very fine.
weight:	About 1000 kg/m³ (63 lb/ft³) at 12 per cent moisture content.
strength:	Heartwood only: comparable to greenheart. Data for Nigerian ebony:

Moisture Content	Bending Strength		Modulus of Elasticity		Compression parallel to grain	
	N/mm²	lbf/in²	N/mm²	1000 lbf/in²	N/mm²	lbf/in²
Green	–	–	–	–	–	–
12 per cent	189	27 400	17 700	2560	92·0	13 350

movement: No information available.

processing

drying: In small dimensions dries fairly rapidly and well, with little tendency to split or distort.
Kiln Schedule E.
Shrinkage: Green to 12 per cent moisture content:
Tangential   about 6·5 per cent ($\frac{13}{16}$ in/ft)
Radial   about 5·5 per cent ($\frac{11}{16}$ in/ft)

working properties:

Heartwood very hard to work with hand and machine tools.

Blunting: Severe.

Sawing: Rip-sawing – Saw type HR 80.
Cross-cutting – Saw type 2 probably most suitable.
Narrow bandsawing – Satisfactory.
Wide bandsawing – Saw type C (tipped teeth).

Machining: Planing – a reduction of cutting angle to 20° required when irregular grain is present. Increase in pressure bar and shoe pressures advised to prevent wood from riding or chattering on cutters.

Nailing: Pre-boring necessary.

Gluing: No information.

wood bending:

Can be bent successfully provided that equipment is capable of withstanding and controlling fairly high end pressures.

Classification – Good.

Ratio radius/thickness for solid bends (steamed):
Supported: 10   Unsupported: 15

Limiting radius for 3·2 mm ($\frac{1}{8}$ in) laminae (unsteamed): 130 mm (5 in).

plywood manufacture: No information.

staining and polishing: Polishes well.

durability and preservation

insect attack: Reported to be moderately resistant to termites in West Africa (*D. crassiflora* highly resistant).

durability of heartwood: Very durable (provisional).

preservative treatment: Probably extremely resistant.

uses

Used for purposes where hardness and fine texture, together with a black colour, are required. Uses include tool handles, especially small tools requiring very hard handles, halves or scales of table cutlery and pocket knives, door knobs, butt ends of billiard cues, facing of tee squares, and various parts of musical instruments (piano and organ keys, organ stops, violin finger boards and pegs, parts of bagpipes, etc.).

EBONY, EAST INDIAN
Diospyros spp.

other names: Indian ebony (*D. melanoxylon*); Ceylon ebony (*D. ebenum*); Andaman marble-wood or zebra wood (*D. marmorata*); Macassar ebony (*D. celebica* and possibly other species).

THE TREE

The various species are generally small- to medium-sized trees having a straight, cylindrical bole of length up to 4·5–6 m (15–20 ft), but often less, diameter 0·3–0·6 m (1–2 ft). India, Ceylon, Andaman Islands, Celebes, according to species.

THE TIMBER properties

colour: Ceylon ebony, *D. ebenum*, is a uniform black wood, rarely with a few irregular light brown streaks. In other species of *Diospyros* light-coloured streaks are characteristically present, e.g. in Macassar ebony pale to medium brown zones contrast with black wood, and in Andaman marble-wood there is a marked contrast between almost white and black wood.

sapwood: Pale in colour, sometimes streaked with black, sharply demarcated from heartwood.

grain: Straight or sometimes irregular or wavy. Texture fine and even.

weight: Ceylon ebony about 1170 kg/m³ (73 lb/ft³), other species somewhat lighter, about 880–1040 kg/m³ (55–65 lb/ft³), seasoned.

strength: The timbers are strong and hard; the black heartwood is much more brittle than the light-coloured sapwood. Ebonies are used largely for decorative purposes where strength is of secondary importance.

movement: No information.

| processing | **drying:** | The black portions of all species are difficult to dry. Ceylon ebony and Andaman marble-wood develop long, fine, deep checks, especially if cut to relatively large dimensions. Logs should be converted to the smallest convenient size as soon as possible after felling and stored under cover. With Indian ebony (*D. melanoxylon*) good results are said to have been obtained by girdling the trees and allowing them to stand for two years before felling, followed by six months drying in plank or scantling. The timber should be well protected against too rapid drying. Sapwood is not difficult to dry. |

Kiln Schedule C.

Shrinkage: No information.

working properties:

The ebonies are extremely hard to work with hand and machine tools and the wood is of a brittle nature. The light-coloured sapwood is rather less hard to work than the heartwood.

Blunting: Severe.

Sawing: Rip-sawing — Saw type HR 80.

Cross-cutting — Saw type 2 most suitable.

Narrow bandsawing — Satisfactory.

Wide bandsawing — Saw type C (tipped teeth).

Machining: In planing, a reduction in cutting angle to 20° is required for material with irregular grain. An increase of pressure bar and shoe pressures is advised to prevent wood riding or chattering on cutters. Turns well.

Nailing: Pre-boring necessary.

Gluing: Macassar ebony is reputed to be difficult.

wood bending: No information.

plywood manufacture: No information.

polishing: Excellent.

durability and preservation

insect attack: Trees and logs are liable to attack by forest longhorn or Buprestid beetles. Reported to be moderately resistant to termites in India and Ceylon.

durability of heartwood:

The black heartwood of Ceylon ebony is very durable. Black heartwood of other species probably similar.

preservative treatment: No information.

uses

Used primarily for decorative purposes. Has good turning properties and is used for special handles, cutlery hafts, handles for pocket knives, brush backs, fancy ware and small turned articles. Parts of stringed instruments such as fingerboards, saddles, pegs, buttons, tailpieces and bow nuts are made from ebony. Used in cabinet work for inlaying.

EKKI
Lophira alata

other names: bongossi (Cameroon); azobé (France and parts of West Africa); kaku (Ghana); eba (Nigeria).

THE TREE

Reaches a height of 55 m (180 ft); usually without buttresses but may have some basal swelling. Long clear bole, diameter up to 1·5–1·8 m (5–6 ft). West Africa and Congo.

THE TIMBER
properties

colour: Dark red or deep chocolate-brown with conspicuous white deposits in the pores.

sapwood: Paler in colour, about 50 mm (2 in) wide.

grain: Usually interlocked. Texture coarse and uneven.

weight: 950–1100 kg/m³ (59–69 lb/ft³) at 12 per cent moisture content.

strength: (Based on limited sampling only.) Slightly weaker than greenheart.

Moisture Content	Bending Strength		Modulus of Elasticity		Compression parallel to grain	
	N/mm²	lbf/in²	N/mm²	1000 lbf/in²	N/mm²	lbf/in²
Green	123	17 800	13 900	2010	68·4	9920
14 per cent	178	25 800	16 900	2450	90·5	13 120

movement: Large.

Moisture content in 90 per cent relative humidity   20 per cent
Moisture content in 60 per cent relative humidity   13·5 per cent
Corresponding tangential movement   2·6 per cent ($\frac{5}{16}$ in/ft)
Corresponding radial movement   2·1 per cent ($\frac{1}{4}$ in/ft)

processing

drying: Extremely refractory. Dries very slowly, and severe splitting and some distortion are likely. Needs to be piled with special care.
Kiln Schedule B.
Shrinkage: Green to 12 per cent moisture content:
    Tangential   about 5·5 per cent ($\frac{11}{16}$ in/ft)
    Radial   about 4·5 per cent ($\frac{9}{16}$ in/ft)

working properties:
Difficult to work with machine and hand tools. Grain usually interlocked, texture coarse.
Blunting: Severe, especially with dry material.
Sawing: Rip-sawing – Saw type HR 80 or TC.
    Cross-cutting – Saw type 2 probably most suitable.
    Narrow bandsawing – Satisfactory.
    Wide bandsawing – Saw type C (tipped teeth).
Machining: Planing – A strong cutting edge required with sharpness angle 40–45°, obtained by reducing clearance angle or cutting angle, or both.
    Timber should be firmly held during all operations to prevent chatter.
    Boring – Tends to char.
Nailing: Pre-boring necessary.
Gluing: Variable.

wood bending: Probably unsuitable for steam bending.

plywood manufacture: No information.

staining and polishing: probably satisfactory.

durability and preservation

insect attack: Reported to be moderately resistant to attack by termites in West Africa.

durability of heartwood: Very durable.

preservative treatment: Extremely resistant.

uses

A very heavy, hard and durable timber. Very suitable for dock and river piling, but not generally available in as large dimensions or long lengths as greenheart. In Africa used untreated for sleepers and for construction work such as bridges, wharfs, etc. Has been employed in the United Kingdom for mine shaft guides and heavy surface structures. As a flooring timber it has high resistance to wear and is suitable for heavy duty flooring where high surface smoothness is not essential. Possesses good resistance to acids and is suitable for filter press plates and frames, especially where conditions are severe or scraping is necessary, but its use is limited by the difficulty of machining.

ELM, ENGLISH
Ulmus procera
and ELM, DUTCH
Ulmus hollandica
var. hollandica

other names: red elm, nave elm (Great Britain).

THE TREE   Reaches a height of 38–45 m (120–150 ft). Clear bole 12–18 m (40–60 ft). Diameter usually 1·0–1·5 m (3–5 ft), but may be 2·5 m (8 ft) or more. Very large trees are often unsound at the centre. Grows in hedgerows, seldom found in forests. Europe, including British Isles. Dutch elm occurs throughout Great Britain and English elm mainly in England and Wales.

THE TIMBER
properties

colour: Heartwood dull brown when dried.

sapwood: Clearly distinguishable from heartwood, especially when freshly felled.

grain: Annual rings distinct, due to large early wood pores, giving the wood a coarse texture. Tends to be cross grained and of irregular growth, producing an attractive figure. Continental elm is usually comparatively straight grained and of more even growth rate than the home-grown timber.

weight: Both species about 550 kg/m³ (34 lb/ft³) at 12 per cent moisture content.

strength: Somewhat higher than obeche. The two species of elm are similar in most strength properties, but Dutch elm is about 40 per cent tougher than English elm.

	Moisture Content	Bending Strength		Modulus of Elasticity		Compression parallel to grain	
		N/mm^2	lbf/in^2	N/mm^2	1000 lbf/in^2	N/mm^2	lbf/in^2
English elm	Green	40	5800	5200	760	16·9	2450
	12%	68	9800	7000	1020	33·9	4920
Dutch elm	Green	44	6400	5400	780	18·8	2710
	12%	74	10 700	7200	1040	34·0	4930

movement: Medium.
Moisture content in 90 per cent relative humidity   22 per cent
Moisture content in 60 per cent relative humidity   13 per cent
Corresponding tangential movement   2·4 per cent ($\frac{9}{32}$ in/ft)
Corresponding radial movement   1·5 per cent ($\frac{3}{16}$ in/ft)

processing   **drying:** Both species dry fairly rapidly but with a very marked tendency to distort. Little tendency to check and split, but some liability for collapse to occur. The timber should be carefully piled with closely spaced sticks and the top of the load should be weighted, e.g. with concrete blocks. The amount of distortion, shrinkage and collapse can be reduced by reconditioning.
Kiln Schedule A.
Shrinkage: Green to 12 per cent moisture content:
Tangential   about 6·5 per cent ($\frac{13}{16}$ in/ft)
Radial   about 4·5 per cent ($\frac{9}{16}$ in/ft)

working properties:

Blunting: Moderate.

Sawing: Irregular grain may cause binding on the saw during conversion.
Rip-sawing – Saw type HR 54.
Cross-cutting – Satisfactory.
Narrow bandsawing – Satisfactory.
Wide bandsawing – Saw type B.

Machining: Satisfactory. Dutch elm generally machines to a better finish than English elm, owing to less wild grain.

Nailing: Satisfactory.

Gluing: Good.

wood bending:

Dutch elm, when free from defects, has very good bending properties. Requires very low end-pressure and bending is little affected by irregular grain. Knots on the inner face tend to induce splitting.

English elm is less suitable because of a pronounced tendency to distort during setting.

	Dutch elm	English elm
Classification	Very good	Very good, but distorts during setting
Ratio radius/thickness for solid bends (steamed):		
Supported	less than 0·5	1·5
Unsupported	9·5	13·5
Limiting radius for 3·2 mm ($\frac{1}{8}$ in) laminae (unsteamed)	100 mm (4 in)	147 mm (5$\frac{4}{8}$ in)

plywood manufacture: Shown by tests to be usable for plywood.

Movement: 1·5 mm plywood from 30 per cent to 90 per cent relative humidity – 0·44 per cent.

Surface splitting on exposure – No information.

staining and polishing: Satisfactory.

durability and preservation

insect attack: Logs liable to attack by forest longhorn or Buprestid beetles. Sapwood liable to attack by powder-post beetles and by the common furniture beetle. Sapwood and heartwood of timber in old buildings in England and Wales may be attacked by the death-watch beetle. The bark of English elm is frequently riddled with tunnels of elm bark beetles, but these cause no damage to the timber.

durability of heartwood: Non-durable.

preservative treatment: Moderately resistant. Sapwood permeable.

uses

A timber of moderate weight and good working and bending properties, and attractive appearance, but its utility and technical performance, especially drying and working characteristics, depend largely on the wood quality.

Used in the furniture industry for cabinet work, chairs and settee frames and is traditionally the preferred timber for seats of Windsor chairs. Used for fancy turnery of many kinds. Extensively employed for manufacture of coffins. In boat-building used for keels and deadwood, stern posts, transoms and rudders; in barge-building for chines, ends of hatch covers and bottoms of canal boats, and in fishing vessels for bottom planks, trawler bobbins and parts of trawler boards. Also finds uses in docks and harbours, e.g. for keel blocks, cappings, wedges, fenders and rubbers. Employed as weatherboarding on farm buildings. As a flooring timber has moderate to low resistance to wear and is best used as blocks.

For some of its uses Dutch elm is preferred to English elm because of its straighter grain.

ELM, JAPANESE
Ulmus spp., principally
Ulmus davidiana var.
japonica and *Ulmus laciniata*

other name: nire (Japan).

THE TREE   Similar in size to elms grown in Britain, but generally of better stem form and more uniform growth than much of the home-grown elm obtained from hedgerows. Japan and North-east Asia.

THE TIMBER
properties   Somewhat variable in character, but often of even, slow growth and relatively free from the growth defects characteristic of English elm. Heartwood dull grey-brown with straight grain and generally rather plain in appearance, resembling American white elm more closely than English elm. Weight about 560 kg/m³ (35 lb/ft³), seasoned, similar to English elm. Strength about half-way between obeche and European beech.

processing   Reported to dry satisfactorily with care, but is rather more liable to split than English elm. Kiln Schedule F. Superior to average English elm in machining and finishing qualities owing to more even growth, straighter grain and relative freedom from defects. Bends well.

durability
and preservation   Heartwood non-durable; probably resistant to preservative treatment.

uses   Generally milder in character and plainer in appearance than English and Dutch elm. Has been used for furniture, including dining tables, sideboards, headboards, and Windsor chair seats but is considered less attractive in appearance than selected English and Dutch elm. Also used for coffins.

ELM, ROCK
Ulmus thomasii

other names: Canadian rock elm (Great Britain); cork elm (Canada); cork elm, cork bark elm, hickory elm (United States).

THE TREE   Reaches a height of 15–21 m (50–70 ft) and diameter of 0·3–0·8 m (1–2½ ft). Straight, clear bole. Parts of eastern Canada and United States.

THE TIMBER
properties

colour:	Light brown.
sapwood:	Not clearly distinguishable from heartwood.
grain:	Straight. Texture moderately fine: rock elm is the finest textured of the elms.
weight:	Heavier than other elms, about 620–780 kg/m³ (39–49 lb/ft³) at 12 per cent moisture content.
pore size:	Rock elm may be distinguished from other commercial elms by the small size and sparse distribution of its early wood pores, which are scarcely visible on a clean-cut end-grain surface without a lens. Those of other elms are large enough to be quite distinct.
strength:	Slightly weaker than European beech.

Moisture Content	Bending Strength		Modulus of Elasticity		Compression parallel to grain	
	N/mm²	lbf/in²	N/mm²	1000 lbf/in²	N/mm²	lbf/in²
Green	69	10 000	7700	1110	27·0	3920
12 per cent	108	15 600	9900	1440	50·5	7320

(Data from U.S. Department of Agriculture).

movement: No information available.

processing

drying: Should be dried with care owing to tendency to check and twist.
Kiln Schedule D.
Shrinkage: Probably large, but no data available.

working properties:
Denser and more difficult to machine than other elms.
Blunting: Moderate.
Sawing: Rip-sawing – Saw type HR 54.
Cross-cutting – Tendency to burn with all saws.
Narrow bandsawing – Satisfactory.
Wide bandsawing – Saw type B or C.
Machining: Satisfactory, but tendency to char in boring.
Nailing: Satisfactory.
Gluing: Good.

wood bending:
Classification – Very good.
Ratio radius/thickness for solid bends (steamed):
Supported: 1·5   Unsupported: 14
Limiting radius for 3·2 mm ($\frac{1}{8}$ in) laminae (unsteamed): 100 mm (4 in).

plywood manufacture:
No information, but there appears to be no technical reason why the species should not be usable.
Movement of plywood and surface splitting – No information.

staining and polishing: Good.

durability and preservation

insect attack: Sapwood liable to attack by powder-post beetles.

durability of heartwood:
Non-durable, but shown by limited tests to be somewhat more resistant to fungus than white elm.

preservative treatment: Probably resistant.

uses

A strong, tough timber. Shipments from Canada may include denser grades of white elm (*U. americana*). It is a good bending timber and is used in boat-building for stern posts, ribs, general framing, keels, rubbing strips and other components, particularly those that are completely submerged in water. Also used for a number of purposes in agricultural implements, and for hubs of wheels, blades of ice hockey sticks and various minor uses. Can be used for underwater parts in dock and wharf construction.

ELM, WHITE
Ulmus americana

other names: American elm (Great Britain); orhamwood (Canada); American elm, soft elm (United States).

THE TREE

Average height 18–25 m (60–80 ft) and diameter 1·0–1·2 m (3–4 ft), but sometimes grows considerably larger. Straight, clear bole. Parts of Canada and United States.

THE TIMBER
properties

colour: Medium brown, often with a reddish tinge.

sapwood: Wide, and of very pale brown colour.

grain: Generally straight but sometimes interlocked. Texture coarse and rather woolly.

weight: Averages about 560 kg/m³ (35 lb/ft³) at 12 per cent moisture content, i.e. considerably lighter in weight than rock elm.

strength: About half-way between obeche and European beech.

Moisture Content	Bending Strength		Modulus of Elasticity		Compression parallel to grain	
	N/mm²	lbf/in²	N/mm²	1000 lbf/in²	N/mm²	lbf/in²
Green	52	7600	7200	1040	20·8	3020
12 per cent	85	12 400	8600	1250	39·5	5730

(Data from US Department of Agriculture)

movement: No information available.

<table>
<tr><td>processing</td><td></td></tr>
</table>

drying: Dries readily with medium shrinkage.
Kiln Schedule F.

working properties:
Blunting: Moderate.
Sawing: Rip-sawing – Saw type HR 54.
Cross-cutting – Satisfactory.
Narrow bandsawing – Satisfactory.
Wide bandsawing – Saw type B.
Machining: Satisfactory.
Nailing: Good.
Gluing: Good.

wood bending: Similar to rock elm.
Classification – Very good.
Ratio radius/thickness for solid bends (steamed):
Supported: 1·7   Unsupported: 13·5
Limiting radius for 3·2 mm ($\frac{1}{8}$ in) laminae (unsteamed): 110 mm ($4\frac{1}{4}$ in).

plywood manufacture:
Shown by tests in Canada to be usable for plywood.
Movement of plywood and surface splitting – No information.

staining and polishing: Satisfactory.

durability and preservation

insect attack: Trees and logs liable to attack by forest longhorn or Buprestid beetles.
durability of heartwood: Non-durable.

preservative treatment: Probably moderately resistant. Sapwood permeable.

uses

Valuable where strength and toughness are required and suitable for many of the purposes for which rock elm is used, provided that its slightly lower strength properties are taken into account. Used to a limited extent in boat-building.

ELM, WYCH
Ulmus glabra

other names: mountain elm, Scotch elm (Great Britain).

THE TREE

Reaches a height of 30–38 m (100–125 ft) and diameter occasionally up to 1·5 m (5 ft). Northern Europe including British Isles.

THE TIMBER properties

colour: Light brown, often with greenish tinge or distinct green streaks.

sapwood: Clearly differentiated from heartwood, especially when freshly felled.

grain: Generally straighter grained and finer textured than English or Dutch elm.

F

weight: Averages about 670 kg/m³ (42 lb/ft³) at 12 per cent moisture content.

strength: Slightly weaker than European beech.

Moisture Content	Bending Strength		Modulus of Elasticity		Compression parallel to grain	
	N/mm²	lbf/in²	N/mm²	1000 lbf/in²	N/mm²	lbf/in²
Green	68	9900	9400	1360	30·4	4410
12 per cent	105	15 300	10 600	1540	49·2	7130

movement: No information available.

processing

drying: Dries fairly well and fairly rapidly, with some tendency to distort, though less than English elm. Not liable to split or collapse. Careful piling with close spacing of sticks is advisable.
Kiln Schedule A.
Shrinkage about the same as in English elm.

working properties:
Has good machining properties which, because of its generally straighter grain, compare favourably with those of other elms.
Blunting: Moderate.
Sawing: Rip-sawing – Saw type HR 54.
Cross-cutting – Satisfactory.
Narrow bandsawing – Satisfactory.
Wide bandsawing – Saw type B.
Machining: Satisfactory in all operations.
Nailing: Good.
Gluing: Good.

wood bending:
In the air-dry condition the wood is very suitable for bending, but when much moisture is present it is liable to buckle and fracture badly. Material must be straight grained to avoid distortion, and even small knots render bending impossible.
Classification – Very good.
Ratio radius/thickness for solid bends (steamed):
Supported: 1·7 Unsupported: 12·5
Limiting radius for 3·2 mm ($\frac{1}{8}$ in) laminae (unsteamed): 117 mm ($4\frac{3}{8}$ in).

plywood manufacture: Shown by tests to be usable for plywood.
Movement: 1·5 mm plywood from 30 per cent to 90 per cent relative humidity – 0·27 per cent.
Surface splitting on exposure to weather: Grade I.

staining and polishing: Satisfactory.

durability and preservation

insect attack: Sapwood liable to attack by powder-post beetles and by the common furniture beetle.

durability of heartwood: Non-durable.

preservative treatment: Resistant. Sapwood permeable.

uses

Used for the same purposes as English or Dutch elm, but is straighter grained and therefore works better. Used in boat-building for planking and framing in dinghies, and for keels and deadwood in larger vessels. For furniture manufacture, English or Dutch elm is commonly preferred because it is milder than wych elm.

ESIA

*Combretodendron
macrocarpum*
(formerly *C. africanum*)

other names: owewe (Nigeria); minzu (Congo Rep.).

THE TREE
Reaches a height of 37 m (120 ft) or more and average diameter of 0·8–1·1 m (2½–3½ ft). Straight, clear, cylindrical bole but sometimes with basal thickening. Tropical West Africa.

**THE TIMBER
properties**
Heartwood reddish-brown, often of plain appearance; some material is marked with darker streaks and, especially when quarter cut to expose the conspicuous ray figure, has some decorative value. Sapwood wide and pale in colour. Grain tends to be interlocked. Texture moderately coarse. Weight about 800 kg/m³ (50 lb/ft³) at 12 per cent moisture content. The freshly-cut timber has a powerful unpleasant odour, which does not persist after drying. Considerably stronger than European beech. Large movement.

processing
Dries slowly with very pronounced tendency to check and split. Appreciable distortion liable to occur and end splitting, surface checking and shakes are likely to prove serious. Virtually impracticable to kiln dry. Has a moderate blunting effect on cutting edges. In rip-sawing the fine sawdust tends to adhere to packings and cause overheating of saws. A thicker gauge plate may be advantageous. In cross-cutting saw type 2 most suitable. In planing a cutting angle of 20° is necessary to prevent tearing. Tendency to char in boring. Other operations satisfactory, though feeding effort is often high. Stains and polishes satisfactorily when filled.

**durability
and preservation**
Has been reported moderately resistant to termites in West Africa. Heartwood durable and probably extremely resistant to preservative treatment. The sapwood is permeable.

uses
Although locally abundant in West Africa, and with good stem form, has not been extensively utilised, mainly because of the degrade which occurs in drying. May be suitable for heavy, rough construction work or railway sleepers in countries of origin. Can be sliced to produce a decorative quartered veneer.

FREIJO

Cordia goeldiana

THE TREE
Reaches a height of about 30 m (100 ft) and diameter 0·6–1·0 m (2–3 ft). Brazil, especially Amazon basin.

**THE TIMBER
properties**

colour:	Golden brown, somewhat similar to teak.
grain:	Usually straight. Texture moderately fine.
weight:	Variable from 400 to 700 kg/m³ (25–44 lb/ft³), averaging about 590 kg/m³ (37 lb/ft³) at 12 per cent moisture content.
strength:	Slightly weaker than European beech.

Moisture Content	Bending Strength		Modulus of Elasticity		Compression parallel to grain	
	N/mm²	lbf/in²	N/mm²	1000 lbf/in²	N/mm²	lbf/in²
Green	–	–	–	–	–	–
12 per cent	97	14 000	12 000	1740	54·3	7880

movement: No information available.

processing **drying:** Dries well with little distortion, but with a slight tendency for end splits to develop.
Kiln Schedule E.

working properties:

Blunting: Moderate.
Sawing: Rip-sawing – Saw type HR 54.
Cross-cutting – Satisfactory.
Narrow bandsawing – Satisfactory.
Wide bandsawing – Saw type B.
Machining: Satisfactory, but sharp cutting edges are required to prevent tearing and grain raising. Timber requires support in end-grain working, as in mortising, boring, etc., to prevent breaking away.
Nailing: Satisfactory.
Gluing: No information.

wood bending:

No exact data available, but its use in Brazil for cooperage indicates that it may be suitable for bends of moderate radius of curvature.
Classification – Poor.
Limiting radius for 3·2 mm ($\frac{1}{8}$ in) laminae (unsteamed): 185 mm ($7\frac{1}{4}$ in).

plywood manufacture: No information available.

staining and polishing: Satisfactory when filled.

durability and preservation   **insect attack:** Sapwood liable to attack by powder-post beetles. Reported to be attacked by dry-wood termites in West Indies.

durability of heartwood: Durable.

preservative treatment: No information available.

uses Somewhat similar in appearance to teak. Has been used in boat-building, including decking. Suitable for furniture, vehicle manufacture, and interior and exterior joinery. Used in Brazil for cooperage but not found satisfactory in Europe owing to the quality of the material shipped.

GABOON
Aucoumea klaineana

other name: okoumé (France).

THE TREE A large tree, height 30–40 m (100–130 ft), occasionally up to 60 m (200 ft). Bole cylindrical, usually slightly curved, diameter 0·8–1·1 m ($2\frac{1}{2}$–$3\frac{1}{2}$ ft), exceptionally up to 2 m ($6\frac{1}{2}$ ft), clear for 25 m (80 ft) or more. Buttresses usually not more than 1 m (3 ft) up the stem. Gaboon, Equatorial Guinea and Congo Rep. (Brazzaville).

THE TIMBER properties

colour: Light salmon-pink to dark pink, toning on exposure to pinkish-brown.

sapwood: Narrow, pale grey in colour, not clearly distinct from heartwood.

grain: Usually shallowly interlocked, sometimes slightly wavy. Texture medium.

weight: Varies from 370 to 560 kg/m³ (23–35 lb/ft³), average about 430 kg/m³ (27 lb/ft³), seasoned.

strength: Somewhat higher than obeche.

movement: No information.

processing **drying:** Dries readily without excessive distortion or checking.
Kiln Schedule E.
No information on shrinkage during drying.

working properties:

Blunting:	Moderate to severe. The timber contains silica.
Sawing:	Rip-sawing — Saw type HR 54 or TC.
	Cross-cutting — Saw type 2 most satisfactory.
	Narrow bandsawing — Satisfactory.
	Wide bandsawing — Saw type A (hard tipping an advantage).
Machining:	A reduced cutting angle required to prevent tearing and blunting.
Nailing:	Satisfactory.
Gluing:	Good.

wood bending: No information.

plywood manufacture:

Widely employed for plywood and readily available in the United Kingdom.

Movement: 4·5 mm plywood from 30 per cent to 90 per cent relative humidity — 0·17 per cent.

Surface splitting on exposure to weather — Grade I.

staining and polishing: Satisfactory.

durability and preservation

insect attack: Logs liable to attack by forest longhorn or Buprestid beetles. Sapwood liable to attack by powder-post beetles. Reported to be non-resistant to termites in West Africa.

durability of heartwood: Non-durable.

preservative treatment: Resistant.

uses

Used principally for manufacture of plywood and blockboard in the growing areas and in several European countries. Logs are graded by an established system before shipment. They vary in diameter from a legal minimum of 0·6 m (2 ft) to 1·8 m (6 ft) or more with an average of about 1·1 m (3½ ft). Veneer can be very variable in quality but the yield of face stock is usually high. Logs are generally converted by rotary peeling, but are occasionally sliced on the quarter for semi-decorative veneer. Gaboon plywood is used for panelling, doors, partitions, etc. The solid timber has been used for cigar boxes, packing cases and interior frame constructions.

GEDU NOHOR
Entandrophragma angolense

other names: gedu lohor, gedu noha (Nigeria); tiama (France and Ivory Coast); edinam (Ghana); kalungi (Congo).

THE TREE

Reaches a height of 50 m (160 ft) and average diameter 1·2–1·5 m (4–5 ft). Broad buttresses. Bole moderately straight and cylindrical, averaging 18–25 m (60–80 ft) in length. West, Central and East Africa.

THE TIMBER properties

colour: Rather dull reddish-brown and plainer in appearance than sapele (*Entandrophragma cylindricum*). Occasional logs are much paler in colour.

sapwood: Lighter in colour, up to 100 mm (4 in) wide.

grain: Interlocked, but not as closely as in sapele and so does not generally produce well-figured wood. Texture moderately coarse.

weight: Appreciably lighter than sapele, resembling African mahogany (*Khaya ivorensis*) in this respect. Average about 540 kg/m³ (34 lb/ft³) at 12 per cent moisture content.

strength: About half-way between obeche and European beech.

Moisture Content	Bending Strength		Modulus of Elasticity		Compression parallel to grain	
	N/mm²	lbf/in²	N/mm²	1000 lbf/in²	N/mm²	lbf/in²
Green	52	7500	6900	1000	25·4	3680
12 per cent	77	11 200	8600	1250	45·2	6550

movement: Small.
Moisture content in 90 per cent relative humidity   22 per cent
Moisture content in 60 per cent relative humidity   13·5 per cent
Corresponding tangential movement   1·6 per cent ($\frac{3}{16}$ in/ft)
Corresponding radial movement   1·2 per cent ($\frac{9}{64}$ in/ft)

processing

drying: Dries fairly rapidly with a marked tendency to distort.
Kiln Schedule A.
Shrinkage: Green to 12 per cent moisture content:
Tangential   about 5·0 per cent ($\frac{5}{8}$ in/ft)
Radial   about 2·5 per cent ($\frac{5}{16}$ in/ft)

working properties: Interlocked grain affects machining operations.
Blunting: Moderate.
Sawing: Rip-sawing – Saw type HR 54.
Cross-cutting – Saw type 2 most satisfactory.
Narrow bandsawing – Satisfactory.
Wide bandsawing – Saw type B.
Machining: Planing – A reduction of the cutting angle to 15° is necessary
to prevent tearing of interlocked grain.
Mortising – A tendency to char.
Other operations – Satisfactory.
Nailing: Satisfactory.
Gluing: Good.

wood bending:
Has very poor bending properties, comparable with utile.
Classification – Poor.

plywood manufacture:
Employed for plywood and usually available in the United Kingdom.
Movement of plywood and surface splitting on exposure to weather – No
information available.

staining and polishing: Good.

durability and preservation

insect attack: Sapwood liable to attack by powder-post beetles. Reported to be
non-resistant to termites in West Africa.

durability of heartwood: Moderately durable.

preservative treatment: Extremely resistant. The sapwood is resistant.

uses A mahogany-type timber, related to sapele but of plainer appearance. Used as an
alternative to mahogany in furniture manufacture. In building work, suitable for
exterior and interior joinery. In boat-building has been used successfully for
planking, cabins and furniture, and suitable also for use in the motor and bus body
trade and in railway carriage construction, and for coffins.

GREENHEART
Ocotea rodiaei

THE TREE Reaches a height of 21–40 m (70–130 ft), occasionally more. Average diameter
about 1·0 m (3 ft). Sometimes has low buttresses. Long, straight, cylindrical bole,
15–25 m (50–80 ft) long. Guyana and part of Surinam.

THE TIMBER
properties

colour: Light to dark olive-green, sometimes marked with brown or black streaks. Local distinction between so-called varieties (black, brown, yellow, etc.) is based on variation in superficial appearance of wood from different trees or even different parts of same tree. There is no reliable evidence that the useful properties of the timber are related to colour variations.

sapwood: Pale yellow or green in colour, shading gradually into the heartwood; 25–50 mm (1–2 in) wide, but may be more in timber from small trees.

grain: Straight or interlocked. Texture fine and even.

weight: About 1030 kg/m³ (64 lb/ft³) at 12 per cent moisture content.

strength: Has exceptionally high-strength properties even when its weight is taken into account. Used as one of the standards of comparison for other species in this book.

Moisture Content	Bending Strength		Modulus of Elasticity		Compression parallel to grain	
	N/mm²	lbf/in²	N/mm²	1000 lbf/in²	N/mm²	lbf/in²
Green	140	20 300	15 900	2310	67·4	9770
12 per cent	181	26 200	21 000	3040	89·9	13 040

movement: Medium.
Moisture content in 90 per cent relative humidity   16 per cent
Moisture content in 60 per cent relative humidity   11 per cent
Corresponding tangential movement   2·0 per cent ($\frac{1}{4}$ in/ft)
Corresponding radial movement   1·5 per cent ($\frac{3}{16}$ in/ft)

processing

drying: Dries very slowly and with considerable degrade, particularly in the thicker sizes. For economic reasons, timber over 25 mm (1 in) in thickness should be partly air dried before kiln drying. Distortion is not serious, but checking and splitting may be severe. Existing shakes are very liable to extend and splitting of knots may occur.
Kiln Schedule B.
Shrinkage: Green to 12 per cent moisture content:
Tangential   about 4·5 per cent ($\frac{9}{16}$ in/ft)
Radial   about 3·0 per cent ($\frac{3}{8}$ in/ft)

working properties:
Interlocked grain, when present, affects many machining operations.
Blunting: Moderate.
Sawing: Rip-sawing – Saw type HR 60 or HR 80.
Cross-cutting – Saw type 2 most suitable.
Narrow bandsawing – Breaking out at the bottom of the cut may occur when working against the grain; otherwise satisfactory.
Wide bandsawing – Saw type C.
Machining: Planing – A cutting angle of 20° is necessary owing to the high density of the wood and interlocked grain.
Other operations – Satisfactory.
Nailing: Unsuitable.
Gluing: Variable.

wood bending:
Very efficient support on the outer face is necessary when bending even to a comparatively large radius of curvature.
Classification – Moderate.
Ratio radius/thickness for solid bends (steamed):
Supported: 18   Unsupported: 36.
Limiting radius for 3·2 mm ($\frac{1}{8}$ in) laminae (unsteamed): 185 mm ($7\frac{1}{4}$ in).

plywood manufacture: Unsuitable on account of its high density.

staining and polishing: Staining rarely necessary. Polishes satisfactorily.

durability and preservation

insect attack: No information available.

durability of heartwood: Very durable.

preservative treatment: Extremely resistant.

uses

A very heavy, hard timber, outstanding in most of its strength properties, and of very high durability and having good resistance to attack by marine borers. Available in very large sizes (generally squares of 300–450 mm (12–18 in) and length up to 17 m (55 ft), but exceptionally squares of 600 mm (24 in) and length over 18 m (60 ft) are obtainable), and therefore suitable for piling, piers, lock gates, and docks and harbour work. For this work all sapwood should be excluded as it is not resistant to marine borer attack. As sawn timber, greenheart has been used for pier decking and hand rails, factory flooring, and in the engineering industry as bearers for engines. Gives good service in chemical plant for vats and filter press plates and frames. Used also for fishing rods and as centre laminae for longbows.

GREVILLEA
Grevillea robusta

other names: silky-oak, African silky-oak (UK).

An Australian species, extensively planted as a shade tree for coffee and tea plantations in Africa, India, Ceylon and other parts of the world; was imported for a while into the United Kingdom, mainly from Tanzania. A medium-sized tree, bole not usually exceeding 9 m (30 ft) in length, diameter about 0·6 m (2 ft).

properties

The timber is similar in appearance to Australian silky-oak (*Cardwellia sublimis*), but is somewhat paler in colour. Heartwood distinctly pink, turning pale brown on exposure. Sapwood white or straw-coloured, moderately well defined, about 38 mm (1½ in) wide. The most prominent feature is the large rays, which on accurately quarter-sawn surfaces produce a well-marked silver-grain figure, not unlike that of true oak. Grain straight or wavy. Texture medium to coarse. The wood is reported to be somewhat knotty. Weight about 560–660 kg/m³ (35–41 lb/ft³), average 610 kg/m³ (38 lb/ft³), seasoned. A soft timber, weak in bending and compression, strength below average in relation to its density. Medium movement.

processing

Dries slowly with slight distortion, some surface checking, and serious splitting in thicker sizes. Saws and machines very easily. Said to be usable for plywood manufacture, but its use as a sliced decorative veneer may be preferable.

durability and preservation

Moderately durable (provisional), and probably moderately resistant to preservative treatment.

uses

A medium-quality joinery and furniture timber. Has been used in the United Kingdom in place of oak for furniture and for flooring in the form of blocks and strips. Used locally in place of softwood for building and shuttering but its low strength properties must be taken into account.

GUAREA
Guarea cedrata and *G. thompsonii*

other names: obobo (Nigeria); bossé (France and Ivory Coast).

    G. cedrata – scented guarea (Great Britain); white guarea, obobonufua (Nigeria).

    G. thompsonii – black guarea, obobonekwi (Nigeria).

These two species of *Guarea* are frequently marketed under a single trade name and for many purposes differences between them are unimportant. It is sometimes desirable, however, to keep the two species distinct and technical differences between them are detailed below.

THE TREE Both species reach a height of about 50 m (160 ft) with diameters of 0·9–1·2 m (3–4 ft) above buttresses. *G. cedrata* has rather heavier buttresses than *G. thompsonii*. Both species have long, straight, cylindrical boles above buttresses. West Africa and Congo; *G. cedrata* also recorded in Uganda.

THE TIMBER
properties

colour: Pinkish-brown, like a pale mahogany.

sapwood: 50–100 mm (2–4 in) wide, pale in colour.

grain: Sometimes straight but frequently interlocked. *G. cedrata* sometimes produces an attractive mottled or curly figure. *G. thompsonii* tends to be straighter in the grain and has a plainer appearance. Texture (both species) fine.

weight: *G. cedrata* about 580 kg/m³ (36 lb/ft³), and *G. thompsonii* about 620 kg/m³ (39 lb/ft³) at 12 per cent moisture content.

silica: Often present in *G. cedrata*; usually absent in *G. thompsonii*.

resin content: Resin present in both species but rather more common in *G. cedrata*.

irritant properties: Fine dust produced during machining or sanding of both species can cause irritation of the skin or of the eyes, nose and throat, but *G. thompsonii* has a more marked irritant effect than *G. cedrata*.

odour: *G. cedrata* has a pleasant cedar-like scent which tends to disappear in time and which is absent in *G. thompsonii*.

strength: Both species comparable to European beech when green. When dry *G. cedrata* is somewhat weaker than European beech and *G. thompsonii* is almost comparable to European beech.

Species	Moisture Content	Bending Strength		Modulus of Elasticity		Compression parallel to grain	
					1000		
		N/mm²	lbf/in²	N/mm²	lbf/in²	N/mm²	lbf/in²
G. cedrata	Green	74	10 800	8900	1290	35·4	5140
	12 per cent	103	14 900	9400	1370	53·2	7720
G. thompsonii	Green	85	12 400	10 600	1540	43·2	6260
	12 per cent	107	15 500	10 800	1570	59·8	8680

movement: (Both species) Small.
Moisture content in 90 per cent relative humidity   19 per cent
Moisture content in 60 per cent relative humidity   13 per cent
Corresponding tangential movement   1·6 per cent ($\frac{3}{16}$ in/ft)
Corresponding radial movement   1·2 per cent ($\frac{9}{64}$ in/ft)

processing

drying: Both species dry fairly rapidly with little tendency to warp, but resin exudation may adversely affect the appearance of the seasoned timber. *G. thompsonii* is liable to split and requires greater care in drying.
Kiln Schedule E for both species.
Shrinkage: Green to 12 per cent moisture content:

		G. cedrata	G. thompsonii
Tangential	about	3·5 per cent ($\frac{7}{16}$ in/ft)	4·0 per cent ($\frac{1}{2}$ in/ft)
Radial	about	2·5 per cent ($\frac{5}{16}$ in/ft)	2·0 per cent ($\frac{1}{4}$ in/ft)

working properties:

G. *cedrata* – Resinous, grain interlocked, with a woolly or fibrous texture.

G. *thompsonii* – Less resinous than G. *cedrata*, grain interlocked.

Blunting: G. *cedrata* – Moderate. G. *thompsonii* – Slight.

Sawing: Rip-sawing – Saw type HR 40.
Cross-cutting – Satisfactory. Saw type 2 most suitable for G. *cedrata*.
Narrow bandsawing – Satisfactory. A blade having a tooth pitch of 8·5 mm recommended for G. *cedrata*.
Wide bandsawing – Saw type B.

Machining: G. *cedrata* – Cutters having a reduced sharpness angle an advantage.
Tendency to char when boring, recessing and mortising. Other operations satisfactory.
G. *thompsonii* – A cutting angle of 20° required to prevent tearing of interlocked grain in planing.

Nailing: G. *cedrata* – Satisfactory.
G. *thompsonii* – Pre-boring desirable.

Gluing: Good.

wood bending:

G. *cedrata* better than G. *thompsonii* for solid bending, but cannot be bent successfully if small knots are present. Inclined to distort during setting.

Classification: G. *cedrata* – Good. G. *thompsonii* – Moderate.

Ratio radius/thickness for solid bends (steamed):

	G. *cedrata*	G. *thompsonii*
Supported	7·5	14
Unsupported	20	36

Limiting radius for 3·2 mm ($\frac{1}{8}$ in) laminae (unstreamed):
G. *cedrata* – 200 mm (8 in)

plywood manufacture:

Employed for plywood and usually available in the United Kingdom.

Movement: 1·5 mm plywood from 30 per cent to 90 per cent relative humidity – G. *cedrata*   0·23 per cent
G. *thompsonii* 0·21 per cent

Surface splitting on exposure to weather:
G. *cedrata*   Grade I
G. *thompsonii* Grade II

staining and polishing:

Satisfactory when filled, but some tendency to resin exudation in G. *cedrata*.

durability and preservation

insect attack: Sapwood rarely liable to attack by powder-post beetles. Has been reported moderately resistant to termites in West Africa.

durability of heartwood: Very durable.

preservative treatment: Extremely resistant. The sapwood is permeable.

uses

A mahogany-type timber of good stability and durability, but with some tendency to resin exudation, particularly in G. *cedrata*. For most purposes differences between the two species are unimportant. Used in furniture (e.g. chairs, drawer sides and rails), interior fittings and high-class joinery. In vehicle construction it is employed for underframes, tail boards, floor boards and sides of trucks and lorries, and for frames and planking in caravans. Has been used for rifle furniture. Should not be used for articles such as instrument cases where resin exudation may be undesirable. Efficient dust extraction is necessary in machining and sanding on account of the irritant properties of the dust.

GUM, AMERICAN RED
Liquidambar styraciflua

other names: gum, sweet gum, bilsted, red gum (heartwood) and sap gum (sapwood) (United States).

An important timber in the United States. Sapwood nearly white, marketed as a distinct commercial timber known as sap gum. Heartwood reddish-brown with a satiny lustre, sometimes having an attractive figure. Grain usually irregular. Texture fine and uniform. Weight about 560 kg/m³ (35 lb/ft³), seasoned. Strength about half-way between obeche and European beech.

Requires care in drying. Easy to work and finishes very smoothly. Glues well. Employed for plywood manufacture but seldom seen in the United Kingdom.

In the USA it has a wide range of uses, e.g. furniture, joinery, doors, plywood, boxes and packing cases.

GUM, SALIGNA
Eucalyptus saligna

other names: Sydney blue gum, blue gum (Australia).

This species is a native of Australia but it has been extensively planted, together with allied species of *Eucalyptus*, in South Africa and many other countries. *Eucalyptus saligna* grown in Australia is a large tree with a straight and symmetrical bole. Heartwood pink to red in colour, grain usually straight or slightly interlocked, texture rather coarse. Weight about 820 kg/m³ (51 lb/ft³), seasoned. Strength slightly higher than European beech. The timber is hard and tough with a tendency to split. It is not difficult to work, takes a good polish and glues well. Has been shown by tests to be usable for plywood but requires care in drying the veneers. Sapwood liable to attack by powder-post beetles. Moderately durable (provisional). It is reported that sapwood is sometimes refractory to preservative treatment. Used as a general purpose hardwood for building purposes, flooring, packing cases and many o ther purposes.

Plantation timber is typically faster grown and considerably lighter in weight, usually 480–620 kg/m³ (30–39 lb/ft³). It is generally rather lower in quality than the Australian timber and is widely used for mining timber, poles, box making, and manufacture of pulp and fibreboard.

GUM, SPOTTED
Eucalyptus maculata

other names: maculata gum (South Africa); macula (Great Britain).

THE TREE
Grows to a height of 25–45 m (80–150 ft) and diameter 1·2–1·5 m (4–5 ft). South-eastern Australia.

THE TIMBER
properties
Varies in colour from pale brown or greyish-brown to dark brown. Sapwood generally about 50 mm (2 in) wide, pale in colour and distinct from heartwood. Grain straight or interlocked and occasionally wavy. Texture coarse but even. Average weight about 1000 kg/m³ (63 lb/ft³), seasoned. A heavy, hard, strong and tough timber; strength properties slightly lower than those of greenheart.

processing
Care is necessary in drying to avoid checking of plain sawn material. Some distortion may be expected in long stock because of the inclined grain which is common in this timber, and slight collapse sometimes occurs. Kiln Schedule C. Not unduly difficult to work, but a reduced cutting angle is required in planing material with wavy or interlocked grain. Pre-boring necessary in nailing. Believed to be satisfactory for gluing and has been shown by tests to be usable for plywood, though it is too heavy for many purposes. Stains and polishes well.

durability and preservation	Sapwood liable to attack by powder-post beetles and by the common furniture beetle. Reported to be non-resistant to termites in Australia. Heartwood moderately durable. Extremely resistant to preservative treatment; the sapwood is permeable.
uses	A good structural timber, used where strength and toughness are required, as in underframing and decking for bridges, and scantlings for road and rail vehicles. Used in Australia for handles of axes and picks. Has been used in Great Britain as cross-arms for transmission and telegraph poles.

HALDU
Adina cordifolia

other names: hnaw (Burma); kwao, kwow (Thailand).

A large tree with a long straight stem, growing in India, Ceylon, Thailand and Burma.

Timber yellow when freshly cut, darkening to yellowish- or reddish-brown. Sapwood yellowish-white, wide, not sharply distinguished from heartwood. Grain fairly straight or somewhat spiral or interlocked. Texture fine and even. Weight about 650 kg/m³ (41 lb/ft³), seasoned. Strength somewhat lower than European beech. Stated to be slightly refractory in drying with some tendency to check and split. Kiln Schedule E. Works fairly easily giving a good finish, and turns well. Heartwood non-durable.

Could be used for furniture, cabinet work and joinery, and has good wearing qualities as a flooring timber. Used in turnery (bobbins, toys, etc.).

HICKORY
Carya spp.

other names: pignut hickory – *Carya glabra*; mockernut hickory – *C. tomentosa*; shellbark hickory – *C. laciniosa*; shagbark hickory – *C. ovata*.

THE TREE	The above species are all marketed as hickory and it is not practicable to distinguish the timber. Average height varies from 18 m (60 ft) to 36 m (120 ft). Bole generally straight and cylindrical and, when well grown, clear of branches for about half the height of the tree. Diameter 0·6–0·9 m (2–3 ft). Eastern United States and south-eastern Canada, distribution varying according to species.

THE TIMBER
properties

colour:	Heartwood brown or reddish-brown, known as red hickory.
sapwood:	Clearly distinguishable from heartwood. Usually wide, very pale in colour. Known as white hickory.
grain:	Typically straight, but may occasionally be wavy or irregular. Texture rather coarse.
weight:	Ranges from 700 to 900 kg/m³ (45–56 lb/ft³), averaging 820 kg/m³ (51 lb/ft³), air dry.
growth rate:	Density and related properties vary according to the rate of growth. Generally material with ring width greater than 1·5 mm (less than 16 rings per inch) is preferred where high density and strength are required.
strength:	Slightly higher than European beech, but exceptionally high in toughness. Exhibits an outstanding combination of high strength, stiffness, hardness and shock resistance. Colour of the timber is no indication of strength.

Moisture Content	Bending Strength		Modulus of Elasticity		Compression parallel to grain		Impact Maximum drop of hammer	
	N/mm²	lbf/in²	N/mm²	1000 lbf/in²	N/mm²	lbf/in²	m	in
Green	74	10 800	10 600	1540	31·5	4570	1·88	74
12 per cent	139	20 200	15 000	2180	61·0	8850	2·23	88

(Data from Canadian Department of Forestry)

movement: No data available.

<table>
<tr><td>processing</td><td></td></tr>
</table>

processing

drying: No precise information available but shrinks considerably in drying. Kiln Schedule E suggested.

working properties: Density variable and grain sometimes irregular.
Blunting: Moderate to severe.
Sawing: Rip-sawing — Saw type HR 54 or HR 60.
Cross-cutting — Saw type 2 most satisfactory.
Narrow bandsawing — Satisfactory.
Wide bandsawing — Saw type B.
Machining: Planing — A cutting angle of 20° recommended with material having irregular grain.
Nailing: Pre-boring necessary.
Gluing: Difficult.

wood bending: Has excellent steam-bending properties.
Classification — Very good.
Ratio radius/thickness for solid bends (steamed):
Supported: 1·8   Unsupported: 15·0
Limiting radius for 3·2 mm ($\frac{1}{8}$ in) laminae (unsteamed): 147 mm ($5\frac{4}{8}$ in).

plywood manufacture:
Said to have been used for plywood. Movement of plywood and surface splitting — No information.

staining and polishing: No information.

durability and preservation

insect attack: Trees and logs liable to attack by forest longhorn or Buprestid beetles. Sapwood liable to attack by powder-post beetles. Reported to be attacked by dry-wood termites in West Indies.

durability of heartwood: Non-durable.

preservative treatment: Moderately resistant.

uses

Has very good strength properties, especially toughness, in relation to its weight, and is used where shock resistance is required, e.g. for handles of striking tools (hammers, picks and axes) and for sports goods (lacrosse sticks, laminae in tennis racquets, baseball bats, backs of long bows, skis, etc.).
Used for tops of heavy sea fishing rods, drum sticks, and for picking sticks in the textile industry. In America used in vehicle building for those parts for which ash is employed in the United Kingdom.

HOLLY, EUROPEAN
Ilex aquifolium

THE TREE

Generally a small tree, about 9 m (30 ft) in height, but may reach 25 m (80 ft). Bole averages 3 m (10 ft) in length, diameter up to 0·6 m (2 ft). Europe, including British Isles, and Western Asia.

THE TIMBER properties

colour: White or greyish-white.
sapwood: Not distinct from heartwood.

	grain:	Tends to be irregular. Texture very fine and even.

grain: Tends to be irregular. Texture very fine and even.

weight: Average about 780 kg/m³ (49 lb/ft³), seasoned.

strength: No precise data available but is known to be harder than most native woods. In the uses to which holly is usually put, other strength properties are not important.

movement: Large.

Moisture content in 90 per cent relative humidity 20 per cent
Moisture content in 60 per cent relative humidity 12 per cent
Corresponding tangential movement 3·8 per cent ($\frac{15}{32}$ in/ft)
Corresponding radial movement   2·0 per cent ($\frac{1}{4}$ in/ft)

processing

drying: Reported to be inclined to distort in drying unless cut to small sizes. Kiln drying not recommended.

Shrinkage: Green to 12 per cent moisture content:
Tangential   about 12·0 per cent ($1\frac{7}{16}$ in/ft)
Radial   about 5·0 per cent ($\frac{5}{8}$ in/ft)

working properties:
Blunting: Moderate.
Sawing: Rip-sawing — Saw type HR 60.
           Cross-cutting — Saw types 1, 2 or 3, but tendency to char when teeth are dull.
           Narrow bandsawing — Tendency to burn, otherwise satisfactory.
           Wide bandsawing — Saw type B or C.
Machining: Planing — A cutting angle of 20° required for material with irregular grain.
           Mortising — Chain most suitable.
           Other operations — Satisfactory apart from tendency to burn when working end-grain. Turns well.
Nailing: Probably requires pre-boring.
Gluing: Good.

wood bending: No information.

plywood manufacture: No information.

staining and polishing: Good.

durability and preservation

insect attack: Trees and logs liable to attack by forest longhorn or Buprestid beetles. Sapwood not liable to attack by powder-post beetles, but may be attacked by the common furniture beetle.

durability of heartwood: Perishable.

preservative treatment: No information.

uses

Available in limited quantities and small sizes only. Suitable for fancy turnery and for inlaid work in furniture. Also used for action parts of harpsichords and clavichords and for billiard cue butts. It is sometimes stained black and employed as a substitute for ebony.

HORNBEAM
Carpinus betulus

THE TREE

Reaches a height of 15–25 m (50–80 ft). Bole fluted, may be clear of branches for half the height of the tree, but in unfavourable localities it may branch low down. Logs from trees grown in Britain usually not more than 6 m (20 ft) long. Diameter 0·9–1·2 m (3–4 ft). Temperate Europe, including Britain (mainly in southern and eastern counties), extending to Asia Minor and Iran. Obtained commercially from European countries, particularly France.

**THE TIMBER
properties**

colour: Dull white, marked with greyish streaks and flecks due to the broad rays.

sapwood: Not easily distinguishable from heartwood.

grain: Commonly cross-grained. Texture fine and even.

weight: Average about 750 kg/m³ (47 lb/ft³), seasoned.

strength: Comparable to European beech.

Moisture Content	Bending Strength		Modulus of Elasticity		Compression parallel to grain	
	N/mm²	lbf/in²	N/mm²	1000 lbf/in²	N/mm²	lbf/in²
Green	66	9600	9700	1400	27·0	3910
12 per cent	119	17 300	11 900	1730	55·0	7980

movement: Large.
Moisture content in 90 per cent relative humidity   21·5 per cent
Moisture content in 60 per cent relative humidity   12 per cent
Corresponding tangential movement   2·8 per cent ($\frac{21}{64}$ in/ft)
Corresponding radial movement   1·8 per cent ($\frac{7}{32}$ in/ft)

processing

drying: Dries well and fairly readily.
Kiln Schedule E.
Shrinkage: Green to 12 per cent moisture content:
Tangential   about 7·0 per cent ($\frac{13}{16}$ in/ft)
Radial   about 5·0 per cent ($\frac{5}{8}$ in/ft)

working properties:
Blunting: Moderate, normal in relation to its density.
Sawing: Rip-sawing – Saw type HR 60.
Cross-cutting – Satisfactory.
Narrow bandsawing – Satisfactory.
Wide bandsawing – Saw type B.
Machining: Satisfactory. Turns well and finishes smoothly.
Nailing: No information.
Gluing: Good.

wood bending:
Very suitable for most types of bend. Small knots are no disadvantage.
The wood discolours slightly in the steaming treatment.
Classification – Very good.
Ratio radius/thickness for solid bends (steamed):
Supported: 4·0   Unsupported: 16·5
Limiting radius for 3·2 mm ($\frac{1}{8}$ in) laminae (unsteamed): 140 mm (5$\frac{1}{2}$ in).

plywood manufacture:
Shown by limited tests to be unsuitable because of its hardness.

staining and polishing: Good.

durability and preservation

insect attack: Logs liable to attack by forest longhorn or Buprestid beetles. Bark and sapwood of logs and converted timber liable to attack by the longhorn beetle *Phymatodes testaceus*. Sapwood may be attacked by the common furniture beetle *Anobium punctatum* and by *Ptilinus pectinicornis*, but not by powder-post beetles.

durability of heartwood: Perishable.

preservative treatment: Permeable.

uses

A hard and tough timber which finishes very smoothly. Used in musical instruments, especially in piano actions and in clavichords and harpsichords. Also for wood cogs and turnery, pulleys, dead-eyes, mallets and wooden pegs. Suitable for drum sticks, shafts of billiard cues, skittles and Indian clubs. As a flooring timber, it has high resistance to wear and is a satisfactory alternative to maple for the light type of industrial floors.

HORSE-CHESTNUT, EUROPEAN
Aesculus hippocastanum

THE TREE May reach a height of about 30 m (100 ft), but is usually somewhat smaller. When grown as a parkland tree the bole rarely exceeds 6 m (20 ft) in length. Diameter up to 1·5–1·8 m (5–6 ft). Europe and Asia. In Britain commonly planted in parkland.

THE TIMBER properties

colour: Creamy-white or yellowish.

sapwood: Not easily distinguished from heartwood.

grain: Inclined to be cross grained or wavy grained. Texture very fine and uniform.

weight: About 510 kg/m³ (32 lb/ft³), seasoned.

strength: Slightly higher than obeche.

Moisture Content	Bending Strength		Modulus of Elasticity		Compression parallel to grain	
	N/mm²	lbf/in²	N/mm²	1000 lbf/in²	N/mm²	lbf/in²
Green	41	5900	5300	770	17·4	2530
12 per cent	71	10 300	6300	910	40·3	5850

movement: Small.
Moisture content in 90 per cent relative humidity 20 per cent
Moisture content in 60 per cent relative humidity 12·5 per cent
Corresponding tangential movement 1·5 per cent ($\frac{3}{16}$ in/ft)
Corresponding radial movement 0·8 per cent ($\frac{3}{32}$ in/ft)

processing

drying: Dries readily with little degrade. No special precautions necessary to preserve the white colour of the wood.
Kiln Schedule H.
Shrinkage: Green to 20 per cent moisture content:
Tangential about 3·0 per cent ($\frac{3}{8}$ in/ft)
Radial about 2·0 per cent ($\frac{1}{4}$ in/ft)

working properties:
Blunting: Slight.
Sawing: Rip-sawing – Saw type HR 54.
Cross-cutting – Satisfactory.
Narrow bandsawing – Satisfactory.
Wide bandsawing – Saw type A.
Machining: All operations satisfactory, but a reduced sharpness angle on cutting edges is advantageous.
Nailing: Satisfactory.
Gluing: Good.

wood bending:
Bends well if free from knots and air dry, but very liable to rupture on the inner compressed face if bent in the green state.
Classification – Good.
Ratio radius/thickness for solid bends (steamed):
Supported: 6 Unsupported: 18
Limiting radius for 3·2 mm ($\frac{1}{8}$ in) laminae (unsteamed): 127 mm (5 in).

plywood manufacture: Shown by tests to be suitable for plywood manufacture.
Movement: 1·5 mm plywood from 30 per cent to 90 per cent relative humidity – 0·22 per cent.
Surface splitting on exposure to weather – Grade II.

staining and polishing: Satisfactory.

durability and preservation

insect attack: Sapwood not liable to attack by powder-post beetles, but may be attacked by the common furniture beetle.
durability of heartwood: Perishable.
preservative treatment: Permeable.

uses

A soft, white wood which machines easily and turns well. Suitable for brush backs and handles for special brushes, general turnery, dairy and kitchen utensils, fruit storage trays and racks, moulders patterns, and hand pieces of tennis, badminton and squash racquets.

HURA
Hura crepitans

other names: assacu (Great Britain and Brazil); possentrie (Surinam); sand box (Great Britain and United States).

Grows in West Indies and tropical America. Reaches its maximum height (60 m, 200 ft) in Surinam. Bole straight and fairly regular, free of branches for 15–30 m (50–100 ft), diameter above buttresses up to 2 m (7 ft).

Colour of wood varies from nearly white to yellowish-brown or olive-grey. Heartwood and sapwood not generally clearly distinguished. Sapwood likely to be wide in immature trees. A soft, light wood, weight usually between 370 and 430 kg/m³ (23–27 lb/ft³), seasoned. Texture moderately fine and even. Strength somewhat higher than obeche.

Seasons fairly well, but slowly, without serious degrade. Kiln Schedule E suggested. Works easily and takes stains well.

Can be used for interior joinery, carpentry work and generally as an alternative to obeche.

IDIGBO
Terminalia ivorensis

other names: black afara (Nigeria); framiré (France and Ivory Coast); emeri (Ghana).

Black afara is a name used in Nigeria for the tree. As a timber name it is confusing and should not be used.

THE TREE

Reaches a height of 45 m (150 ft). Clean straight bole, frequently fluted, length up to 21 m (70 ft) above buttresses. Diameter 0·9–1·2 m (3–4 ft) or more. West Africa.

THE TIMBER properties

colour: Yellowish or light yellowish-brown, more rarely light pinkish-brown.
sapwood: 25–50 mm (1–2 in) wide. Not usually easily distinguishable from heartwood, but may be somewhat paler in colour.
grain: Fairly straight with local irregularities, sometimes slightly interlocked, producing an irregular stripe figure when quarter-sawn. Texture medium to fairly coarse. Growth rings unusually distinct and marked by comparatively dense zones which give the flat sawn timber a characteristic appearance suggesting plain oak.
weight: Very variable, partly owing to prevalence of lightweight brittleheart. Range 370–740 kg/m³ (23–46 lb/ft³) at 12 per cent moisture content, but usually between 480 and 620 kg/m³ (30–39 lb/ft³), averaging 540 kg/m³ (34 lb/ft³).
brittleheart: Liable to occur in the inner or core-wood. Has below average density and strength, and sometimes develops a pinkish colour after exposure to light. Natural compression failures (thunder-shakes) commonly occur in brittleheart.

G

chemical staining:
Liable to stain if in contact with iron under damp conditions. Contains yellow colouring matter which may stain moist fabrics in contact with the wood.

corrosive properties:
Has slightly acidic properties and, when moist, tends to promote corrosion of metals, especially iron and steel.

strength: About half-way between European beech and obeche, but material containing brittleheart is appreciably weaker and is unsuitable for purposes where strength is important.

Moisture Content	Bending Strength		Modulus of Elasticity		Compression parallel to grain	
	N/mm^2	lbf/in^2	N/mm^2	$1000\,lbf/in^2$	N/mm^2	lbf/in^2
Green	–	–	–	–	–	–
12 per cent	83	12 100	9300	1350	47·8	6930

movement: Small.
Moisture content in 90 per cent relative humidity   18 per cent
Moisture content in 60 per cent relative humidity   12 per cent
Corresponding tangential movement   1·0 per cent ($\frac{1}{8}$ in/ft)
Corresponding radial movement   0·6 per cent ($\frac{5}{64}$ in/ft)

processing

drying: Dries rapidly and well with little checking or distortion and knots split only slightly.
Kiln Schedule J.
Shrinkage: Green to 12 per cent moisture content:
            Tangential   about 3·0 per cent ($\frac{3}{8}$ in/ft)
            Radial   about 1·5 per cent ($\frac{3}{16}$ in/ft)

working properties:
Slightly interlocked grain affects machining properties. Material containing brittleheart and compression failures liable to crumble when cut on end grain.
Blunting: Slight.
Sawing: Rip-sawing – Saw type HR 54.
            Cross-cutting – Satisfactory.
            Narrow bandsawing – Satisfactory.
            Wide bandsawing – Saw type B.
Machining: Planing – A 20° cutting angle advisable to prevent tearing.
            Boring and hollow square chisel mortising – Burning may occur due to poor chip extraction.
            Other operations – Generally satisfactory but a tendency to break away when working on end-grain.
Nailing: Satisfactory with care.
Gluing: Good.

wood bending: Generally unsuitable for steam bending purposes.
Classification – Very poor.
Limiting radius for 3·2 mm ($\frac{1}{8}$ in) laminae (unsteamed): 190 mm ($7\frac{1}{2}$ in).

plywood manufacture: Shown by tests to be usable for plywood.
Movement: 1·5 mm plywood from 30 per cent to 90 per cent relative humidity – 0·14 per cent.
Surface splitting on exposure to weather – Grade I.

staining and polishing: Satisfactory when filled.

durability and preservation

insect attack: Sapwood liable to attack by powder-post beetles. Reported to be moderately resistant to termites in West Africa.

durability of heartwood: Durable.

preservative treatment: Extremely resistant. Sapwood moderately resistant.

uses   An attractive wood of medium weight, combining ease of working, durability and outstanding stability. Used in furniture and high-class joinery for both interior and exterior work. Material containing brittleheart should be avoided where strength is important. As flooring it has a good appearance and moderate resistance to wear and is suitable for domestic buildings, but is too soft for industrial uses.

When used out of doors, iron or steel fittings should be well protected to avoid staining of the timber and corrosion of the metal. The timber contains a yellow dye-stuff which may stain damp fabrics (see above). It is not therefore recommended for kitchen equipment.

ILOMBA
Pycnanthus angolensis

other names: akomu (Nigeria); otie (Ghana); walele (Ivory Coast); pycnanthus (Great Britain).

THE TREE   A medium-sized tree, 25–35 m (80–115 ft) in height. Long, clear bole, straight and cylindrical, only slightly buttressed, diameter 0·6–1·0 m (2–3 ft), or occasionally more. Tropical Africa.

THE TIMBER
properties

colour: Heartwood rather plain, greyish-white to pinkish-brown, occasionally with yellowish or mauve markings.

sapwood: Wide, not clearly demarcated from heartwood. Very liable to discoloration if extraction and conversion are delayed.

grain: Generally straight. Texture moderately coarse, but even.

odour: Freshly sawn timber often has an unpleasant odour, which disappears when it is dry.

weight: Average about 510 kg/m³ (32 lb/ft³), seasoned.

strength: Somewhat higher than obeche, but tends to be brittle. No precise data available.

movement: No information.

processing

drying: Requires careful drying. Marked tendency to split, and may distort very badly. Difficult to dry in thicker sizes.
Kiln Schedule C.
Shrinkage: No information.

working properties:
Blunting: No information.
Sawing and Machining: Satisfactory. Gives a good finish.
Nailing: Satisfactory, though with some tendency to split.
Gluing: Good.

wood bending: No information.

plywood manufacture:
Employed for plywood and generally available in the United Kingdom. No information on movement of plywood or surface splitting on exposure.

durability and preservation

insect attack: Sapwood liable to attack by powder-post beetles. Reported to be non-resistant to termites in West Africa.

durability of heartwood:
Perishable. Rapid extraction and conversion are necessary to avoid degrade from insect and fungal attack.

preservative treatment: Permeable.

uses   A general utility wood, similar in many respects to South American virola and East African mtambara. Sawn timber can be used for interior joinery, interior parts of furniture, mouldings, etc. Logs are sent to continental Europe where they are used mainly for plywood manufacture as an alternative to gaboon, but rapid extraction and shipment, and protection against insect and fungal attack are necessary. Also used for plywood manufacture in the United Kingdom.

IMBUYA
Phoebe porosa

other name: embuia (Brazil).

THE TREE
A fairly large tree, reaching a maximum height of 40 m (130 ft), diameter of bole about 2 m (6–7 ft). Southern Brazil.

THE TIMBER
properties
Heartwood somewhat variable in colour, from yellowish or olive to chocolate-brown, and may be plain or figured. Grain may be straight, curly or wavy. Texture rather fine. When fresh has a spicy, resinous odour, which is largely lost in drying. Weight about 660 kg/m³ (41 lb/ft³), seasoned. Strength somewhat lower than European beech.

processing
Considered easy to dry; Kiln Schedule E recommended. Saws and machines satisfactorily. Easy to work and finishes smoothly, but a cutting angle of 20° may be required to prevent tearing during planing. Turns satisfactorily. Glues well. Employed for plywood manufacture, but seldom seen in the United Kingdom.

durability and preservation
Heartwood reported to be durable.

uses
Used in Brazil for high-grade flooring, furniture and interior joinery. Has been recommended for gun stocks. As a flooring timber, has a moderate to high resistance to wear. In Europe it is used as veneer for furniture and interior work.

'INDIAN LAUREL'
Terminalia alata,
T. crenulata and
T. coriacea

other names: taukkyan (Burma); asna, mutti, sain (India).

THE TREE
Commonly reaches a height of 30 m (100 ft), but may grow to a larger size in favourable localities. Bole clean and straight, average diameter about 1·0 m (3 ft), length up to 21 m (70 ft), but more commonly 12–15 m (40–50 ft). Widely distributed in India and Burma.

THE TIMBER
properties

colour: Very variable from light brown with few markings or finely streaked with darker lines, to dark brown with irregular darker streaks, producing an attractive figure. Well-figured logs are obtainable from Southern India and Burma. There is some evidence that they are furnished by *Terminalia crenulata*.

sapwood: Reddish-white, sharply defined.

grain: Fairly straight or irregular. Texture rather coarse.

weight: Varies from about 740 to 940 kg/m³ (46–60 lb/ft³), average 850 kg/m³ (53 lb/ft³), seasoned.

strength: Comparable to European beech.

Moisture Content	Bending Strength		Modulus of Elasticity		Compression parallel to grain	
	N/mm²	lbf/in²	N/mm²	1000 lbf/in²	N/mm²	lbf/in²
Green	76	11 000	10 900	1580	37·7	5470
12 per cent	98	14 170	12 300	1790	55·1	7990

(Data from Indian Forest Records)

movement: No information.

processing　　**drying**: Rather difficult to dry, especially in large dimensions. Should be dried slowly and evenly to avoid surface checking, distortion and splitting. It has been found that the timber air dries well if trees are felled during or just after the rains, converted without delay, and the timber stacked for drying under shade. Kiln drying is stated to give more satisfactory results than air drying, but there is a marked tendency for distortion, splitting and checking to occur.

Kiln Schedule C.

Shrinkage: Green to 12 per cent moisture content:
Tangential　　about 6·5 per cent ($\frac{13}{16}$ in/ft)
Radial　　　　about 4·0 per cent ($\frac{1}{2}$ in/ft)

working properties:
Blunting:　Moderate.
Sawing:　　Rip-sawing – Saw type HR 60.
　　　　　　Cross-cutting – Saw type 2.
　　　　　　Narrow bandsawing – Satisfactory.
　　　　　　Wide bandsawing – Saw type C.
Machining: In planing and moulding a cutting angle of 20° is recommended to prevent tearing where grain is interlocked.
Nailing:　 Difficult. Pre-boring necessary.
Gluing:　　Difficult.

wood bending: No information.

plywood manufacture:
Has been tested in India and found to be unsuitable for plywood because of splitting of the veneer during peeling.

staining and polishing: Satisfactory when filled.

durability and preservation　　**insect attack**: Trees and logs liable to attack by forest longhorn or Buprestid beetles and sapwood liable to attack by powder-post beetles. Has been reported to be highly resistant to termites in Malaya and moderately resistant in India.

durability of heartwood: Moderately durable (provisional).

preservative treatment: Resistant. The sapwood is probably permeable.

uses　　Has been used for furniture and cabinet work, panelling, doors, staircases, etc. It yields a very attractive decorative veneer. Employed in India and Burma for building purposes, boat-building, posts and pit props. It is one of the most valuable woods of India.

IROKO
Chlorophora excelsa
and *C. regia*

other names: mvule (East Africa); odum (Ghana); kambala (Zaire); tule, intule (Mozambique); moreira (Angola).

THE TREE　　*Chlorophora excelsa* reaches a height of about 50 m (160 ft). Buttresses small or absent. Bole unbranched for 21 m (70 ft) or more. Diameter up to about 2·5 m (8–9 ft). Widely distributed in Central Africa.

Chlorophora regia, generally similar to *C. excelsa*, but the tree may be rather smaller. Restricted to West Africa.

THE TIMBER properties　　A timber possessing many of the desirable features of teak.

colour:　Yellowish-brown or brown, deepening to dark brown, with lighter markings associated with the vessel lines and particularly conspicuous on flat-sawn surfaces.

sapwood:　Pale in colour, about 50–75 mm (2–3 in) wide. Clearly distinguished from heartwood.

grain: Typically interlocked and sometimes irregular. Texture rather coarse. May be distinguished from teak by its somewhat coarser texture and commonly interlocked grain. When dry it lacks the characteristic smell and greasy feel of teak.

weight: About 640 kg/m³ (40 lb/ft³) at 12 per cent moisture content, similar to teak.

irritant properties:
The fine dust produced in machining occasionally causes irritation of the skin.

other features:
Hard deposits, composed of calcium carbonate and commonly known as stone, liable to occur. They are often completely hidden but can sometimes be detected by the darker colour of the wood surrounding them.

strength: Slightly lower than European beech or teak.

Moisture Content	Bending Strength		Modulus of Elasticity		Compression parallel to grain	
	N/mm²	lbf/in²	N/mm²	1000 lbf/in²	N/mm²	lbf/in²
Green	74	10 700	8300	1200	35·3	5120
12 per cent	90	13 100	9400	1360	54·5	7910

movement: Small.
Moisture content in 90 per cent relative humidity   15 per cent
Moisture content in 60 per cent relative humidity   11 per cent
Corresponding tangential movement   1·0 per cent ($\frac{1}{8}$ in/ft)
Corresponding radial movement   0·5 per cent ($\frac{1}{16}$ in/ft)

processing

drying: Dries well and fairly rapidly without much degrade. Little splitting or distortion occurs but there is some tendency for stick marks to appear during drying.
Kiln Schedule E.
Shrinkage: Green to 12 per cent moisture content:
Tangential   about 2·0 per cent ($\frac{1}{4}$ in/ft)
Radial   about 1·5 per cent ($\frac{3}{16}$ in/ft)

working properties:
Interlocked grain affects machining operations. Occasional deposits of stone (calcium carbonate) severely damage cutting edges.
Blunting: Moderate, but severe when deposits are present.
Sawing: Rip-sawing — Saw type HR 54 or TC.
Cross-cutting — Saw type 3 least suitable for long runs.
Narrow bandsawing — Satisfactory.
Wide bandsawing — Saw type C, possibly with tipped or hardened teeth.
Machining: Planing — A reduction of cutting angle to 15° required to overcome tearing of interlocked grain.
Other operations — Satisfactory.
Nailing: Satisfactory.
Gluing: Good.

wood bending:
Classification — Moderate.
Ratio radius/thickness for solid bends (steamed):
Supported: 15   Unsupported: 18
Limiting radius for 3·2 mm ($\frac{1}{8}$ in) laminae (unsteamed): 210 mm (8$\frac{1}{4}$ in).

plywood manufacture:
Employed for plywood and sometimes available in the United Kingdom. No information on movement of plywood or surface splitting on exposure to weather.

staining and polishing: Good when filled.

durability and preservation

insect attack: Sapwood liable to attack by powder-post beetles. Has been reported highly resistant to termites in East, West and South Africa.

durability of heartwood: Very durable.

preservative treatment: Extremely resistant. Sapwood permeable.

uses

A strong, very durable timber of attractive appearance and small movement, suitable for many of the purposes for which teak is used, including exterior and interior joinery, bench tops and draining boards. Used also in boat-building, apart from bent parts, and in vehicle building. It is a valuable structural timber suitable for piling and marine work. Satisfactory for domestic flooring but not recommended for heavy duty flooring because of a tendency to splinter.

IRONBARK
Eucalyptus spp.

The trade name ironbark covers the timber of the Australian trees *Eucalyptus paniculata*, *E. drepanophylla* and *E. siderophloia* (grey ironbarks) and *E. crebra* and *E. sideroxylon* (red ironbarks).

THE TREE

The size of the tree varies with the species. Average height about 18 m (60 ft), diameter 0·6–0·9 m (2–3 ft). *E. paniculata* and *E. sideroxylon* may reach a height of up to 30 m (100 ft) and diameter 0·9–1·2 m (3–4 ft). Eastern Australia.

THE TIMBER properties

The grey ironbarks are greyish-brown in colour and the red ironbarks brown to reddish-brown or dark red. Sapwood pale in colour, about 25 mm (1 in) wide. Grain usually interlocked. Texture fine and even. A very hard, dense and strong timber. Weight varies from about 1020 to 1180 kg/m³ (64–74 lb/ft³), seasoned. Strength comparable to greenheart.

processing

Stated to be very refractory in drying. In air drying care must be taken to restrict air circulation through the pile sufficiently to maintain a high relative humidity, since the wood is extremely prone to check. It is stated that thin quarter-sawn boards can be kiln dried rapidly without checking, but in general the timber requires very mild drying conditions. Air drying before kiln drying is recommended. Kiln Schedule B. Hard to work in all machine operations. Has a severe blunting effect on cutting edges. Requires care in sawing; in cross-cutting timber requires support at saw exit. In planing a reduced cutting angle is required to prevent tearing. All operations on cross-grain tend to cause breaking out. Unlikely to be suitable for bending or for plywood manufacture.

durability and preservation

Heartwood very durable. Extremely resistant to preservative treatment. The sapwood is permeable.

uses

A heavy, strong and durable timber used in Australia for railway sleepers, bridge work, fencing, and heavy construction work, where resistance to decay is required.

JACAREUBA
Calophyllum brasiliense
SANTA MARIA
Calophyllum brasiliense
var. *rekoi*

The trade name jacareuba refers to the Brazilian timber and Santa Maria to the Central American variety.

THE TREE

Reaches a height of 30–45 m (100–150 ft). Bole straight, cylindrical and un-buttressed, clear for about 20 m (65 ft), diameter 1·0–1·5 m (3–5 ft). Parts of Brazil, Central America and West Indies.

colour: Pink to brick-red.

sapwood: Paler in colour, usually 25–50 mm (1–2 in) wide, distinct from heartwood, but often without a sharp line of demarcation.

grain: Interlocked. Texture medium and fairly uniform. On flat-sawn surfaces dark lines of soft tissue produce a streaked appearance.

weight: Ranges from 540 to 700 kg/m³ (34–44 lb/ft³), average 590 kg/m³ (37 lb/ft³), at 12 per cent moisture content.

strength: Comparable to European beech. Figures in table refer to the timber Santa Maria.

Moisture Content	Bending Strength		Modulus of Elasticity		Compression parallel to grain	
	N/mm²	lbf/in²	N/mm²	1000 lbf/in²	N/mm²	lbf/in²
Green	77	11 100	10 100	1470	37·9	5490
12 per cent	108	15 700	11 800	1710	60·2	8730

movement: Medium.
Data for Santa Maria:
Moisture content in 90 per cent relative humidity   21·0 per cent
Moisture content in 60 per cent relative humidity   12·5 per cent
Corresponding tangential movement   2·8 per cent ($\frac{21}{64}$ in/ft)
Corresponding radial movement   1·7 per cent ($\frac{13}{64}$ in/ft)

processing

drying: Dries slowly with appreciable distortion. Knots tend to split appreciably but little checking is likely to occur.
Kiln Schedule A.
Shrinkage: Green to 12 per cent moisture content:
Tangential   about 5·5 per cent ($\frac{11}{16}$ in/ft)
Radial   about 3·0 per cent ($\frac{3}{8}$ in/ft)

working properties:
Interlocked grain requires care in machining. Brown gum streaks, when present, cause rapid blunting of cutting edges.
Blunting: Moderate, but occasionally severe (see above).
Sawing: Rip-sawing – Saw type HR 54.
Cross-cutting – Satisfactory.
Narrow bandsawing – Satisfactory.
Wide bandsawing – Saw type B.
Machining: Planing – A reduction of cutting angle to 15–20° necessary to prevent tearing.
Drilling – Three-wing drills most suitable; timber requires ample support.
Nailing: Tends to split unless pre-bored.
Gluing: Good.

wood bending:
Tests on a small amount of Santa Maria indicate that severe buckling and fibre rupture occur.
Classification – Moderate.
Ratio radius/thickness for solid bends (steamed):
Supported: 20   Unsupported: 56
Limiting radius for 3·2 mm ($\frac{1}{8}$ in) laminae (unsteamed): No information.

plywood manufacture: Reported to be unsuitable.

staining and polishing: Satisfactory.

durability and preservation

insect attack: Reported to be moderately resistant to termites in West Indies and Borneo.

durability of heartwood: Durable.

preservative treatment: Extremely resistant. Sapwood permeable.

uses	A timber of medium weight but difficult to dry without degrade. Could be employed for constructional work, but owing to its poor drying properties it needs careful handling if used for joinery and furniture.	

JARRAH
Eucalyptus marginata

THE TREE

Reaches a height of 30–45 m (100–150 ft). Diameter 1·0–1·5 m (3–5 ft). South-western Australia.

THE TIMBER
properties

colour: Uniform pinkish to dark red. Often a rich dark red mahogany hue, turning to deep brownish-red with age and exposure to light. Timber of some logs is marked by short, dark brown radial flecks (pencil marks) which are scattered over the end-grain. On flat-sawn surfaces the markings appear as short boat-shaped flecks and on quartered surfaces as small dark patches. They are not detrimental to the timber and may enhance its decorative value.

sapwood: Pale in colour, usually very narrow in old trees.

grain: Usually fairly straight but sometimes interlocked or wavy. Texture even, moderately coarse. The timber may contain gum veins or pockets.

weight: Ranges from 690 to 1040 kg/m³ (43–65 lb/ft³), average about 800 kg/m³ (50 lb/ft³), at 12 per cent moisture content.

strength: Comparable to European beech.

Moisture Content	Bending Strength		Modulus of Elasticity		Compression parallel to grain	
	N/mm²	lbf/in²	N/mm²	1000 lbf/in²	N/mm²	lbf/in²
Green	72	10 400	9600	1390	37·2	5390
12 per cent	118	17 100	12 100	1760	63·5	9210

movement: Medium.
Moisture content in 90 per cent relative humidity 21·5 per cent
Moisture content in 60 per cent relative humidity 14·0 per cent
Corresponding tangential movement 2·6 per cent ($\frac{5}{16}$ in/ft)
Corresponding radial movement 1·8 per cent ($\frac{7}{32}$ in/ft)

processing

drying: Partial air drying before kiln drying recommended. Low temperatures and high humidities advisable when kiln drying from the green condition, especially in thicker sizes. Distortion is the principal cause of degrade in kiln drying. Checking not serious with narrow boards such as flooring, but with wide and thick flat-cut stock care must be exercised. Kiln Schedule C.
Shrinkage: Green to 12 per cent moisture content:
Tangential about 8·0 per cent (1 in/ft)
Radial about 5·0 per cent ($\frac{5}{8}$ in/ft)

working properties: Grain sometimes wavy or interlocked.
Blunting: Moderate.
Sawing: Rip-sawing – Saw type HR 60.
Cross-cutting – Satisfactory.
Narrow bandsawing – Satisfactory.
Wide bandsawing – Saw type C.
Machining: Planing – Cutting angle should be reduced to 15° to prevent tearing out.
Other operations – Satisfactory.
Nailing: Difficult.
Gluing: Good.

wood bending:
>Satisfactory bends of moderate radius of curvature can be made if the material is reasonably straight grained.
>Classification – Moderate.
>Ratio radius/thickness for solid bends (steamed):
>>Supported: 17·5   Unsupported: 39
>Limiting radius for 3·2 mm ($\frac{1}{8}$ in) laminae (unsteamed): 173 mm ($6\frac{4}{8}$ in).

plywood manufacture:
>Shown by tests to be usable for plywood. No information on movement of plywood or surface splitting on exposure.

staining and polishing: Staining is rarely necessary. Polishes well.

durability and preservation

insect attack: Sapwood rarely attacked by powder-post beetles. Heartwood reported to be highly resistant to termites in Australia.

durability of heartwood: Very durable.

preservative treatment: Extremely resistant. Sapwood permeable.
>*Note:* Some specimens of the timber absorb preservatives readily and their vessels are penetrated deeply, although the fibres are not to any extent penetrated.

uses

A hard and heavy timber having good strength properties. Used for decking and underframing of piers, jetties and bridges, also for piling and rubbers in dock and harbour work. As a flooring timber, has high resistance to wear and is suitable for normal conditions of pedestrian traffic, but is inclined to splinter under heavy traffic.

JELUTONG
Dyera costulata
and *D. lowii*

THE TREE

A large tree, reaching a height of 60 m (200 ft). Bole straight and cylindrical, free from buttresses, up to 27 m (90 ft) long. Diameter up to 2·5 m (8 ft). South-east Asia (Malaya, Sabah, Brunei, Sarawak).

THE TIMBER
properties

colour: White or straw coloured, but may be discoloured by staining fungi after tapping for latex.

sapwood: Not distinct from heartwood.

grain: Almost straight. Texture fine and even. Plain appearance.

latex traces:
>Rows of slit-like passages (latex traces or latex canals), 13–38 mm ($\frac{1}{2}$–$1\frac{1}{2}$ in) high, may occur at intervals along the grain but can be eliminated when converting the timber into relatively small dimensions.

weight: About 420–500 kg/m³ (26–31 lb/ft³), average 460 kg/m³ (29 lb/ft³) at 12 per cent moisture content.

strength: Comparable to obeche.

Moisture Content	Bending Strength		Modulus of Elasticity		Compression parallel to grain	
	N/mm²	lbf/in²	N/mm²	1000 lbf/in²	N/mm²	lbf/in²
Green	41	5900	7400	1080	21·9	3170
16 per cent	53	7700	7600	1100	28·1	4070

(Data from Forest Research Institute, Malaya)

movement: Small.
>Moisture content in 90 per cent relative humidity   20 per cent
>Moisture content in 60 per cent relative humidity   12 per cent
>Corresponding tangential movement   1·9 per cent ($\frac{7}{32}$ in/ft)
>Corresponding radial movement   0·9 per cent ($\frac{7}{64}$ in/ft)

processing	**drying:**	Dries easily with little tendency to check or distort, but staining may be troublesome. In air drying ample circulation of air is necessary to ensure quick drying of the surface. It may be difficult to extract moisture from the core of thick stock.

drying: Dries easily with little tendency to check or distort, but staining may be troublesome. In air drying ample circulation of air is necessary to ensure quick drying of the surface. It may be difficult to extract moisture from the core of thick stock.

Kiln Schedule H.

Shrinkage: Green to 12 per cent moisture content:
Tangential about 3·0 per cent ($\frac{3}{8}$ in/ft)
Radial about 2·0 per cent ($\frac{1}{4}$ in/ft)

working properties:
Blunting: Slight.
Sawing: Rip-sawing – Saw type HR 54.
Cross-cutting – Satisfactory.
Narrow bandsawing – Satisfactory.
Wide bandsawing – Saw type A.
Machining: Satisfactory.
Nailing: Satisfactory.
Gluing: Good.

wood bending: No information available.

plywood manufacture:
Reported to have been used as plywood for chests containing rubber.

staining and polishing: Satisfactory.

durability and preservation

insect attack: Sapwood liable to attack by powder-post beetles. Reported to be non-resistant to termites in Malaya.

durability of heartwood: Non-durable.

preservative treatment: Probably permeable.

uses

A plain, stable, lightweight and easily worked timber, but some planks are marred by latex ducts. Has been accepted as an alternative to yellow pine for pattern making. Its ease of working renders it suitable for handicraft work and it has also been used for drawing-boards.

The trees are an important source of latex, which is tapped and used in the manufacture of chewing gum.

JEQUITIBA
Cariniana spp.

other name: jequitiba rosa (Brazil)

THE TREE

A large tree reaching a height of 30–38 m (100–125 ft), bole clear of branches for 18–25 m (60–80 ft), diameter 0·9–1·2 m (3–4 ft). Brazil and northern South America.

THE TIMBER
properties

A rather plain timber, yellowish, pinkish or reddish-brown in colour, sometimes with darker streaks. Sapwood greyish to pale brown, not usually well defined. Grain typically straight, texture fine to medium. Weight rather variable, average about 580 kg/m³ (36 lb/ft³), seasoned. Strength somewhat lower than European beech.

processing

No information on drying; Kiln Schedule D suggested. Working properties generally satisfactory with only a slight blunting effect on cutting edges. In planing there is a tendency to grain raising if cutters are dull. Tends to split in nailing and may require pre-boring. Glues well and is employed in plywood manufacture, but seldom seen in the United Kingdom.

durability and preservation

Heartwood durable. Reported to be extremely resistant to preservative treatment. The sapwood is permeable.

Widely used in South America for general construction, carpentry, interior construction and cabinet making. Used in ship-building in place of the mahoganies. Could be used as an alternative to oak, but is somewhat lighter in weight.

KABUKALLI
Goupia glabra

other names: cupiuba (Brazil); kopie (Surinam); goupi (French Guiana).

Grows in Guyana and neighbouring countries, reaching a height of about 37 m (120 ft), with a long trunk. Heartwood light reddish-brown, of plain appearance, darkening on exposure. Distinct, but not sharply demarcated from thick brownish or pinkish sapwood. Grain irregularly interlocked, texture medium to coarse and rather harsh. The fresh wood has an unpleasant odour, most of which is lost on drying. Weight about 830 kg/m³ (52 lb/ft³), seasoned. Strength about half-way between European beech and greenheart.

Generally satisfactory in sawing; Saw type HR 54 recommended for rip-sawing. Has a moderate blunting effect on cutting edges. Owing to wild grain care is required in machining. In planing a cutting angle of 15° is required to prevent tearing on quarter-sawn surfaces and in moulding there is a tendency for arrises to chip. A large amount of dust is produced during recessing, boring and turning. Liable to split when nailed. Glues well but is unsuitable for plywood manufacture because of splitting of logs. Stains and polishes satisfactorily when filled.

Essentially a heavy and durable construction timber.

KAPUR
Dryobalanops spp.

other names: 'Borneo camphorwood' (Great Britain); kapor (Sabah); kapoer (Indonesia).

Several species of *Dryobalanops* are marketed as kapur. Malayan shipments are mainly *Dryobalanops aromatica*; in Sabah and Sarawak *D. lanceolata* and *D. beccarii* with some *D. aromatica* are important species.

THE TREE Reaches a height of 60 m (200 ft) or sometimes more. Bole straight, usually 27–30 m (90–100 ft) above well-developed buttresses. Diameter usually 1·0–1·5 m (3–5 ft) but may be more. South-east Asia (Malaya, Sarawak, Sabah).

THE TIMBER
properties

colour: Light reddish-brown to deep reddish-brown, uniform.

sapwood: Pale yellow-brown or pinkish, distinct from heartwood.

grain: Straight or shallowly interlocked. Texture fairly coarse but even. Fine resin ducts are present but resin does not exude from the surface of the wood.

odour: The freshly-cut wood has a characteristic camphor-like odour which gradually disappears.

chemical staining:
The wood is liable to stain in contact with iron or iron compounds under damp conditions. It may also promote corrosion of some metals due to its acidic character. The sapwood sometimes contains a yellow colouring matter which may stain damp fabrics.

weight: Usually between 720 and 800 kg/m³ (45–50 lb/ft³) at 12 per cent moisture content. Average 770 kg/m³ (48 lb/ft³). *D. oblongifolia* from Malaya and some Sabah species tend to be slightly lighter in weight, averaging 700 kg/m³ (44 lb/ft³).

strength: Slightly higher than European beech.

Species	Moisture Content	Bending Strength		Modulus of Elasticity		Compression parallel to grain	
		N/mm^2	lbf/in^2	N/mm^2	1000 lbf/in^2	N/mm^2	lbf/in^2
Dryobalanops lanceolata	Green	88	12 800	11 000	1590	42·9	6220
(Sabah)	12 per cent	126	18 300	13 000	1890	69·6	10 100
Dryobalanops beccarii	Green	81	11 700	10 900	1580	41·2	5980
(Sabah)	12 per cent	117	16 900	13 300	1930	66·4	9630

movement: Medium.
Moisture content in 90 per cent relative humidity   17·5–18 per cent
Moisture content in 60 per cent relative humidity   12–12·5 per cent
Corresponding tangential movement   2·1–2·3 per cent ($\frac{1}{4}$–$\frac{9}{32}$ in/ft)
Corresponding radial movement   1·0–1·2 per cent ($\frac{1}{8}$–$\frac{9}{64}$ in/ft)

processing

drying: D. lanceolata dries rather slowly but very well, apart from cup. Shakes tend to increase.
Kiln Schedule H.
Shrinkage: Green to 12 per cent moisture content:
Tangential   about 8·0 per cent (1 in/ft)
Radial about 3·5 per cent ($\frac{7}{16}$ in/ft)
D. beccarii dries rather slowly with a tendency to cup and twist. A little end-splitting may develop in 25 mm (1 in) thick material. Much more difficult to dry in thicker dimensions without serious end-splitting and excessive surface checking.
Kiln Schedule G, but relative humidities 10 per cent higher at each stage of schedule are advisable for timber above 38 mm (1$\frac{1}{2}$ in) thick.
Shrinkage: Green to 12 per cent moisture content:
Tangential   about 6·5 per cent ($\frac{13}{16}$ in/ft)
Radial   about 3·0 per cent ($\frac{3}{8}$ in/ft)

working properties: Texture coarse, grain slightly interlocked.
Blunting: Moderate, but occasionally severe.
Sawing: Rip-sawing – Saw type HR40 or TC.
Cross-cutting – Saw type 2 most suitable. Tendency to break out at bottom of cut.
Narrow bandsawing – Satisfactory, but also some tendency to break out at bottom of cut.
Wide bandsawing – Saw type B.
Machining: Standard machining conditions satisfactory but general finish rather fibrous and blunting of cutting edges is considerable.
Nailing: Satisfactory.
Gluing: No information.

wood bending: Steaming is accompanied by some resin exudation.
Classification – Moderate.
Ratio radius/thickness for solid bends (steamed):
Supported: 17   Unsupported: 30
Limiting radius for 3·2 mm ($\frac{1}{8}$ in) laminae (unsteamed): 173 mm (6$\frac{4}{5}$ in).

plywood manufacture:
Shown by tests to be usable and might be important in local markets.

staining and polishing: Satisfactory.

durability and preservation

insect attack: Sapwood liable to attack by powder-post beetles. Reported to be non-resistant to termites in Malaya.

durability of heartwood: Very durable.

preservative treatment: Extremely resistant. Sapwood permeable.

A strong and very durable timber but rather difficult to machine. Similar in some respects to keruing but is of a more uniform character and more stable, and is non-resinous. A good constructional timber, particularly suitable for estate and farm buildings and has been used for wharf decking. Also used for trailer bottoms, and suitable for exterior joinery, e.g. sills and thresholds, external stairways, window and door frames, cladding, outdoor seats, etc.

KARRI
Eucalyptus diversicolor

THE TREE
A large tree, reaching a height of 45–60 m (150–200 ft) or sometimes more. Clear bole, length 25–30 m (80–100 ft), diameter 1·8–3·0 m (6–10 ft). South-western Australia.

THE TIMBER
properties

colour: Reddish-brown, resembling jarrah in appearance but generally paler in colour.

grain: Interlocked, producing a striped figure. Texture even, moderately coarse.

weight: Average about 880 kg/m³ (55 lb/ft³) at 12 per cent moisture content.

strength: About half-way between European beech and greenheart.

Moisture Content	Bending Strength		Modulus of Elasticity		Compression parallel to grain	
	N/mm²	lbf/in²	N/mm²	1000 lbf/in²	N/mm²	lbf/in²
Green	77	11 200	13 400	1940	37·6	5450
12 per cent	139	20 200	17 900	2590	74·5	10 800

(Data from **Division of Building Research, CSIRO, Forest Products Laboratory, Australia**)

movement: Large.
Moisture content in 90 per cent relative humidity   19·5 per cent
Moisture content in 60 per cent relative humidity   12·5 per cent
Corresponding tangential movement   3·0 per cent ($\frac{3}{8}$ in/ft)
Corresponding radial movement   2·3 per cent ($\frac{9}{32}$ in/ft)

processing

drying: Has a pronounced tendency to check; checks become very deep in thick pieces under adverse conditions. Checking is most severe on the tangential face but also occurs on quarter-sawn material. Distortion is liable to occur in thin stock. In kiln drying from the green condition low temperatures and high humidities are advisable. Partial air drying before kiln drying is advised.
Kiln Schedule C.
Shrinkage: Green to 15 per cent moisture content:
Tangential   about 10·0 per cent ($1\frac{3}{16}$ in/ft)
Radial   about   5·0 per cent ($\frac{5}{8}$ in/ft)

working properties: Wavy and interlocked grain affects machining operations.
Blunting: Moderate to severe.
Sawing: Rip-sawing – Saw type HR 60.
Cross-cutting – Satisfactory, but saw type 3 not suitable for production runs.
Narrow bandsawing – Satisfactory.
Wide bandsawing – Saw type C.
Machining: In planing and moulding a cutting angle of 15° is required to avoid picking up.
Nailing: Difficult.
Gluing: Good.

wood bending:
>>> Limiting data indicate moderately good steam bending properties but the timber cannot be bent if small knots are present.
>>> Classification — Moderate.
>>> Ratio radius/thickness for solid bends (steamed):
>>>>> Supported: 8·0   Unsupported: 12·5
>>> Limiting radius for thin laminae (unsteamed): No information.

plywood manufacture:
>>> Shown by tests to be usable for plywood and might be important in local markets. No information on movement of plywood, or surface splitting on exposure to weather.

staining and polishing: Satisfactory.

durability and preservation

insect attack: Sapwood not liable to attack by powder-post beetles.

durability of heartwood: Durable.

preservative treatment: Extremely resistant. Sapwood permeable.

uses

A heavy, hard timber, having very good strength properties. Similar to jarrah in many respects and very suitable for use in bridges, joists, rafters, beams, etc. Inferior to jarrah for underground use and for situations in contact with water, e.g. dock and harbour works.

KEMPAS
Koompassia malaccensis

THE TREE

Reaches a height of 55 m (180 ft). Bole columnar, with large buttresses, clear for 25–27 m (80–90 ft). Malaya, Sumatra and Borneo.

THE TIMBER properties

colour: Brick red, darkening on exposure to orange-red or red-brown, with numerous yellow-brown streaks due to soft tissue associated with the pores.

sapwood: White or pale yellow, clearly defined, about 50 mm (2 in) wide in large trees and wider in young trees.

grain: Interlocked, sometimes wavy. Texture coarse but even. Streaks or veins of hard stone-like tissue occur irregularly and may be 6 mm ($\frac{1}{4}$ in) wide radially, several inches tangentially, and may extend several feet along the grain. This abnormal tissue is a source of mechanical weakness and may give rise to degrade in drying.

brittleheart: Occasionally present.

corrosion: The timber is slightly acidic and may promote the corrosion of some metals.

weight: 770–1000 kg/m³ (48–62 lb/ft³), average 880 kg/m³ (55 lb/ft³), seasoned.

strength: About half-way between European beech and greenheart.

Moisture Content	Bending Strength		Modulus of Elasticity		Compression parallel to grain	
	N/mm²	lbf/in²	N/mm²	1000 lbf/in²	N/mm²	lbf/in²
Green	105	15 300	15 600	2260	56·7	8230
15 per cent	128	18 600	17 400	2520	68·1	9880

(Data from Forest Research Institute, Malaya)

movement: No data available.

<table>
<tr><td>processing</td><td>drying:</td><td>Dries fairly well, but zones of abnormal tissue which occasionally occur may give rise to serious splitting.
Kiln Schedule E.
Shrinkage: No information.</td></tr>
</table>

working properties:

Machining properties affected by interlocked grain and fibrous texture.

Blunting: Moderate to severe.

Sawing: Rip-sawing – Saw type HR40.
Cross-cutting – Saw type 1 or 2 satisfactory.
Narrow bandsawing – Satisfactory.
Wide bandsawing – Saw type C.

Machining: Planing – Cutting angle should be reduced to 20° to avoid tearing.

Nailing: Pre-boring advisable.

Gluing: No information.

wood bending: No information.

plywood manufacture: No information.

staining and polishing: Satisfactory.

durability and preservation

insect attack: Sapwood liable to attack by powder-post beetles. Recorded as non-resistant to termites in Malaya. Reported to be attacked by dry-wood termites in Borneo.

durability of heartwood: Durable.

preservative treatment: Probably resistant.

uses A heavy constructional timber but somewhat difficult to work. In Malaya has been used after preservative treatment for railway sleepers.

KERUING, GURJUN, YANG, APITONG and ENG
Dipterocarpus spp.

Timber of the keruing type is produced by more than seventy species of *Diptero-carpus*. It is known by distinctive names according to its country of origin. The most important commercially are:

keruing – Malaya, Sarawak, Sabah, Indonesia
gurjun – India and Burma
yang – Thailand
apitong – Philippines
eng *or* in – Burma (see below).

The general character of the timber from the different countries is similar, but gurjun from Burma and yang from Thailand, which are produced by comparatively few species, tend to be more uniform in properties than keruing from Malaya and Sabah, which is derived from a large number of species.

THE TREE
Trees vary in size according to species and locality. Commonly 30–60 m (100–200 ft) in height, with a clear, straight bole up to 21 m (70 ft) long, diameter 1·0–1·8 m (3–6 ft). Some species have buttresses.

THE TIMBER
properties

In comparison with many tropical timbers these timbers are characterised by straightness of general grain direction (which may nevertheless be interlocked).

colour: Varies from pinkish-brown to dark brown, sometimes with a purple tint. Rather plain in appearance.

sapwood: Grey in colour, 50–75 mm (2–3 in) wide, usually well defined.

grain: Straight or shallowly interlocked. Texture moderately coarse but even.

resin exudation:

Some of the timber exudes resin which mars the surface. Resin exudation is very variable but appears to be more common in keruing than in gurjun or yang.

weight: Varies between 640 and 910 kg/m³ (40–57 lb/ft³), but mostly in the range 720 to 800 kg/m³ (45–50 lb/ft³), seasoned. Probably no significant difference in average density between keruing and gurjun, but keruing tends to be more variable because it is derived from a larger number of species. Eng is about 20 per cent heavier than keruing or gurjun, and correspondingly higher in its strength properties.

strength: On average somewhat higher than European beech.
Average values for keruing from Malaya and Sabah:

Moisture Content	Bending Strength		Modulus of Elasticity		Compression parallel to grain	
	N/mm^2	lbf/in^2	N/mm^2	$1000\,lbf/in^2$	N/mm^2	lbf/in^2
Green	79	11 400	13 500	1960	39·2	5690
12 per cent	131	19 000	16 000	2320	71·0	10 300

movement: Medium to large.

	Gurjun	Keruing
Moisture content in 90 per cent relative humidity	20 per cent	20 per cent
Moisture content in 60 per cent relative humidity	12 per cent	12·5 per cent
Corresponding tangential movement	3·3 per cent ($\frac{13}{32}$ in/ft)	2·5 per cent ($\frac{5}{16}$ in/ft)
Corresponding radial movement	2·0 per cent ($\frac{1}{4}$ in/ft)	1·5 per cent ($\frac{3}{16}$ in/ft)

processing

drying: Generally dries slowly, even at high temperatures, and may be difficult to dry uniformly from the centre outwards, especially in quarter-sawn stock. Distortion, especially cupping, is often considerable and slight collapse may occur. Some material is liable to exude gummy resin during drying, more particularly at high temperatures.
Kiln Schedule D.
Shrinkage: Green to 12 per cent moisture content:
Tangential   about 11·5 to 7·5 per cent ($1\frac{3}{8}$ to $\frac{15}{16}$ in/ft)
Radial   about   5·5 to 2·5 per cent ($\frac{5}{8}$ to $\frac{5}{16}$ in/ft)

working properties: Very variable according to species.
Blunting: Moderate to severe. The timber sometimes contains silica, causing rapid blunting of cutting edges.
Sawing: Generally satisfactory apart from rapid blunting of saws with some material. Tungsten carbide-tipped saws advisable when sawing dry timber.
Machining: Can be machined to a clean, but slightly fibrous finish if straight grained. Reduction of cutting angle to 20° recommended when planing material with interlocked grain. Resin adhering to tools, machines and fences sometimes troublesome.
Nailing: Satisfactory.
Gluing: Variable.

wood bending:
Limited tests indicate that gurjun is unsuitable for bending. Keruing from Sabah shown to have moderate bending properties. Severe resin exudation accompanies steaming.
Ratio radius/thickness for solid bends (steamed):
    Supported: 16–17   Unsupported: 29–37
Limiting radius for 3·2 mm ($\frac{1}{8}$ in) laminae (unsteamed):
    147–165 mm ($5\frac{4}{5}$–$6\frac{1}{2}$ in)

H

plywood manufacture:
Employed for plywood and usually available in the United Kingdom. The lighter weight logs are generally used.

Movement of plywood – No information.

Surface splitting on exposure to weather – Grade I.

staining and polishing: Care required when resin is present.

durability and preservation

insect attack: Sapwood liable to attack by powder-post beetles. Gurjun reported to be moderately resistant to termites and keruing non-resistant to termites in Malaya.

durability of heartwood: Moderately durable.

preservative treatment:
Varies from moderately resistant to resistant. Sapwood moderately resistant.

uses Used for many purposes for which oak was formerly used, e.g. for frames, flooring and sides of road vehicles, and in boat-building. Suitable for general construction work, but liable to exude resin if used for exterior joinery exposed to the sun. The resin penetrates most paints and varnishes. Satisfactory as flooring in domestic and public buildings if very resinous material is avoided, but not suitable for heavy-duty flooring.

KOKKO
Albizia lebbek

other name: siris (India).

THE TREE Varies considerably in size according to locality. Reaches its greatest size (height 18–30 m (60–100 ft), diameter 0·6–1·1 m (2–3½ ft)), in the Andaman Islands. Occurs also in India and Burma.

THE TIMBER properties Heartwood medium brown in colour, often having a handsome figure owing to irregular dark markings and interlocked grain. Sapwood wide, white or yellowish-white, distinct from heartwood. Grain somewhat irregularly interlocked, texture very coarse but even. Weight variable, average about 620 kg/m³ (39 lb/ft³), seasoned. Strength somewhat lower than European beech.

processing Moderately refractory in drying; care should be taken to avoid too rapid drying. End-splitting and surface checking may occur in air drying and thick material dries very slowly. End coating of logs is recommended. Presents little difficulty in kiln drying. Kiln Schedule E recommended. Working properties generally satisfactory, but has a woolly texture and interlocked grain. In rip-sawing there is a tendency to bind with thin gauge saws. A cutting angle of 20° is recommended in planing. Shown by tests to be unsuitable for plywood manufacture because of interlocked grain and alternate bands of hard and soft tissue, but has been sliced satisfactorily for decorative veneer. Finishes well if the grain is filled.

durability and preservation Heartwood moderately durable. Probably moderately resistant to preservative treatment.

uses Seen in this country mainly as veneer, which has a rich brown colour and handsome appearance. Has been used in the past in the solid for furniture and may be of interest on account of its walnut-like appearance. Might be acceptable for some of the purposes for which iroko is used, though it is not quite as stable. Used in India for structural purposes, boat-building and as a decorative timber for furniture, interior woodwork, etc.

KRABAK or MERSAWA
Anisoptera spp.

other names: Several species of *Anisoptera* are marketed together. The timber is known as krabak in Thailand, mersawa in Malaya, Sabah, Brunei and Sarawak, kaunghmu in Burma and palosapis in the Philippines.

THE TREE Reaches a height of 45 m (150 ft). Bole fairly straight, free of branches for 25–30 m (80–100 ft). Diameter 1·0–1·5 m (3–5 ft). South-east Asia.

THE TIMBER
properties

colour: Pale yellow or yellow-brown, sometimes with a pinkish tinge, darkening on exposure.

sapwood: Not normally clearly defined from the heartwood, but may be rather paler in colour. About 50 mm (2 in) wide. Very liable to discoloration by staining fungi.

grain: Interlocked, usually not very deeply. Texture moderately coarse but even. Plain appearance apart from a slight ribbon figure on quarter-sawn surfaces.

weight: Varies between species from about 510 to 740 kg/m³ (32–46 lb/ft³), average about 640 kg/m³ (40 lb/ft³), seasoned. The timber from Burma is more uniform in quality and somewhat lighter in weight (560 kg/m³, 35 lb/ft³) than krabak or mersawa.

strength: No precise information available but probably slightly higher than European beech.

movement: Medium.
Moisture content in 90 per cent relative humidity   18 per cent
Moisture content in 60 per cent relative humidity   11 per cent
Corresponding tangential movement   2·8 per cent ($\frac{11}{32}$ in/ft)
Corresponding radial movement   1·2 per cent ($\frac{9}{64}$ in/ft)

processing

drying: Dries very slowly; it is difficult to extract the moisture from the core of thick planks. Little degrade apart from slight distortion. Considerable variation in the drying rate has been found.
Kiln Schedule E.
Shrinkage: Green to 12 per cent moisture content:
Tangential   about 6·5 per cent ($\frac{25}{32}$ in/ft)
Radial   about 3·0 per cent ($\frac{3}{8}$ in/ft)

working properties: Interlocked grain affects machining properties.
Blunting: Severe.
Sawing: Rip-sawing – Saw type HR40 or TC. Type TC essential when dry.
Cross-cutting – Saw type 1 or 2.
Narrow bandsawing – An increased tooth pitch required.
Wide bandsawing – Saw type B.
Machining: Planing – Planes fairly cleanly with sharp cutters but dull cutters produce a fibrous finish. A reduction of cutting angle to 20° is advisable.
Nailing: No information.
Gluing: No information.

wood bending:
Buckles severely when bent to any appreciable extent, even when supported by a strap. Severe resin exudation accompanies steaming.
Classification – Poor.
Limiting radius for 3·2 mm ($\frac{1}{8}$ in) laminae (unsteamed): 180 mm (7 in).

plywood manufacture:
Employed in plywood and usually available in the United Kingdom. No information on movement of plywood or surface splitting on exposure.

109

staining and polishing: Satisfactory.

<table>
<tr><td style="vertical-align:top">durability
and preservation</td><td>insect attack: Sapwood liable to attack by powder-post beetles. Reported to be non-resistant to termites in Malaya.

durability of heartwood: Moderately durable.

preservative treatment: Probably moderately resistant.</td></tr>
<tr><td style="vertical-align:top">uses</td><td>A timber of moderate weight, but rather slow to dry and has a severe blunting effect on saws and cutters. Can be used for general construction, interior joinery and vehicle bodies. Used in its country of origin for ships' planking and furniture. Plywood not recommended for concrete shuttering because it retards the setting of cement.</td></tr>
</table>

KUROKAI
Protium decandrum

THE TREE Grows to a height of 27 m (90 ft). Bole up to 18 m (60 ft) long, sometimes fluted but usually of good form, diameter above buttresses up to 1 m (3 ft). Guyana.

THE TIMBER
properties Heartwood pinkish-brown. Sapwood paler in colour, not clearly demarcated from heartwood. The appearance of the timber is sometimes marred by the presence of dark reddish-brown streaks running in the general direction of the grain. Grain straight or shallowly interlocked, texture fine and even. Average weight about 660 kg/m³ (41 lb/ft³), seasoned. Strength comparable to European beech. Medium movement.

processing Dries fairly rapidly with a marked tendency to cup and twist and for original shakes to extend during drying. Kiln Schedule C. Has a moderate blunting effect on cutting edges. In cross-cutting and narrow bandsawing timber requires support to prevent breaking out. Interlocked grain necessitates care in machining, and in moulding and drilling the timber requires support at tool exits. Liable to split in nailing. Suitable for making solid bends of moderate radius of curvature if straight grained and free from knots. Has been shown by tests to be suitable for plywood manufacture.

durability and preservation Reported to be attacked by dry-wood termites in the West Indies. Heartwood non-durable and extremely resistant to preservative treatment. Sapwood moderately resistant.

uses Used to a limited extent in Guyana for carpentry, but the difficulties arising in processing, together with its lack of durability and resistance to preservative treatment, limit its utilisation in the United Kingdom. Probably best suited to plywood manufacture.

LANCEWOOD
Oxandra lanceolata

other name: asta (United States).

Generally a small tree, growing in the West Indies. Older trees develop a dark heartwood, but only the pale yellow sapwood is in demand and young trees are therefore preferred. The timber is straight grained, fine textured, and weighs about 980 kg/m³ (61 lb/ft³), seasoned. It is a hard, heavy wood, noted for its strength and resilience. It is fairly hard to work and has a tendency to chatter in sawing. In planing, extra pressure is required to prevent riding on cutters. Finishes well in other operations, particularly in turning. Takes stains and polishes very well. The timber is non-durable.

When available, used for archery bows, the top joints of fishing rods, and small tool handles of exceptional hardness. Also used for small parts in organ-building and in textile machinery.

LIGNUM VITAE
Guaiacum spp.
principally *G. officinale*

THE TREE

A small tree, reaching a height of 9 m (30 ft). Diameter usually 250–300 mm (10–12 in), occasionally 450–750 mm (18–30 in). The timber is usually marketed as bolts up to 3 m (10 ft) long. West Indies, Central America and parts of northern South America.

THE TIMBER
properties

One of the hardest and heaviest timbers known to commerce.

colour: Dark greenish-brown or nearly black.

sapwood: Yellowish, sharply defined from heartwood. Varies in width according to species, being narrow in *Guaiacum officinale* and wider in *G. sanctum* (Bahamas) which comes in smaller billets, often with a large proportion of sapwood.

grain: Interlocked. Texture fine and uniform.

weight: Varies from 1150 to 1300 kg/m³ (72–82 lb/ft³), average about 1230 kg/m³ (77 lb/ft³), seasoned.

strength: Outstanding in strength properties, particularly hardness, but splits easily in a tangential plane. No precise data available.

movement: Medium.
Moisture content in 90 per cent relative humidity   20·5 per cent
Moisture content in 60 per cent relative humidity   14 per cent
Corresponding tangential movement   2·5 per cent ($\frac{5}{16}$ in/ft)
Corresponding radial movement   No data

processing

drying: Reported to be refractory in drying; under the tropical sun logs are liable to check and become ring-shaken, at least at the ends.
Kiln Schedule B.

working properties:
Difficult to work with hand tools and very hard to saw and machine.
Blunting: Moderate.
Sawing: Difficult, with a tendency to vibration.
Rip-sawing – Saw type HR 80.
Cross-cutting – Saw type 1 or 2.
Machining: Planing and moulding – Pressure bar, shoe and clamp loadings should be increased to prevent cutters riding.
Turning – Excellent.
Nailing: Not recommended.
Gluing: Difficult.

wood bending: Unsuitable for bending.

plywood manufacture: Unsuitable because of its weight.

staining and polishing: Polishes well. Staining not usually necessary.

durability and preservation

insect attack: Logs liable to attack by forest longhorn or Buprestid beetles.

durability of heartwood: Very durable.

preservative treatment: Extremely resistant.

uses

Owing to the self-lubricating properties of the wood, which are associated with its high oil content, lignum vitae is the most satisfactory wood for ships' propeller bushes and bearings. Used in the textile industry for cotton gins, polishing sticks and rollers. Used also for mallet heads, dead-eyes, etc., and in the sports goods industry for woods used in bowls.

Verawood or Maracaibo lignum vitae (*Bulnesia arborea*) from Venezuela is closely related to true lignum vitae and is sometimes employed as a substitute.

LIME, EUROPEAN
Tilia spp., principally
T. vulgaris

THE TREE Reaches an average height of 25–30 m (80–100 ft), occasionally up to 40 m (130 ft). Bole may be 15 m (50 ft) long, but when grown in the open it branches much nearer the ground. Diameter up to 1·2 m (4 ft). Europe including British Isles.

THE TIMBER
properties

colour: Uniform white or pale yellow, turning pale brown on exposure.
sapwood: Not visually distinguishable from heartwood.
grain: Straight. Texture fine and uniform.
weight: Average about 540 kg/m³ (34 lb/ft³), seasoned.
strength: About half-way between obeche and European beech.

Moisture Content	Bending Strength		Modulus of Elasticity		Compression parallel to grain	
	N/mm²	lbf/in²	N/mm²	1000 lbf/in²	N/mm²	lbf/in²
Green	54	7900	9200	1330	26·1	3780
12 per cent	92	13 300	11 200	1620	47·6	6910

movement: Medium.
Moisture content in 90 per cent relative humidity   22 per cent
Moisture content in 60 per cent relative humidity   11·5 per cent
Corresponding tangential movement   2·5 per cent ($\frac{5}{16}$ in/ft)
Corresponding radial movement   1·3 per cent ($\frac{5}{32}$ in/ft)

processing

drying: Dries well and fairly rapidly with some tendency to distort.
Kiln Schedule H.
Shrinkage: Green to 12 per cent moisture content:
Tangential   about 7·5 per cent ($\frac{15}{16}$ in/ft)
Radial   about 5·0 per cent ($\frac{5}{8}$ in/ft)

working properties:
Blunting: Slight.
Sawing: Rip-sawing – Saw type HR 54.
Cross-cutting – Satisfactory.
Narrow bandsawing – Satisfactory.
Wide bandsawing – Saw type A.
Machining: Planing and moulding – Machines well but a reduced sharpness angle on cutters is recommended and cutting edges should be kept sharp because of the soft nature of the timber.
Turning – Good.
Nailing: Satisfactory.
Gluing: Good.

wood bending:
Suitable for solid bends of moderate radius of curvature. Bending properties are not much improved by supporting the convex face with a strap and end-pressure device.
Classification – Moderate.
Ratio radius/thickness for solid bends (steamed):
Supported: 14   Unsupported: 16
Limiting radius for 3·2 mm ($\frac{1}{8}$ in) laminae (unsteamed): 178 mm (7 in).

plywood manufacture: Shown by tests to be suitable.
Movement: 4·5 mm plywood from 30 per cent to 90 per cent relative humidity – 0·14 per cent.
Surface splitting on exposure to weather – Grade I.

staining and polishing: Satisfactory.

durability and preservation

insect attack: Sapwood liable to attack by the common furniture beetle.
durability of heartwood: Perishable.
preservative treatment: Permeable.

uses

A soft, white wood of fine texture and easy to machine. Has been used by brush makers, mainly for flat paint brushes, and generally in turnery for such articles as toys and bobbins. The hat block manufacturers regard it as an alternative to alder. Would be more extensively used if graded material were available in larger quantities.

LINGUE
Persea lingue

A Chilean tree, growing to an average height of 17 to 18 m (55–60 ft), with a straight bole.

The timber is rather variable in colour, commonly pale reddish-brown. Grain interlocked. Inclined to be knotty. Resembles guarea in grain and texture. Weight about 510 to 620 kg/m³ (32–39 lb/ft³), average 560 kg/m³ (35 lb/ft³), seasoned. Strength somewhat lower than European beech, especially in hardness and resistance to shock loads.

Easy to work and has only a slight blunting effect on cutting edges. Satisfactory in sawing, but in planing a cutting angle of 20° and sharp cutting edges are necessary to produce a good finish on account of interlocked grain. Can be stained and polished satisfactorily. Has been reported to be suitable for steam bending.

Sapwood liable to attack by powder-post beetles. Heartwood moderately durable. Used in Chile for furniture, joinery, interior construction and flooring. It should be suitable for some of the purposes for which birch is employed.

LOURO INAMUI
Ocotea barcellensis

other name: louro inamuhy.

A medium-sized tree, 18 to 30 m (60–100 ft) high. Brazil and Venezuela. Heartwood has a uniform pale golden-brown colour, not clearly differentiated from sapwood. Grain slightly interlocked, giving a broad stripe figure on quartered surfaces. Texture medium. The freshly-cut wood has a strong fragrant odour which is largely lost after drying. It is also oily but the oil evaporates, leaving the surface dry. Weight about 670 kg/m³ (42 lb/ft³), seasoned.

Works well and finishes smoothly. In planing quarter-sawn surfaces a cutting angle of 20° is advisable owing to interlocked grain. The timber has a reputation for stability and durability. Used in South America for carpentry and should be suitable for furniture and high-class joinery.

LOURO, RED
Ocotea rubra

other names: determa (Guyana); wane (Surinam); louro vermelho (Brazil).

THE TREE

A large tree, up to 30 to 40 m (100–130 ft) in height, average diameter 0·8 m (2½ ft), though occasionally up to 1·2 m (4 ft). Bole well formed, cylindrical, un-buttressed and free of branches for 18 to 25 m (60–80 ft). The Guianas, Surinam and Northern Brazil.

THE TIMBER properties

Pale reddish-brown with a subdued golden lustre. Grain straight to irregular, texture rather coarse. Bears some similarity to a dense grade of African mahogany. Average weight about 620 kg/m³ (39 lb/ft³), seasoned. Strength somewhat lower than European beech.

<table>
<tr><td>processing</td><td>Dries at a moderate rate with a tendency to check and split. Thick stock dries rather slowly. Kiln Schedule E. Working properties satisfactory; has only a slight blunting effect on cutting edges. In machining has a tendency to grain raising if cutters are dull. Stains and polishes well after filling. Used in Brazil for planking of boats, and therefore probably suitable for bending to a moderate radius of curvature. Glues well. No information on plywood manufacture.</td></tr>
</table>

**durability
and preservation**

Heartwood durable, probably extremely resistant to preservative treatment. Sapwood probably moderately resistant.

uses

Used locally for furniture, interior joinery and for planking in boat-building. Also employed in railway and road vehicle construction and for ships' spars and masts. Has not been imported into the United Kingdom in large quantities, but should be useful in carpentry and joinery.

MAHOGANY, AFRICAN
Khaya ivorensis and
K. anthotheca

other names: Nigerian, Ghana, Ivory Coast, Benin, Lagos, Degema, Takoradi Grand Bassam mahogany, etc., according to origin (Great Britain); Lagos wood, Benin wood, ogwango (Nigeria); acajou d'Afrique, ngollon (France); khaya (United States). Krala, acajou blanc (*K. anthotheca* only).

African mahogany consists largely of *K. ivorensis* and, from some countries *K. anthotheca*. Small quantities of *K. grandifoliola* are sometimes shipped from certain areas. Consignments are commonly classified according to country of origin, port of shipment, or district from which they are derived. The latter sometimes gives a clue to the species, which may be of value in view of certain differences in properties noted below.

The timber of *K. grandifoliola* differs appreciably in some of its properties from *K. ivorensis* and *K. anthotheca* and is described separately.

THE TREE

K. ivorensis reaches a height of 55–60 m (180–200 ft), usually rather less. Sharp buttresses extend only 1·2–1·5 m (4–5 ft) up the bole. Clear bole up to 25–27 m (80–90 ft). Diameter up to 1·5–1·8 m (5–6 ft). West Africa (Ivory Coast to Gaboon). *K. anthotheca* grows to a height of 55 m (180 ft), not generally of such a good shape as *K. ivorensis*. Buttresses may be 2·7 m (9 ft) high. Diameter of bole 1·2 m (4 ft). Parts of West Africa, extending eastwards to Uganda.

**THE TIMBER
properties**

colour: Pink when freshly sawn, darkening upon exposure to reddish-brown. Seldom shows the yellowish-brown colour found in the paler shades of American mahogany.

sapwood: Creamy-white or yellowish, up to 50 mm (2 in) wide, not always sharply demarcated from heartwood.

grain: Sometimes straight, more usually interlocked producing a stripe or roe figure on quarter-sawn material. Occasional logs produce highly decorative timber. Texture variable, typically moderately coarse.

brittleheart: Present in some logs. Cross fractures variously termed thundershakes, cross-breaks and heart-breaks (actually natural compression failures) frequently occur in trees containing brittleheart and may be more common in figured logs than in plain logs.

weight: Average about 700 kg/m³ (44 lb/ft³) green, 530 kg/m³ (33 lb/ft³) seasoned.

irritant properties: The dust from *K. anthotheca* may cause irritation of the skin in some individuals. Other species of *Khaya* do not possess irritant properties.

strength: About half-way between obeche and European beech.

114

Species	Moisture Content	Bending Strength		Modulus of Elasticity		Compression parallel to grain	
		N/mm^2	lbf/in^2	N/mm^2	1000 lbf/in^2	N/mm^2	lbf/in^2
K. ivorensis (Nigeria and Ghana)	Green	54	7800	7400	1080	26·8	3890
	12 per cent	78	11 300	9000	1300	46·4	6730
K. anthotheca (Ghana)	Green	60	8700	7700	1120	28·4	4120
	12 per cent	83	12 100	9200	1330	45·9	6660
(Uganda)	Green	53	7700	7400	1080	25·4	3680
	12 per cent	83	12 000	9000	1310	44·3	6430

movement: Small. Data for *K. ivorensis*.

Moisture content in 90 per cent relative humidity   20 per cent
Moisture content in 60 per cent relative humidity   13·5 per cent
Corresponding tangential movement   1·5 per cent ($\frac{3}{16}$ in/ft)
Corresponding radial movement   0·9 per cent ($\frac{7}{64}$ in/ft)
K. anthotheca is very similar.

processing

drying: Dries fairly rapidly with little degrade. Where strongly developed tension wood is present serious distortion may occur during drying.
Kiln Schedule F.
Shrinkage: Green to 12 per cent moisture content:
Tangential   about 4·5 per cent ($\frac{9}{16}$ in/ft)
Radial   about 2·5 per cent ($\frac{5}{16}$ in/ft)

working properties:

Machining properties affected by interlocked grain and by woolliness in some material. Tension wood and brittleheart sometimes present.
Blunting: Moderate.
Sawing: Rip-sawing – Saw type HR54.
Cross-cutting – Satisfactory.
Narrow bandsawing – Satisfactory.
Wide bandsawing – Saw type B.
Machining: Planing – A cutting angle of 20° recommended to prevent tearing of interlocked grain.
Moulding – Square block causes most tearing. French head not satisfactory on woolly material.
Other operations generally satisfactory, except on woolly material.
Nailing: Satisfactory.
Gluing: Good.

wood bending:

Cannot be bent without severe buckling and fibre rupture occurring.
No advantage in using a supporting strap.
Classification – Very poor, but *Khaya anthotheca* from East Africa may be moderately satisfactory.
Limiting radius for 3·2 mm ($\frac{1}{8}$ in) laminae (unsteamed):
K. ivorensis   236 mm ($9\frac{1}{4}$ in)
K. anthotheca (West Africa)   173 mm ($6\frac{4}{8}$ in)
K. anthotheca (East Africa)   147 mm ($5\frac{4}{8}$ in)

plywood manufacture:
Employed for plywood and usually available in the United Kingdom.

	K. ivorensis	K. anthotheca
Movement of 4·5 mm plywood, 30 per cent to 90 per cent relative humidity	0·14 per cent	No information
Surface splitting on exposure to weather	Grade I	Grade I

staining and polishing: Good.

durability and preservation

insect attack: Trees and logs liable to attack by forest longhorn or Buprestid beetles. Sapwood liable to attack by powder-post beetles and by the common furniture beetle. Has been reported to be non-resistant to termites in West Africa. Said to be attacked by dry-wood termites in West Indies.

durability of heartwood: Moderately durable.

preservative treatment: Extremely resistant. Sapwood moderately resistant.

uses

A wood of medium density and pleasing appearance, having good working properties and small movement. Used extensively in the furniture industry for reproduction furniture, contract furniture, office desks, cabinet work, etc. Suitable for interior parts such as rails, shelf lipping, divisions, cabinet interiors and drawer sides. In boat-building, suitable for almost all parts of boats except steamed bent framing. Used chiefly for planking and general joinery and for keels, hogs, transomes, stems and many other items. Commonly laminated, e.g. for stems and frames, and used in veneer form in the cold moulding process. Very suitable for use in racing craft where weight is important. Employed in the joinery trade for panelling, general interior joinery, doors, bank fittings, etc. In motor vehicles, used for mouldings, shells and internal fittings in vans, ambulances and caravans. Used for many other purposes where a good quality, medium weight hardwood is required.

MAHOGANY, AFRICAN
Khaya grandifoliola

The timber of *K. grandifoliola* is not always marketed separately from *K. ivorensis* and *K. anthotheca*, but a separate description is provided because it differs appreciably from these two species in some of its properties.

THE TREE

Generally a smaller tree than *K. ivorensis* or *K. anthotheca*, reaching a height of 30–40 m (100–130 ft). Bole sometimes twisted or branching low, diameter above buttresses about 0·9–1·2 m (3–4 ft). Drier parts of West Africa.

THE TIMBER
properties

Heartwood variable in colour from pink to reddish-brown, generally somewhat darker than *K. ivorensis*. Sapwood straw-coloured or pink, up to 50 mm (2 in) wide, not sharply defined. Grain interlocked, generally irregular in direction, sometimes very wild. Texture moderately coarse. Weight very variable from 560 to 770 kg/m³ (35–48 lb/ft³), average about 670 kg/m³ (42 lb/ft³) at 12 per cent moisture content. Appreciably stronger and harder than *K. ivorensis*, approaching European beech in its strength properties. Small movement.

processing

Dries rather slowly but fairly well with little checking or distortion, but cup may sometimes occur. Moderate blunting effect on cutting edges, and appreciably harder to work than *K. anthotheca* and does not machine so cleanly. Material with wild grain is difficult to machine to a good surface. Tends to split in nailing. Finishes satisfactorily. Less suitable for plywood manufacture than other species of *Khaya*, because of its weight and irregular grain.

durability and preservation

Trees and logs liable to attack by forest longhorn and Buprestid beetles and sapwood liable to attack by powder-post beetles. Heartwood durable (provisional). Extremely resistant to preservative treatment; sapwood moderately resistant.

uses Generally darker in colour and appreciably heavier, harder and more resistant to decay than *K. ivorensis* or *K. anthotheca*. Suitable for heavier types of furniture, interior and exterior joinery, shop fittings, etc. Contains so much distorted grain that it is unlikely to be suitable for rotary-cut veneers.

MAHOGANY, AMERICAN
Swietenia spp.,
principally
S. macrophylla

other names: British Honduras, Costa Rica, Guatemala, Honduras, Mexican, Nicaraguan, Brazilian, Peruvian mahogany, etc., according to origin (Great Britain); aguano (Brazil and Peru), caoba (Peru).

Cuban or Spanish mahogany from the West Indies is *Swietenia mahagoni*.

THE TREE
Grows to an average height of 30 m (100 ft). Bole straight and cylindrical, diameter 1·2–1·8 m (4–6 ft) above heavy buttresses. Clear of branches for 12–18 m (40–60 ft). Central America, particularly Mexico and Honduras, and parts of northern South America.

THE TIMBER
properties

colour: Varies from light reddish- or yellowish-brown to dark reddish-brown, with a high natural lustre. The heavier material is generally darker in colour. Darkens upon exposure.

grain: Tends to be interlocked, but there is a fair proportion of plain, straight-grained timber. Irregularities of grain produce a variety of figure — fiddleback, blister, stripe or roe, curl, mottle, etc. Characteristically, flat-sawn timber shows a growth ring figure which distinguishes it from most African mahogany. Texture moderately coarse, but generally finer than African mahogany.

weight: Averages about 540 kg/m³ (34 lb/ft³), seasoned, and is appreciably lighter and softer than Cuban mahogany.

strength: About half-way between obeche and European beech.

Moisture Content	Bending Strength		Modulus of Elasticity		Compression parallel to grain	
	N/mm²	lbf/in²	N/mm²	1000 lbf/in²	N/mm²	lbf/in²
Green	–	–	–	–	–	–
12 per cent	83	12 100	8800	1280	44·2	6410

movement: Small.
Moisture content in 90 per cent relative humidity   19 per cent
Moisture content in 60 per cent relative humidity   12·5 per cent
Corresponding tangential movement   1·3 per cent ($\frac{5}{32}$ in/ft)
Corresponding radial movement   1·0 per cent ($\frac{1}{8}$ in/ft)

processing

drying: Dries fairly rapidly and well without much checking or distortion. Kiln Schedule F.
Shrinkage: Green to 12 per cent moisture content:
            Tangential   about 3·0 per cent ($\frac{3}{8}$ in/ft)
            Radial   about 2·0 per cent ($\frac{1}{4}$ in/ft)

working properties:

Blunting: Slight.

Sawing: Rip-sawing — Saw type HR 54.
Cross-cutting — Satisfactory.
Narrow bandsawing — Satisfactory.
Wide bandsawing — Saw type B.

Machining: Planing and moulding — Generally finishes well but some material tends to produce a woolly finish, and the use of sharp cutters is then essential.
Other operations — Satisfactory.

Nailing: Satisfactory.

Gluing: Good.

wood bending:

Tests on timber grown in India indicate that it possesses moderately good bending properties.
Classification — Moderate.
Ratio radius/thickness for solid bends (steamed):
Supported: 12   Unsupported: 28

plywood manufacture:

Employed for plywood but seldom seen in the United Kingdom. No information on movement of plywood or surface splitting on exposure to weather.

staining and polishing: Good.

durability and preservation

insect attack: Sapwood liable to attack by powder-post beetles and by the common furniture beetle. Reported to be attacked by dry-wood termites in West Indies.

durability of heartwood: Durable.

preservative treatment: Extremely resistant.

uses

A high-class timber, generally regarded as superior in quality to other mahoganies, and valued especially for its small dimensional movement, lack of distortion and good finishing qualities. Uses are mainly confined to special purpose high-quality work on account of its relatively high cost and limited availability. Used in the furniture industry mainly for reproduction cabinet work, chairs, and for special contract work. Suitable for panelling and interior joinery and used occasionally for window and door cappings in private cars. In boats, proved satisfactory over many years for planking, deck housings, cabin fittings, etc. Suitable for models, especially where fine detail or sharp edges are required, and for purposes such as pattern making when long-term reliability is required.

MAHOGANY, DRY-ZONE
Khaya senegalensis

other names: bisselon, bissilongo (Portuguese Guinea).

Grows to a height of about 30 m (100 ft). Diameter of bole 1 m (3 ft), but not so well shaped as in other African mahoganies. West Africa.

The timber is a heavy African mahogany, resembling *Khaya grandifoliola* more closely than *K. ivorensis* or *K. anthotheca*. Heartwood pink-brown, darkening to deep red-brown with a purplish tinge, darker in colour than ordinary commercial African mahogany. Sapwood only slightly paler and browner than heartwood and not very distinct from it. Grain interlocked. Average weight approaching 800 kg/m³ (50 lb/ft³), seasoned.

Dries fairly well; Kiln Schedule F. Has a moderate blunting effect on cutting edges but works well, though harder and less easy to work than *K. ivorensis*. A reduced cutting angle required in machining on account of interlocked grain. Unsuitable for plywood manufacture because of its weight.

Suitable for veneer and for interior decoration and most other purposes for which African mahogany is used. Some timber is of excellent quality, especially suitable for purposes where mahogany of firm texture and a natural dark reddish-brown colour is required.

MAHOGANY, MOZAMBIQUE
Khaya nyasica

other names: mbaua, umbaua (Mozambique); Nyasaland mahogany (UK).

THE TREE

Grows to a height of up to 30 m (100 ft). Bole straight and cylindrical, diameter 1·0–1·5 m (3–5 ft), clear for 15–18 m (50–60 ft). East and Central Africa.

THE TIMBER
properties

Heartwood bright pink to pale red when freshly sawn, darkening to a lustrous golden brown on exposure. Tends to be somewhat darker than *Khaya ivorensis*. Sapwood 50–75 mm (2–3 in) wide, paler than heartwood when freshly cut, but similar to it when dry. Grain usually interlocked and may be wavy or irregular. Texture medium. Average weight about 580 kg/m³ (36 lb/ft³) air dry. Strength properties generally similar to *K. ivorensis* and *K. anthotheca*. Small movement.

processing

Dries fairly rapidly with only slight distortion unless tension wood is present. Kiln Schedule F. Has a moderate blunting effect on cutting edges. Tendency for saws to bind in rip-sawing and to burn in cross-cutting. Machining generally satisfactory but a reduced cutting angle is advisable. Pre-boring necessary in nailing. Glues well and has been shown by tests to be suitable for plywood manufacture.

durability and preservation

Heartwood moderately durable (provisional). Extremely resistant to preservative treatment; sapwood moderately resistant.

uses

Generally similar in properties to *K. ivorensis* and *K. anthotheca* and suitable for the same purposes as those for which these timbers are used.

MAKARATI
Burkea africana

other name: mukarati.

A small- to medium-sized tree, height 12–15 m (40–50 ft), bole up to 4·5 m (15 ft) long, diameter about 0·6 m (2 ft). Widely distributed in tropical Africa.
Heartwood dark brown to reddish-brown. Sapwood white or yellowish, 25 mm (1 in) wide, distinct from heartwood. Has a fine texture and interlocked grain, giving a stripe figure. A hard and heavy timber, average weight about 900 kg/m³ (56 lb/ft³), seasoned, but stated to be appreciably lower in Mozambique and Uganda (about 720 kg/m³, 45 lb/ft³).

Stated to dry moderately rapidly with little distortion or splitting. Kiln Schedule B. Difficult to work because of its hardness. Tends to tear in planing but can be polished to a good finish.

Has high wearing qualities and forms a good flooring timber. Small quantities have been imported from Mozambique, mainly for flooring. Also used locally for mining timbers, heavy construction and sleepers.

MAKORÉ
Tieghemella heckelii
(formerly *Mimusops heckelii*)

other names: baku (Ghana); a similar timber from Cameroon and Gaboon, produced by a related species, is known as douka.

THE TREE

Reaches a height of 37–45 m (120–150 ft). Bole long, clean, straight and cylindrical, free from buttresses. Diameter generally about 1·2 m (4 ft), but may be up to 2·7 m (9 ft). West Africa.

THE TIMBER   Large trees are reported to be liable to shatter in felling, but logs received in this country are generally of good shape, sound in the heart and free from defects.

properties

colour: Varies from pinkish- or purplish-brown to dark red, with a high natural lustre.

sapwood: Paler in colour, 50–75 mm (2–3 in) wide.

grain: Often straight, giving timber of fairly plain appearance, but the grain of selected timber is figured with a broken stripe or mottle, and has a very decorative moiré or watered silk appearance. Occasionally the wood is handsomely marked with irregular veins of darker colour. Texture fine.

weight: Average about 620 kg/m³ (39 lb/ft³) at 12 per cent moisture content.

chemical staining:
              Liable to stain in contact with iron or iron compounds under moist conditions.

irritant properties:
              The fine dust produced in machining is liable to cause irritation of the nose, eyes and throat and other ill effects in some individuals.

strength: Slightly lower than European beech.

Moisture Content	Bending Strength		Modulus of Elasticity		Compression parallel to grain	
	N/mm²	lbf/in²	N/mm²	1000lbf/in²	N/mm²	lbf/in²
Green	75	10 900	8200	1190	36·5	5300
12 per cent	101	14 700	10 100	1470	53·3	7730

movement: Small.
           Moisture content in 90 per cent relative humidity   19 per cent
           Moisture content in 60 per cent relative humidity   13 per cent
           Corresponding tangential movement   1·8 per cent ($\frac{7}{32}$ in/ft)
           Corresponding radial movement   1·1 per cent ($\frac{1}{8}$ in/ft)

processing

drying: Dries at a moderate rate with little degrade. Distortion generally slight but some twisting may occur in a small proportion of a load. A little splitting tends to develop around knots.
          Kiln Schedule H.
          Shrinkage: Green to 12 per cent moisture content:
                      Tangential   about 4·5 per cent ($\frac{9}{16}$ in/ft)
                      Radial   about 3·0 per cent ($\frac{3}{8}$ in/ft)

working properties:
          Contains silica, causing blunting of cutting edges, especially in dry wood.
          Blunting: Severe.
          Sawing: Rip-sawing – Saw type TC, with 20–25° hook.
                      Cross-cutting – Tungsten carbide-tipped saw necessary.
                      Narrow bandsawing – Not a commercial proposition.
                      Wide bandsawing – Type B; tipped or hardened teeth required for dry timber.
          Machining: Planing – a cutting angle of 20° required.
                      Boring – tendency to char.
                      Machining generally difficult due to the blunting effect on tools.
          Nailing: Tendency to split.
          Gluing: Good.

wood bending:
Heartwood suitable for bends of moderate radius of curvature. Sapwood tends to buckle and rupture when bent even to a moderate extent and is generally unsatisfactory.
Classification — moderate (heartwood only).
Ratio radius/thickness for solid bends (steamed):
Supported: 12   Unsupported: 18
Limiting radius for 3·2 mm ($\frac{1}{8}$ in) laminae (unsteamed): 158 mm ($6\frac{1}{8}$ in).

plywood manufacture:
Employed for plywood and usually available in the United Kingdom.
Movement: 4·5 mm plywood from 30 per cent to 90 per cent relative humidity — 0·20 per cent.
Surface splitting on exposure — no information.

staining and polishing:
Excellent when a small amount of filler is used.

durability and preservation

insect attack: Sapwood liable to attack by powder-post beetles. Has been reported to be highly resistant to termites in West Africa.

durability of heartwood: Very durable.

preservative treatment: Extremely resistant. Sapwood moderately resistant.

uses

A timber having some resemblance to a close-grained mahogany. Has good stability and high natural durability and is somewhat heavier than African mahogany. Used for furniture and decorative work, both in the solid and as veneer, and for high-class joinery and interior fittings. In vehicle construction satisfactory as framing for commercial vehicles and used in railway carriages for interior decorative framing and as veneers for coach panelling. In boat-building its durability is a valuable property and it is suitable for keels, dead wood, planking and frames. Marine plywood of makoré is also widely used in boat-building. Other uses include exterior doors, sills, thresholds and flooring, where it is suitable for normal pedestrian traffic.

Would probably be more widely used for furniture and similar purposes if it were not for the disadvantages of the irritant dust formed in machining and sanding and the blunting effect on tools.

MANSONIA
Mansonia altissima

other names: aprono (Ghana); ofun (Nigeria); bété (Ivory Coast and Cameroon).

THE TREE

Reaches a height of 30 m (100 ft). Bole cylindrical with narrow buttresses, diameter above buttresses 0·6–0·8 m ($2–2\frac{1}{2}$ ft). West Africa.

THE TIMBER properties

colour: Rather variable, yellowish- to dark greyish-brown, often with a purple tinge, relieved by lighter and darker bands. The colour fades on exposure.

sapwood: White, 25–38 mm ($1–1\frac{1}{2}$ in) wide.

grain: Generally straight. Texture fine and even.

weight: Average about 590 kg/m³ (37 lb/ft³), seasoned.

irritant properties:
The fine dust produced in machining and sanding operations is liable to cause dermatitis and irritation of the nose, eyes and throat among men working with the timber.

strength: Comparable to European beech (based on a limited number of tests).

Moisture Content	Bending Strength		Modulus of Elasticity		Compression parallel to grain	
	N/mm^2	lbf/in^2	N/mm^2	$1000\,lbf/in^2$	N/mm^2	lbf/in^2
Green	90	13 000	9700	1400	44·1	6400
12 per cent	122	17 700	10 900	1580	58·6	8500

movement: Medium.

Moisture content in 90 per cent relative humidity   20 per cent
Moisture content in 60 per cent relative humidity   12 per cent
Corresponding tangential movement   2·3 per cent ($\frac{9}{32}$ in/ft)
Corresponding radial movement   1·3 per cent ($\frac{5}{32}$ in/ft)

processing

drying: Dries fairly rapidly and well, but shakes are inclined to extend and knots tend to split appreciably; very little distortion.
Kiln Schedule H.
Shrinkage: Green to 12 per cent moisture content:
Tangential   about 3·0 per cent ($\frac{3}{8}$ in/ft)
Radial   about 1·5 per cent ($\frac{3}{16}$ in/ft)

working properties:
Blunting: Moderate.
Sawing: Rip-sawing – Saw type HR54.
Cross-cutting – All types satisfactory.
Narrow bandsawing – Satisfactory.
Wide bandsawing – Saw type B.
Machining: Good.
Nailing: Satisfactory.
Gluing: Good.

bending properties:
Generally good, but considerable variation occurs in bending properties even between pieces from the same part of a tree, hence only recommended for bends of moderate radius of curvature. Cannot be successfully bent if knots, even of small diameter, are present. Tends to split at setting temperatures of about 65°C. Less tendency for buckling and fracture during bending if the material is bent in the green rather than the air-dried condition.
Classification – Good.
Ratio radius/thickness for solid bends (steamed):
Supported: 10   Unsupported: 15·5
Limiting radius for 3·2 mm ($\frac{1}{8}$ in) laminae (unsteamed): 110 mm ($4\frac{1}{4}$ in).

plywood manufacture:
Employed for plywood but seldom seen in the United Kingdom. No information on movement or surface splitting.

staining and polishing: Satisfactory.

durability and preservation

insect attack: Sapwood rarely liable to attack by powder-post beetles. Reported to be highly resistant to termites in West Africa.

durability of heartwood: Very durable.

preservative treatment: Extremely resistant. Sapwood permeable.

uses

A very suitable timber for furniture and joinery manufacture, but the irritant properties of the fine dust limit its use. Bears a general resemblance to American walnut and may be used for similar purposes. Suitable for high quality cabinet work and chairs, radio and television cabinets and fancy turnery. Used in the motor body industry for high-class work such as window edges, dashboards, etc.

MAPLE, JAPANESE
Acer spp.

Consists principally of *Acer mono*, but possibly also other species.
Under favourable conditions the tree grows to a height of about 12–15 m (40–50 ft), diameter 0·6 m (2 ft).

Timber very similar in properties to rock maple, but usually slightly darker in colour and a little lighter in weight. Heartwood pinkish-brown in colour, not clearly demarcated from sapwood. Weight about 610–710 kg/m³ (38–44 lb/ft³), air dry. Dries without undue difficulty. Used in Japan for plywood manufacture.

Has a high resistance to abrasive action and is equivalent to rock maple as a flooring timber.

MAPLE, ROCK
Acer saccharum

other names: hard maple (Great Britain, Canada, United States); sugar maple (Canada); white maple (sapwood) (United States).

Black maple (*Acer nigrum*) is similar in properties to *A. saccharum* and is commonly sold under the name of rock maple or hard maple.

THE TREE

Sometimes reaches a height of 40 m (130 ft), more usually 25–27 m (80–90 ft). Clear bole up to 21 m (70 ft), but often less. Diameter 0·6–1·0 m (2–3 ft). Canada and northern and eastern states of USA.

THE TIMBER
properties

colour: Creamy white, generally with a reddish tinge. Occasional large trees have a dark brown heart.

sapwood: Light coloured, not sharply defined from heartwood.

grain: Usually straight but sometimes curly or wavy. Fine brown lines marking the growth rings give a distinctive figure on plain-sawn surfaces. Texture fine and even.

weight: Average about 720 kg/m³ (45 lb/ft³) at 12 per cent moisture content.

strength: Comparable to European beech.

Moisture Content	Bending Strength		Modulus of Elasticity		Compression parallel to grain	
	N/mm²	lbf/in²	N/mm²	1000 lbf/in²	N/mm²	lbf/in²
Green	74	10 700	11 000	1590	32·6	4730
15 per cent	121	17 600	13 200	1910	58·5	8490

(Data from Canadian Department of Forestry)

movement: Medium.
Moisture content in 90 per cent relative humidity   21 per cent
Moisture content in 60 per cent relative humidity   12·5 per cent
Corresponding tangential movement   2·6 per cent ($\frac{5}{16}$ in/ft)
Corresponding radial movement   1·8 per cent ($\frac{7}{32}$ in/ft)

processing

drying: Stated to dry slowly but without undue difficulty.
Kiln Schedule E.
Shrinkage: Green to 12 per cent moisture content:
Tangential   about 5·0 per cent ($\frac{5}{8}$ in/ft)
Radial   about 2·5 per cent ($\frac{5}{16}$ in/ft)

working properties:

Blunting:	Moderate.
Sawing:	Satisfactory, but with a tendency to tooth vibration. Rip-sawing – Saw type HR60. Cross-cutting – Satisfactory. Narrow bandsawing – Satisfactory. Wide bandsawing – Saw type B.
Machining:	Planing – a cutting angle of 20° required. Other operations – generally good results, but a tendency to ride on cutters and to burn on end-grain working.
Nailing:	Difficult.
Gluing:	Good.

wood bending:

No exact data available on minimum bending radius, but appears to be a very good bending wood. Requires stronger and more efficient bending apparatus than would be necessary for well-known bending timbers such as beech or ash.

Classification – Good.

plywood manufacture:

Employed in plywood but seldom seen in the United Kingdom. No information on movement or surface splitting.

staining and polishing: Good.

durability and preservation

insect attack: Sapwood liable to attack by the common furniture beetle *Anobium punctatum* and by the furniture beetle *Ptilinus pectinicornis*. Growth defects known as pith flecks caused by insects are sometimes present in heartwood and sapwood.

durability of heartwood: Non-durable.

preservative treatment: Resistant. Sapwood permeable.

uses

A timber with good strength properties and resistance to wear and which finishes and turns well. Suitable for furniture and panelling. An excellent flooring timber with high resistance to abrasion, wearing smoothly without surface disintegration, suitable for heavy industrial traffic, roller skating rinks, dance halls, squash courts, bowling alleys, etc. The timber is also used for rollers in textile machinery, shoe lasts, parts of piano actions, and in sports goods. Some trees have a fine figure ('birds eye') and curly grain and furnish a very decorative veneer.

MAPLE, SOFT
Acer saccharinum

other names: silver maple (Canada and United States).

Red maple (*Acer rubrum*) is similar in properties to *A. saccharinum* and is commonly sold with that species as soft maple.

THE TREE

Commonly reaches a height of 21–27 m (70–90 ft), sometimes up to 38 m (125 ft). Straight clear bole, diameter 0·6–1·2 m (2–4 ft). Canada and eastern United States.

THE TIMBER properties

colour:	Creamy-white.
sapwood:	Not easily distinguishable from heartwood.
grain:	Straight. Timber generally similar in appearance to rock maple but less lustrous and growth rings comparatively indistinct. Rays are narrower and less conspicuous than in rock maple and pith flecks are frequently present.

weight:	About 540 kg/m³ (34 lb/ft³), seasoned. Red maple (*Acer rubrum*) is somewhat heavier, about 610 kg/m³ (38 lb/ft³), seasoned.
strength:	About half-way between obeche and European beech.

Moisture Content	Bending Strength		Modulus of Elasticity		Compression parallel to grain	
	N/mm²	lbf/in²	N/mm²	1000 lbf/in²	N/mm²	lbf/in²
Green	50	7200	8500	1240	21·0	3040
15 per cent	86	12 500	10 500	1530	43·1	6250

(Data from Canadian Department of Forestry)

movement: No information available.

processing

drying: No exact information available, but probably similar to rock maple. Kiln Schedule E.

working properties:
	Blunting:	Moderate.
	Sawing:	Rip-sawing – Saw type HR54.
		Cross-cutting – Satisfactory.
		Narrow bandsawing – Satisfactory.
		Wide bandsawing – Saw type B.
	Machining:	Satisfactory. The timber is less hard and works more easily than rock maple.
	Nailing:	Satisfactory with care.
	Gluing:	Variable.

wood bending: No information available.

plywood manufacture:
Shown by tests to be usable for plywood. No information on movement or surface splitting.

staining and polishing: Good.

durability and preservation

insect attack: No information.

durability of heartwood: Non-durable.

preservative treatment: Moderately resistant. Sapwood permeable.

uses

Resembles rock maple, but is lighter in weight, softer and lower in strength properties. Suitable for furniture and joinery work and for turnery, but less suitable than rock maple for flooring.

MECRUSSE
Androstachys johnsonii

Occurs in Mozambique and Rhodesia. A tall straight tree, height up to 36 m (120 ft), average diameter 0·5 m (1½ ft). Logs straight and cylindrical. Heartwood brown or light brown, slightly pinkish, sometimes with darker markings. Sapwood narrow and pale in colour. Grain characteristically curly. Texture very fine and even. A hard, heavy and strong timber, but with some tendency to split under stress. Weight about 1000 kg/m³ (62 lb/ft³), seasoned. Medium movement.

Dries slowly and surface checks readily. Not difficult to saw or machine although hard and heavy. Turns well and finishes and polishes excellently. Heartwood reported to be very durable and resistant to termites. Used in East Africa mainly for heavy construction work such as bridges, posts and piling, harbour work and railway sleepers. In Britain used as a flooring timber.

MENGKULANG
Heritiera spp., principally
H. simplicifolia (formerly
Tarrietia simplicifolia)

other names: kembang (Sabah); chumprak (Thailand) and lumbayan (Philippines) are very similar timbers to mengkulang.

THE TREE A medium-sized to large tree, 30–45 m (100–150 ft) in height. Bole well formed, 0·6–1·0 m (2–3 ft) in diameter above buttresses. Malaysia.

THE TIMBER
properties

colour: Varies from medium pink to red-brown or dark red-brown, sometimes with dark streaks on longitudinal surfaces. Similar in appearance to West African niangon.

sapwood: 50–125 mm (2–5 in) wide, paler in colour than heartwood, not always sharply defined.

grain: Typically interlocked and sometimes somewhat irregular in direction. Quartered surfaces have a broad stripe figure and, when accurately cut, a conspicuous ray figure, very much like that of niangon. Texture moderately coarse.

weight: Somewhat variable, but generally in the range 640–720 kg/m³ (40–45 lb/ft³), seasoned. Limited tests indicate that kembang from Sabah may be somewhat lighter in weight than Malaysian mengkulang.

strength: Comparable to European beech.

Moisture Content	Bending Strength		Modulus of Elasticity		Compression parallel to grain	
	N/mm²	lbf/in²	N/mm²	1000lbf/in²	N/mm²	lbf/in²
Green	81	11 800	11 700	1700	39·8	5770
12 per cent	117	17 000	13 200	1920	61·2	8880

movement: Small.
Moisture content in 90 per cent relative humidity   16·5 per cent
Moisture content in 60 per cent relative humidity   12·0 per cent
Corresponding tangential movement   1·5 per cent ($\frac{3}{16}$ in/ft)
Corresponding radial movement   0·9 per cent ($\frac{7}{64}$ in/ft)

processing

drying: Dries rapidly and well, apart from some tendency to surface checking which may vary from log to log.
Kiln Schedule H.
Shrinkage: Green to 12 per cent moisture content:
Tangential   about 7·0 per cent ($\frac{13}{16}$ in/ft)
Radial   about 3·0 per cent ($\frac{3}{8}$ in/ft)

working properties: Machining properties affected by interlocked grain.
Blunting: Severe.
Sawing: Rip-sawing – Saw type HR40 or TC.
Cross-cutting – Saw type 1, 2 or TC-tipped for long runs.
Narrow bandsawing – Increased tooth pitch necessary.
Wide bandsawing – Saw type B. Tipped or hardened teeth required for dry material.

Machining: Planing — A cutting angle of 20° is recommended for quarter-sawn material.
Moulding — French head not suitable.
Recessing — Rapid blunting produces dull cutting edges which tend to char.
Nailing: Satisfactory.
Gluing: No information.

wood bending: No information.

plywood manufacture:
Employed for plywood and usually available in the United Kingdom.
Movement of plywood — No information.
Surface splitting on exposure — No information.

staining and polishing: Satisfactory when filled.

durability and preservation

durability of heartwood: Non-durable.
preservative treatment: Resistant. Sapwood moderately resistant.

uses

A medium-weight red wood, useful as an alternative to timbers of similar type from West Africa. Has an attractive appearance and good strength properties but presents some difficulty in sawing and working and its durability is lower than that of niangon and utile. Best used as a general purpose wood for carpentry, interior joinery and construction purposes. Can be rotary peeled and is used for commercial plywood production in south-east Asia.

MERANTI, SERAYA and LAUAN
principally *Shorea* spp.

A large number of species of the genus *Shorea* growing in south-east Asia produce timber known according to its country of origin as meranti, seraya or lauan. Generally, the name meranti is applied to timbers from Malaya, Sarawak and Indonesia, seraya to timbers from Sabah, and lauan to timbers from the Philippines. The timbers vary in colour and density and are conveniently grouped as follows:
1 Light red meranti, light red seraya, white lauan (in part).
   Note: white lauan also includes species of *Parashorea* and *Pentacme*.
2 Dark red meranti, dark red seraya, red lauan.
3 Yellow meranti, yellow seraya.
4 White meranti, melapi (Sabah).*
The commercial timber in each of these groups consists of a number of species and the descriptions that follow are based, in the main, on a few of the more important species in each group. However, because of the number of species included, the commercial timbers are somewhat variable in character and the properties (e.g. strength) of some pieces in a parcel may vary appreciably from those given. In particular, it may be found that some timber within a group may have a different durability classification from that stated.

*Note that white seraya (*Parashorea* spp.) is not the same type of timber as white meranti (*Shorea* spp.).

MERANTI, DARK RED, DARK RED SERAYA and RED LAUAN
Shorea spp.

A group of timbers which are darker in colour and usually somewhat heavier than light red meranti and light red seraya. The commercial timber may include a number of species of *Shorea*, but the most important in both Malaya and Sabah is *S. pauciflora*, sometimes known in Malaya as nemesu and in Sabah as oba suluk.

THE TREE

A large tree. *S. pauciflora* in Sabah may reach a height of nearly 70 m (225 ft), diameter 1·5 m (5 ft), with a tall, well-shaped bole above large buttresses. Trees of other species generally rather smaller in size.

THE TIMBER
properties

The following description refers to timber of *S. pauciflora*. Timbers of other species in this group are similar to it in general character.

colour: Medium to dark red-brown, commonly with conspicuous white dammar or resin streaks.

sapwood: 25–65 mm (1–2½ in) wide, pink in colour, rather poorly defined.

grain: Interlocked, giving a broad stripe figure on quartered surfaces. Texture rather coarse. Brittleheart not usually troublesome.

weight: Appreciably heavier than light red meranti, generally in the range 580 to 770 kg/m³ (36–48 lb/ft³), average about 670 kg/m³ (42 lb/ft³), seasoned.

strength: Appreciably higher than light red meranti, and approaching European beech.

Data for dark red seraya (*Shorea pauciflora*) from Sabah:

Moisture Content	Bending Strength		Modulus of Elasticity		Compression parallel to grain	
	N/mm²	lbf/in²	N/mm²	1000 lbf/in²	N/mm²	lbf/in²
Green	68	9900	9700	1400	33·9	4920
12 per cent	92	13 300	11 400	1650	52·9	7670

The density of the material tested was appreciably lower than the average for this timber. The strength values of timber of average density may be expected to be somewhat higher than the figures quoted.

movement: Small.
Moisture content in 90 per cent relative humidity   20 per cent
Moisture content in 60 per cent relative humidity   13 per cent
Corresponding tangential movement   2·0 per cent (¼ in/ft)
Corresponding radial movement   1·0 per cent (⅛ in/ft)

processing

drying: Dries more slowly than light red meranti, especially in thicker sizes. Some tendency to distort, and risk of splitting and checking in thick material.
Kiln Schedule F.
Shrinkage: Green to 12 per cent moisture content:
Tangential   about 5·5 per cent ($\frac{11}{16}$ in/ft)
Radial   about 3·0 per cent (⅜ in/ft)

working properties:
Blunting: Slight.
Sawing: Sawn surfaces fibrous, otherwise satisfactory.
Rip-sawing – Saw type HR 54.
Cross-cutting – Satisfactory.
Narrow bandsawing – Satisfactory.
Wide bandsawing – Saw type B.
Machining: Satisfactory provided cutters are sharp. Fibrous finish may be obtained on end-grain. Support can be provided with advantage in boring and mortising.
Nailing: Good.
Gluing: Good.

wood bending:
Severe buckling occurs at comparatively large radii of curvature, even when using a supporting strap. Pronounced tendency for bends to distort during drying.
Classification – Poor.
Limiting radius for 3·2 mm (⅛ in) laminae (unsteamed): 165 mm (6½ in).

plywood manufacture:
Used for plywood manufacture and generally available in the United Kingdom.
Movement of plywood — No information.
Surface splitting on exposure to weather — Grade I.

staining and polishing: Satisfactory.

durability and preservation

insect attack: Sapwood liable to attack by powder-post beetles.

durability of heartwood: Moderately durable to durable.

preservative treatment:
Varies from resistant to extremely resistant. Sapwood usually moderately resistant.

uses

The timbers are attractive and are heavier, stronger and more durable than light red meranti and light red seraya and somewhat less variable in character than these timbers. Suitable for more exacting purposes in construction, exterior and interior joinery, shopfitting and boat-building. As flooring timber, suitable for moderate conditions of wear.

MERANTI, LIGHT RED, LIGHT RED SERAYA, WHITE LAUAN
Shorea spp.

THE TREE

The timber is derived from about a dozen species of *Shorea*. Trees reach a large size, height up to 60 m (200 ft), with well shaped boles, 27–30 m (90–100 ft) long, 1·0 m or more (3–4 ft) in diameter, sometimes with buttresses.

THE TIMBER properties

colour: Variable in colour from very pale pink to mid-red.

sapwood: Distinct from heartwood, paler in colour, usually with a greyish tinge, 25–65 mm (1–2½ in) wide.

grain: Usually shallowly interlocked, producing a broad stripe figure on quartered surfaces. Texture coarse but even.

brittleheart: Commonly present in some species.

weight: Variable, from about 400 to 640 kg/m³ (25–40 lb/ft³), average about 510 kg/m³ (32 lb/ft³), seasoned.

strength: On average about half-way between obeche and European beech. Data for light red meranti from Malaya and light red seraya from Sabah:

Timber	Moisture Content	Bending Strength		Modulus of Elasticity		Compression parallel to grain	
		N/mm²	lbf/in²	N/mm²	1000 lbf/in²	N/mm²	lbf/in²
Light red meranti	Green	63	9100	9700	1400	32·0	4640
	12 per cent	88	12 700	10 500	1520	50·1	7260
Light red seraya*	Green	50	7200	7200	1040	25·7	3730
	12 per cent	71	10 300	8500	1230	41·1	5960

*The density of the light red seraya tested was appreciably lower than the average for this timber. The strength values of timber of average density may be expected to be somewhat higher than the figures quoted.

movement: Small.

Moisture content in 90 per cent relative humidity 19·5 per cent
Moisture content in 60 per cent relative humidity 12·5 per cent
Corresponding tangential movement 1·6–1·9 per cent ($\frac{3}{16}$ to $\frac{1}{4}$ in/ft)
Corresponding radial movement   0·7–0·8 per cent
                                                                (about $\frac{3}{32}$ in/ft)

processing

drying: Variable, depending on species. Timbers generally dry fairly rapidly, with little distortion but with a marked tendency to cup. Thick material may be rather slow to dry with a tendency to surface checking.
Kiln Schedule F generally suitable.
Shrinkage: Green to 12 per cent moisture content:
Tangential   about 6·0–7·0 per cent ($\frac{3}{4}$ to $\frac{7}{8}$ in/ft)
Radial   about 2·0–3·5 per cent ($\frac{1}{4}$ to $\frac{7}{16}$ in/ft)

working properties:
Blunting: Slight.
Sawing: Rip-sawing – Saw type HR 54.
Cross-cutting – Satisfactory, but gives a fibrous finish.
Narrow bandsawing – Satisfactory.
Wide bandsawing – Saw type A.
Machining: Planing – Generally satisfactory with standard cutters provided they are kept sharp.
Other operations – Generally satisfactory with sharp cutting edges. Brittleheart should be avoided.
Nailing: Satisfactory.
Gluing: Good.

wood bending:
Cannot be steam bent appreciably without severe buckling, even when a supporting strap is used.
Classification – Very poor.
Limiting radius for 3·2 mm ($\frac{1}{8}$ in) laminae (unsteamed):
about 150–200 mm (6–8 in)

plywood manufacture:
Used for plywood manufacture and generally available in the United Kingdom.
Movement of plywood – No information.
Surface splitting on exposure to weather – Grade I.

staining and polishing: Satisfactory when filled.

durability and preservation

insect attack: Sapwood liable to attack by powder-post beetles.

durability of heartwood: Non-durable to moderately durable.

preservative treatment:
Varies from resistant to extremely resistant. Sapwood usually moderately resistant.

uses Widely used in Malaya and Sabah and exported in large quantities to the United Kingdom, Australia, Japan and other countries. Used for a wide range of light structural work and interior joinery, and for general purposes. Has been employed as a mahogany-like wood for interior parts of furniture. Extensively used for plywood manufacture.

MERANTI, WHITE
Shorea spp.

other name: melapi (Sabah).

Timber produced by trees of several species in the *Anthoshorea* group of *Shorea*.
Note: White meranti is *not* the equivalent of white seraya from Sabah.

THE TREE Medium-sized to large trees, some species reaching 60 m (200 ft) in height, diameter 1·0–1·5 m (3–5 ft). Bole well formed, generally with buttresses.

THE TIMBER **properties**	Heartwood almost white when freshly sawn, darkening to pale yellow-brown on exposure. Sapwood about 50–65 mm (2–2½ in) wide, distinguishable from heartwood when dry. Grain interlocked, texture moderately coarse but even. The presence of silica in the form of minute grains in the wood is a characteristic feature of the white meranti group. Weight variable, average about 660 kg/m³ (41 lb/ft³), seasoned. Little information available on strength properties; probably slightly superior to light red meranti.
processing	Dries well without serious degrade. Kiln Schedule F. Has a severe blunting effect on cutting edges owing to presence of silica in the wood. In sawing, tungsten carbide-tipped saws and increased tooth pitch are recommended. A cutting angle of 20° is advisable in planing. Recessing difficult due to rapid blunting of cutter. Stains and polishes satisfactorily when filled.
durability **and preservation**	Heartwood moderately durable (provisional). Varies from resistant to extremely resistant to impregnation.
uses	Not generally favoured for constructional and joinery purposes owing to the difficulty in sawing and machining. Its main use is for veneer and plywood manufacture, since the presence of silica does not affect the peeling process.

MERANTI, YELLOW, YELLOW SERAYA
Shorea spp.

Yellow meranti and yellow seraya, from Malaya and Sabah respectively, are produced by about a dozen species of *Shorea* and are similar in general character to red meranti and red seraya, but do not have the red tint of these timbers.

THE TREE	Medium-sized to large trees, often 60 m (200 ft) or more in height. Generally buttressed, with straight cylindrical boles up to 1·5 m (5 ft) in diameter above buttresses.

THE TIMBER
properties

colour:	Dull yellow or yellow-brown, darkening somewhat on exposure.
sapwood:	Paler in colour than heartwood, often with a greyish tinge, 50–75 mm (2–3 in) wide.
grain:	Shallowly interlocked. Texture moderately coarse, but finer than that of red merantis.
weight:	Variable but generally in the range 480 to 670 kg/m³ (30–42 lb/ft³), seasoned.
chemical staining:	
	Liable to discoloration if in contact with iron under moist conditions.
brittleheart:	Sometimes present.
strength:	About half-way between obeche and European beech. Data for yellow seraya from Sabah (average of four species):

Moisture Content	Bending Strength		Modulus of Elasticity		Compression parallel to grain	
	N/mm²	lbf/in²	N/mm²	1000 lbf/in²	N/mm²	lbf/in²
Green	62	9000	8600	1250	29·9	4340
12 per cent	84	12 200	9700	1410	45·7	6630

movement: Small.
Data for *Shorea faguetiana*:
Moisture content in 90 per cent relative humidity   18·5 per cent
Moisture content in 60 per cent relative humidity   12·0 per cent
Corresponding tangential movement   2·8 per cent ($\frac{11}{32}$ in/ft)
Corresponding radial movement   0·8 per cent ($\frac{3}{32}$ in/ft)

processing **drying:** In thicknesses up to 50 mm (2 in) dries slowly but well, apart from a tendency to cup. In thicker sizes moisture movement is slow with some risk of honeycombing, and a tendency for existing shakes to extend. Kiln Schedule J.

Shrinkage: Green to 12 per cent moisture content:
Tangential about 6·5–7·5 per cent ($\frac{13}{16}$ to $\frac{15}{16}$ in/ft)
Radial about 2·5 per cent ($\frac{5}{16}$ in/ft)

working properties:
Blunting: Moderate.
Sawing: Rip-sawing – Saw type HR 54.
Cross-cutting – Satisfactory.
Narrow bandsawing – Satisfactory.
Wide bandsawing – Saw type B.
Machining: Planing – A cutting angle of 20° required to prevent tearing of quarter-sawn stock, due to interlocked grain.
Boring and mortising – Support of workpiece advisable.
Nailing: Good.
Gluing: Good.

wood bending:
Better for steam bending than light red seraya and suitable for bends of moderate radius of curvature.
Classification – Moderate.
Ratio radius/thickness for solid bends (steamed):
Supported: 18   Unsupported: 31
Limiting radius for 3·2 mm ($\frac{1}{8}$ in) laminae (unsteamed): 150 mm (6 in).

plywood manufacture: Shown by tests to be suitable.
Movement of plywood – No information.
Surface splitting on exposure to weather – Grade I.

staining and polishing: Satisfactory when filled.

durability and preservation

insect attack: Sapwood liable to attack by powder-post beetles. Reported to be non-resistant to termites in Malaya.

durability of heartwood: Moderately durable.

preservative treatment: Extremely resistant. Sapwood moderately resistant.

uses A non-resinous, easy working wood, somewhat harder and heavier than light red meranti. Suitable for light building construction, interior joinery and fittings and general utility purposes. As flooring timber, best suited to light conditions of traffic. Used for plywood manufacture in Malaya.

MERBAU
Intsia palembanica
and *I. bijuga*

A large tree with a rather short, thick bole, often fluted, diameter about 1·5 m (5 ft). South-east Asia and islands of south-west Pacific.

properties Heartwood yellowish to orange-brown when freshly cut, darkening to medium to dark red-brown on exposure. Sapwood pale yellow, 50–75 mm (2–3 in) wide, sharply defined from heartwood. Grain interlocked, sometimes wavy, producing a ribbon figure on radial faces. Texture rather coarse but even. Liable to stain in contact with iron under damp conditions. Weight about 740–900 kg/m³ (46–56 lb/ft³), average 800 kg/m³ (50 lb/ft³), seasoned. Strength comparable to or slightly lower than European beech. Stated to be lacking in resilience. Small movement.

processing Dries well and fairly rapidly with little degrade and low shrinkage. Kiln Schedule C. Working properties rather variable. Has a moderate blunting effect on cutting edges. Cuts cleanly in most operations but a cutting angle of 20° is advantageous in planing material with interlocked grain. Pre-boring advisable in nailing. Finishes well but requires considerable filling. Heartwood reported to be durable.

uses   A timber of good appearance and generally free from defects. Similar in general character to afzelia and suitable for similar purposes, e.g. structural work and panelling, high-class joinery, and furniture. The heavier material is satisfactory for good-quality flooring.

MISSANDA
Erythrophleum guineense
and *E. ivorense*

other names: tali (Ivory Coast); erun, sasswood (Nigeria); potrodom (Ghana); kassa (Zaire); muave (Zambia).

Reaches a height of up to 30 m (100 ft). Widespread in tropical Africa.
Heartwood reddish-brown, becoming rich, dark brown when dried. Sapwood yellowish-white. Grain interlocked, texture coarse. A hard, heavy timber, weighing about 900 kg/m³ (56 lb/ft³), seasoned. Strength somewhat lower than greenheart. Small movement.

Stated to dry slowly with a tendency to distort. Kiln Schedule D. Difficult to work; tungsten carbide-tipped saws and cutters are recommended. Finishes well and takes a good polish. Heartwood very durable.

As flooring it has a high resistance to wear and has been used successfully in schools and other public buildings. Suitable for heavy-duty flooring where smoothness of surface is not essential. Used in Africa for heavy construction work, door frames, bridge decking, railway sleepers, etc.

MORA
Mora excelsa and
MORABUKEA
Mora gonggrijpii

THE TREE   Trees of both species reach a height of 30–45 m (100–150 ft), or occasionally more, with buttresses up to 4·5 m (15 ft) up the stem. Clear bole above buttresses 18–25 m (60–80 ft) in length, usually cylindrical but sometimes flattened. The Guianas, Trinidad and eastern Venezuela.

THE TIMBER
properties   Mora and morabukea are similar in general appearance. Heartwood varies from chocolate-brown to reddish-brown, occasionally pinkish-brown particularly in morabukea. The timbers are rather plain in appearance. Sapwood yellowish to pale brown, commonly 50–300 mm (2–6 in) wide. Grain generally interlocked and somewhat irregular. Texture coarse. Weight generally between 900 and 1100 kg/m³ (56–68 lb/ft³), average about 1020 kg/m³ (64 lb/ft³), seasoned. Strength comparable to greenheart. Large movement.

processing   Dries very slowly with appreciable degrade. Existing shakes tend to extend during drying, surface checking develops readily and degrade from cupping and twisting may be serious. Initial longitudinal distortion tends to increase markedly. Kiln Schedule B. Hard to work because of its density and interlocked grain. Morabukea is slightly milder than mora but their machining characteristics are similar. Has a moderate to severe blunting effect on cutting edges. In sawing, resin builds up on teeth and teeth tend to vibrate. In machining, has a high resistance to cutting and charring may occur in some operations. In planing, a cutting angle of 20° is required to overcome tearing on quarter-sawn stock. Pre-boring is necessary in nailing. Bending properties moderately good, but not comparable with those of a good bending timber such as beech. Unsuitable for plywood manufacture because of its weight. Staining and polishing satisfactory.

durability
and preservation   Sapwood liable to attack by powder-post beetles. Reported to be resistant to termites in the West Indies. Heartwood durable and extremely resistant to preservative treatment. Sapwood permeable or moderately resistant.

uses   These timbers are best suited for heavy constructional work, e.g. piling, jetties and foreshore work. Used locally for building work, boat-building, railway sleepers, paving blocks, etc.

MTAMBARA
Cephalosphaera usambarensis

THE TREE   A large tree, commonly 45 m (150 ft), sometimes up to 60 m (200 ft) in height. Bole straight and cylindrical, 15–25 m (50–80 ft) long and 1·2–1·8 m (4–6 ft) in diameter above buttresses. Tanzania.

THE TIMBER
properties   Heartwood pale pinkish-brown with a faint orange tint, rather plain in appearance. Sapwood somewhat paler in colour but not sharply demarcated from heartwood. Grain generally straight, texture moderately fine and even. Weight about 590 kg/m^3 (37 lb/ft^3), seasoned. Strength properties generally somewhat lower than European beech, but comparable to beech in stiffness. Medium movement.

processing   Reported to be very susceptible to staining fungi and should be given an anti-stain treatment before being stacked for air drying. Kiln dries rapidly with little degrade; negligible splitting and little distortion apart from cup. Kiln Schedule J suggested. Saws and machines without difficulty. Can be sanded to a smooth finish and takes polish well. Nails well. Can be peeled and is used for plywood in East Africa.

durability and preservation   Heartwood perishable. Both heartwood and sapwood are moderately resistant to preservative treatment.

uses   A timber of plain appearance with a combination of moderately light weight, pale colour, straight grain and good working properties. A general purpose utility wood, suitable for interior joinery and for use in the furniture industry.

MUBURA
Parinari excelsa

THE TREE   Grows to a height of up to 45 m (150 ft). Bole straight and cylindrical with buttresses up to 3 m (10 ft) high. Diameter above buttresses 0·9–1·2 m (3–4 ft). Mature trees have some tendency to develop heart rot. Widely distributed in tropical Africa.

THE TIMBER
properties   Heartwood pale reddish-brown, darkening on exposure. Sapwood yellowish-white, distinct from heartwood. Grain usually interlocked and irregular in direction. Texture moderately coarse. Deposits of silica, often 1 per cent or more by weight, commonly occur in the ray cells as crystalline aggregates. Average weight about 740 kg/m^3 (46 lb/ft^3). The timber is hard, moderately heavy and comparable in strength properties to European beech. Large movement.

processing   A difficult timber to kiln dry from green. Dries very slowly with a marked tendency to distort and split. With thick material it may be more satisfactory to air dry before kiln drying. Kiln Schedule B, but with material over 38 mm (1$\frac{1}{2}$ in) in thickness an increase in relative humidity of 10 per cent is recommended. Has a severe blunting effect on cutting edges due to its high silica content and its hardness. Tungsten carbide-tipped saws are necessary for satisfactory sawing, particularly with seasoned timber. Machining is also difficult with carbon steel or high-speed steel tools owing to severe blunting. Pre-boring is advisable in nailing. Has moderately good bending properties if the steamed material is adequately supported with a metal strap and end-pressure device. Can be glued satisfactorily but is unsuitable for plywood manufacture.

durability and preservation   Heartwood non-durable. Moderately resistant to preservative treatment.

134

uses
Its use is restricted by the difficulty in drying the timber and by its severe blunting effect on saws and tools. It is not used in the United Kingdom and the best prospect lies in using it locally in large constructional sizes, converted when green. Used to some extent in Africa for bridge beams and decking, mining timbers, railway sleepers and other types of heavy construction.

MUERI
Pygeum africanum

The tree varies in size according to area of growth, reaching a maximum height of over 30 m (100 ft) in Kenya and Uganda. Diameter up to 1·0 m (3 ft). Tropical and southern Africa.

Heartwood pale red when freshly cut, darkening on exposure to rich dark red. Sapwood pale pink, not clearly demarcated from heartwood. Grain somewhat interlocked. Texture medium to fine. Weight about 720 kg/m³ (45 lb/ft³), air dry. Strength comparable to European beech.

Very refractory in drying; dries slowly, liable to split, distort and collapse. Kiln drying requires a long period on a mild schedule (Kiln Schedule C) and is probably not economic. Working properties generally satisfactory. Difficult to nail on account of its hardness and tendency to split if nailed near edges. Polishes well. Stated to be unsuitable for plywood manufacture.

Used in Africa for lorry bodies, bridge decking and building work where a strong, tough wood is required. Has good wearing qualities and should give satisfactory service as a flooring timber if carefully dried.

MUHIMBI
Cynometra alexandri

other name: muhindi (Uganda).

THE TREE
Reaches a height of 36–46 m (120–150 ft). The tree has large plank-like buttresses and larger trees are usually hollow. Bole straight and cylindrical in young trees but older trees are frequently gnarled, with a clear bole rarely exceeding 12 m (40 ft), diameter 0·8 m (2½ ft). Central and East Africa.

THE TIMBER
properties

colour: Dull reddish-brown with darker markings.

sapwood: Pale in colour, sharply defined, 50–75 mm (2–3 in) wide.

grain: Usually interlocked and variable in direction. Texture fine. The timber is generally of plain appearance but figured logs occasionally occur. The appearance of the timber is sometimes marred by the presence of scar tissue extending along the grain and occasionally containing a white chalk-like deposit.

weight: Varies from 830 to 1020 kg/m³ (52–64 lb/ft³), seasoned, average about 900 kg/m³ (56 lb/ft³).

strength: About half-way between European beech and greenheart, except in stiffness in which it is similar to beech.

Moisture Content	Bending Strength		Modulus of Elasticity		Compression parallel to grain	
	N/mm²	lbf/in²	N/mm²	1000 lbf/in²	N/mm²	lbf/in²
Green	94	13 700	9900	1430	48·5	7030
12 per cent	151	21 900	14 100	2050	71·7	10 400

135

movement: Medium.

Moisture content in 90 per cent relative humidity   20 per cent
Moisture content in 60 per cent relative humidity   13 per cent
Corresponding tangential movement   2·8 per cent ($\frac{11}{32}$ in/ft)
Corresponding radial movement   1·3 per cent ($\frac{5}{32}$ in/ft)

processing

drying: Dries slowly, especially in thick sizes, with a tendency to serious surface checking and extension of existing shakes. Some end-splitting and splitting of knots occurs, but distortion generally is slight.
Kiln Schedule B.
Shrinkage: Green to 12 per cent moisture content:
Tangential   about 4·5 per cent ($\frac{9}{16}$ in/ft)
Radial   about 2·5 per cent ($\frac{5}{16}$ in/ft)

working properties:
Grain moderately to severely interlocked, affecting machining properties.
Blunting: Severe.
Sawing: Rip-sawing — Saw type HR60 or TC. Build-up of resin occurs on saw teeth. Plate saws two gauges thicker than standard are recommended to avoid tooth vibration.
Cross-cutting — Saw type 3 not suitable.
Narrow bandsawing — Tendency to burn.
Wide bandsawing — Saw type C; tipped or hardened teeth advisable on dry material.
Machining: Planing — A cutting angle of 15° required to prevent tearing. Rapid blunting of cutting edges soon causes the timber to ride on cutters.
Moulding — French head produces least tearing, but feeding effort is high.
Boring — All drills cause some burning and deposition of gum is troublesome. Two-wing twist flute most suitable.
Recessing — Difficult.
Mortising — Hollow square chisel difficult, chain tends to snatch.
Nailing: Pre-boring necessary to avoid splitting and buckling of nails.
Gluing: No information.

wood bending:
Appears to be weak in tension when steamed but can be bent to a moderate radius of curvature, even if small knots are present, if adequately supported with a metal strap and end-pressure device.
Classification — Moderate.
Ratio radius/thickness for solid bends (steamed):
Supported: 16·5   Unsupported: 37
Limiting radius for 3·2 mm ($\frac{1}{8}$ in) laminae (unsteamed): 185 mm ($7\frac{1}{4}$ in).

plywood manufacture: Tested and found to be unsuitable because of its weight.

staining and polishing: Satisfactory with a little filler.

**durability
and preservation**

insect attack: Reported to be highly resistant to termites in East Africa.

durability of heartwood: Durable.

preservative treatment: Variable, from moderately resistant to resistant.

uses

A hard and heavy timber of fine texture, with very good resistance to abrasion. Too heavy for general carpentry and joinery but very suitable for flooring, particularly industrial and other heavy-duty flooring.

MUHUHU
Brachylaena hutchinsii

THE TREE

Height up to 25 m (80 ft), but commonly only 9–18 m (30–60 ft). Bole fluted and twisted, 2–6 m (6–20 ft) long, diameter 0·5–0·6 m (1½–2 ft). East Africa.

THE TIMBER
properties

colour: Yellow-brown when freshly cut, darkening on exposure to medium brown.

sapwood: Greyish-white, distinct from heartwood, 25–38 mm (1–1½ in) wide.

grain: Typically closely interlocked or sometimes wavy. Striped appearance, due to closely interlocked grain. Texture very fine and even.

odour: Has a pleasant spicy scent.

weight: Varies from 830 to 1000 kg/m³ (52–62 lb/ft³), average about 910–960 kg/m³ (57–60 lb/ft³), seasoned.

strength: Slightly lower than European beech in most strength properties but has a high resistance to indentation and to abrasion.

Moisture Content	Bending Strength		Modulus of Elasticity		Compression parallel to grain	
	N/mm^2	lbf/in^2	N/mm^2	$1000\,lbf/in^2$	N/mm^2	lbf/in^2
Green	92	13 300	8600	1250	53·6	7770
12 per cent	112	16 200	10 100	1460	70·3	10 200

movement: Small.
Moisture content in 90 per cent relative humidity   16 per cent
Moisture content in 60 per cent relative humidity   11·5 per cent
Corresponding tangential movement   1·4 per cent ($\frac{11}{64}$ in/ft)
Corresponding radial movement   1·1 per cent ($\frac{1}{8}$ in/ft)

processing

drying: Dries fairly rapidly in 25 mm (1 in) thickness with some tendency for surface checking and end-splitting to develop. Thicker timber is disproportionately more difficult to dry, the drying being slower and accompanied by serious surface checking in some instances.
Kiln Schedule B.
Shrinkage: Green to 12 per cent moisture content:
Tangential   about 3·0 per cent ($\frac{3}{8}$ in/ft)
Radial   about 2·0 per cent ($\frac{1}{4}$ in/ft)

working properties:
Interlocked and sometimes irregular grain presents some difficulty in machining. Gum tends to build up on tools.
Blunting: Moderate.
Sawing: Rip-sawing – Saw type HR 40. A large tooth pitch or fast feed speed required to produce coarse sawdust and prevent packing and overheating of saws.
Cross-cutting – Satisfactory.
Narrow bandsawing – Satisfactory.
Wide bandsawing – Saw type B.
Machining: Planing – A cutting angle of 20° required to plane quarter-sawn material satisfactorily.
Moulding – French head most suitable.
Mortising – Hollow square chisel tends to char.
Turning – Very good finish produced.
Other operations – Satisfactory.
Nailing: Pre-boring necessary.
Gluing: No information.

wood bending:
Can be bent to a moderate radius of curvature unless pin knots are present.
Classification – Moderate.
Ratio radius/thickness for solid bends (steamed):
Supported: 18·0   Unsupported: 36·0
Limiting radius for 3·2 mm ($\frac{1}{8}$ in) laminae (unsteamed): 262 mm (10$\frac{3}{8}$ in).

137

plywood manufacture: Tested and found to be unsuitable because of its weight.

staining and polishing: Very good.

<table>
<tr><td align="right">durability
and preservation</td><td>insect attack: Reported to be moderately resistant to termites in East Africa.</td></tr>
</table>

insect attack: Reported to be moderately resistant to termites in East Africa.

durability of heartwood: Very durable.

preservative treatment: Extremely resistant.

uses

An attractive, hard-wearing and very durable timber, outstanding as a flooring timber and suitable for high-quality floors in hotels, public buildings, etc., where there is heavy pedestrian traffic. Also suitable for floors in factories, warehouses, etc., where it is exposed to industrial traffic including trucking. Used in East Africa for carving and turnery.

The timber is available mainly as flooring blocks and strips. Although suitable for heavy and durable construction work, it is unlikely to be available in the sizes required for such purposes.

MUNINGA
Pterocarpus angolensis

THE TREE

other names: ambila (Mozambique); mukwa (Rhodesia); kiatt, kajat, kajaten-hout (South Africa); mninga (Tanzania).

THE TREE Grows to a height of 21 m (70 ft), but more usually 15 m (50 ft). Bole 4–8 m (15–25 ft) long, usually straight when mature. Diameter about 0·6 m (2 ft). South Central Africa.

THE TIMBER
properties

colour: Pale to medium golden brown or chocolate-brown, often with darker or redder streaks, toning down to an attractive golden brown and retaining its irregular darker markings.

sapwood: Pale in colour, clearly demarcated from heartwood, variable in width but commonly about 38 mm ($1\frac{1}{2}$ in) wide.

grain: Rarely straight, often irregularly interlocked producing an attractive figure. Texture medium to fairly coarse, somewhat uneven. Small white spots occur irregularly in some logs – they are caused by a natural component of the wood and are generally more obvious in veneers than in solid timber.

weight: Rather variable, ranging from 480 to 780 kg/m³ (30–49 lb/ft³), average 620 kg/m³ (39 lb/ft³) at 12 per cent moisture content. Timber from Rhodesia is rather softer and lighter in weight, averaging 540 kg/m³ (34 lb/ft³).

strength: Somewhat lower than European beech, especially in stiffness.

Moisture Content	Bending Strength		Modulus of Elasticity		Compression parallel to grain	
	N/mm²	lbf/in²	N/mm²	1000 lbf/in²	N/mm²	lbf/in²
Green	85	12 300	7600	1100	40·6	5890
12 per cent	94	13 700	8400	1220	57·1	8280

movement: Small.
Moisture content in 90 per cent relative humidity   13 per cent
Moisture content in 60 per cent relative humidity   10 per cent
Corresponding tangential movement   0·6 per cent ($\frac{5}{64}$ in/ft)
Corresponding radial movement   0·5 per cent ($\frac{1}{16}$ in/ft)

processing	**drying:** Dries very well though rather slowly, especially in thicker sizes of quartered material. Little tendency for the timber to split or distort and knots split only slightly.

processing   **drying:** Dries very well though rather slowly, especially in thicker sizes of quartered material. Little tendency for the timber to split or distort and knots split only slightly.
Kiln Schedule J.
Shrinkage: Green to 12 per cent moisture content:
       Tangential   about 1·5 per cent ($\frac{3}{16}$ in/ft)
       Radial   about 1·0 per cent ($\frac{1}{8}$ in/ft)

working properties: Grain generally interlocked, affecting machining operations.
Blunting: Moderate.
Sawing: Rip-sawing – Saw type HR 54.
             Cross-cutting – Satisfactory.
             Narrow bandsawing – Satisfactory.
             Wide bandsawing – Saw type B.
Machining: Generally satisfactory but care required when working end-grain or breaking through.
             Planing – A cutting angle of 20° required for material with interlocked grain.
             Turning – Good.
Nailing: Satisfactory with care; thin gauge nails an advantage.
Gluing: Good.

wood bending:
Some buckling and fibre rupture liable to occur. No great advantage gained by using a supporting strap.
Classification – Moderate.
Ratio radius/thickness for solid bends (steamed):
       Supported: 16·5   Unsupported: 18
Limiting radius for 3·2 mm ($\frac{1}{8}$ in) laminae (unsteamed): 173 mm (7 in).

plywood manufacture:
Employed in plywood but seldom seen in United Kingdom.
Movement of plywood – No information.
Surface splitting on exposure to weather – Grade I.

staining and polishing: Good.

durability and preservation   **insect attack:** Sapwood liable to attack by powder-post beetles. Has been reported highly resistant to termites in East and South Africa.

durability of heartwood: Very durable.

preservative treatment: Resistant. Sapwood moderately resistant.

uses   A timber of attractive appearance, small dimensional movement and good durability. Used for furniture, panelling and high-class joinery. As a flooring timber, it is moderately resistant to wear and is suitable only for pedestrian traffic. Widely used as decorative veneer.

MUSINE
Croton megalocarpus

An East African tree growing to a height of 30 m (100 ft) or sometimes more. Clean cylindrical bole, length 12–18 m (40–60 ft), diameter 0·6–1·0 m (2–3 ft), free of buttresses.

Wood yellowish-grey or brownish-grey, sometimes with dark brown streaks near the centre of the tree. Straight grained and of medium texture. Weight about 720 kg/m³ (45 lb/ft³), seasoned. Large movement.

Reported to be rather difficult to dry without distortion and splitting. Kiln Schedule C suggested. Sawing satisfactory, moderately difficult to machine. Reported to be non-durable.

Has been imported mainly as a flooring timber. Has high resistance to wear and is suitable for most conditions of traffic including heavy-duty flooring. Used in Kenya for general construction.

K

MUSIZI
Maesopsis eminii

THE TREE Reaches an average height of 27 m (90 ft), occasionally 36–43 m (120–140 ft). Bole straight and cylindrical, only slightly buttressed. Average length of clear bole 21 m (70 ft), diameter up to 1·2 m (4 ft). West and Central Africa.

THE TIMBER
properties

colour: Variable in colour, yellowish-green when freshly cut, darkening on exposure to golden brown or dark brown.

sapwood: Nearly white, up to 75 mm (3 in) wide.

grain: Typically interlocked, producing a well-marked stripe on quarter-sawn surfaces. Pin knots sometimes present causing local irregularities of the grain. Texture moderately coarse and even.

weight: About 460 kg/m³ (28–29 lb/ft³), seasoned.

strength: About half-way between obeche and European beech.

Moisture Content	Bending Strength		Modulus of Elasticity		Compression parallel to grain	
	N/mm²	lbf/in²	N/mm²	1000 lbf/in²	N/mm²	lbf/in²
Green	55	8000	8100	1170	28·5	4140
12 per cent	76	11 000	9200	1340	46·0	6670

movement: Small.

Moisture content in 90 per cent relative humidity   18 per cent
Moisture content in 60 per cent relative humidity   12 per cent
Corresponding tangential movement   1·5 per cent ($\frac{3}{16}$ in/ft)
Corresponding radial movement   1·0 per cent ($\frac{1}{8}$ in/ft)

processing

drying: Dries fairly rapidly and very well. Slight distortion occurs but no tendency to split or check and knots remain sound.
Kiln Schedule F.
Shrinkage: Green to 12 per cent moisture content:
Tangential   about 4·0 per cent ($\frac{1}{2}$ in/ft)
Radial   about 2·5 per cent ($\frac{5}{16}$ in/ft)

working properties:
Grain interlocked, texture woolly, requiring sharp cutting edges to produce a good finish.
Blunting: Slight.
Sawing: Rip-sawing – Saw type HR 54.
Cross-cutting – Satisfactory.
Narrow bandsawing – Satisfactory.
Wide bandsawing – Saw type A.
Machining: Planing – A cutting angle of 20° necessary to prevent tearing of quarter-sawn material.
Recessing – Care required to prevent tearing and woolly finish.
Drilling – Fibrous finish, support required at tool exit.
Mortising – Requires support to avoid crumbling on end-grain.
Nailing: Good.
Gluing: Good.

wood bending:
Severe buckling and fibre rupture occur with only a small degree of bending.
Classification – Very poor.

plywood manufacture: Employed for plywood.
No information on movement or surface splitting on exposure.

staining and polishing: Satisfactory when filled.

durability and preservation

insect attack: Said to be non-resistant to termites in East and West Africa.

durability of heartwood: Non-durable.

preservative treatment: Permeable.

uses

A useful lightweight general-purpose hardwood, suitable for interior use. Has good nailing properties and can be used for boxes and packing cases.

MUTENYE
Guibourtia arnoldiana

other names: benge, libengi.

THE TREE

Height up to 27 m (90 ft). Bole irregular, length 9–18 m (30–60 ft), diameter 0·4–0·8 m ($1\frac{1}{4}$–$2\frac{1}{2}$ ft). West Central Africa.

THE TIMBER
properties

Heartwood pale yellowish-brown to medium brown, sometimes with a faint reddish tinge, and marked by grey or almost black veining. Similar in appearance to *Guibourtia ehie* but a little finer textured. Grain interlocked and sometimes wavy, giving the wood a highly decorative appearance. Weight about 800–960 kg/m³ (50–60 lb/ft³), seasoned. A moderately hard timber, with medium movement.

processing

Reported to have a high shrinkage in drying. Kiln Schedule C suggested. Satisfactory in sawing, but planing of quarter-sawn material presents some difficulty because of interlocked and sometimes wavy grain. A satisfactory finish can be produced with care. Only suitable for bends of moderate radius of curvature; not tolerant of pin knots and inclined to distort during steaming. There is a risk of resin exudation in the steaming process. Limiting radius for 3·2 mm ($\frac{1}{8}$ in) laminae (un-steamed): 178 mm (7 in). Can be sliced to produce a high-quality veneer.

durability and preservation

Reported to be moderately resistant to termites in West Africa. Heartwood moderately durable (provisional).

uses

An attractive wood, somewhat heavy to find extensive use as solid timber, but suitable for turnery, flooring and some furniture parts. Used mainly as veneer for furniture and interior decorative work.

NIANGON
Tarrietia utilis

other name: nyankom (Ghana).

THE TREE

Reaches a height of 30 m (100 ft), but occasionally up to 40 m (130 ft). Bole usually straight and cylindrical, length 20 m (65 ft) or sometimes more, diameter 0·6–1·0 m (2–3 ft). When grown on swampy ground the trees develop a twisted, irregular stem. Plank buttresses often develop into stilt roots. West Africa from Sierra Leone to Ghana.

A similar but darker and somewhat heavier wood produced by *T. densiflora* is known as ogoué in Gaboon.

**THE TIMBER
properties**

colour: Varies from pale pink to reddish-brown, darkening on exposure.

sapwood: Paler in colour, commonly about 75 mm (3 in) wide, not clearly demarcated from heartwood.

grain: Generally interlocked and sometimes wavy, producing an irregular stripe figure on quarter-sawn boards. Texture rather coarse with conspicuous ray figure. The timber has a characteristic greasy feel.

weight: Very variable from 510 to 750 kg/m³ (32–47 lb/ft³), average 620 kg/m³ (39 lb/ft³), seasoned.

strength: About half-way between obeche and European beech.

Moisture Content	Bending Strength		Modulus of Elasticity		Compression parallel to grain	
	N/mm²	lbf/in²	N/mm²	1000 lbf/in²	N/mm²	lbf/in²
Green	70	10 200	8400	1220	36·5	5300
12 per cent	90	13 000	9500	1380	51·7	7500

movement: Medium.
Moisture content in 90 per cent relative humidity   18·5 per cent
Moisture content in 60 per cent relative humidity   12·5 per cent
Corresponding tangential movement   2·0 per cent ($\frac{15}{64}$ in/ft)
Corresponding radial movement   1·1 per cent ($\frac{1}{8}$ in/ft)

processing

drying: Dries fairly rapidly and well. Distortion unlikely to be appreciable, but a small proportion of the timber may tend to twist. Slight end-splitting and surface checking may develop.
Kiln Schedule E.
Shrinkage: Green to 12 per cent moisture content:
    Tangential   about 4·5 per cent ($\frac{9}{16}$ in/ft)
    Radial   about 2·5 per cent ($\frac{5}{16}$ in/ft)

working properties: Steeply interlocked grain affects machining properties.
Blunting: Moderate.
Sawing: Rip-sawing – Saw type HR 40. A long tooth pitch required to produce coarse sawdust and prevent packing and overheating of saw blade.
    Cross-cutting – Pronounced tendency to breaking out at bottom of cut.
    Narrow bandsawing – Satisfactory.
    Wide bandsawing – Saw type B.
Machining: Planing – A cutting angle of 15° required to prevent tearing of quarter-sawn material.
    Moulding – Tendency to tear. French head most suitable.
    Other operations – Satisfactory.
Nailing: Tends to split.
Gluing: Good.

wood bending: Bending properties vary considerably but are not impaired by knots.
Classification – Moderate.
Ratio radius/thickness for solid bends (steamed):
    Supported: 18   Unsupported: 30
Limiting radius for 3·2 mm ($\frac{1}{8}$ in) laminae (unsteamed): 188 mm (7$\frac{1}{2}$ in).

plywood manufacture:
Employed for plywood and usually available in the United Kingdom.
Movement: 4·5 mm plywood from 30 per cent to 90 per cent relative humidity – 0·22 per cent.
Surface splitting on exposure to weather – Grade I.

staining and polishing: Satisfactory when filled.

durability and preservation

insect attack: Sapwood liable to attack by powder-post beetles.

durability of heartwood: Durable.

preservative treatment: Extremely resistant. Sapwood resistant.

uses

A useful general purpose timber for furniture, carpentry and joinery. Has also been used in boat-building, greenhouses, etc.

NYATOH
principally *Palaquium* spp. and *Payena* spp.

Timber of a number of species of *Palaquium* and *Payena* from Malaya and the south-east Asian islands of weight up to 880 kg/m³ (55 lb/ft³) is known as nyatoh. Timber exceeding 880 kg/m³ in weight is known as bitis in Malaya and nyatoh batu in Sabah.

THE TREE

Trees often reach a large size, 30 m (100 ft) or more in height and up to 1·0 m (3 ft) in diameter.

**THE TIMBER
properties**

The timber is variable in properties because it is produced by a number of different species. Heartwood pale pink to red-brown, sometimes with darker streaks. Sapwood paler in colour, up to 75 mm (3 in) wide, not sharply defined. Grain straight or shallowly interlocked, sometimes slightly wavy. Texture moderately fine. Has some resemblance to makoré and sometimes shows an attractive moiré or 'watered silk' figure. Weight variable, but generally between 640 and 720 kg/m³ (40–45 lb/ft³), seasoned. Strength properties variable but on average comparable to European beech.

processing

Reported to dry rather slowly but with some tendency to end split and distort. Kiln Schedule E. Ease of working depends on species, since some species are siliceous and cause rapid blunting of cutting edges. Non-siliceous wood saws easily and can be planed to a smooth surface. Peels easily and well and can be used for plywood.

durability and preservation

Sapwood liable to attack by powder-post beetles. Reported to be non-durable to moderately durable and very resistant to preservative treatment.

uses

An attractive wood, but variable in colour and weight and in its working properties. May be expected from its appearance, and fine, even texture to be suitable for interior joinery and fittings and for furniture. It is a potential plywood timber and for this purpose siliceous species may be acceptable. Occasional logs having figured wood may be a source of decorative veneer.

OAK, AMERICAN RED
Quercus spp.

THE TREE

Commercial American red oak is made up of a number of species, principally *Quercus rubra* (northern red oak) and *Q. falcata* var. *falcata* (southern red oak).

**THE TIMBER
properties**

Resembles other oaks in appearance but heartwood usually has a reddish tinge. Generally coarser in texture and has a less attractive silver grain figure than American white oak, due to its smaller rays. Exhibits considerable variation in structure and quality, depending on species and conditions of growth. Southern red oak is typically of more rapid growth than northern red oak and produces a rather harder, heavier and coarser textured wood. Average weight about 770 kg/m³ (48 lb/ft³), seasoned, similar to American white oak. Strength rather variable, but similar on average to American white oak and slightly lower than European beech. Medium movement.

processing	Drying properties similar to other oaks. Kiln Schedule C. Working properties vary according to the density of the wood. Gluing variable. Has very good bending properties.
durability and preservation	Heartwood non-durable. Moderately resistant to preservative treatment.
uses	Considered inferior to white oak for high-class furniture and decorative work. Used for flooring, vehicle construction, interior joinery, furniture and veneer, but unsuitable for exterior work because of its lack of durability. Not suitable for tight cooperage because of its porosity. Has been planted on a small scale in Great Britain. Tests on the timber have shown that it is generally similar to timber from America.

OAK, AMERICAN WHITE
Quercus spp., principally *Q. alba*, *Q. prinus*, *Q. lyrata* and *Q. michauxii*

other names:	*Quercus alba*, true white oak; *Q. prinus*, chestnut oak; *Q. lyrata*, overcup oak; *Q. michauxii*, swamp chestnut oak (United States).
THE TREE	Under favourable conditions reaches a height of 30 m (100 ft), but under less favourable conditions may be of poor form and only 15 m (50 ft) high. Well-grown trees have a straight, clear bole, length 12–15 m (40–50 ft), diameter 0·9–1·2 m (3–4 ft). Eastern half of United States and south-eastern Canada, the distribution varying according to species.
THE TIMBER	Similar in many respects to European oak.
properties	
colour:	Rather variable from pale yellow-brown to mid-brown.
sapwood:	Almost white in colour, distinct from heartwood.
grain:	Generally straight. Quarter-sawn material has a characteristic ornamental silver grain due to the broad rays. Structure and quality vary widely according to the conditions of growth. Oak from the northern Appalachian area is usually slow grown producing a comparatively lightweight, mild type of wood. Oak from the southern States is typically fast grown with correspondingly wide growth rings, producing a harder and tougher timber.
weight:	Slightly heavier than European oak. Average weight about 750 kg/m³ (47 lb/ft³) at 12 per cent moisture content.
chemical staining:	If the timber comes into contact with iron or iron compounds in presence of moisture, blue-black stains are liable to appear due to reaction between the iron and tannin present in the wood.
corrosive properties:	A somewhat acidic timber which tends to promote corrosion of metals especially iron and steel in contact with it under damp conditions. Metals, e.g. lead, may also be attacked if exposed to vapours from undried oak. Metals used in association with oak should be protected by painting or galvanising.
strength:	Slightly lower than European beech.

Species	Moisture Content	Bending Strength		Modulus of Elasticity		Compression parallel to grain	
					1000		
		N/mm²	lbf/in²	N/mm²	lbf/in²	N/mm²	lbf/in²
Q. alba	Green	60	8700	8100	1170	25·5	3700
	12 per cent	110	16 000	11 500	1670	53·2	7720
Q. prinus	Green	58	8400	8800	1280	25·2	3650
	12 per cent	97	14 000	10 300	1490	48·9	7090
Q. lyrata	Green	58	8400	7400	1080	24·1	3500
	12 per cent	92	13 300	9200	1330	44·4	6440
Q. michauxii	Green	62	9000	8700	1260	25·4	3680
	12 per cent	101	14 600	11 400	1660	52·1	7550

(Data from US Department of Agriculture)

movement: Medium.
Moisture content in 90 per cent relative humidity   21 per cent
Moisture content in 60 per cent relative humidity   12·5 per cent
Corresponding tangential movement   2·8 per cent ($\frac{11}{32}$ in/ft)
Corresponding radial movement   1·3 per cent ($\frac{5}{32}$ in/ft)

processing

drying: Dries relatively slowly with a tendency to check, split and honeycomb. Kiln Schedule C.
Shrinkage: Green to 12 per cent moisture content:
Tangential   about 5·5 per cent ($\frac{11}{16}$ in/ft)
Radial   about 3·0 per cent ($\frac{3}{8}$ in/ft)

working properties:
Working properties vary according to species and also with rate of growth. Slowly grown timber is softer and more easily worked than the fast grown, tougher material.
Blunting: Moderate.
Sawing: Rip-sawing — Saw type HR 54, or type HR 60 for dense material.
Cross-cutting — Satisfactory.
Narrow bandsawing — Satisfactory.
Wide bandsawing — Saw type B.
Machining: Planing — A cutting angle of 20° advisable.
Other operations — Satisfactory.
Nailing: Pre-boring advisable.
Gluing: Variable.

wood bending:
An excellent bending timber. Material free from defects, drying checks, etc., can be bent after steaming to very small radii of curvature. The steamed wood is liable to stain if it comes into contact with iron or steel. Classification — Very good.
Ratio radius/thickness for solid bends (steamed):
Supported: 0·5   Unsupported: 13
Limiting radius for 3·2 mm ($\frac{1}{8}$ in) laminae (unsteamed): 137 mm ($5\frac{1}{2}$ in).

plywood manufacture: No information.

staining and polishing: Good.

<table>
<tr><td style="vertical-align:top">durability
and preservation</td><td>insect attack:</td><td>Logs liable to severe attack by ambrosia (pinhole-borer) beetles, and trees and logs liable to attack by forest longhorn or Buprestid beetles. Reported to be attacked by dry-wood termites in West Indies.</td></tr>
</table>

durability of heartwood: Durable.

preservative treatment: Extremely resistant. Sapwood moderately resistant.

uses Milder to work than European oaks and suitable for furniture, cabinet making and joinery. A good timber for constructional work owing to its strength and durability. A good flooring timber, suitable for parquet and strip flooring. Suitable for tight cooperage and used particularly for manufacture of whisky casks.

OAK, EUROPEAN
Quercus robur
(Q. pedunculata) and
Quercus petraea
(Q. sessiliflora)

(Hybrids between these two species are common.)

other names: English, French, Yugoslavian oak, etc., according to origin (Great Britain).
Q. robur pedunculate oak (Great Britain).
Q. petraea durmast oak, sessile oak (Great Britain).

THE TREE

Reaches a height of 18 to 30 m (60–100 ft), varying according to soil and locality. Diameter of bole about 1·2 to 1·8 m (4–6 ft), occasionally more. Forms a straight, clear bole, sometimes up to 15 m (50 ft) in length, when grown under forest conditions, but carries lower branches when grown in the open. Occurs in pure stands, and in mixed woods where it is often the dominant species.

Both species grow throughout Europe south of about 63°N, and in Asia Minor and North Africa. Large quantities are grown in France, Germany, Austria, Czechoslovakia and the Balkan States. Occurs throughout the British Isles, but is commonest in the South and Midlands.

THE TIMBER
properties

There is no inherent difference between timber of the two species.

colour: Yellowish brown. A yellow stain (golden oak) caused by a harmless surface mould is sometimes noticeable during drying, but is not permanent.

sapwood: Light in colour, usually 25 to 50 mm (1–2 in) wide, distinct from heartwood.

grain: Generally straight, but varying with growth conditions. The characteristic ornamental silver grain, due to the broad rays, is seen in quarter-sawn material.

Structure and quality are both affected by growth conditions. Typical material has alternating zones of large-pored early wood and dense late wood, but in slow-grown timber the late wood zones may be almost eliminated, the timber being consequently soft and light in weight. Oak from central Europe is often of slow, even growth, uniform colour and straight grain, while that from northern countries is characteristically harder and tougher. Home-grown oak is very variable in quality, but timber from well-grown trees compares favourably with that grown on the continent.

weight: Variable according to origin and character of growth. Timber of slow growth from Central Europe, such as Slavonian oak, averages about 670 kg/m³ (42 lb/ft³) and home-grown timber, which is usually of more vigorous growth, about 720 kg/m³ (45 lb/ft³) at 12 per cent moisture content.

corrosive properties:

A somewhat acidic timber which tends to promote corrosion of metals, especially iron and steel, in contact with it under damp conditions. Metals exposed to vapours from undried oak may also be attacked. Corrosion of lead under these conditions can be very severe. Metals which are not corrosion resistant should be painted or galvanised.

chemical staining:

Blue-black stains, formed by reaction of iron with the tannin in oak, are liable to appear on the timber when it is in contact with iron or iron compounds in presence of moisture.

strength: Slightly lower than European beech.
Data for home-grown oak:

Moisture Content	Bending Strength		Modulus of Elasticity		Compression parallel to grain	
	N/mm^2	lbf/in^2	N/mm^2	$1000\,lbf/in^2$	N/mm^2	lbf/in^2
Green	59	8500	8300	1210	27·6	4000
12 per cent	97	14 000	10 100	1460	51·6	7490

movement: Medium.
Moisture content in 90 per cent relative humidity   20 per cent
Moisture content in 60 per cent relative humidity   12 per cent
Corresponding tangential movement   2·5 per cent ($\frac{5}{16}$ in/ft)
Corresponding radial movement   1·5 per cent ($\frac{3}{16}$ in/ft)

processing

drying: Dries very slowly with a marked tendency to split and check, particularly in the early stages of drying. In air drying thin piling sticks should be used and some end protection is advisable. In kiln drying there is a considerable risk of honeycombing developing later in the process if drying is forced, and distortion may be appreciable. A yellow stain sometimes develops during the drying process but gradually fades in service.
Kiln Schedule C.
Shrinkage: Green to 12 per cent moisture content:
            Tangential   about 7·5 per cent ($\frac{15}{16}$ in/ft)
            Radial   about 4·0 per cent ($\frac{1}{2}$ in/ft)

working properties: Working properties vary with density.
Blunting: Moderate. More rapid with denser than with milder timber.
Sawing: Rip-sawing – Saw type HR 54, or HR 60 for denser material.
            Cross-cutting – Satisfactory.
            Narrow bandsawing – Satisfactory.
            Wide bandsawing – Saw type B.
Machining: Generally satisfactory. In planing a reduction in cutting angle to 20° is required for irregular and cross-grained material.
Nailing: Difficult. Pre-boring advisable.
Gluing: Good.

wood bending:

A very good bending wood. It is advisable to reduce the moisture content to about 25 per cent for steam bending purposes, since green material is liable to rupture on the inner face. Rather large forces are induced during the bending operation. Rapid drying should be avoided during the setting process. The steamed wood may become stained if it comes into contact with iron or steel.
Classification – Very good.
Ratio radius/thickness for solid bends (steamed):
        Supported: 2   Unsupported: 13
Limiting radius for 3·2 mm ($\frac{1}{8}$ in) laminae (unsteamed): 147 mm ($5\frac{3}{4}$ in).

plywood manufacture:
Tested and found to be suitable for plywood. Widely used as decorative veneer.
Movement of plywood – No information.
Surface splitting on exposure – No information.

staining and polishing: Good.

durability and preservation

insect attack: Logs and green planks liable to attack by ambrosia (pinhole-borer) beetles. Logs may be attacked by forest longhorn or Buprestid beetles, and bark and sapwood of logs and converted timber by the longhorn beetle *Phymatodes testaceus*. Sapwood is susceptible to attack by powder-post beetles and by the common furniture beetle and sapwood and heartwood of timber in old buildings in England and Wales liable to be attacked by the death-watch beetle.

durability of heartwood: Durable.

preservative treatment: Extremely resistant. The sapwood is permeable.

uses

Oak is one of the most widely used hardwoods in the United Kingdom. Good quality oak is used for furniture (particularly reproduction and church furniture) and for panelling, high-class interior joinery and carving. It is also sliced to produce decorative veneers for furniture and panelling. Used in the building industry for exterior work such as sills, fascias and doors, where its combination of durability and decorative appearance is valuable. Makes an attractive floor, although not suitable for heavy industrial wear.

Widely used for fencing, gates, leading edges of pallets and for mining timber; these uses provide an outlet for lower grades of oak than those required for furniture and panelling.

Its impermeability combined with good bending properties make it suitable for tight cooperage. It is also used in boat-building, mainly for keels and framing, and in vehicle building for bearers and for flooring in heavy trucks.

Oak from central European countries is somewhat milder and easier to work than that from northern Europe. Home-grown oak, grade for grade, is comparable to other northern European oak, though supplies of good quality material are small in proportion to the total quantity of home-grown oak available, owing to the extensive use of this timber over the centuries.

OAK, HOLM
Quercus ilex

other name: evergreen oak (Great Britain).

Grows to an average height of 15–18 m (50–60 ft), with a short bole, diameter up to 1·8 m (6 ft). Mediterranean region; planted as an ornamental tree in the United Kingdom.

The timber differs from that of the common commercial oaks which are deciduous trees. Heartwood is a pleasing shade of brown, not sharply defined from the paler sapwood. When cut on the quarter shows the silver grain figure of all true oaks, the rays being exceptionally broad and conspicuous. Annual rings comparatively indistinct. Wood very close-textured, hard and compact. Weight generally about 800–960 kg/m³ (50–60 lb/ft³), seasoned, or sometimes higher.

Liable to distort and check in drying. Usually harder to convert than the common oaks and has a moderate blunting effect on cutting edges. Saw type HR 60 recommended in rip-sawing and type C in wide bandsawing. Tendency to pick up in planing owing to irregular grain; reduction in cutting angle to 20° necessary to produce a good finish. Polishes well. Pre-boring advisable in nailing. Unsuitable for plywood manufacture because of its weight.

Can be used for some of the same purposes as European oak, taking into account its higher density. In Mediterranean countries considered inferior to European oak in its general properties and generally used for rough work.

OAK, JAPANESE
Quercus spp.,
principally *Q. mongolica*
var. *grosseserrata*

**THE TIMBER
properties**

An oak of the white oak group, somewhat paler in colour than American white oak or European oak. Variations in growth rate and quality occur as with other oaks, but timber imported into the United Kingdom is typically of slow, even growth and is fairly uniform in character. Timber from the North Island of Japan (Hokkaido) is slower grown and milder than that from the Main Island (Honshu). Grain generally straight and free from knots. Average weight about 660 kg/m³ (41 lb/ft³), somewhat lower than other commercial oaks. Lower in strength properties than European oak. Medium movement.

processing

Dries without undue difficulty. Kiln Schedule C. Milder and easier to work than other types of oak owing to its lower density. Steam bending properties very good. Ratio radius/thickness for solid bends (steamed): supported 1·5, unsupported 12·5. Limiting radius for 3·2 mm ($\frac{1}{8}$ in) laminae (unsteamed): 137 mm ($5\frac{1}{2}$ in). Employed in plywood manufacture but seldom seen in the United Kingdom.

**durability
and preservation**

Heartwood durable (provisional), but timber may contain a rather large proportion of sapwood.

uses

Used for furniture, interior fittings, joinery, flooring blocks, boat-building and as veneer.

OAK, PERSIAN
Quercus castaneaefolia

The timber is classed as a red oak and is similar to the better grades of American red oak. Weight about 770 kg/m³ (48 lb/ft³), seasoned, similar to American red oak. Like other red oaks it is less durable than the white oaks. It differs from most species of red oak in having the pores of the heartwood more or less blocked by tyloses. This renders the wood relatively impermeable to liquids and it can therefore be used for barrel staves.

Has been imported mainly for production of staves for tight cooperage. Probably suitable for flooring and furniture, though differing somewhat from other oaks in appearance.

OAK, TURKEY
Quercus cerris

THE TREE

Grows to a height of 37 m (120 ft), diameter 1·8 m (6 ft) or sometimes more. Southern Europe and south-western Asia. Has been introduced into southern England.

**THE TIMBER
properties**

colour: Heartwood generally similar to other oaks, sometimes with a reddish tint.

sapwood: Wide, up to 150 to 180 mm (6–7 in), but may be less in forest-grown as distinct from parkland trees.

grain: Generally straight. When cut on the quarter the wood shows the silver-grain figure characteristic of true oaks. Owing to the often rapid rate of growth, a growth ring figure is usually a conspicuous feature.

weight: Somewhat heavier than the common commercial oaks, generally between 780 and 870 kg/m³ (49–54 lb/ft³), seasoned.

strength: Home-grown timber shown by tests to be comparable to European beech and slightly stronger than European oak.

Data for home-grown Turkey oak:

Moisture Content	Bending Strength		Modulus of Elasticity		Compression parallel to grain	
	N/mm^2	lbf/in^2	N/mm^2	$1000\,lbf/in^2$	N/mm^2	lbf/in^2
Green	70	10 100	10 100	1460	28·5	4140
12 per cent	130	18 800	11 200	1620	57·6	8350

movement: Large.

Moisture content in 90 per cent relative humidity   21 per cent
Moisture content in 60 per cent relative humidity   12 per cent
Corresponding tangential movement   3·3 per cent ($\frac{13}{32}$ in/ft)
Corresponding radial movement   1·3 per cent ($\frac{5}{32}$ in/ft)

processing

drying: Dries very slowly with considerable degrade. Difficult to dry without checking and somewhat severe distortion, particularly cup.

Kiln Schedule B.

Shrinkage: Green to 12 per cent moisture content:
Tangential   about 10·5 per cent ($1\frac{1}{4}$ in/ft)
Radial   about 4·5 per cent ($\frac{9}{16}$ in/ft)

working properties:
Rather hard to work, but finishes cleanly in most operations. Generally comparable to American red oak in behaviour.

Home-grown material may be difficult to machine owing to prevalence of knots. Care required in nailing.

wood bending:
Clear material has very good bending properties, but unsuitable for bending if knots are present. Tendency for splitting and distortion to occur immediately after conversion of the timber and when bends are dried and set to shape.

Classification – Very good.

Ratio radius/thickness for solid bends (steamed):
          Supported: 1·5–3·5   Unsupported: 11–12
Limiting radius for 3·2 mm ($\frac{1}{8}$ in) laminae (unsteamed): 115 mm ($4\frac{1}{2}$ in).

plywood manufacture: Unsuitable for plywood manufacture.

staining and polishing: Satisfactory.

durability and preservation

insect attack: Logs liable to attack by forest longhorn or Buprestid beetles. Sapwood liable to attack by powder-post beetles.

durability of heartwood: Moderately durable.

preservative treatment: Extremely resistant; the sapwood is permeable.

uses

Due to its poor seasoning properties and large shrinkage in drying, together with limited natural durability, economic utilisation of the timber is difficult. Suitable only for rough work such as temporary construction, shuttering or mining timber. Might be used in small sizes for bent work. Sliced veneers might be produced from selected large-size trees.

OBECHE or WAWA
Triplochiton scleroxylon

other names: arere (Nigeria); ayous (France and the Cameroon); samba (France and the Ivory Coast).
Wawa is recognised as an alternative standard name for timber from Ghana.

THE TREE

A large tree, reaching a height of 45–55 m (150–180 ft). Narrow buttresses extend up to 6 m (20 ft) up the bole. Bole straight and cylindrical, free from branches up to 25 m (80 ft). Diameter above buttresses 0·9–1·5 m (3–5 ft). West Africa.

THE TIMBER
properties

colour: Nearly white to pale straw-coloured. Machined surfaces have a high natural lustre.

sapwood: Commonly 75–100 mm (3–4 in) wide; no clear distinction between sapwood and heartwood.

grain: Typically interlocked, giving a characteristic striped appearance to quartered stock. Texture moderately coarse but even.

brittleheart: May be present in large logs.

weight: About 380 kg/m³ (24 lb/ft³), at 12 per cent moisture content.

odour: Has a disagreeable odour when fresh but this does not usually persist after drying.

strength: Used in this book as a standard for comparison of other timbers.

Moisture Content	Bending Strength		Modulus of Elasticity		Compression parallel to grain	
	N/mm²	lbf/in²	N/mm²	1000 lbf/in²	N/mm²	lbf/in²
Green	37	5400	4600	660	18·5	2680
12 per cent	54	7900	5500	800	28·2	4090

movement: Small.
Moisture content in 90 per cent relative humidity   19 per cent
Moisture content in 60 per cent relative humidity   12 per cent
Corresponding tangential movement   1·3 per cent ($\frac{5}{32}$ in/ft)
Corresponding radial movement   0·8 per cent ($\frac{3}{32}$ in/ft)

processing

drying: Dries very rapidly and very well. Practically no tendency to split or for existing shakes to extend, but slight distortion may occur and knots split a little.
Kiln Schedule L, or perhaps an even more severe treatment.
Shrinkage: Green to 12 per cent moisture content:
   Tangential   about 3·0 per cent ($\frac{3}{8}$ in/ft)
   Radial   about 2·0 per cent ($\frac{1}{4}$ in/ft)

working properties:
Blunting: Slight.
Sawing: Rip-sawing – Saw type SR.
   Cross-cutting – Satisfactory.
   Narrow bandsawing – Satisfactory.
   Wide bandsawing – Saw type A.
Machining: Sharp cutters with a reduced sharpness angle recommended. Generally satisfactory, but a tendency to crumble when end-grain is worked and for chipping to occur at tool break-through (as in mortising).
Nailing: Can be easily nailed, but holding qualities are poor.
Gluing: Good.

wood bending:

Suitable for producing solid bends of moderate radius of curvature. Slight wrinkling may occur on the edges of bends, and no advantage is gained by using a supporting strap.

Classification – Moderate.

Ratio radius/thickness for solid bends (steamed):

Supported: 18   Unsupported: 17

Limiting radius for 3·2 mm ($\frac{1}{8}$ in) laminae (unsteamed): 152 mm (6 in).

plywood manufacture:

Employed for plywood and usually available in the United Kingdom.

Movement: 1·5 mm plywood from 30 per cent to 90 per cent relative humidity – 0·20 per cent.

Surface splitting on exposure to weather – Grade II.

staining and polishing: Satisfactory, but careful filling required.

durability and preservation

insect attack: Sapwood liable to attack by powder-post beetles. Reported to be non-resistant to termites in West Africa.

durability of heartwood: Non-durable.

preservative treatment: Resistant. Sapwood permeable.

uses

Widely used for purposes for which a light, easily worked hardwood is required and where its lack of durability and low strength are not important. Examples are in furniture for interior rails, drawer sides and some types of cabinet framing; interior joinery; sliderless soundboards in organs where stability is particularly important, and model making. Also used for plywood manufacture.

ODOKO
Scottellia coriacea

THE TREE

Reaches a height of 30 m (100 ft). Bole long and straight, but not quite cylindrical, with slight fluting in the lower part. Diameter 0·3–0·6 m (1–2 ft). West Africa.

THE TIMBER
properties

colour: Pale yellow to biscuit-coloured, with a clean appearance.

sapwood: About 100 mm (4 in) wide, not clearly distinct from heartwood.

grain: Generally straight but may be slightly interlocked. Accurately quarter-sawn stock has an attractive silver-grain figure due to the presence of broad rays. Texture fine.

weight: Average about 620 kg/m³ (39 lb/ft³), seasoned.

strength: Comparable to, or slightly higher than, European beech.

Moisture Content	Bending Strength		Modulus of Elasticity		Compression parallel to grain	
	N/mm^2	lbf/in^2	N/mm^2	$1000\,lbf/in^2$	N/mm^2	lbf/in^2
Green	83	12 100	11 300	1640	38·6	5600
12 per cent	117	16 900	12 800	1860	63·6	9220

movement: Medium.

Moisture content in 90 per cent relative humidity   21 per cent

Moisture content in 60 per cent relative humidity   13 per cent

Corresponding tangential movement   2·8 per cent ($\frac{11}{32}$ in/ft)

Corresponding radial movement   1·5 per cent ($\frac{3}{16}$ in/ft)

processing	**drying:** Dries fairly rapidly, but has a pronounced tendency to split and existing shakes are liable to extend. Little distortion occurs and knots remain sound. Kiln Schedule E. Shrinkage: Green to 12 per cent moisture content: Tangential about 5·0 per cent ($\frac{5}{8}$ in/ft) Radial about 2·5 per cent ($\frac{5}{16}$ in/ft)

working properties:

Blunting: Moderate.

Sawing: Rip-sawing — Saw type HR54.
Cross-cutting — Satisfactory.
Narrow bandsawing — Satisfactory.
Wide bandsawing — Saw type B.

Machining: A good finish normally obtained, but on true quarter-sawn faces the rays tend to flake slightly. Has a rather brittle nature and care is needed to prevent chipping away at the tool exit in certain operations.

Nailing: Tends to split. Pre-boring advisable.

Gluing: Good.

wood bending:

Limited tests indicate that fibre rupture and buckling occur at comparatively large radii of curvature, and that the timber is liable to fracture on the convex face of bends.
Classification — Poor.

plywood manufacture: Tested and found suitable for manufacture of plywood.
Movement: 4·5 mm plywood from 30 per cent to 90 per cent relative humidity — 0·22 per cent.
Surface splitting on exposure to weather — No information.

durability and preservation

insect attack: Reported to be non-resistant to termites in West Africa.

durability of heartwood: Non-durable.

preservative treatment: Permeable.

uses

Similar in some respects to beech, but whiter, and susceptible to discoloration by staining fungi. Suitable for domestic woodware and for use in the furniture and joinery industries. As a flooring timber it has a high resistance to wear and is suitable for most normal conditions of traffic.

OGEA
Daniellia ogea and
D. thurifera

other names: oziya, daniellia (Nigeria); incenso or insenso (Portuguese Guinea); fara (Ivory Coast).

THE TREE
Reaches a height of 30–45 m (100–150 ft) or more. Bole long, clear, straight and cylindrical, length 15–30 m (50–100 ft). Unbuttressed or with very short rounded buttresses. Diameter 1·2–1·5 m (4–5 ft). West Africa.

THE TIMBER properties

colour: Pale pinkish- to reddish-brown, with occasional darker streaks.

sapwood: Almost white to straw-coloured and distinct from heartwood. Very wide, commonly 100–180 mm (4–7 in).

grain: Shallowly interlocked. Texture rather coarse and inclined to be woolly. The heartwood is apt to be somewhat gummy but this does not appear to be a serious defect.

brittleheart: Prevalent near the centre of the stem.

weight: Varies from 420 to 580 kg/m³ (26–36 lb/ft³), seasoned. The darker coloured heartwood is appreciably heavier than the pale coloured sapwood.

strength: About half-way between obeche and European beech.

Moisture Content	Bending Strength		Modulus of Elasticity		Compression parallel to grain	
	N/mm²	*lbf/in²*	*N/mm²*	*1000 lbf/in²*	*N/mm²*	*lbf/in²*
Green	–	–	–	–	–	–
12 per cent	81	11 800	9100	1320	41·6	6030

movement: Medium.

    Moisture content in 90 per cent relative humidity 20 per cent

    Moisture content in 60 per cent relative humidity 12 per cent

    Corresponding tangential movement 2·0 per cent ($\frac{1}{4}$ in/ft)

    Corresponding radial movement 1·0 per cent ($\frac{1}{8}$ in/ft)

processing

drying: Dries fairly rapidly with little degrade. Slight distortion may take place and collapse may occur on thick material but these are unlikely to be severe.

Kiln Schedule J.

Shrinkage: Green to 12 per cent moisture content:

    Tangential   about 4·5 per cent ($\frac{9}{16}$ in/ft)

    Radial   about 1·5 per cent ($\frac{3}{16}$ in/ft)

working properties:

Machining properties affected by interlocked grain and woolly texture.

Blunting: Slight.

Sawing: When a large proportion of woolly material is present, saws appear to blunt more rapidly, otherwise satisfactory.

Rip-sawing – Saw type HR 54.

Cross-cutting – Considerable breaking out occurs at bottom of cut.

Narrow bandsawing – Considerable breaking out occurs at bottom of cut.

Wide bandsawing – Saw type A.

Machining: Planing – Sharp cutters with reduced sharpness angle required to prevent woolly finish.

Moulding – Difficult.

On end-grain working either the wood crumbles or the fibres are bent and are not severed cleanly.

Nailing: Good.

Gluing: Good.

wood bending:

When bent even to a large radius of curvature, severe buckling and fibre rupture occur. No advantage in using a supporting strap.

Classification – Very poor.

Limiting radius for 3·2 mm ($\frac{1}{8}$ in) laminae (unsteamed): 214 mm ($8\frac{2}{8}$ in).

plywood manufacture:

Employed in plywood but seldom seen in United Kingdom.

Movement: 4·5 mm plywood from 30 per cent to 90 per cent relative humidity – 0·16 per cent.

Surface splitting on exposure to weather – Grade II.

staining and polishing: Satisfactory when filled.

durability and preservation

insect attack: Sapwood liable to attack by powder-post beetles. Reported to be non-resistant to termites in West Africa.

durability of heartwood: Perishable.

preservative treatment: Moderately resistant. Sapwood permeable.

uses
Only small quantities of the timber have been imported. The unusually wide sapwood, which is likely to become affected by sapstain if conversion is delayed, and its susceptibility to degrade after conversion are drawbacks to its use. Could be used for interior parts of cabinet furniture and for manufacture of light crates and packing cases. A decorative veneer can be produced from the wood.

OKAN
Cylicodiscus gabunensis

other name: denya (Ghana).

THE TREE
A large tree, reaching an average height of 55–60 m (180–200 ft). Diameter up to 2·5–3·0 m (8–10 ft), average exploitable diameter 0·9–1·2 m (3–4 ft). Short buttresses, rarely more than 1·0 m (3 ft) high. Bole very straight and cylindrical, clear of branches for 25 m (80 ft). West Africa.

THE TIMBER
properties

colour: Yellow to golden brown, often with a slight greenish tinge, darkening on exposure to reddish-brown.

sapwood: Pale pinkish shade, distinct from heartwood, width 50–75 mm (2–3 in).

grain: Typically interlocked. Texture moderately coarse.

odour: Has a disagreeable odour when fresh.

weight: Average about 960 kg/m³ (60 lb/ft³) at 12 per cent moisture content.

strength: About half-way between European beech and greenheart.

Moisture Content	Bending Strength		Modulus of Elasticity		Compression parallel to grain	
	N/mm²	lbf/in²	N/mm²	1000 lbf/in²	N/mm²	lbf/in²
Green	101	14 700	12 800	1850	56·7	8230
12 per cent	140	20 300	16 100	2330	85·4	12 380

movement: No information.

processing

drying: Dries slowly with a marked tendency to split and check. Distortion generally not serious.
Kiln Schedule B.
Shrinkage: Green to 12 per cent moisture content:
Tangential   about 3·5 per cent ($\frac{7}{16}$ in/ft)
Radial   about 3·0 per cent ($\frac{3}{8}$ in/ft)

working properties:
Difficult to obtain a clean finish in some operations owing to pronounced interlocked grain.
Blunting: Fairly severe.
Sawing: Rip-sawing – Saw type HR 60.
Cross-cutting – Saw types 1 and 2.
Narrow bandsawing – Satisfactory.
Wide bandsawing – Saw type C.
Machining: Planing – A cutting angle of 10° required for satisfactory planing of quarter-sawn material.
Moulding – Rather difficult.
Other operations – Satisfactory.
Nailing: Pre-boring necessary.
Gluing: No information.

L

wood bending: Unsuitable for bending owing to buckling and fibre fracture. No exact data available.

plywood manufacture: Unsuitable because of its weight.

staining and polishing: Satisfactory when filled.

durability and preservation

insect attack: Sapwood liable to attack by powder-post beetles. Reported to be highly resistant to termites in West Africa.

durability of heartwood: Very durable.

preservative treatment: Extremely resistant. Sapwood resistant.

uses

A hard and heavy timber, difficult to machine and very durable. Most suitable for piling and wharf decking as it can be used without preservative treatment. Has very high resistance to wear and is suitable for heavy-duty flooring in factories and warehouses but the difficulty in machining must be taken into account when it is considered for flooring.

OKWEN
Brachystegia spp.

other names: brachystegia (Nigeria); meblo (Ivory Coast); naga (Cameroon and France).

Okwen is produced by four species of *Brachystegia* in West Africa:
- *B. nigerica*
- *B. eurycoma*
- *B. kennedyi*
- *B. leonensis* (does not occur in Nigeria).

Brachystegia nigerica is appreciably higher in density than the other species and differs correspondingly from them in its technical properties. Information below refers mainly to *B. nigerica* and *B. kennedyi*.

THE TREE

Large trees, growing to a height of 36–45 m (120–150 ft). Bole may be buttressed, diameter above buttresses about 1·0 m (3–4 ft) or sometimes more. West Africa.

THE TIMBER properties

colour: Light to dark brown.

sapwood: Pale in colour, white or yellowish, well defined from heartwood. Very wide (often about 150 mm, 6 in).

grain: Sometimes straight but more usually deeply interlocked. Quarter-sawn surfaces commonly show a pronounced stripe or roe figure, especially in *B. kennedyi*. Texture medium to coarse.

weight: Variable, from about 530 to 770 kg/m³ (33–48 lb/ft³), seasoned. *B. nigerica* averages about 700 kg/m³ (44 lb/ft³) and *B. kennedyi* about 540 kg/m³ (34 lb/ft³).

strength: Somewhat lower than European beech. *B. nigerica* is appreciably stronger than *B. kennedyi*.

Species	Moisture Content	Bending Strength		Modulus of Elasticity		Compression parallel to grain	
		N/mm²	lbf/in²	N/mm²	1000 lbf/in²	N/mm²	lbf/in²
B. nigerica	Green	79	11 400	8800	1280	39·4	5720
	12 per cent	105	15 200	10 500	1530	57·0	8270
B. kennedyi	Green	–	–	–	–	–	–
	12 per cent	85	12 300	8200	1190	43·0	6230

(Data on *B.kennedyi* from Nigerian Department of Forest Research)

movement: (*B. nigerica*) Medium.

Moisture content in 90 per cent relative humidity   20 per cent
Moisture content in 60 per cent relative humidity   13 per cent
Corresponding tangential movement   2·1 per cent ($\frac{1}{4}$ in/ft)
Corresponding radial movement   1·1 per cent ($\frac{1}{8}$ in/ft)

processing

drying: Dries slowly but fairly well; distortion is the main cause of degrade. Slight tendency for end-splitting and surface checking to occur. Cup may occasionally become serious and slight collapse is possible. *B. kennedyi* stated to dry more rapidly than *B. nigerica* with relatively little degrade.

Kiln Schedule E.

Shrinkage: Green to 12 per cent moisture content:
              Tangential   about 3·5 per cent ($\frac{7}{16}$ in/ft)
              Radial   about 2·5 per cent ($\frac{5}{16}$ in/ft)

working properties: Grain often steeply interlocked.

Blunting: Moderate.

Sawing: Rip-sawing – Saw type HR 54. A tendency to pinch the saw.
Cross-cutting – Saw types 1 and 2. Extensive chipping out often occurs at saw exit.
Narrow bandsawing – Satisfactory apart from chipping out at saw exit.
Wide bandsawing – Saw type B.

Machining: *B. nigerica:*
Planing – Difficult to produce a satisfactory finish even with a reduced cutting angle owing to steeply interlocked grain.
Moulding – Finish poor.
Recessing – Blunting of tool quite rapid.
Mortising – Hollow square chisel difficult to feed and does not clear chips well.
B. kennedyi has generally satisfactory machining properties.

Nailing: Pre-boring advisable.

Gluing: No information.

wood bending:
Considerable variation may occur in bending properties. Pronounced tendency for the timber to distort during steaming and bending operations. Fracturing of the convex face can only be prevented by applying high initial end pressures.
Classification – *B. nigerica:* Moderate.
Ratio radius/thickness for solid bends (steamed):
          Supported: 13   Unsupported: 34
Limiting radius for 3·2 mm ($\frac{1}{8}$ in) laminae (unsteamed): 188 mm ($7\frac{2}{8}$ in).

plywood manufacture: Tested and found to be suitable.
No information on movement of plywood or surface splitting during weathering.

staining and polishing:
B. nigerica not suitable for finishing treatments owing to the difficulty in obtaining a satisfactory smooth surface. *B. kennedyi* finishes satisfactorily after filling.

durability and preservation

insect attack: Sapwood liable to attack by powder-post beetles.

durability of heartwood: Moderately durable.

preservative treatment: Extremely resistant. Sapwood permeable.

Very little of the timber has been imported into this country but it is comparable to oak in weight and strength properties and should be suitable for many of the uses to which oak is put. Suitable for vehicle manufacture and general construction where great durability is not essential, parquet flooring, etc.

B. kennedyi is lighter in weight and presents less difficulty than *B. nigerica* in drying and machining. It is suitable for a variety of uses including joinery, flooring, and general utility purposes.

OLIVE, EAST AFRICAN
Olea hochstetter.

other name: musharagi (Kenya).

THE TREE

Grows to a height of up to 25 m (80 ft), but is often smaller. Bole free from buttresses, but heavily fluted and rarely straight. Length of clear bole 4·5–9 m (15–30 ft), diameter 0·45–0·75 m ($1\frac{1}{2}$–$2\frac{1}{2}$ ft). East Africa.

THE TIMBER
properties

colour: Pale brown to mid-brown with dark grey-brown streaks and irregular markings, giving a handsome appearance.

sapwood: Pale yellow or cream, without characteristic markings, 25–50 mm (1–2 in) wide.

grain: Straight or shallowly interlocked. Texture fine and even.

weight: 830–1020 kg/m³ (52–64 lb/ft³), average 880 kg/m³ (55 lb/ft³) at 12 per cent moisture content.

strength: About half-way between European beech and greenheart, but with rather low resistance to splitting.

Moisture Content	Bending Strength		Modulus of Elasticity		Compression parallel to grain	
	N/mm²	lbf/in²	N/mm²	1000 lbf/in²	N/mm²	lbf/in²
Green	105	15 300	13 700	1980	48·8	7080
12 per cent	174	25 300	17 400	2530	84·1	12 200

movement: Large.
Moisture content in 90 per cent relative humidity   19 per cent
Moisture content in 60 per cent relative humidity   12·5 per cent
Corresponding tangential movement   2·9 per cent ($\frac{23}{64}$ in/ft)
Corresponding radial movement   1·7 per cent ($\frac{13}{64}$ in/ft)

processing

drying: Dries slowly with a tendency to check and split. Internal checking or honeycombing may develop in thick material if too rapid drying is attempted.
Kiln Schedule E.
Shrinkage: Green to 12 per cent moisture content:
Tangential   about 6·5 per cent ($\frac{13}{16}$ in/ft)
Radial   about 4·0 per cent ($\frac{1}{2}$ in/ft)

working properties:

Grain slightly interlocked, affecting machining properties.

Blunting: Moderate.

Sawing: Rip-sawing – Saw type HR 60 or HR 80. Dust tends to build up in packings and consequently to overheat saw.
Cross-cutting – Saw type 1 tends to vibrate. Type 2 most suitable.
Narrow bandsawing – Satisfactory.
Wide bandsawing – Saw type C.

Machining: Planing – A 20° cutting angle and increased load on pressure bars and shoes required to obtain a smooth, clean finish. Resistance to feeding is high.
Moulding – French head and collars most suitable. A tendency to chatter.
Mortising – Hollow square chisel mortiser difficult to feed. Other operations – Satisfactory.

Nailing: Requires pre-boring.

Gluing: No information.

wood bending:

Considerable variation in bending properties. Sapwood may be bent to a smaller radius of curvature than heartwood. Slight resin exudation accompanies steaming.

Classification – Moderate.

Ratio radius/thickness for solid bends (steamed):
Supported: 11·5   Unsupported: 30

Limiting radius for 3·2 mm ($\frac{1}{8}$ in) laminae (unsteamed): 188 mm ($7\frac{3}{8}$ in).

plywood manufacture: Unsuitable because of its weight.

staining and polishing: Good.

durability and preservation

insect attack: Logs liable to attack by ambrosia (pinhole-borer) beetles.

durability of heartwood: Moderately durable.

preservative treatment: Moderately resistant. Sapwood permeable.

uses

A hard and heavy timber with an attractive appearance and good resistance to abrasion. Suitable for furniture, panelling and turnery. As a flooring timber, it has high resistance to wear and makes a high-grade, decorative floor, suitable for public buildings, etc. In East Africa the less decorative wood is used in vehicle building and for tool handles.

OMU
Entandrophragma candollei

The name 'heavy sapele' or 'heavy mahogany' has been applied because the green logs tend to sink in water, but the seasoned wood has about the same density as sapele.

THE TREE

A large tree, growing to a height of 50 m (160 ft) or more, diameter up to 2 m (7 ft). Bole straight and cylindrical, up to 27 m (90 ft) long, with rounded buttresses extending about 3 m (10 ft) up the trunk. West Africa.

THE TIMBER properties

Compared with sapele and utile the timber is darker in colour, coarser textured, and generally somewhat less attractive in appearance. Heartwood dull brown, often with a purplish tinge, darkening on exposure. Sapwood pale pink, 50 to 75 mm (2–3 in) wide, sharply defined from heartwood. Texture rather coarse. Average weight about 640 kg/m³ (40 lb/ft³), seasoned. Strength slightly lower than European beech. Medium movement.

processing Dries rather slowly with a marked tendency to distort. Kiln Schedule A. Quartered material 25 mm (1 in) thick has been found to dry well to Kiln Schedule J. Shrinkage, green to 12 per cent moisture content: tangential about 6·0 per cent (¾ in/ft), radial about 4·0 per cent (½ in/ft). Has a moderate blunting effect on cutting edges. Satisfactory in sawing. Tends to tear in planing unless the cutting angle is reduced to 20°. Tendency to burn in boring. Other operations satisfactory. Nailing properties fairly good. Employed in plywood manufacture and selected logs are sliced for their decorative striped figure. Stains and polishes well when filled.

durability and preservation Heartwood moderately durable. Resistant to preservative treatment.

uses Suitable for many of the same purposes as sapele and utile, though somewhat less attractive in appearance and less durable than utile.

OPEPE
Nauclea diderrichii
(formerly
*Sarcocephalus
diderrichii*)

other names: kusia, kusiaba (Ghana); badi (Ivory Coast); bilinga (Cameroon).

THE TREE
A large tree, up to about 50 m (160 ft) in height. Free from buttresses, but old trees have a short basal thickening. Bole long and cylindrical, length up to 25–30 m (80–100 ft), diameter up to 1.5 m (5 ft). West Africa.

THE TIMBER
properties

colour: Distinctive yellow or orange-yellow.

sapwood: Whitish or pale yellow, clearly defined from heartwood, width about 50 mm (2 in).

grain: Usually interlocked or irregular, enhancing the attractive appearance of the wood. A reasonable proportion of straight-grained material can probably be obtained by grading. Texture fairly coarse, owing to the rather large pores.

weight: Average about 740 kg/m³ (46 lb/ft³) at 12 per cent moisture content.

strength: Slightly higher than European beech.

Moisture Content	Bending Strength		Modulus of Elasticity		Compression parallel to grain	
	N/mm²	lbf/in²	N/mm²	1000 lbf/in²	N/mm²	lbf/in²
Green	94	13 700	11 900	1720	51·6	7490
12 per cent	120	17 400	13 400	1940	71·7	10 400

movement: Small.
Moisture content in 90 per cent relative humidity   18 per cent
Moisture content in 60 per cent relative humidity   12 per cent
Corresponding tangential movement   1·8 per cent ($\frac{7}{32}$ in/ft)
Corresponding radial movement   0·9 per cent ($\frac{7}{64}$ in/ft)

processing **drying:** Quarter-sawn material dries fairly quickly with very little checking or distortion. Flat-sawn timber is more refractory; considerable checking and splitting may occur, and serious distortion sometimes develops. In thicker sizes the timber dries rather slowly.
Kiln Schedule E.
Shrinkage: No information.

working properties:

Machining properties affected by interlocked irregular grain and coarse texture.

Blunting: Moderate.

Sawing: Rip-sawing — Saw type HR 54.
Cross-cutting — Satisfactory.
Narrow bandsawing — Satisfactory.
Wide bandsawing — Saw type B.

Machining: Planing — Flat-sawn material planes to a smooth finish but on quarter-sawn faces a cutting angle of 10° is required to prevent tearing. Other operations satisfactory.

Nailing: Pre-boring necessary to prevent splitting unless thin gauge nails are used.

Gluing: Good.

wood bending: Shown by tests to have poor steam bending properties.
Classification — Poor.
Limiting radius for 3·2 mm ($\frac{1}{8}$ in) laminae (unsteamed): 277 mm ($10\frac{3}{4}$ in).

plywood manufacture: No information.

staining and polishing: Satisfactory when filled.

durability and preservation

insect attack: Sapwood liable to attack by powder-post beetles. Reported to be moderately resistant to termites in West Africa.

durability of heartwood: Very durable.

preservative treatment: Moderately resistant. Sapwood permeable.

uses

A strong, stable and durable wood, particularly suitable for use in large sizes for structural purposes such as piling, wharf and jetty decking and other dock and marine work. Has been used in place of oak in boat-building (except for bent parts) and for general construction work and is suitable for exterior joinery, provided some surface checking is tolerated. Has been used for railway sleepers in the United Kingdom and West Africa. Makes an attractive flooring for normal conditions of pedestrian traffic.

OVANGKOL
Guibourtia ehie

other names: ehie, anokye, hyeduanini (Ghana); amazoue, amazakoue (Ivory Coast).

Height generally about 30 m (100 ft), sometimes up to 45 m (150 ft). Bole straight and cylindrical above buttresses, diameter about 0·8 m (2½ ft). On older stems there is a tendency for narrow, slightly raised horizontal rings to be formed; this feature also occurs in ogea (*Daniellia ogea*). West Africa.

Heartwood yellow-brown to chocolate coloured, with grey to almost black stripes. Similar in appearance to Queensland walnut, but somewhat paler in colour. Sapwood pale in colour, distinct from heartwood, sometimes fairly wide. Grain interlocked, texture moderately coarse. When fresh the timber has a strong smell but this disappears on drying. Average weight 800 kg/m³ (50 lb/ft³) at 12 per cent moisture content.

No information on drying. May present a moderate resistance in sawing and machining and care is necessary in finishing quartered stock because of interlocked grain. Reported to be highly resistant to termites in West Africa. Heartwood moderately durable (provisional).

An attractive wood which should be suitable for cabinet work, high-grade furniture, interior decorative work, turnery, flooring, etc., provided that it can be dried satisfactorily. As veneer it should provide a useful addition to the range of walnut-like woods.

PADAUK, AFRICAN
Pterocarpus soyauxii

other names: barwood, camwood.

THE TREE Reaches a height of 30 m (100 ft), with a straight, cylindrical bole, length 12–18 m (40–60 ft), diameter 0·6–1·0 m (2–3 ft). Central and west tropical Africa.

THE TIMBER
properties Heartwood very distinctive when freshly cut, vivid red, toning down to a medium to dark purple-brown on exposure. Sapwood wide 100–200 mm, (4–8 in), pale in colour and sharply defined from heartwood. Grain straight or somewhat interlocked, texture very coarse. Weight about 640–800 kg/m³ (40–50 lb/ft³), seasoned. Has good strength properties, particularly in compression and static bending. Movement exceptionally small.

processing Dries fairly rapidly and very well with a minimum of degrade. Kiln Schedule J. Although a heavy wood, generally easy to machine. Finishes very well.

durability and preservation Heartwood very durable. Moderately resistant to preservative treatment.

uses Has an attractive appearance together with high-strength properties, durability and outstanding stability. Used for high-class joinery, fancy turnery and carvings, and for tool and knife handles and spirit levels. Has high resistance to abrasion and makes an excellent flooring timber of good appearance, suitable for heavy pedestrian traffic in public buildings. Suitable for floors where under-floor heating is installed because of its dimensional stability. Has also been used as decorative veneer.

PADAUK, ANDAMAN
Pterocarpus dalbergioides

other names: padauk (Great Britain); Andaman redwood, vermilion wood (United States).

THE TREE Reaches a height of 25–37 m (80–120 ft). Buttresses may be very large. Bole straight and cylindrical above buttresses, diameter 0·75–0·9 m (2½–3 ft), occasionally more. Bole clear for 12 m (40 ft) or more. Andaman Islands.

THE TIMBER
properties

colour: A handsome wood, rather variable in colour, but mainly of a rich crimson hue, varying through shades of red to brown and often having darker red or black streaks. The markings often produce an attractive figure. The colour darkens on exposure to dark reddish-brown. Some trees produce a pale red or yellowish timber which differs from the darker wood only in appearance.
sapwood: Greyish, narrow.
grain: Generally interlocked. Texture variable, from medium to rather coarse.
weight: Average about 770 kg/m³ (48 lb/ft³) at 12 per cent moisture content.
strength: Comparable to European beech.

Moisture Content	Bending Strength		Modulus of Elasticity		Compression parallel to grain	
	N/mm²	lbf/in²	N/mm²	1000 lbf/in²	N/mm²	lbf/in²
Green	88	12 800	10 300	1500	48·8	7080
12 per cent	105	15 300	11 200	1630	61·5	8920

(Data from Indian Forest Records)

movement: No information.

| processing | **drying:** If converted green it may develop fine surface splits, but it is believed that this can be overcome by girdling the trees and allowing them to die before felling, when it air dries very well with little degrade. Somewhat liable to develop fine wavy surface checks if cut to broad sections. For storage in the log or the square, ends should be protected against rapid drying particularly in dry climates. Kiln dries well without much checking or distortion. |

drying: If converted green it may develop fine surface splits, but it is believed that this can be overcome by girdling the trees and allowing them to die before felling, when it air dries very well with little degrade. Somewhat liable to develop fine wavy surface checks if cut to broad sections. For storage in the log or the square, ends should be protected against rapid drying particularly in dry climates. Kiln dries well without much checking or distortion.
Kiln Schedule F.
No data on shrinkage.

working properties: Machining properties affected by interlocked grain.
Blunting: Moderate.
Sawing: Rip-sawing — Saw type HR 60.
              Cross-cutting — Saw type 1 or 2.
              Narrow bandsawing — Satisfactory.
              Wide bandsawing — Saw type B.
Machining: Planing — A cutting angle of 15° required to prevent tearing of quarter-sawn material.
              Turning — Excellent.
Nailing: Difficult.
Gluing: Good.

wood bending: No information.

plywood manufacture: Found by tests in India to be suitable.
No information on movement of plywood or surface splitting during weathering.

staining and polishing: Satisfactory when filled.

durability and preservation

insect attack: Reported to be moderately resistant to termites in India.

durability of heartwood: Very durable.

preservative treatment: Moderately resistant. Sapwood probably permeable.

uses

Used in joinery, especially for fittings such as bank counters, where wear is likely to take place. Also suitable for exterior joinery and for boat-building, except for steam-bent parts. For flooring, suitable for normal conditions of traffic where a decorative floor is required. Used in the East for vehicle framing, building and furniture.

PADAUK, BURMA
Pterocarpus macrocarpus

other names: mai pradoo, pradoo (Thailand).

THE TREE

A medium-sized tree, height about 18–25 m (60–80 ft). Bole up to 8 m (25 ft) long, often straight and cylindrical, but sometimes crooked or forked. Diameter 0·6–0·9 m (2–3 ft). Burma and Thailand.

THE TIMBER
properties

Heartwood yellowish-red to brick-red, streaked with darker lines. Less variable in colour than Andaman padauk. Tones down on exposure to an attractive golden brown. Sapwood grey and narrow. Grain interlocked, producing a narrow ribbon-grain figure. Texture moderately coarse. Weight about 850 kg/m³ (53 lb/ft³), seasoned. A heavy, hard and strong timber, considerably stronger than Andaman padauk.

processing

Air dries fairly well with little distortion or splitting. May develop surface checks in drying if converted green. If stored in log form the ends of logs should be well protected against rapid drying. In kiln drying, dries rather slowly with relatively little degrade. Kiln Schedule F. Has a moderate blunting effect on cutting edges. Satisfactory in sawing. In machining a cutting angle of 15° is recommended in planing to produce a good finish. Turns well. Pre-boring advisable for nailing. Glues satisfactorily and polishes well when filled.

Sapwood liable to attack by powder-post beetles. Heartwood very durable and extremely resistant to preservative treatment.

uses

Harder and stronger and rather more difficult to work than Andaman padauk, and generally suitable for similar purposes. As decorative flooring it has high resistance to wear and is suitable for normal conditions of traffic. Used in Burma for under-framing of vehicles and for shafts and furniture.

PALDAO
Dracontomelum dao

other names: dao (Philippines).

Other species of *Dracontomelum* with timbers similar to paldao and known as New Guinea walnut, Pacific walnut, Papuan walnut and loup occur in New Guinea and neighbouring islands.

A tall tree, 30 m (100 ft) or more in height, with a straight bole, clear up to 20 m (65 ft), but buttresses may extend up to 6 m (20 ft).

A decorative timber with a walnut-like appearance, resembling Queensland walnut more closely than European walnut. Heartwood greyish-brown with a greenish-yellow cast and irregular, dark brown to almost black banding. Sapwood very wide, pale and featureless. Grain interlocked or sometimes wavy, texture medium. Weight about 740 kg/m³ (46 lb/ft³), seasoned. A fairly strong wood, notably in toughness and bending.

Used mainly in the form of veneer. Can be successfully rotary peeled but is more usually sliced to display the figure. Glues satisfactorily, takes a good finish and polishes well. The timber is reported to have a moderate shrinkage on drying with a tendency to distort. Working properties generally satisfactory. Heartwood non-durable (provisional).

An attractive wood with a walnut-like appearance. Suitable as veneer for many purposes for which walnut is used, such as high-grade furniture, shop fitting, interior panelling and other decorative work.

PANGA PANGA
Millettia stuhlmannii

The timber wengé (*Millettia laurentii*) from Zaire is generally similar in appearance and properties to panga panga.

THE TREE

Height not usually exceeding 18 m (60 ft), diameter about 0·6 m (2 ft). Logs 3–4·5 m (10–15 ft) long, generally straight and cylindrical but occasionally with heart rot, shakes and pockets of ingrown bark. East tropical Africa.

**THE TIMBER
properties**

Heartwood dark brown with alternate bands of dark and lighter coloured tissue, giving a characteristic decorative figure. Sapwood yellowish-white, 25–50 mm (1–2 in) wide, distinct from the dark heartwood. Grain straight, texture rather coarse and not very uniform. Weight about 830–1000 kg/m³ (52–62 lb/ft³), seasoned. Has good strength properties and high resistance to abrasion. Small movement.

processing

Stated to dry well but very slowly, with little degrade. Kiln Schedule E suggested. Has a moderate blunting effect on cutting edges. A reduced cutting angle recommended in planing. Turns well to a clean finish. Difficult to nail. Unsuitable for plywood manufacture because of its weight.

**durability
and preservation**

Heartwood durable (provisional). Extremely resistant to preservative treatment.

uses   Very suitable for use as flooring strips or blocks. Has a pleasing appearance and high resistance to wear and is best suited to normal conditions of pedestrian traffic, as in public buildings, hotels, showrooms and boardrooms. May also be used for interior and exterior joinery and general construction.

PEAR
Pyrus communis

Trees are usually 9–12 m (30–40 ft) high, occasionally 18 m (60 ft). Diameter 0·3–0·6 m (1–2 ft). The timber usually comes from old orchard trees. Europe and western Asia.

The timber is typically a pinkish-brown colour. Grain straight, texture fine and even. Weight about 700 kg/m³ (44 lb/ft³), seasoned.

Dries slowly with a definite tendency to distort. Kiln Schedule A. Moderately hard to saw and has a moderate blunting effect on cutting edges. In planing a cutting angle of 20° is recommended. Turns very well and gives good results with stains and polishes.

Has excellent turning properties and is employed for fancy turnery such as bowls, the backs of brushes, etc. More readily available in continental Europe where it is used for recorders and, when stained black, for violin finger-boards.

PEROBA ROSA
Aspidosperma spp.
principally *A. peroba*

other name: red peroba (Great Britain).

THE TREE   Height variable from 15 m (50 ft) to 38 m (125 ft), average about 27 m (90 ft). Bole straight and well-formed, average diameter 0·8 m (2½ ft) sometimes up to 1·2–1·5 m (4–5 ft). Brazil.

THE TIMBER
properties

colour:	Tan to rose-red, often streaked with purple or brown and becoming brownish-yellow to medium brown on exposure.
sapwood:	Yellowish, paler than heartwood, but not sharply demarcated.
grain:	Very variable. Texture fine and uniform.
weight:	700–850 kg/m³ (44–53 lb/ft³), average about 750 kg/m³ (47 lb/ft³), seasoned.
strength:	Generally somewhat lower than European beech, but comparable to beech in compressive strength.

Moisture Content	Bending Strength		Modulus of Elasticity		Compression parallel to grain	
	N/mm²	lbf/in²	N/mm²	1000 lbf/in²	N/mm²	lbf/in²
Green	79	11 500	8300	1210	39·8	5770
12 per cent	88	12 800	9900	1440	56·9	8250

movement: No information.

processing   **drying:** Dries without much splitting, but some distortion may develop. Kiln Schedule E.
No information on shrinkage during drying.

working properties: Grain very variable and irregular.

      Blunting: Moderate.

      Sawing: Rip-sawing — Saw type HR 54.

                    Cross-cutting — Satisfactory.

                    Narrow bandsawing — Satisfactory.

                    Wide bandsawing — Saw type B.

      Machining: Planing — A cutting angle of 20° required to produce a good finish due to irregular grain.

                    Other operations — Generally satisfactory but timber requires support at tool exit on end-grain working, e.g. boring, mortising.

      Gluing: Good.

wood bending: No information.

plywood manufacture: No information.

staining and polishing: Satisfactory.

durability and preservation

insect attack: Reported to be attacked by dry-wood termites in West Indies.

durability of heartwood: Durable.

preservative treatment:

Reported to be extremely resistant. Sapwood permeable.

uses

Probably most suitable for construction work requiring strength and durability, and could also be useful for exterior joinery. Less suitable than white peroba for vat-making owing to grain distortion and tendency for colouring matter to be extracted in some processes.

PEROBA, WHITE
Paratecoma peroba

other names: ipé peroba, peroba amarella, peroba branca, peroba de campos (Brazil).

THE TREE

Reaches a maximum height of about 40 m (130 ft). Bole straight and symmetrical, clear of branches for about 27 m (90 ft), diameter up to 1·5 m (5 ft). Brazil.

THE TIMBER properties

colour: Variable, light olive with a yellowish, greenish or reddish hue, sometimes indistinctly striped.

sapwood: White or yellowish, sharply demarcated from heartwood.

grain: Commonly interlocked giving rise to a narrow stripe or roe figure. Wavy grain also commonly occurs. Texture fine.

weight: 690–830 kg/m³ (43–52 lb/ft³), average 750 kg/m³ (47 lb/ft³), seasoned.

irritant properties:

The fine dust produced in machining operations is reported to cause skin irritations in some workers.

strength: Comparable to European beech.

Moisture Content	Bending Strength		Modulus of Elasticity		Compression parallel to grain	
	N/mm²	lbf/in²	N/mm²	1000 lbf/in²	N/mm²	lbf/in²
Green	–	–	–	–	–	–
12 per cent	112	16 200	11 400	1650	63·8	9260

movement: No information.

drying: Dries readily and well with negligible splitting. Distortion not generally serious though fairly severe twisting may occur in a few pieces.
Kiln Schedule D.
Shrinkage: Green to 12 per cent moisture content:
Tangential about 3·5 per cent ($\frac{7}{16}$ in/ft)
Radial about 2·0 per cent ($\frac{1}{4}$ in/ft)

working properties: No information.
Gluing: Good.

wood bending: No information.

plywood manufacture: Known to be used in plywood.

staining and polishing: Satisfactory.

durability and preservation

insect attack: No information.

durability of heartwood: Very durable.

preservative treatment: Resistant.

uses

Has been used for lorry and motor body frames and shipbuilding, including decking and flooring. Suitable for manufacture of vats and tanks for use with foodstuffs and chemicals. Used in Brazil for good quality joinery and furniture.

PERSIMMON
Diospyros virginiana

THE TREE

Usually a small- or medium-sized tree, but occasionally reaching a height of 30 m (100 ft), bole diameter 0·5–0·8 m (1$\frac{1}{2}$–2$\frac{1}{2}$ ft). North America (middle and southern States of USA).

THE TIMBER
properties

The timber consists almost entirely of pale-coloured sapwood; the heartwood is confined to a small central core of brown or black wood.

grain: Straight. Texture fine and even.

weight: Average about 830 kg/m³ (52 lb/ft³), seasoned.

strength: Slightly higher than European beech.

Moisture Content	Bending Strength		Modulus of Elasticity		Compression parallel to grain	
	N/mm²	lbf/in²	N/mm²	1000 lbf/in²	N/mm²	lbf/in²
Green	72	10 500	8800	1280	29·9	4330
12 per cent	128	18 600	13 000	1880	65·6	9520

(Data from US Forest Products Laboratory)

movement: Large.
Moisture content in 90 per cent relative humidity   24 per cent
Moisture content in 60 per cent relative humidity   13·5 per cent
Corresponding tangential movement   3·4 per cent ($\frac{27}{64}$ in/ft)
Corresponding radial movement   2·0 per cent ($\frac{1}{4}$ in/ft)

processing drying: Dries fairly rapidly with some tendency to check.
Kiln Schedule C.
Shrinkage: Green to 12 per cent moisture content:
Tangential about 6·5 per cent ($\frac{9}{16}$ in/ft)
Radial about 4·5 per cent ($\frac{5}{16}$ in/ft)

working properties: No information. Finishes very smoothly.
Gluing: Difficult.

wood bending: No information.

plywood manufacture: Unsuitable because of its weight.

staining and polishing: No information.

durability
and preservation

insect attack: Sapwood liable to attack by powder-post beetles.

durability of heartwood: No information.

preservative treatment: No information.

uses

A hard and heavy timber of fine texture, giving a very smooth finish. Used mainly for making shuttles for the textile industry and in the sports goods trade for heads of golf clubs.

PLANE, EUROPEAN
Platanus hybrida
(formerly
Platanus acerifolia)

other names: English plane, French plane, etc., according to origin, London plane, lacewood (quartered wood only) (Great Britain).

THE TREE

Grows to a height of 30 m (100 ft). Clear bole, length about 9 m (30 ft), diameter 0·9–1·2 m (3–4 ft). Europe, including United Kingdom. The tree is seldom planted for timber.

THE TIMBER
properties

colour: Resembles beech in colour, but is distinguished by the numerous broad rays which are very conspicuous when the wood is cut on the true quarter, showing reddish-brown against the light coloured background and producing a distinctive and highly decorative figure.

sapwood: Not usually distinct from heartwood, but some logs contain an irregular core of darker coloured wood.

grain: Usually straight; texture fine and even.

weight: Average about 620 kg/m³ (39 lb/ft³), seasoned.

strength: About half-way between obeche and European beech.

Moisture Content	Bending Strength		Modulus of Elasticity		Compression parallel to grain	
	N/mm²	lbf/in²	N/mm²	1000 lbf/in²	N/mm²	lbf/in²
Green	54	7900	6400	930	24·2	3510
12 per cent	–	–	–	–	41·6	6040

movement: No information.

processing

drying: Dries fairly rapidly without much splitting but with a tendency to distort. Kiln Schedule E.
Shrinkage: Green to 12 per cent moisture content:
Tangential   about 9·0 per cent (1⅛ in/ft)
Radial   about 4·0 per cent (½ in/ft)

working properties:

Blunting: Moderate.

Sawing: A slight tendency to bind on saws.
Rip-sawing – Saw type HR 54.
Cross-cutting – Satisfactory.
Narrow bandsawing – Satisfactory.
Wide bandsawing – Saw type B.

Machining: Generally satisfactory.
Planing – A small reduction in cutting angle may be advantageous. On true quarter-sawn faces the large rays tend to flake when planed unless cutters are sharp.

Nailing: Good.

Gluing: Good.

wood bending:

Classification – Very good.
Ratio radius/thickness for solid bends (steamed):
Supported: 2·0   Unsupported: 17·0
Limiting radius for 3·2 mm ($\frac{1}{8}$ in) laminae (unsteamed): 135 mm ($5\frac{1}{4}$ in).

plywood manufacture: No information.

staining and polishing: Satisfactory with care.

durability and preservation

insect attack: Sapwood liable to attack by the common furniture beetle.

durability of heartwood: Perishable.

preservative treatment: No information.

uses

A decorative wood, used for inlaying cigarette and other fancy boxes, and decorative work generally. Also used as veneer, particularly for panelling.

POPLAR
Populus spp.

other names: The main species of poplar growing throughout Great Britain and Europe are:
Black Italian poplar (*P canadensis* var. *serotina*)
European aspen (*P. tremula*)
Black poplar (*P. nigra*)
Grey poplar (*P. canescens*)
White poplar (*P. alba*)
Many hybrids between the different species also occur.

THE TREE

The black poplar and grey poplar reach a height of 30–35 m (100–115 ft). Aspen and white poplar are smaller trees reaching about 18–25 m (60–80 ft) in height. Diameter usually 0·9–1·2 m (3–4 ft), but can be larger.

Home-grown poplar is largely derived from the black poplars, of which black Italian poplar is the most important. Poplar imported from France and Belgium is also mainly of the black poplar group. White poplar and grey poplar are less important as timber.

THE TIMBER
properties

The timber is seldom marketed by species and the following properties may be taken as generally applicable to all species, but the quality of the timber varies considerably according to the conditions of growth.

colour: Light in colour, white, greyish, pale brown or reddish. White poplar and grey poplar have a distinct reddish-brown heartwood.

sapwood: Not usually distinct from heartwood except in white poplar and grey poplar.

grain: Usually straight and inclined to be woolly. Texture fine and even, due to the uniform structure and lack of contrast between early wood and late wood. Aspen from northern Europe is of particularly good quality, being typically straight grained, nearly white in colour and of finer texture than the faster-growing varieties.

weight: Generally in the range 380–530 kg/m³ (23–33 lb/ft³), average about 450 kg/m³ (28 lb/ft³), seasoned.

strength: Somewhat higher than obeche.

Species	Moisture Content	Bending Strength		Modulus of Elasticity		Compression parallel to grain	
		N/mm^2	lbf/in^2	N/mm^2	1000 lbf/in^2	N/mm^2	lbf/in^2
Black Italian	Green	41	6000	6800	990	19·3	2800
poplar	12 per cent	72	10 400	8600	1250	37·4	5430
Grey poplar	Green	44	6400	7200	1040	20·1	2920
	12 per cent	76	11 000	9500	1380	36·9	5350

movement: Medium.
Moisture content in 90 per cent relative humidity 22 per cent
Moisture content in 60 per cent relative humidity 13 per cent
Corresponding tangential movement 2·8 per cent ($\frac{11}{32}$ in/ft)
Corresponding radial movement 1·2 per cent ($\frac{9}{64}$ in/ft)

processing

drying: Generally dries well and fairly rapidly, but local pockets of moisture are apt to remain in the timber. Knots are inclined to split.
Kiln Schedule E.
Shrinkage: (Data for black Italian poplar)
Green to 12 per cent moisture content:
Tangential about 5·5 per cent ($\frac{11}{16}$ in/ft)
Radial about 2·0 per cent ($\frac{1}{4}$ in/ft)

working properties:
Blunting: Slight.
Sawing: Occasional tendency to bind on saws.
Rip-sawing – Saw type HR 54.
Cross-cutting – Satisfactory.
Narrow bandsawing – Satisfactory.
Wide bandsawing – Saw type A.
Machining: Texture is often woolly. Cutters with reduced sharpness angle and sharp cutting edges required to produce a good finish. Black Italian poplar has the best working properties.
Nailing: Satisfactory.
Gluing: Good.

wood bending:
Severe buckling on the concave face occurs when wood is bent to any appreciable extent. Bending properties not improved by use of a supporting strap and end-pressure device. Generally unsuitable for solid bending.
Classification – Very poor.
Limiting radius for 3·2 mm ($\frac{1}{8}$ in) laminae (unsteamed): 160 mm ($6\frac{1}{4}$ in).

plywood manufacture:
Employed for plywood and usually available in United Kingdom.
Movement: 1·5 mm plywood from 30 per cent to 90 per cent relative humidity:
P. robusta 0·18 per cent
P. serotina 0·16 per cent
Surface splitting on exposure to weather – Group I.

staining and polishing:
Generally satisfactory, though staining may sometimes be patchy.

durability and preservation

insect attack: Trees and logs liable to attack by forest longhorn or Buprestid beetles, and trees liable to attack by wood-boring caterpillars (Cossidae). Sapwood not liable to attack by powder-post beetles.

170

durability of heartwood: Perishable (all species).

preservative treatment:

> Black Italian poplar moderately resistant; the sapwood which constitutes a large proportion of the tree is permeable.
> Grey and black poplar resistant; sapwood permeable.
> No information on other species.

uses   A plain, pale-coloured and lightweight timber, very tough for its weight, and less liable to splinter than softwoods. It is the main timber used for manufacture of matches and is widely used for making chip baskets. It is also used for shelving and other interior joinery and for toys and non-decorative turnery. Employed also for plywood manufacture.

PTERYGOTA
Pterygota bequaertii and *P. macrocarpa*

other names: ware, awari (Ghana); poroposo (Nigeria); koto (Ivory Coast).

THE TREE   *Pterygota macrocarpa* grows to a height of about 37 m (120 ft), *P. bequaertii* a little smaller. Bole reasonably straight but with fairly heavy buttresses, extending up to 6 m (20 ft) above ground level. Diameter above buttresses 0·5–1·2 m (1½–4 ft). West Africa.

THE TIMBER properties

colour: Creamy white.

sapwood: Pale in colour, little different from heartwood.

grain: Shallowly interlocked, commonly with small knot clusters. Texture moderately coarse. The high rays give the timber a striking fleck figure on accurately quarter-sawn stock.

weight: Rather variable, from 530–750 kg/m³ (33–47 lb/ft³) at 12 per cent moisture content; the average for *P. bequaertii*, 650 kg/m³ (41 lb/ft³) is somewhat higher than that for *P. macrocarpa*, 560 kg/m³ (35 lb/ft³).

strength: *P. bequaertii* comparable to European beech; *P. macrocarpa* somewhat lower.

Species	Moisture Content	Bending Strength		Modulus of Elasticity		Compression parallel to grain	
					1000		
		N/mm²	lbf/in²	N/mm²	lbf/in²	N/mm²	lbf/in²
P. bequaertii	Green	73	10 600	8800	1270	35·4	5130
	12 per cent	111	16 100	11 500	1670	57·9	8400
P. macrocarpa	Green	57	8300	7400	1080	26·7	3870
	12 per cent	85	12 300	9200	1340	43·4	6300

movement: Medium.
Moisture content in 90 per cent relative humidity   19·5 per cent
Moisture content in 60 per cent relative humidity   11·5 per cent
Corresponding tangential movement   3·0 per cent (⅜ in/ft)
Corresponding radial movement   1·5 per cent ($\frac{3}{16}$ in/ft)

M

| processing | **drying:** | Dries fairly rapidly. Slight tendency for surface checking and extension of original shakes to take place. Distortion generally small but moderate cupping may occasionally occur. Tendency for 'sticker stain' to develop. Kiln Schedule H. |

drying: Dries fairly rapidly. Slight tendency for surface checking and extension of original shakes to take place. Distortion generally small but moderate cupping may occasionally occur. Tendency for 'sticker stain' to develop. Kiln Schedule H.

Shrinkage: Green to 12 per cent moisture content:
Tangential   about 5·0 per cent ($\frac{5}{8}$ in/ft)
Radial   about 2·0 per cent ($\frac{1}{4}$ in/ft)

working properties:
(Data for *Pterygota kamerunensis*) Texture coarse, grain moderately interlocked.

Blunting: Moderate.

Sawing: Rip-sawing – Saw type HR 54.
Cross-cutting – All saw types satisfactory.
Narrow bandsawing – Satisfactory.
Wide bandsawing – Saw type B.
Breaking out at bottom of cut may be pronounced in all sawing operations.

Machining: Planing – Reduction of cutting angle to 20° required to prevent tearing of quarter-sawn surfaces. Cutters must be maintained in a sharp condition to prevent fibrous finish. Other operations – Generally satisfactory, but fibrous finish often obtained.

Nailing: Tendency to split when nailed near edges, otherwise satisfactory.

Gluing: Good.

wood bending:
Considerable variation in bending properties but, in general, the wood buckles when bent appreciably, and requires careful end-pressure control to prevent distortion during bending. Cannot be bent satisfactorily if pin knots are present. Inclined to split and check immediately after steaming treatment.
Classification – Very poor.
Limiting radius for 3·2 mm ($\frac{1}{8}$ in) laminae (unsteamed): 155 mm (6 in).

plywood manufacture: Found by tests to be unsuitable.

staining and polishing: Satisfactory when filled.

durability and preservation

insect attack: Sapwood liable to attack by powder-post beetles. Reported to be non-resistant to termites in West Africa.

durability of heartwood: Perishable.

preservative treatment: Permeable.

uses

Requires rapid extraction and conversion to be marketed in clean condition, free from stain and insect damage. Suitable as an alternative to beech or ramin for use in furniture manufacture, interior joinery and carpentry and as a general utility construction timber for interior work.

PUNAH
Tetramerista glabra

Grows in Sumatra, Borneo and Malaya. Diameter of logs commonly 0·6–1·0 m (2–3 ft). Timber straw-coloured or light brown, often tinged with pink. Sapwood paler in colour than heartwood, 38–75 mm (1$\frac{1}{2}$–3 in) wide, not always distinct in the log, but sharply differentiated in seasoned timber. Grain straight or spiral, not generally interlocked. Texture moderately coarse and even. Average weight 720 kg/m³ (45 lb/ft³), seasoned. Probably comparable in strength properties to European beech, but lower in shock resistance and hardness.

Dries rapidly but requires care to avoid cupping and end-splitting. Saws relatively easily for its weight and planes satisfactorily. Finishes satisfactorily but considerable filling is required.

Suitable for general construction work, carcassing and joinery out of contact with the ground.

172

PURPLEHEART
Peltogyne spp.

other name: amaranth (United States).

THE TREE Varies in size and in habit of growth according to locality. Commonly reaches a height of 38–45 m (125–150 ft). Buttresses may be present. Bole straight and cylindrical, clear for 15 m (50 ft) or more. Diameter up to 1·2 m (4 ft), more usually 0·6–0·9 m (2–3 ft). Central America and northern South America.

THE TIMBER
properties

colour: Dull brown when freshly cut, rapidly changing to purple on exposure to light and gradually toning down in course of time to dark purplish-brown. Considerable variation in colour occurs between species and probably within a species.

sapwood: Whitish or cream coloured, 50–100 mm (2–4 in) wide.

grain: Generally straight, sometimes wavy or interlocked. Texture moderate to fine.

weight: Variable, from about 800–1000 kg/m³ (50–63 lb/ft³), average about 860 kg/m³ (54 lb/ft³), seasoned.

strength: About half-way between European beech and greenheart.

Moisture Content	Bending Strength		Modulus of Elasticity		Compression parallel to grain	
	N/mm²	*lbf/in²*	*N/mm²*	*1000 lbf/in²*	*N/mm²*	*lbf/in²*
Green	*105*	*15 200*	*14 000*	*2030*	*56·5*	*8190*
12 per cent	*147*	*21 300*	*16 700*	*2420*	*78·5*	*11 380*

movement: Small.
Moisture content in 90 per cent relative humidity   16 per cent
Moisture content in 60 per cent relative humidity   11 per cent
Corresponding tangential movement   1·8 per cent ($\frac{7}{32}$ in/ft)
Corresponding radial movement   1·1 per cent ($\frac{1}{8}$ in/ft)

processing **drying:** Dries well and fairly rapidly with little degrade. Some difficulty in extracting moisture from the centre of thicker planks.
Kiln Schedule E.
Shrinkage: Green to 12 per cent moisture content:
Tangential   about 4·5 per cent ($\frac{9}{16}$ in/ft)
Radial   about 2·0 per cent ($\frac{1}{4}$ in/ft)

working properties:
(Data for *Peltogyne pubescens*) Grain often wavy and **interlocked.** Build up of gum on tools is troublesome.
Blunting: Moderate to severe.
Sawing: Rip-sawing – Saw type HR 60.
Cross-cutting – Saw type 1 or 2.
Narrow bandsawing – Satisfactory.
Wide bandsawing – Saw type C.
Machining: Planing – Reduction of cutting angle to 15° required for satisfactory planing of material with wavy or interlocked grain.
Mortising – Chain mortiser most suitable, but **support** required at tool exit.
Drilling – Difficult, with a tendency to burn.
Nailing: Satisfactory with care.
Gluing: No information.

wood bending:
        Classification – Moderate.
        Ratio radius/thickness for solid bends (steamed):
                Supported: 18   Unsupported: 30
        Limiting radius for 3·2 mm ($\frac{1}{8}$ in) laminae (unsteamed): 203 mm (8 in).

plywood manufacture: Unsuitable because of its weight.

staining and polishing: Good.

durability and preservation

insect attack: Sapwood liable to attack by powder-post beetles.

durability of heartwood: Very durable.

preservative treatment: Extremely resistant. Sapwood permeable.

uses

Possesses high strength and very good durability and is an excellent structural timber suitable for heavy outdoor constructional work such as bridges and dock work. As flooring, it has high wearing qualities and is suitable for most conditions of traffic. Has a limited outlet for decorative purposes, e.g. small turned articles, and can be sliced for veneer which is used on a small scale for decorative inlay. Has been used successfully in chemical plant for vats and filter press plates and frames.

PYINKADO
Xylia xylocarpa
(formerly
X. dolabriformis)

THE TREE

Grows to a height of 30–37 m (100–120 ft), diameter 0·8–1·2 m (2½–4 ft). Bole straight, fairly cylindrical and free of branches for 12 m (40 ft) in well-grown trees. In less favourable districts trees are smaller and the bole is of poor form. Burma and parts of India.

THE TIMBER properties

colour: Dull reddish-brown, with darker streaks due to the presence of a dense fibrous zone in each growth ring, and locally speckled with dark, gummy exudations.

sapwood: Pale reddish-white, narrow.

grain: Variable, from straight to broadly interlocked, sometimes wavy. Texture moderately fine and even.

weight: About 980 kg/m³ (61 lb/ft³), seasoned.

strength: Somewhat lower than greenheart.

Moisture Content	Bending Strength		Modulus of Elasticity		Compression parallel to grain	
	N/mm²	lbf/in²	N/mm²	1000 lbf/in²	N/mm²	lbf/in²
Green	113	16 400	14 600	2120	57·4	8320
12 per cent	145	21 000	16 100	2340	79·5	11 530

(Data from Forest Research Institute, Dehra Dun, India)

movement: Medium.
        Moisture content in 90 per cent relative humidity 19·5 per cent
        Moisture content in 60 per cent relative humidity 13·5 per cent
        Corresponding tangential movement 2·1 per cent ($\frac{1}{4}$ in/ft)
        Corresponding radial movement 1·5 per cent ($\frac{3}{16}$ in/ft)

processing

drying: Not unduly refractory but may have a tendency to surface check and split and to distort. Air dries slowly, without degrade.
Kiln Schedule C.
No information on shrinkage during drying.

working properties:

Wavy and interlocked grain affects machining properties. Contains varying amounts of resin.

Blunting: Severe.

Sawing: Difficult to saw when green. Tendency to tooth vibration when dry.

Rip-sawing – Saw type HR 60 or TC.

Cross-cutting – Saw type 1 or 2.

Wide bandsawing – Saw type C.

Machining: General – Tends to ride on cutters, requiring increased pressure to hold the timber firmly. Requires support at tool exit when working end-grain, e.g. cross-cutting, drilling, etc.

Planing – A cutting angle of 20° required to prevent tearing.

Nailing: Unsuitable for nailing.

Gluing: No information.

wood bending: No information.

plywood manufacture: Unsuitable because of its weight.

staining and polishing:

Excellent except when resin content is exceptionally high.

durability and preservation

insect attack: Trees and logs liable to attack by forest longhorn or Buprestid beetles. Reported to be moderately resistant to termites in India and Malaya.

durability of heartwood: Very durable.

preservative treatment:

Extremely resistant. Sapwood probably moderately resistant.

uses

Suitable for heavy, structural work, especially in contact with the ground or water, as in piles, bridge girders, decking and dock work. As a flooring timber it has high resistance to abrasion and is comparable with maple, and makes a decorative floor suitable for public buildings. Flat-sawn stock wears more smoothly than rift-sawn, which may have zones of interlocked grain.

QUARUBA
Vochysia spp.

other names: quaruba branca, quaruba vermelha (Brazil); yemeri, *V. hondurensis* (British Honduras); iteballi, *V. surinamensis* and *V. tetraphylla* (Guyana); kwari, *V. guianensis* and other species (Surinam).

THE TREE

Very variable in size, reaching a height of 27–40 m (90–130 ft) in Brazil and Guyana and sometimes more in British Honduras. Bole straight and cylindrical, unbuttressed, diameter commonly 0·6–1·0 m (2–3 ft), clear of branches for 15–18 m (50–60 ft). Tropical America.

THE TIMBER properties

The timber from Brazil (quaruba) is rather more variable in all respects than that from British Honduras (yemeri). Heartwood pale pinkish-brown. Sapwood wide, lighter in colour than heartwood but not always sharply demarcated from it. Grain straight or slightly interlocked; texture moderately coarse, uniform and inclined to be woolly and fibrous. Average weight about 480–510 kg/m³ (30–32 lb/ft³), seasoned. The green timber is reported to have an exceptionally high moisture content. Strength properties somewhat lower than European beech, but rather high in relation to its weight. Yemeri from British Honduras appears to be rather lower in strength than quaruba from Brazil.

processing

Dries fairly rapidly but with a marked tendency to distort, particularly in the form of twist and cup. Some collapse may occur in thicker stock. Kiln Schedule A. Has a moderate blunting effect on cutting edges. Satisfactory in sawing. Machining generally satisfactory, but dull cutters cause grain raising and a fibrous finish is obtained on end-grain in most operations. Takes nails well. Glues well and is stated to be suitable for plywood manufacture. Stains and polishes satisfactorily, but water stains should not be used as they tend to raise the grain.

durability and preservation

Sapwood liable to attack by powder-post beetles. Reported to be non-resistant to dry-wood termites in the West Indies. Brazilian quaruba (quaruba branca and quaruba vermelha) is moderately durable and probably resistant to preservative treatment. Yemeri is non-durable and permeable to preservatives.

uses

A timber of medium weight, but difficult to dry without degrade. Could be used for lightweight structural work where its coarse texture and woolly finish are not objectionable.

'QUEENSLAND MAPLE'
Flindersia brayleyana and *F. pimenteliana*

other names: silkwood, maple silkwood (Australia); 'Australian maple' (Great Britain).

THE TREE

Grows to a height of up to 30 m (100 ft), diameter of bole 0·9–1·2 m (3–4 ft). Northern Queensland.

THE TIMBER properties

An attractive wood, unrelated to and unlike true maple. Heartwood brownish-pink with a silky lustre, toning down to pale brown on exposure. Grain often interlocked and may also be wavy or curly, producing a wide range of figure. Texture medium and uniform. Average weight about 550 kg/m³ (34 lb/ft³), seasoned. Strength properties high for its weight, but somewhat lower than European beech.

processing

Stated to air dry and kiln dry satisfactorily, though with some tendency to collapse, and distortion is sometimes troublesome especially in dense stock with interlocked grain. Kiln Schedule C. Has a moderate blunting effect on cutting edges. Satisfactory in sawing and machining, but a cutting angle of 20° is recommended in planing, particularly with quarter-sawn material. Nails satisfactorily and glues well. Reported to be suitable for plywood manufacture and can be sliced to give a decorative veneer. Polishes well.

durability and preservation

Sapwood not liable to attack by powder-post beetles. Heartwood moderately durable (provisional).

uses

An attractive wood, sometimes highly figured, used in Australia for cabinet work and interior joinery. In the United Kingdom used mainly in the form of decorative veneer.

'QUEENSLAND WALNUT'
Endiandra palmerstonii

other names: walnut bean (Australia); oriental wood (United States).

THE TREE

Reaches a height of 37–43 m (120–140 ft). The tree is buttressed. Bole above buttresses usually well shaped and unbranched for 25 m (80 ft), diameter up to 1·8 m (6 ft). Northern Queensland.

THE TIMBER properties

colour: Bears a general resemblance to plain European walnut but stripes are more regular. Colour varies from pale to dark brown, often with pinkish, greyish-green or blackish streaks. More lustrous than European walnut.

grain:	Generally interlocked and frequently rather wavy, giving a chequered or broken stripe on the quarter. Texture medium and even.
odour:	When freshly cut has a disagreeable odour, which largely disappears when the wood is dried.
weight:	Varies from 600 to 770 kg/m³ (37–48 lb/ft³), average 680 kg/m³ (42 lb/ft³), seasoned.
silica:	The timber commonly contains deposits of silica which occur as crystalline aggregates in the ray cells.
strength:	Somewhat lower than European beech. No precise data available.
movement:	No information.

processing

drying: Kiln dries fairly rapidly in thinner sizes without checking but with some tendency to warp, and slight collapse may occur. It may be necessary to apply end coating to long boards to prevent end-splitting. Thicker material is liable to split unless quarter-sawn.

Kiln Schedule E.

Shrinkage: Green to 12 per cent moisture content:
Tangential about 6·5 per cent ($\frac{13}{16}$ in/ft)
Radial about 4·0 per cent ($\frac{1}{2}$ in/ft)

working properties: Grain occasionally irregular and interlocked.

Blunting: Severe due to presence of silica.

Sawing: Rip-sawing – Saw type TC.
Cross-cutting – Tungsten carbide-tipped saws recommended.
Narrow bandsawing – Increased tooth pitch necessary.
Wide bandsawing – Saw type B with increased tooth pitch and heavy swage.

Machining: Planing – Best quality high-speed steel cutters or tungsten carbide-tipped cutters required, with a cutting angle of 20° and the largest sharpness angle possible while still giving clearance.
Other operations – Satisfactory provided cutting edges are strong and are kept sharp.

Gluing: Satisfactory.

wood bending:

Suitable only for bends of comparatively large radius of curvature; liable to buckle if bent to smaller radii.

Classification – Moderate.

Ratio radius/thickness for solid bends (steamed):
Supported: 20·0   Unsupported: 44·0

Limiting radius for 3·2 mm ($\frac{1}{8}$ in) laminae (unsteamed): 188 mm ($7\frac{2}{8}$ in).

plywood manufacture:

No information on plywood manufacture but used as decorative veneer.

staining and polishing: Good.

durability and preservation

insect attack: Sapwood not liable to attack by powder-post beetles.

durability of heartwood: Non-durable.

preservative treatment: No information.

uses

Used in high-grade furniture, shopfitting, interior decorative work and panelling. Forms an attractive decorative veneer. As a flooring timber it is moderately resistant to wear and is suitable for pedestrian traffic.

RAMIN
Gonystylus spp.,
principally *G. bancanus*

other name: melawis (Malaya).

THE TREE

A tall tree with a straight, cylindrical bole, unbuttressed but sometimes slightly fluted at base. Average diameter of bole 0·6 m (2 ft), sometimes up to 1·1 m (3½ ft). Free from branches for 15 to 18 m (50–60 ft). South-east Asia, especially Sarawak.

THE TIMBER properties

colour: White to pale straw-coloured, without any outstanding features. Freshly sawn timber is liable to stain but this can be avoided by anti-stain dipping treatment.

sapwood: Similar in colour to heartwood and not readily distinguished from it.

grain: Straight or shallowly interlocked. Texture moderately fine and even.

weight: About 640 to 720 kg/m³ (40–45 lb/ft³), average about 660 kg/m³ (41 lb/ft³), seasoned.

odour: Under certain conditions timber may have an unpleasant smell when freshly cut. This does not generally persist after drying but may become noticeable again if the timber is rewetted.

irritant properties: It is reported that workers handling logs develop skin irritation, probably due to penetration of the skin by sharp-pointed bark fibres remaining on the log. Washing the hands with soap and water is apparently an effective remedy.

strength: Generally slightly higher than European beech, but slightly less tough and hard and weaker in shear and in resistance to splitting.

Moisture Content	Bending Strength		Modulus of Elasticity		Compression parallel to grain	
	N/mm²	lbf/in²	N/mm²	1000 lbf/in²	N/mm²	lbf/in²
Green	71	10 300	10 100	1470	38·7	5620
12 per cent	134	19 400	14 000	2030	72·4	10 500

movement: Large
Moisture content in 90 per cent relative humidity   20 per cent
Moisture content in 60 per cent relative humidity   12 per cent
Corresponding tangential movement   3·1 per cent (3/8 in/ft)
Corresponding radial movement   1·5 per cent (3/16 in/ft)

processing

drying: Dries readily with little distortion but with a tendency to end-splitting and surface checking. End-splitting may become serious when drying timber in thicker sizes. When kiln drying timber greater than 38 mm (1½ in) in thickness a relative humidity 10 per cent higher than that stated in the Schedule (instead of the normal 5 per cent relative humidity increase) should be maintained during the early stages of drying. An initial high temperature, high humidity treatment (i.e. 3 hours at 70°C (160°F), 100 per cent relative humidity) is advisable in some instances to prevent discoloration due to mould growth. A strong unpleasant odour may be evolved during kiln drying.
Kiln Schedule C.
Shrinkage: Green to 12 per cent moisture content:
Tangential   about 5·0 per cent (5/8 in/ft)
Radial   about 2·5 per cent (5/16 in/ft)

working properties: Grain slightly interlocked.
Blunting: Moderate.
Sawing: Rip-sawing – Saw type HR 54.
Cross-cutting – Satisfactory.
Narrow bandsawing – Satisfactory.
Wide bandsawing – Saw type B.
Machining: Planing – A cutting angle of 20° recommended to avoid tearing.
Boring – Straight fluted drills most suitable.
General – Requires support at tool exit.
Nailing: Marked tendency to split.
Gluing: Good.

wood bending: Cannot be bent appreciably without buckling.
Classification – Very poor.
Limiting radius for 3·2 mm (⅛ in) laminae (unsteamed): 230 mm (9·0 in).

178

plywood manufacture: Shown by tests to be suitable.
Movement: 4·5 mm plywood from 30 per cent to 90 per cent relative humidity – 0·13 per cent.
Surface splitting on exposure to weather – Grade II.

staining and polishing: Satisfactory with a small amount of filler.

durability and preservation

insect attack: Sapwood liable to attack by powder-post beetles. Reported to be attacked by dry-wood termites in Borneo.

durability of heartwood: Perishable.

preservative treatment: Permeable.

uses

A plain, pale-coloured wood, slightly lighter in weight than beech. A good utility timber used in furniture, interior joinery, mouldings, small handles, wooden toys, and other domestic articles.

RAULI
Nothofagus procera

THE TREE

Reaches a maximum height of 40 m (130 ft), and average diameter of 0·8 m (2½ ft), with a clear bole of length about 18 m (60 ft). Chile; was once very abundant but stands have been depleted by heavy felling.

THE TIMBER
properties

Resembles a very mild beech, but distinctly red in colour and lacks the very conspicuous rays of the true beeches (*Fagus* spp.).

colour: Uniform, reddish-brown to cherry red.

grain: Straight. Texture fine and uniform.

weight: Average about 540 kg/m³ (34 lb/ft³), seasoned.

strength: Somewhat lower than European beech.

Moisture Content	Bending Strength		Modulus of Elasticity		Compression parallel to grain	
	N/mm²	lbf/in²	N/mm²	1000 lbf/in²	N/mm²	lbf/in²
Green	–	–	–	–	–	–
12 per cent	92	13 300	9200	1330	49·7	7210

movement: Probably small.
Moisture content in 90 per cent relative humidity   19 per cent
Moisture content in 60 per cent relative humidity   12 per cent
Corresponding radial movement   1·0 per cent (⅛ in/ft)

processing

drying: Appears to dry rather slowly but well and with little degrade.
Kiln Schedule E.
Shrinkage: Green to 12 per cent moisture content:
Tangential   about 4·5 per cent ($\frac{9}{16}$ in/ft)
Radial   about 2·5 per cent ($\frac{5}{16}$ in/ft)

working properties:
Blunting: Slight.
Sawing: Rip-sawing – Saw type HR 54.
Cross-cutting – Saw types 1, 2 and 3.
Narrow bandsawing – Satisfactory.
Wide bandsawing – Saw type A.
Machining: Satisfactory. Finishes cleanly and is comparable in working properties with a mild grade of European beech.
Gluing: Good.

wood bending:
Suitable only for manufacture of solid bends of moderate radius of curvature and not suitable for most types of bends used in making furniture. Cannot be bent if pin knots are present. No advantage appears to be gained by using a supporting strap.
Classification — Moderate.
Ratio radius/thickness for solid bends (steamed):
Supported: 16·5   Unsupported: 16·5
Limiting radius for 3·2 mm ($\frac{1}{8}$ in) laminae (unsteamed): 191 mm ($7\frac{1}{2}$ in).

plywood manufacture:
Employed in plywood but seldom seen in the United Kingdom.

staining and polishing: Good.

durability and preservation

insect attack: Sapwood liable to attack by powder-post beetles.

durability of heartwood: Durable (provisional).

preservative treatment: Probably moderately resistant.

uses
Widely used in Chile for furniture, cabinet work, doors, window frames, flooring and other purposes. Has been used satisfactorily in the United Kingdom as a substitute for beech, but is lighter in weight and should be regarded as an alternative to beech only where strength properties, including hardness, are unimportant.

RHODESIAN COPALWOOD
Guibourtia coleosperma

other names: muxibe, mussive, musibi (Angola).

A medium-sized tree, height about 15–18 m (50 to 60 ft), bole 8–12 m (25–40 ft) long, diameter 0·5–0·6 m ($1\frac{1}{2}$–2 ft). Southern Central Africa.

Heartwood pinkish-brown when freshly sawn, with darker, almost purple veins or stripes. Darkens on exposure to a rich mahogany red-brown and veining becomes much less conspicuous. Less highly figured than bubinga. Sapwood yellowish-white, distinct from heartwood. Grain typically interlocked, texture moderately fine. Weight about 800 kg/m³ (50 lb/ft³), seasoned. No information on drying. Reported to saw without difficulty. Somewhat hard to work but machines to a fine finish and polishes well. Reported to be moderately durable. Extremely resistant to preservative treatment.

A timber of good appearance and wearing properties. Makes an attractive and hard-wearing floor. Used in Africa for furniture and panelling. Somewhat heavy for furniture in Britain but would be suitable for shop and bank fittings and for turnery.

'RHODESIAN TEAK'
Baikiaea plurijuga

other names: Zambesi redwood, umgusi, mukushi, mukusi (Rhodesia).

THE TREE
A much-branched tree, growing to a height of 15 to 18 m (50–60 ft). Clear bole varies in length from 3·0 to 4·5 m (10–15 ft), diameter about 0·8 m ($2\frac{1}{2}$ ft). Rhodesia and Zambia.

THE TIMBER properties

colour: A handsome reddish-brown wood, sometimes marked with irregular black lines or flecks.

sapwood: Pale in colour, sharply demarcated from heartwood.

grain: Straight or slightly interlocked. Texture fine and even, giving a smooth surface.

chemical staining:

Liable to stain in contact with iron or iron compounds under damp conditions owing to the presence of tannin in the wood.

weight: Average about 900 kg/m³ (56 lb/ft³), seasoned.

strength: A heavy, hard timber having high resistance to abrasion. No precise data on strength properties available.

movement: Small.

Moisture content in 90 per cent relative humidity   18 per cent
Moisture content in 60 per cent relative humidity   11·5 per cent
Corresponding tangential movement   1·6 per cent ($\frac{3}{16}$ in/ft)
Corresponding radial movement   1·0 per cent ($\frac{1}{8}$ in/ft)

processing

drying: Dries well but slowly. Distortion unlikely to be appreciable, and degrade from extension of initial shakes or from splitting of knots not generally serious.
Kiln Schedule D.
Shrinkage: Green to 12 per cent moisture content:
Tangential   about 2·5 per cent ($\frac{5}{16}$ in/ft)
Radial   about 1·5 per cent ($\frac{3}{16}$ in/ft)

working properties:
Blunting: Severe.
Sawing: Considerable build-up of resin on saw teeth, particularly when timber is green.
Rip-sawing — Saw type HR 60 with increased set when green. Saw type TC when dry.
Cross-cutting — Saw type 1 and 2 or tungsten carbide tipped.
Narrow bandsawing — Increased set required.
Wide bandsawing — Stellite tipped.
Machining: Planing — A cutting angle of 20° recommended. Tendency to ride on cutters unless material is held firmly.
Turning — Excellent.
General — A tendency to char in many operations, and blunts cutting edges rapidly.
Nailing: Not suitable.
Gluing: Good.

wood bending:

Tendency to buckle during bending and cannot be bent successfully if small knots are present. Fractures of a brittle type are likely to occur and bending properties are considerably improved by supporting bends with a metal strap. Slight resin exudation accompanies steaming.
Classification — Moderate.
Ratio radius/thickness for solid bends (steamed):
Supported: 13   Unsupported: 25
Limiting radius for 3·2 mm ($\frac{1}{8}$ in) laminae (unsteamed): 165 mm ($6\frac{1}{2}$ in).

plywood manufacture: Unsuitable because of its weight.

staining and polishing: Good.

durability and preservation

insect attack: Logs liable to attack by forest longhorn or Buprestid beetles and sapwood liable to attack by powder-post beetles. Reported to be moderately resistant to termites in Rhodesia.

durability of heartwood: Very durable.

preservative treatment:

Extremely resistant. Sapwood probably moderately resistant.

uses   A hard, stable, durable wood, used mainly for flooring. Makes a highly decorative floor with high resistance to wear, suitable for all conditions of traffic including heavy-duty flooring.

ROBINIA
Robinia pseudoacacia

other names: false acacia (Great Britain); black locust (United States).

THE TREE

Grows to a height of 24–27 m (80–90 ft), though often less in Great Britain. Bole often twisted or fluted, diameter commonly about 0·6 m (2 ft), occasionally up to 0·9 m (3 ft). The tree is inclined to fork close to the ground giving a very short clean bole. It is liable to wind-break. Native of North America, introduced into Europe (including Britain), Asia, North Africa and New Zealand.

THE TIMBER
properties

On the Continent the wood is usually cut from young trees of rapid growth, often of coppice origin. Old trees are often rotten at the heart.

colour: Greenish when freshly cut, turning golden brown on exposure.

sapwood: Pale in colour, about 13 mm ($\frac{1}{2}$ in) wide, clearly demarcated from heartwood.

grain: Usually straight. Texture somewhat coarse owing to the contrast between the large-pored early wood and dense late wood.

weight: Variable, from about 540 to 860 kg/m³ (34–54 lb/ft³), average about 720 kg/m³ (45 lb/ft³), seasoned.

strength: A heavy, hard timber having good strength properties and high toughness, comparable to that of ash.

movement: No information.

processing

drying: Dries slowly with a marked tendency to distort.
Kiln Schedule A.
No information on shrinkage during drying.

working properties:
Resistance to cutting varies considerably between the soft early wood and dense late wood.
Blunting: Moderate.
Sawing: Rip-sawing – Saw type HR 60.
Cross-cutting – Satisfactory.
Narrow bandsawing – Satisfactory.
Wide bandsawing – Saw type B.
Machining: Satisfactory.
Nailing: Difficult.
Gluing: Good.

wood bending:
Has very good bending properties, equal to those of timbers such as beech and ash. Not prone to fail at knots, and bends equally well in either the green or air-dried state. The steamed material stains if allowed to come into contact with iron or steel.
Classification – Very good.
Ratio radius/thickness for solid bends (steamed):
Supported: 1·5   Unsupported: 11
Limiting radius for thin laminae (unsteamed): No data.

plywood manufacture: No information.

staining and polishing: Satisfactory.

durability
and preservation

insect attack: Sapwood liable to attack by powder-post beetles and by the common furniture beetle. Reported to be highly resistant to termites in Central America.

durability of heartwood: Durable.

preservative treatment: Extremely resistant.

uses

A hard, heavy, tough timber with good bending properties and good durability. Used for stakes, posts, gates and other purposes where the timber is exposed to the weather. More widely used on the Continent than in Great Britain.

ROSEWOOD, BRAZILIAN
principally *Dalbergia nigra*

other names: Rio rosewood, Bahia rosewood (United Kingdom).

THE TREE

Grows to a height of 38 m (125 ft), but has a short, irregular bole, often buttressed, diameter about 1·0 m (3–4 ft). Old trunks are often hollow. Logs after removal of sapwood are rarely more than 0·5 m (1½ ft) in diameter. Heartwood of young trees is brown and not attractive. High-quality timber with rich purplish-black markings is only obtained from old, often defective stems.

THE TIMBER
properties

Heartwood varies from various shades of brown to chocolate or violet, irregularly streaked with black. Sapwood pale in colour, sharply but irregularly demarcated from heartwood. Grain generally straight, sometimes wavy. Texture rather coarse. Weight varies from about 750 to 900 kg/m³ (47–56 lb/ft³), average about 850 kg/m³ (53 lb/ft³), seasoned. No information on strength properties.

processing

Not unduly difficult to work, finishing very smoothly. Some specimens are too oily to take a high polish. Shown by tests to have very good bending properties if straight-grained and free from knots.

> Ratio radius/thickness for solid bends (steamed):
> Supported: 4·5   Unsupported: 17·9
> Limiting radius for 3·2 mm (⅛ in) laminae (unsteamed): 158 mm (6⅛ in).

durability and preservation

Heartwood reported to be very durable.

uses

Used in furniture manufacture, cabinet making and piano cases, and for handles for certain tools and instruments. Forms an attractive decorative veneer.

The South American timber *Machaerium scleroxylon* resembles Brazilian rosewood in appearance and has been used as a substitute for it. It should be noted, however, that *M. scleroxylon* possesses irritant properties and the dust produced in machining operations is liable to cause severe skin irritation among men working with it.

ROSEWOOD, HONDURAS
Dalbergia stevensonii

THE TREE

A medium-sized tree, height 15–30 m (50–100 ft), diameter about 1·0 m (3 ft). Bole often fluted and forks at about 6–8 m (20–25 ft) from the ground. British Honduras.

THE TIMBER
properties

Heartwood pinkish or purplish-brown with irregular black markings, giving the wood a varied and attractive figure. Sapwood pale in colour, quickly turning yellow, width 25–50 mm (1–2 in). Grain generally straight, texture medium to rather fine. One of the heaviest rosewoods, weight about 930–1100 kg/m³ (58–68 lb/ft³), seasoned. Has good strength properties and is denser and tougher than Brazilian rosewood, but is used mainly for purposes where strength properties are of minor importance.

processing

Reported to air dry slowly with a marked tendency to check. Kiln Schedule C. Has a moderate blunting effect on cutting edges. Satisfactory in sawing. In planing a cutting angle of 20° is advantageous when interlocked or wavy grain is present. Timber tends to ride over cutters. Excellent for turning and finishes well.

durability and preservation

Reported to be moderately resistant to termites in Honduras. Heartwood very durable (provisional).

uses Harder and heavier than Indian rosewood and rather more difficult to work. Used for the handles of knives and small tools and for parts of musical instruments; especially suitable for xylophone keys.

ROSEWOOD, INDIAN
Dalbergia latifolia

other names: Bombay blackwood (India); East Indian rosewood (Great Britain).

THE TREE Varies considerably in size according to locality. Under good conditions reaches a height of 25 m (80 ft). Bole fairly straight, clean and cylindrical, average length about 6 m (20 ft), sometimes up to 15 m (50 ft). Maximum diameter of bole about 1·5 m (5 ft), more commonly 0·8 m ($2\frac{1}{2}$ ft) and in some regions only 0·3 m (1 ft). India.

THE TIMBER
properties

colour: A handsome medium to dark purplish-brown wood with darker streaks terminating the growth zones and giving an attractive figure on plain-sawn surfaces.

sapwood: Pale yellowish-white, narrow.

grain: Narrowly interlocked, producing an inconspicuous ribbon grain figure. Texture uniform and moderately coarse.

weight: Average about 850 kg/m³ (53 lb/ft³) at 12 per cent moisture content.

strength: Slightly higher than European beech and particularly hard for its weight.

Moisture Content	Bending Strength		Modulus of Elasticity		Compression parallel to grain	
	N/mm²	lbf/in²	N/mm²	1000 lbf/in²	N/mm²	lbf/in²
Green	67	9700	7700	1110	32·4	4700
12 per cent	121	17 500	11 400	1660	65·2	9450

(Data from Forest Research Institute, Dehra Dun, India)

movement: Small.
Moisture content in 90 per cent relative humidity   13·5 per cent
Moisture content in 60 per cent relative humidity   9·5 per cent
Corresponding tangential movement   1·0 per cent ($\frac{1}{8}$ in/ft)
Corresponding radial movement   0·7 per cent ($\frac{3}{32}$ in/ft)

processing Green conversion appears to be satisfactory and the heart, which usually contains shakes, should be boxed.

drying: Air dries fairly rapidly with no appreciable degrade, but requires protection against too rapid drying. Characteristic defects are slight surface checking and end-splitting, and extension of any shakes that may be present.
Kiln dries well but rather slowly and the colour is reported to improve during this process.
Kiln Schedule E.
No information on shrinkage during drying.

working properties:
Blunting: Moderate.
Sawing: Rip-sawing – Saw type HR 60.
Cross-cutting – Saw type 1 or 2.
Narrow bandsawing – Satisfactory.
Wide bandsawing – Saw type C.
Machining: Planing – A cutting angle of 25° most satisfactory and gives a smooth surface.
Other operations – Satisfactory.
Nailing: Not suitable.
Gluing: Good.

wood bending: No information.

plywood manufacture:
Generally unsuitable because of its weight, but reported to yield a commercial plywood in India. Valued as a decorative veneer.

staining and polishing: Satisfactory when filled.

durability and preservation

insect attack: Sapwood liable to attack by powder-post beetles. Reported to be moderately resistant to termites in India.

durability of heartwood: Very durable.

preservative treatment: No information.

uses

A decorative wood used for high-class furniture and cabinet work, for which it is highly valued. Often used in the form of veneer. Used also for parts of musical instruments, including guitars, and small turned articles.

SAPELE
Entandrophragma cylindricum

other names: aboudikro (France and Ivory Coast); sapelli (France and Cameroon).

THE TREE

Height generally about 45 m (150 ft), but may reach 60 m (200 ft). Broad, low buttresses, or sometimes free from buttresses. Bole straight and cylindrical, clear for 30 m (100 ft) or more. West Africa.

THE TIMBER
properties

colour: A medium to fairly dark reddish-brown wood, typically with a well-marked stripe or roe figure which shows to advantage on accurately quarter-cut veneers. Occasional logs with wavy grain yield veneers with a highly decorative fiddle-back figure.

sapwood: Whitish or pale yellow, 75–100 mm (3–4 in) wide.

grain: Interlocked and sometimes wavy. Texture fairly fine. The timber is liable to ring or cup shakes.

odour: Has a pronounced cedar-like scent when freshly cut, which gradually diminishes.

weight: About 560–690 kg/m³ (35–43 lb/ft³), average about 620 kg/m³ (39 lb/ft³), seasoned.

strength: Comparable to European beech.

Moisture Content	Bending Strength		Modulus of Elasticity		Compression parallel to grain	
	N/mm^2	lbf/in^2	N/mm^2	$1000\,lbf/in^2$	N/mm^2	lbf/in^2
Green	74	10 700	9600	1390	36·0	5220
12 per cent	111	16 100	11 700	1700	58·6	8500

movement: Medium.
Moisture content in 90 per cent relative humidity   20·5 per cent
Moisture content in 60 per cent relative humidity   13·5 per cent
Corresponding tangential movement   1·8 per cent ($\frac{7}{32}$ in/ft)
Corresponding radial movement   1·3 per cent ($\frac{5}{32}$ in/ft)

processing	**drying:**	Dries fairly rapidly but with a marked tendency to distort. Kiln Schedule A.

drying: Dries fairly rapidly but with a marked tendency to distort.
Kiln Schedule A.
Shrinkage: Green to 12 per cent moisture content:
Tangential   about 4·5 per cent ($\frac{9}{16}$ in/ft)
Radial   about 2·5 per cent ($\frac{5}{16}$ in/ft)

working properties: Grain interlocked, affecting machining properties.
Blunting: Moderate.
Sawing: Rip-sawing — Saw type HR 40 or HR 54.
Cross-cutting — Saw type 1 or 2 satisfactory. Type 3 tends to char.
Narrow bandsawing — Satisfactory.
Wide bandsawing — Saw type B.
Machining: Planing — Optimum cutting angle 15° on account of interlocked grain. Tendency to chip bruising.
Other operations — Satisfactory.
Nailing: Good.
Gluing: Good.

wood bending:
Buckles and ruptures severely when bent even to a comparatively large radius of curvature.
Classification — Poor.
Limiting radius for 3·2 mm ($\frac{1}{8}$ in) laminae (unsteamed): 160 mm ($6\frac{1}{4}$ in).

plywood manufacture:
Employed for plywood and usually available in the United Kingdom.
Movement: 1·5 mm plywood from 30 per cent to 90 per cent relative humidity — 0·33 per cent.
Surface splitting on exposure to weather — Grade I.

staining and polishing:
Care required when staining. Gives an excellent polished finish.

durability and preservation

insect attack: Sapwood liable to attack by powder-post beetles. Reported to be moderately resistant to termites in West Africa.

durability of heartwood: Moderately durable.

preservative treatment: Resistant. Sapwood moderately resistant.

uses A timber of the mahogany type, widely used for furniture, plywood, joinery, shop-fittings, etc. As flooring it has moderate to high wearing qualities and makes a good decorative floor for domestic and public buildings. The striped figure is valued for decorative veneer.

SATINWOOD, CEYLON
Chloroxylon swietenia

other name: East Indian satinwood (Great Britain).

THE TREE A medium-sized tree, usually about 14–15 m (45–50 ft) in height. Bole straight and cylindrical, about 3 m (10 ft) long, diameter usually about 0·3 m (1 ft), or sometimes more. Central and Southern India and Ceylon.

THE TIMBER properties Heartwood light yellow or golden yellow, darkening in time to a soft brown. Remarkably lustrous. Sapwood not clearly distinct from heartwood, but the outer wood is paler in colour than the inner wood. Grain interlocked, producing a narrow ribbon-grain figure, often broken or variously mottled. Dark gum veins are sometimes present and are liable to develop into splits. Texture fine and even. Average weight about 980 kg/m³ (61 lb/ft³), seasoned. A dense, hard timber, somewhat higher in strength than European beech, but used mainly for decorative purposes where its strength properties are relatively unimportant.

186

processing In drying, reported to have a tendency to surface checking with some distortion. Surface checking is least in material from girdled trees and seasoning in the log is said to give good results. Satisfactory air drying results if the timber is carefully stacked and protected against too rapid drying. Kiln dries well, with little degrade. Kiln Schedule C. Has a moderate blunting effect on cutting edges. In rip-sawing saw type HR 80 of stout gauge should be used. In planing a cutting angle of 15° is required to avoid tearing, and increased pressure is necessary to prevent riding over cutters. Turns well. Difficult to glue. Unsuitable for plywood manufacture because of its weight, but used as a decorative veneer. Stains and polishes well with a little filler.

durability and preservation Reported to be non-resistant to termites in India. Heartwood provisionally durable, and extremely resistant to preservative treatment.

uses A decorative timber suitable for furniture, cabinet work and interior joinery. Turns well and is made into fancy goods. It is principally used as a decorative veneer.

SATINWOOD, WEST INDIAN
Fagara flava

other names: Jamaican satinwood (Great Britain); San Domingan satinwood (United States).

Grows to a maximum height of 12 m (40 ft), diameter 0·5 m (1½ ft). West Indies, Bermuda, the Bahamas and southern Florida. Reaches its best development in Jamaica.

The timber has a lustrous, creamy or golden yellow colour, darkening with exposure. Grain interlocked or irregular, often with a roe or mottle figure. The characteristic scent of coconut oil is very distinct when the wood is freshly worked. Average weight about 880 kg/m³ (55 lb/ft³), seasoned. Texture fine and even.

Has a moderate blunting effect on cutting edges and is slightly less hard and resistant to cutting than Ceylon satinwood. Has a tendency to ride over cutters unless higher loads and pressures are used. In planing a cutting angle of 20° is recommended where irregular grain is present. Turns excellently and finishes well in other operations. It is reported that the fine dust produced in machining operations is liable to cause dermatitis among men working with the timber. The timber is non-durable.

Suitable for high-class cabinet work, inlays, fancy goods, furniture and interior decorative work, and for turnery.

SEPETIR
Sindora spp.

other names: petir (Sarawak); makata (Thailand); gu (Indo-China); supa (Philippines).

The timber of *Pseudosindora palustris*, derived principally from Sarawak, is also known as sepetir. It is described separately under the name swamp sepetir.

THE TREE Several species of *Sindora* growing in south-east Asia produce timber known as sepetir. The trees generally have clean cylindrical boles, free from buttresses diameter about 0·9–1·2 m (3–4 ft).

THE TIMBER
properties Generally similar to swamp sepetir but has a more decorative appearance. Heartwood golden brown, darkening on exposure. Dark brown or black streaks are sometimes present, producing handsomely figured wood. Sapwood light greyish-brown, sometimes with a pinkish tinge, very variable in width, 75 to 300 mm (3–12 in). Grain sometimes variable in direction. Texture moderately fine and even.

Weight generally about 640 to 720 kg/m³ (40–45 lb/ft³) at 12 per cent moisture content. The timber has a characteristic spicy smell.

processing

Drying appears to be very variable. Rather hard to work, but gives a good finish in most operations. Liable to split when nailed. Stains and polishes well.

durability and preservation

Sapwood liable to attack by powder-post beetles. Reported to be attacked by dry-wood termites in Malaya. Heartwood is durable, and extremely resistant to preservative treatment. The sapwood is permeable.

uses

Similar to swamp sepetir; suitable for light construction and joinery, but sapwood is very susceptible to insect attack. Some material has an attractive appearance and may be suitable for furniture manufacture. Figured logs yield a handsome veneer, suitable for panelling or furniture.

SEPETIR, SWAMP
Pseudosindora palustris

other names: sepetir (Sarawak and Sabah); petir, sepetir paya (Sarawak). Sepetir is the British standard name for *Pseudosindora palustris*, found in Sarawak and Sabah, and for several species of *Sindora* occurring in south-east Asia. *P. palustris* is the chief source of sepetir in Sarawak and is here described separately under the name 'swamp sepetir'.

THE TREE

A tall tree with a straight, cylindrical bole, unbuttressed, diameter up to 1·2 m (4 ft), but usually much smaller, about 0·6 m (2 ft). Sarawak and Sabah.

THE TIMBER properties

colour: Pale pink with pale brown veining when fresh, darkening on exposure to a rich reddish-brown with slightly darker veining. It is redder and plainer than sepetir from Malaya (*Sindora* spp.).

sapwood: Straw-coloured at first, turning pink, commonly about 100 mm (4 in) wide.

grain: Straight or shallowly interlocked. Texture moderately fine and even. The wood sometimes has a slightly greasy feel.

weight: Varies from 590 to 820 kg/m³ (37–51 lb/ft³), average about 670 kg/m³ (42 lb/ft³) at 12 per cent moisture content.

strength: Slightly higher than European beech.

Moisture Content	Bending Strength		Modulus of Elasticity		Compression parallel to grain	
	N/mm²	lbf/in²	N/mm²	1000 lbf/in²	N/mm²	lbf/in²
Green	81	11 700	10 100	1470	39·2	5690
12 per cent	125	18 100	12 700	1840	63·8	9250

movement: Small.
Moisture content in 90 per cent relative humidity 17·0 per cent
Moisture content in 60 per cent relative humidity 12·5 per cent
Corresponding tangential movement 1·2 per cent ($\frac{9}{64}$ in/ft)
Corresponding radial movement 0·9 per cent ($\frac{7}{64}$ in/ft)

processing

drying: Dries rather slowly but very well. Tendency for end-splitting to develop but distortion in all forms is small.
Kiln Schedule G.
Shrinkage: Green to 12 per cent moisture content:
Tangential about 3·0 per cent ($\frac{3}{8}$ in/ft)
Radial about 2·0 per cent ($\frac{1}{4}$ in/ft)

working properties:

Grain slightly interlocked affecting machining properties.

Blunting: Moderate.

Sawing: Resin builds up on blades.

Rip-sawing – Saw type HR54.

Cross-cutting – Saw types 1 and 2.

Narrow bandsawing – Satisfactory.

Wide bandsawing – Saw type B.

Machining: Timber requires careful support where breaking through on end-grain, i.e. in cross-cutting, boring and mortising.

In boring and hollow square chisel mortising chips do not clear well and charring may occur.

Nailing: Pronounced tendency to split.

Gluing: No information.

wood bending:

Liable to buckle during bending, and resin exudation accompanies steaming. Fractures of a brittle type are likely to occur and bending properties are improved by use of a supporting strap.

Classification – Moderate.

Ratio radius/thickness for solid bends (steamed):

Supported: 18   Unsupported: 37

Limiting radius for 3·2 mm ($\frac{1}{8}$ in) laminae (unsteamed): 180 mm (7 in).

plywood manufacture:

Employed in plywood but seldom seen in the United Kingdom.

Movement of plywood – No information.

Surface splitting on exposure to weather – Grade I.

staining and polishing: Satisfactory.

durability and preservation

insect attack: No information.

durability of heartwood: Durable.

preservative treatment:

Heartwood extremely resistant. Sapwood moderately resistant.

uses

A durable and stable timber which works fairly readily and takes a good polish. Suitable for general joinery and carpentry.

SERAYA, WHITE
Parashorea spp.

other name: bagtikan (Philippines), also included with white lauan.

White seraya is produced by two species of *Parashorea* and is available commercially only from Sabah and the Philippine Islands.

NOTE: White seraya is *not* the same as Malayan white meranti.

THE TREE

A very large tree, 45–60 m (150–200 ft) in height. Straight, cylindrical bole, up to 30 m (100 ft) long, diameter 1·0–1·5 m (3–5 ft) or more above large buttresses.

THE TIMBER properties

Tends to be more uniform in character than other timbers of this group, being the product of only two species.

colour: Straw-coloured or very pale brown, sometimes with a pinkish tint.

sapwood: Paler in colour and sometimes greyish, not always clearly defined, usually 65–75 mm (2$\frac{1}{2}$–3 in) wide.

grain: Shallowly interlocked, producing a broad stripe figure on quarter-cut surfaces. Texture moderately coarse.

brittleheart: Frequently present near the centre of logs.

weight: Average about 530 kg/m³ (33 lb/ft³), seasoned.

strength: About half-way between obeche and European beech, and rather higher than light red meranti.

Moisture Content	Bending Strength		Modulus of Elasticity		Compression parallel to grain	
	N/mm²	lbf/in²	N/mm²	1000 lbf/in²	N/mm²	lbf/in²
Green	60	8700	9100	1320	31·0	4500
12 per cent	80	11 600	10 100	1460	48·4	7020

movement: Small.
Moisture content in 90 per cent relative humidity 18·5 per cent
Moisture content in 60 per cent relative humidity 11·5–12·0 per cent
Corresponding tangential movement 1·6–2·3 per cent ($\frac{13}{64}-\frac{9}{32}$ in/ft)
Corresponding radial movement   0·8–1·1 per cent ($\frac{3}{32}-\frac{9}{64}$ in/ft)

processing

drying: Dries fairly rapidly and well with little degrade apart from an occasional tendency to cup.
Kiln Schedule J.

working properties:
Blunting: Slight.
Sawing: Generally satisfactory, but giving a woolly finish.
Rip-sawing – Saw type HR 54.
Cross-cutting – Satisfactory.
Narrow bandsawing – Satisfactory.
Wide bandsawing – Saw type B.
Machining: Machines easily, but due to fibrous nature requires sharp cutting edges. Finish tends to be woolly and rough on end-grain.
Planing – A cutting angle of 20° required to prevent tearing of quarter-sawn material.
Nailing: Good.
Gluing: Good.

wood bending:
Severe buckling occurs at comparatively large radius of curvature.
Classification – Very poor.
Limiting radius for 3·2 mm ($\frac{1}{8}$ in) laminae (unsteamed):
about 190 mm ($7\frac{1}{2}$ in).

plywood manufacture:
Employed for plywood and generally available in the United Kingdom.
Movement: 4·5 mm plywood from 30 per cent to 90 per cent relative humidity – 0·12 per cent.
Surface splitting on exposure to weather – Grade II.

staining and polishing: Satisfactory when filled.

durability and preservation

insect attack: Trees and logs liable to attack by forest longhorn or Buprestid beetles. Sapwood liable to attack by powder-post beetles.

durability of heartwood: Non-durable to moderately durable.

preservative treatment: Extremely resistant. Sapwood moderately resistant.

uses A lightweight, pale, easy-working, fairly plain and somewhat coarse-textured wood. Suitable for interior joinery, light construction work and flooring for domestic purposes. For most of its uses brittleheart should be excluded. Has been used for ships' decking, but is only recommended for decks under cover. Its most important use is in the manufacture of plywood.

SERRETTE
Byrsonima coriacea
var. *spicata*

THE TREE

Generally reaches a height of 30 to 36 m (100–120 ft), occasionally more. Diameter of bole 0·6 to 0·9 m (2–3 ft). Trees are liable to split badly when felled and the larger trees are sometimes attacked by termites. West Indies and Central America.

THE TIMBER
properties

A pale to dark reddish-brown timber, sometimes with a greyish tint and with grey or black markings. Sapwood wide, paler in colour than heartwood. Grain mostly straight or slightly interlocked. Quarter-sawn surfaces show a slight roe or stripe figure. Flat-sawn surfaces are of plain appearance. Texture moderately fine. Average weight about 740 kg/m³ (46 lb/ft³), seasoned. Slightly higher in strength than European beech. Medium movement.

processing

Dries rather slowly and quite well up to 25 mm (1 in) thickness. Slight checking may occur and distortion tends to be troublesome, the worst defect being cup. Thicker timber is considerably more difficult to dry. Kiln Schedule E. Has a moderate blunting effect on cutting edges. Satisfactory in sawing and machining but has a tendency to char in boring. In nailing care is necessary to prevent splitting. Wood bending properties very variable, and only recommended for bends of moderate radius of curvature. Shown by tests to be unsuitable for plywood manufacture.

durability
and preservation

Reported to be attacked by dry-wood termites in West Indies. Heartwood moderately durable. Heartwood and sapwood are both moderately resistant to impregnation.

uses

A heavy timber with good strength properties. Can be used for constructional work in situations where its durability is adequate. As a flooring timber it wears smoothly and should be suitable for light industrial types of floor.

SILKY-OAK, AUSTRALIAN
Cardwellia sublimis

other names: The name silky-oak is also sometimes applied to *Grevillea robusta*, see Grevillea, page 82.

A large tree, height up to 37 m (120 ft), with a straight bole, 1·2 m (4 ft) in diameter. North Queensland.

A decorative timber showing a well-marked silver-grain on quartered surfaces, not unlike that of true oak. Heartwood pinkish or reddish brown, similar to American red oak, turning browner with age. Grain generally straight, texture coarse and even. Narrow lines of gum ducts are occasionally present. Average weight about 530 kg/m³ (33 lb/ft³), seasoned.

Rate of drying stated to be very variable even within a charge. Has little tendency to check or collapse but severe cupping may occur in wide, plain-sawn boards dried at the top of a kiln charge. Kiln Schedule E. Saws and machines satisfactorily and has only a slight blunting effect on cutting edges. A cutting angle of 20° is recommended in planing. Nails well. Reported to have very good bending properties. Glues well and is employed in plywood manufacture but seldom seen in the United Kingdom. Sapwood is liable to attack by powder-post beetles. Heartwood moderately durable (provisional).

A decorative timber employed for furniture, joinery and panelling. As a flooring timber it possesses a high resistance to wear. Widely used in Australia for furniture, house panelling and office fittings.

'SILVER BEECH'
Nothofagus menziesii

other name: 'Southland beech'.

In addition to *Nothofagus menziesii*, two other New Zealand species of *Nothofagus*, *N. fusca* ('red beech') and *N. truncata* ('hard beech' or 'clinker beech') are of potential importance.

THE TREE   Grows to a height of about 30 m (100 ft). Trunk commonly buttressed, diameter 0·6–1·5 m (2–5 ft). New Zealand.

THE TIMBER
properties

colour: Uniform pinkish-brown or sometimes a pale salmon colour. Similar in appearance to the closely related rauli (*Nothofagus procera*) and bears a superficial resemblance to European beech but lacks the characteristic large rays of the latter.

sapwood: Consists of a narrow outer zone, paler in colour than the heartwood, and a zone intermediate in colour which for most practical purposes should be regarded as sapwood.

grain: Generally straight, sometimes curly. Texture fine and even.

weight: Rather variable. Timber from South Island, the source of most of the export timber, weighs about 530 kg/m³ (33 lb/ft³), seasoned. Timber from North Island is generally considerably harder and heavier. Red beech and hard beech are appreciably heavier than silver beech.

strength: Somewhat lower than European beech.

Source	Moisture Content	Bending Strength		Modulus of Elasticity		Compression parallel to grain	
		N/mm²	lbf/in²	N/mm²	1000 lbf/in²	N/mm²	lbf/in²
South Island	Green	55	8000	8300	1200	24·5	3560
(Wallace)	12 per cent	89	12 900	10 800	1560	43·6	6330
North Island	Green	66	9500	9200	1330	30·2	4380
(Rotorua)	12 per cent	98	14 200	10 500	1520	51·0	7390

(Data from New Zealand Forest Service)

movement: No information.

processing

drying: Thin material dries fairly easily but difficulty in drying increases rapidly with thickness. Some tendency for end-splitting to develop but distortion is comparatively slight.
Kiln Schedule E.
Shrinkage: From 30 per cent to 12 per cent moisture content:
Tangential   about 6·0 per cent ($\frac{3}{4}$ in/ft)
Radial   about 3·0 per cent ($\frac{3}{8}$ in/ft)

working properties:
Blunting: Moderate.
Sawing: Rip-sawing – Saw type HR 54.
             Cross-cutting – Satisfactory.
             Narrow bandsawing – Satisfactory.
             Wide bandsawing – Saw type B.
Machining: Satisfactory, but a cutting angle of 20° recommended in planing material with irregular grain.
Nailing: Satisfactory.
Gluing: No information.

wood bending:
> Limited tests indicate good bending properties. However, slight buckling is sometimes produced on bends of comparatively large radius.
> Classification – Good.
> Ratio radius/thickness for solid bends (steamed):
> > Supported: 10   Unsupported: 20
> Limiting radius for 3·2 mm ($\frac{1}{8}$ in) laminae (unsteamed): No information.

plywood manufacture: No information.

staining and polishing: Satisfactory.

durability and preservation

insect attack: Reported in New Zealand to be liable to attack by the common furniture beetle and by powder-post beetles.

durability of heartwood: Non-durable.

preservative treatment: Extremely resistant.

uses

Suitable for flooring, furniture, turnery and other purposes for which European beech is used. Also employed in New Zealand for butter boxes, cheese crates and other food containers and in motor body work.

STERCULIA, BROWN
Sterculia rhinopetala

other names: aye (Nigeria); wawabima (Ghana).

THE TREE

Reaches a height of about 37 m (120 ft). Narrow buttresses, extending up the bole to about 3 m (10 ft). Bole straight and cylindrical, length about 21 m (70 ft). West Africa.

THE TIMBER
properties

colour: Variable in colour from pale to deep reddish-brown. Owing to the dark colour of the numerous high rays, accurately quarter-sawn material has a striking figure in contrast to the rather plain appearance of flat-sawn timber.

sapwood: Straw-coloured, clearly demarcated from heartwood, generally about 38 to 63 mm ($1\frac{1}{2}$–$2\frac{1}{2}$ in) wide.

grain: Sometimes straight, more usually interlocked. Texture rather coarse and fibrous.

weight: Varies widely from 530 to 1020 kg/m³ (33–64 lb/ft³), average about 820 kg/m³ (51 lb/ft³), seasoned.

strength: Slightly higher than European beech.

Moisture Content	Bending Strength		Modulus of Elasticity		Compression parallel to grain	
	N/mm²	lbf/in²	N/mm²	1000 lbf/in²	N/mm²	lbf/in²
Green	87	12 600	10 800	1560	42·5	6170
12 per cent	145	21 000	14 100	2040	69·6	10 100

movement: Large.
> Moisture content in 90 per cent relative humidity   21 per cent
> Moisture content in 60 per cent relative humidity   13 per cent
> Corresponding tangential movement   3·2 per cent ($\frac{25}{64}$ in/ft)
> Corresponding radial movement   1·5 per cent ($\frac{3}{16}$ in/ft)

processing

drying: Dries very slowly with severe cupping, though other forms of distortion are slight. Appreciable checking, end-splitting and extension of original shakes may occur, and slight collapse is likely.
> Kiln Schedule B.
> Shrinkage: Green to 12 per cent moisture content:
> > Tangential   about 9·5 per cent ($1\frac{1}{4}$ in/ft)
> > Radial   about 5·0 per cent ($\frac{5}{8}$ in/ft)

working properties: Texture coarse.

Blunting:	Moderate.
Sawing:	If fine sawdust is produced saws rapidly overheat. Rip-sawing – Saw type SR with 15° hook angle. Cross-cutting – Saw types 1 and 2. Narrow bandsawing – Satisfactory. Wide bandsawing – Saw type C.
Machining:	Generally satisfactory, but a fibrous finish may be produced particularly on end-grain.
Nailing:	Pre-boring required.
Gluing:	No information.

wood bending:

Clear material suitable for bends of moderate radius of curvature, but the timber cannot be bent satisfactorily if small knots are present.
Classification – Moderate.
Ratio radius/thickness for solid bends (steamed):
Supported: 12   Unsupported: 14
Limiting radius for 3·2 mm ($\frac{1}{8}$ in) laminae (unsteamed): 115 mm ($4\frac{1}{2}$ in).

plywood manufacture: Shown by tests to be suitable.
Movement: 4·5 mm plywood from 30 per cent to 90 per cent relative humidity – 0·12 per cent.
Surface splitting on exposure to weather – Grade I.

staining and polishing: Satisfactory when filled.

durability and preservation

insect attack: Sapwood liable to attack by powder-post beetles. Reported to be moderately resistant to termites in Nigeria.

durability of heartwood: Moderately durable.

preservative treatment: Extremely resistant. Sapwood moderately resistant.

uses

A heavy timber with good strength properties, especially toughness, but rather difficult to season. Can be used for constructional work in situations where its durability is adequate, and possibly as an alternative to ash where toughness is a particular asset, as, for instance, in tool handles. Has been used as facing for plywood and blockboard, and for flooring blocks.

STERCULIA, YELLOW
Sterculia oblonga

other names: okoko (Nigeria); bongele, ekonge (Cameroon).

THE TREE

Reaches a height of 24–37 m (80–120 ft). Sharp buttresses extend 3·0–3·6 m (10–12 ft) up the stem. Bole straight, cylindrical and clear of branches up to 15–21 m (50–70 ft), diameter 0·5–1·0 m (about 1½–3 ft). West Africa.

THE TIMBER properties

colour:	Creamy-white to light yellowish-brown. The numerous high rays give the timber a striking fleck figure on accurately quarter-sawn stock.
sapwood:	Pale in colour, not clearly distinct from heartwood. Commonly 100–200 mm (4–8 in) wide.
grain:	Shallowly interlocked. Texture rather coarse.
weight:	Varies from 690 to 830 kg/m³ (43–52 lb/ft³), average about 780 kg/m³ (49 lb/ft³), seasoned.
odour:	Freshly-sawn timber has a strong, disagreeable odour but this does not persist after the timber has been thoroughly dried.
strength:	Slightly higher than European beech.

194

Moisture Content	Bending Strength		Modulus of Elasticity		Compression parallel to grain	
	N/mm²	lbf/in²	N/mm²	1000 lbf/in²	N/mm²	lbf/in²
Green	81	11 700	10 300	1500	38·7	5610
12 per cent	123	17 900	13 700	1980	67·2	9750

movement: Medium.
Moisture content in 90 per cent relative humidity   18·5 per cent
Moisture content in 60 per cent relative humidity   11·5 per cent
Corresponding tangential movement   3·0 per cent ($\frac{3}{8}$ in/ft)
Corresponding radial movement   1·3 per cent ($\frac{5}{32}$ in/ft)

processing

drying: Dries slowly, with marked tendency for surface checking to develop and for shakes to extend. End-splitting may be troublesome. Longitudinal distortion is slight, but cup tends to be a serious defect especially in pieces which suffer slight collapse.
Kiln Schedule C.
Shrinkage: Green to 12 per cent moisture content:
Tangential   about 6·5 per cent ($\frac{13}{16}$ in/ft)
Radial   about 3·5 per cent ($\frac{7}{16}$ in/ft)

working properties: Texture very coarse.
Blunting: Moderate.
Sawing: Generally satisfactory apart from fibrous finish.
Rip-sawing – Saw type HR 54.
Cross-cutting – Satisfactory.
Narrow bandsawing – Satisfactory.
Wide bandsawing – Saw type B.
Machining: Generally satisfactory apart from fibrous finish produced in many operations.
Planing – A small amount of tearing may occur but can be mainly avoided by reducing the cutting angle to 20°.
Nailing: Satisfactory.
Gluing: Good.

wood bending:
Generally inferior in its bending properties to brown sterculia. Inclined to distort severely during bending and cannot be bent satisfactorily if small knots are present. Very little improvement in bending properties gained by using a supporting strap.
Classification – Moderate.
Ratio radius/thickness for solid bends (steamed):
Supported: 17   Unsupported: 18·5
Limiting radius for 3·2 mm ($\frac{1}{8}$ in) laminae (unsteamed): 163 mm ($6\frac{2}{8}$ in).

plywood manufacture: Shown by tests to be suitable.
Movement: 4·5 mm plywood from 30 per cent to 90 per cent relative humidity – 0·16 per cent.
Surface splitting on exposure to weather – Grade II.

staining and polishing: Satisfactory with a little filler.

durability and preservation

insect attack: Liable to attack by powder-post beetles. Reported to be non-resistant to termites in Nigeria.

durability of heartwood: Non-durable.

preservative treatment: Extremely resistant. Sapwood permeable.

uses

A timber with good strength properties, but difficult to season and lacking in natural durability. May be suitable for construction work where durability is not important.

SUCUPIRA
Bowdichia spp.

other names: black sucupira, sapupira.

In addition to *Bowdichia* spp. (principally *B. nitida*) the name sucupira is also used in Brazil for species of *Diplotropis*, which have timber very like that of *Bowdichia*, and for *Ferreirea spectabilis*, which has a yellowish-brown wood, better known as yellow sucupira.

THE TREE

Species of *Bowdichia* are medium-sized to large trees, up to 45 m (150 ft) high and 1·2 m (4 ft) in diameter. South America, mainly Brazil.

THE TIMBER
properties

Heartwood dark chocolate-brown in colour with conspicuous paler markings, giving a decorative appearance which may be enhanced on quartered surfaces by a stripe figure. Sapwood whitish, sharply demarcated from heartwood. Grain interlocked, sometimes irregular. Texture moderately coarse. Average weight about 1000 kg/m³ (62 lb/ft³), seasoned. Has high strength properties, approaching those of greenheart.

processing

No information on drying characteristics; has medium shrinkage on drying. Difficult to work on account of its high density and interlocked and irregular grain, but can be finished to a smooth surface and stains and polishes satisfactorily. Glues well but is unsuitable for plywood manufacture because of its weight.

durability
and preservation

Heartwood very durable (provisional).

uses

A heavy, strong, durable wood, best suited for structural purposes and in general too hard and heavy for purposes requiring much fabrication. As flooring it splits badly under heavy pedestrian traffic but should be satisfactory under less exacting conditions. Has a decorative appearance and, although somewhat coarse textured, is of interest for turned work and as a veneer for inlays in high-class furniture.

SYCAMORE
Acer pseudoplatanus

other names: sycamore plane, great maple (Great Britain); plane (Scotland).

THE TREE

Grows to a height of over 30 m (100 ft), diameter 1·5 m (5 ft). Under forest conditions will form a straight, cylindrical bole 15 or 18 m (50 or 60 ft) long. Europe including British Isles, and West Asia.

Note: Field maple (*Acer campestre*) and Norway maple (*Acer platanoides*) also grow in Europe, including British Isles, and produce timber which is very similar in properties to sycamore and may be used for similar purposes.

THE TIMBER
properties

colour:	White or yellowish white with a natural lustre, especially marked on quarter-sawn stock. Slowly dried timber assumes a light brown colour and is known as weathered sycamore.
sapwood:	Not distinct from heartwood.
grain:	Usually straight, but sometimes curly or wavy producing an attractive fiddleback figure.* Texture fine and even.
weight:	Average about 610 kg/m³ (38 lb/ft³), seasoned.
strength:	Somewhat lower than European beech.

*Figured sycamore is the traditional wood for violin backs.

Moisture Content	Bending Strength		Modulus of Elasticity		Compression parallel to grain	
	N/mm^2	lbf/in^2	N/mm^2	$1000\,lbf/in^2$	N/mm^2	lbf/in^2
Green	66	9500	8400	1220	27·5	3990
12 per cent	99	14 300	9400	1370	48·2	6990

movement: Medium.
Moisture content in 90 per cent relative humidity   23 per cent
Moisture content in 60 per cent relative humidity   13·5 per cent
Corresponding tangential movement   2·8 per cent ($\frac{11}{32}$ in/ft)
Corresponding radial movement   1·4 per cent ($\frac{11}{64}$ in/ft)

processing

drying: Air dries well but is very much inclined to stain. Rapid drying of the surface (by using large piling sticks or artificial heat treatment before piling) somewhat reduces the tendency to stain. End stacking of boards separated to allow ample ventilation is common commercial practice and assists in rapid drying of the surface. Very small pieces of wood are sometimes used instead of sticks to hold adjacent boards apart; this helps to prevent the appearance of sticker marks.
Kiln dries well and rapidly, but the temperature should be kept low (below 49°C, 120°F) if darkening of the timber is to be avoided.
Kiln Schedule A normally recommended. Kiln Schedule E suitable if colour is not important.
Shrinkage: Green to 12 per cent moisture content:
Tangential   about 5·5 per cent ($\frac{11}{16}$ in/ft)
Radial   about 2·5 per cent ($\frac{5}{16}$ in/ft)

working properties:
Blunting: Moderate.
Sawing: Rip-sawing – Saw type HR 54 satisfactory, but a tendency to bind on saw.
Cross-cutting – Tendency to burn, otherwise satisfactory.
Narrow bandsawing – Tendency to burn.
Wide bandsawing – Saw type B.
Machining: Planing – Satisfactory when straight grained, but with irregular and wavy-grained material a reduction in cutting angle to 15° is necessary to prevent tearing.
Tendency to burn in drilling and when cutters are dull.
Turns well.
Nailing: Good.
Gluing: Good.

wood bending:
Clear timber bends very well, but the presence of wavy grain and knots, which are sometimes prevalent, greatly detracts from its good bending qualities.
Classification – Very good.
Ratio radius/thickness for solid bends (steamed):
Supported: 1·5   Unsupported: 14·5
Limiting radius for 3·2 mm ($\frac{1}{8}$ in) laminae (unsteamed): 102 mm (4 in).

plywood manufacture: Shown by tests to be suitable.
Movement: 4·5 mm plywood from 30 per cent to 90 per cent relative humidity – 0·15 per cent.
Surface splitting on exposure to weather – Grade II.

staining and polishing: Very good.

durability and preservation

insect attack: Sapwood liable to attack by the common furniture beetle and by *Ptilinus pectinicornis*, but only rarely attacked by powder-post beetles.
durability of heartwood: Perishable.
preservative treatment: Permeable.

uses	A pale, fine-textured wood used largely in turnery for manufacture of bobbins, brush handles, domestic and dairy utensils, and textile rollers. Some logs have an attractive figure and are used in the veneer trade. Figured sycamore is used for the backs of violins.

TALLOWWOOD
Eucalyptus microcorys

THE TREE	Grows to a height of 30–45 m (100–150 ft), average diameter 1·0–1·5 m (3–5 ft). Eastern Australia.
THE TIMBER **properties**	A yellowish-brown timber having a greasy nature to which the name tallowwood refers. Grain usually interlocked. Texture moderately coarse and even. Weight about 980 kg/m³ (61 lb/ft³), seasoned. A hard, strong and tough timber, but its resistance to splitting is relatively low. Medium movement.
processing	Stated to be very refractory in drying and liable to check severely. It is recommended that the timber should be air dried before kiln drying. Kiln Schedule C. Has a severe blunting effect on cutting edges but otherwise satisfactory in sawing. In planing, picking up occurs owing to the interlocked grain and a cutting angle of 20° is necessary to produce smooth surfaces. Finishes cleanly in other operations and turns well. Reported to be difficult to glue and does not stain readily but other finishing treatments are satisfactory. Reported to have moderately good bending properties.
durability **and preservation**	Sapwood liable to attack by powder-post beetles. Has been reported moderately resistant to termites in Australia. Heartwood very durable. Extremely resistant to preservative treatment.
uses	A heavy, strong and very durable wood. If available, could be used for piling, dock work, bridging, etc. Has a very high resistance to abrasive action and is suitable for most types of floors, but tends to splinter or roughen under severe conditions of traffic. Used in Australia in the construction of heavy wagons and vehicles and for telephone and electric transmission poles and cross-arms.

'TASMANIAN MYRTLE'
Nothofagus cunninghamii

	other names: 'myrtle beech', 'Tasmanian beech' (Australia).
THE TREE	Commonly reaches a height of 30 m (100 ft), exceptionally up to 60 m (200 ft). Clear bole up to 12 m (40 ft) long, and 0·6–1·0 m (2–3 ft) or more in diameter. Tasmania and Victoria.
THE TIMBER **properties**	Generally similar to New Zealand 'silver beech', but somewhat heavier. Pink or reddish-brown in colour with a narrow, paler sapwood, separated from heartwood by a zone intermediate in colour. Grain straight or slightly interlocked, occasionally wavy. Texture fine and uniform. Average weight about 720 kg/m³ (45 lb/ft³), seasoned. Comparable in strength to European beech.
processing	Very variable in its drying characteristics. Light-coloured timber (straw to pale pink) is softer than the darker timber and gives little trouble. Darker (red) timber requires very careful drying to avoid serious internal checking. Definite collapse occurs in the red myrtle but this may be largely removed by reconditioning. Preliminary air drying to about 30 per cent moisture content is recommended. Kiln Schedule C. Has a moderate blunting effect on cutting edges. Satisfactory in sawing and machining apart from a tendency to burn in cross-cutting and boring. Glues well and is said to be suitable for plywood manufacture. Has good bending properties. Takes stains and polishes well.

198

durability and preservation	Sapwood liable to attack by powder-post beetles. Heartwood non-durable.	
uses	Generally similar in properties to European beech and suitable for similar purposes, except where a light colour is required. Used in Australia for food containers, including butter boxes, and in motor body work.	

'TASMANIAN OAK'
*Eucalyptus delegatensis,
E. obliqua* and
E. regnans

other name: 'Australian oak'.

In Australia these three species are marketed as separate timbers. The standard trade names adopted there are:

> *Eucalyptus delegatensis*, 'Alpine ash'.
> *E. obliqua*, 'messmate stringybark'.
> *E. regnans*, 'mountain ash'.

For the export trade they are commonly marketed together.

THE TREE

Trees of these three species reach a height of 60–90 m (200–300 ft) with a long, clear, straight bole, diameter up to 2 m (7 ft) or more, but often 1·0–1·2 m (3–4 ft). South-eastern Australia and Tasmania.

THE TIMBER
properties

colour: Varies from a pale colour, commonly with a pinkish tint to a light brown, similar to European oak.

sapwood: Paler in colour, indistinct, 25–38 mm (1–1½ in) wide.

grain: Usually straight, occasionally interlocked or wavy. Texture coarse. Hard gum or kino veins are sometimes present in all three species but appear to be most frequent in *E. obliqua*. The timber bears a general resemblance to plain-sawn European oak but lacks the characteristic silver grain of true oak.

weight: Varies over a rather wide range. The wood of *E. obliqua* tends to be heavier than that of the other two species.

Species	Density Range		Average Density	
	kg/m³	lb/ft³	kg/m³	lb/ft³
E. obliqua	670–990	42–62	780	49
E. regnans	560–690	35–43	620	39
E. delegatensis	580–770	36–48	640	40

strength: Slightly higher than European beech. Data in the table are average values for the three species.

Moisture Content	Bending Strength		Modulus of Elasticity		Compression parallel to grain	
	N/mm²	lbf/in²	N/mm²	1000 lbf/in²	N/mm²	lbf/in²
Green	71	10 300	11 700	1700	34·8	5050
12 per cent	119	17 300	14 500	2110	64·7	9390

movement: (reconditioned timber) Medium.
Moisture content in 90 per cent relative humidity   17·5 per cent
Moisture content in 60 per cent relative humidity   12 per cent
Corresponding tangential movement   2·1 per cent (¼ in/ft)
Corresponding radial movement   1·4 per cent ($\frac{11}{64}$ in/ft)

| processing | **drying:** | Dries readily and comparatively quickly but with some tendency to develop surface checks during the early stages of drying and with a slight tendency to distort. Very prone to collapse and to check internally. Air drying before kiln drying is recommended and a high temperature reconditioning treatment during the final stages of drying is effective in removing collapse. |

drying: Dries readily and comparatively quickly but with some tendency to develop surface checks during the early stages of drying and with a slight tendency to distort. Very prone to collapse and to check internally. Air drying before kiln drying is recommended and a high temperature reconditioning treatment during the final stages of drying is effective in removing collapse.

Kiln Schedule C.

Shrinkage: Green to 12 per cent content followed by reconditioning:
Tangential $10\cdot0$–$6\cdot5$ per cent ($1\frac{3}{16}$ to $\frac{13}{16}$ in/ft)
Radial $5\cdot0$–$4\cdot0$ per cent ($\frac{5}{8}$ to $\frac{1}{2}$ in/ft)

working properties:
Blunting: Moderate.
Sawing: Rip-sawing – Saw type HR 54.
Cross-cutting – Saw types 1, 2 and 3.
Narrow bandsawing – Satisfactory.
Wide bandsawing – Saw type B.
Machining: Satisfactory.
Nailing: Satisfactory.
Gluing: Good.

wood bending:
Some buckling and fracture of the fibres occurs in bending; *Eucalyptus obliqua* is a better bending timber than *E. regnans*.
Classification – *E. obliqua* moderate, *E. regnans* poor.
Ratio radius/ thickness for solid bends (steamed):
E. obliqua – Supported: 16   Unsupported: 24
E. regnans – Supported: 30   Unsupported: 30

plywood manufacture:
Employed for plywood but seldom seen in the United Kingdom.
No information on movement of plywood or surface splitting on exposure.

staining and polishing: Good.

durability and preservation

insect attack: Sapwood liable to attack by powder-post beetles.

durability of heartwood: Moderately durable.

preservative treatment: Resistant. Sapwood permeable.

uses

Timbers of medium density and good appearance, suitable for a wide range of uses including structural timber in buildings, joinery, cladding, furniture, and boxes and cases, and extensively used for these purposes in Australia. As flooring, rather variable in resistance to wear because of the range of species making up the timber, but timber in the higher density range should be suitable for normal conditions of traffic.

TAWA
Beilschmiedia tawa

THE TREE

Generally 18–25 m (60–80 ft) in height. Bole about 9 m (30 ft) long and 0·5–0·8 m ($1\frac{1}{2}$–$2\frac{1}{2}$ ft) in diameter. New Zealand.

THE TIMBER
properties

A uniformly pale-coloured timber, normally with no visible difference between sapwood and heartwood, similar in texture and general appearance to sycamore. Heartwood of large logs sometimes discoloured by blackish streaks. Grain straight, texture moderately fine. Average weight about 640 kg/m³ (40 lb/ft³), seasoned. Comparable in strength properties to European beech. Movement probably small.

processing

Kiln dries fairly readily, but with some tendency to checking. Kiln Schedule E. Has a moderate blunting effect on cutting edges. Satisfactory in sawing. In planing a cutting angle of 20° is required to plane wavy-grained heartwood satisfactorily. Requires support during drilling and chain mortising to prevent excessive breaking out. Tends to split in nailing. Glues well. Employed in plywood manufacture but seldom seen in the United Kingdom. Takes stains and polishes well.

durability and preservation	Sapwood liable to attack by powder-post beetles and by the common furniture beetle. Heartwood non-durable.
uses	As a flooring timber it has high resistance to wear and has been imported into the United Kingdom for this purpose. Used extensively in New Zealand for flooring and for furniture, interior joinery and many other purposes.

TCHITOLA
Oxystigma oxyphyllum
(formerly *Pterygopodium oxyphyllum*)

other names: Iolagbola (Nigeria); kitola (Congo Dem. Rep.); tola mafuta, tola chimfuta (Angola).

The name tola is also used in the Portuguese African territories and in Europe for the timber known in Britain as agba (*Gossweilerodendron balsamiferum*).

THE TREE	A large tree, up to 45 m (150 ft) in height. Bole straight, cylindrical and unbuttressed, diameter 0·6–1·0 m (2–3 ft). Tropical Africa.
THE TIMBER **properties**	Heartwood reddish-brown with an attractive figure, having some resemblance to walnut. Typically darker in colour than agba. Sapwood 60–100 mm ($2\frac{1}{2}$–4 in) wide, paler in colour than heartwood and well defined from it. A gummy wood, but the gum occurs principally in the sapwood. Grain straight or shallowly interlocked, texture moderately coarse. Average weight about 610 kg/m^3 (38 lb/ft^3), seasoned. Comparable in strength to European beech. Small movement.
processing	It is said to air dry with little distortion or splitting. Satisfactory in sawing and planing apart from some tendency for gum to accumulate on tools. Nails and glues well. Employed for plywood and generally available in the United Kingdom. Takes finishes satisfactorily.
durability and preservation	Reported to be moderately resistant to termites in West Africa. Heartwood non-durable (provisional) and probably permeable to preservatives.
uses	Used in Britain mainly as a decorative veneer. Has a walnut-like appearance and is used for furniture, radio and television cases and similar cabinet work. Bleeding of gum from veneers may occasionally be troublesome. Used extensively in South Africa for manufacture of a general-purpose plywood.

TEAK
Tectona grandis

THE TREE	Very variable in size and form according to locality and conditions of growth. In favourable localities may reach a height of 40–45 m (130–150 ft), with a clear bole up to 25–27 m (80–90 ft) but more usually 9–11 m (30–35 ft). Diameter of bole up to 1·8–2·4 m (6–8 ft), but generally less (0·9–1·5 m, 3–5 ft). In drier regions trees are smaller. In older trees the bole is more fluted and buttressed at the base. Indigenous to India, Burma, Thailand, Indo-China and Java. Has been planted in India, East and West Africa, West Indies and elsewhere.
THE TIMBER **properties**	An outstanding timber on account of its many valuable properties including durability, strength, moderate weight, relative ease of working, dimensional stability and pleasing appearance.
colour:	Usually golden brown, darkening on exposure. Sometimes figured with dark markings. The best Burma teak is uniform in colour with few markings. Timber from the dry zone areas of India has a more figured appearance.

sapwood: Light to pale yellowish-brown, narrow to medium width.

grain: Often straight but sometimes wavy. Texture coarse and uneven.

weight: Varies from about 610 to 690 kg/m³ (38–43 lb/ft³), average about 640 kg/m³ (40 lb/ft³), at 12 per cent moisture content.

irritant properties:
The fine dust produced in machining operations is liable to cause irritation of the skin in some individuals.

strength: Comparable to, or slightly lower than, European beech. No marked difference between natural and plantation-grown timber. Teak grown in Nigeria and Trinidad has been shown by tests to have strength properties at least as high as those of Burma-grown teak of similar density.

Origin	Moisture Content	Bending Strength		Modulus of Elasticity		Compression parallel to grain	
					1000		
		N/mm²	lbf/in²	N/mm²	lbf/in²	N/mm²	lbf/in²
Burma	Green	84	12 200	8800	1280	42·8	6210
	12 per cent	106	15 400	10 000	1450	60·4	8760
Nigeria Western Region	Green	90	13 000	8900	1290	41·2	5970
	12 per cent	111	16 100	11 200	1620	57·2	8290
Nigeria Northern Region	Green	83	12 000	8900	1290	37·9	5490
	12 per cent	100	14 500	10 100	1470	51·9	7530

movement: Small.
Moisture content in 90 per cent relative humidity   15 per cent
Moisture content in 60 per cent relative humidity   10 per cent
Corresponding tangential movement   1·2 per cent ($\frac{9}{64}$ in/ft)
Corresponding radial movement   0·7 per cent ($\frac{3}{32}$ in/ft)

processing

drying: Dries well but rather slowly. Particular care required in determining initial and final moisture contents, as large variations in drying rates occasionally occur. The timber is very liable to change colour, but the colour becomes uniform within a reasonable time after kiln drying.
Kiln Schedule H.
Shrinkage: Green to 12 per cent moisture content:
Tangential   about 2·5 per cent ($\frac{5}{16}$ in/ft)
Radial   about 1·5 per cent ($\frac{3}{16}$ in/ft)

working properties:
Blunting: Severe (variable).
Sawing: Rip-sawing – Saw type HR40 or TC.
Cross-cutting – Saw types 1 and 2 or tungsten carbide-tipped.
Narrow bandsawing – 118 teeth/m (36 teeth/ft) recommended.
Wide bandsawing – Saw type B with increased tooth pitch or stellite-tipped teeth.
Machining: General – Tungsten carbide cutters are advantageous. Power feeding of material and reduced cutter speeds recommended.
Planing – Reduction of cutting angle to 20° advisable.
Mortising – Difficult with hollow square chisel mortiser.
Nailing: Pre-boring recommended.
Gluing: Good on freshly-machined or newly-sanded surfaces.

wood bending:

Considerable variation in bending qualities within a consignment. Tendency to buckle on the concave face. Suitable only for bends of moderate radius of curvature.

Classification – Moderate.

	Burma	Teak from W. Nigeria	N. Nigeria
Ratio radius/thickness for solid bends (steamed): Supported	18	10	18
Unsupported	35	26	35
Limiting radius for 3·2 mm ($\frac{1}{8}$ in) laminae (unsteamed)	160 mm ($6\frac{1}{4}$ in)	147 mm ($5\frac{4}{8}$ in)	165 mm ($6\frac{1}{2}$ in)

plywood manufacture:

Employed in plywood and usually available in the United Kingdom. No information on movement of plywood or surface splitting on exposure to weather.

staining and polishing: Satisfactory.

durability and preservation

insect attack: Trees liable to attack by wood-boring caterpillars (Cossidae), and sapwood liable to attack by powder-post beetles. Reported to be highly resistant to termites in India, West Africa, United States and Malaya.

durability of heartwood: Very durable.

preservative treatment: Extremely resistant.

uses

Well known for its good technical properties, particularly stability and durability, and has a wide range of uses. Extensively used in ship building, especially for decking, deck houses, rails, bulwarks, hatches, weather doors, etc., for boat planking, particularly for use in tropical waters, and for furniture and interior fittings of boats. Teak is a good timber for exterior joinery and is also used extensively in furniture manufacture and for garden furniture. As flooring, it has moderate to low resistance to wear and is best used rift-sawn for light to moderate traffic. Has good resistance to a wide variety of chemicals and has been used for scrubbing towers, vats, fume ducts, etc., in industrial chemical plant.

THINGAN
Hopea odorata

A large tree, up to 45 m (150 ft) in height, diameter about 1·2 m (4 ft). Bole straight and cylindrical, free of branches for 25 m (80 ft). Andaman Islands and south-east Asia.

A pale yellowish-brown timber, darkening on exposure to an attractive golden brown, and marked on flat-sawn surfaces by prominent lines of vertical canals with white contents. Sapwood paler in colour and moderately well defined. Grain sometimes straight but more usually interlocked. Texture moderately fine. Average weight about 770 kg/m³ (48 lb/ft³), seasoned. Comparable in strength to European beech. Small movement.

Dries slowly with small shrinkage; liable to surface checking and end-splitting when dried too rapidly. Kiln Schedule C suggested. Somewhat difficult to saw, but finishes to a good surface. Has been employed for plywood manufacture for use in tea chests.

As a flooring timber it has high resistance to wear but is somewhat inferior to maple and is best suited for normal conditions of traffic. It has shown promise as an alternative to maple for shoe and boot lasts and for rollers in the textile industry. In Burma it is widely used for constructional purposes and for boat-building.

THITKA
Pentace burmanica

other name: kashit.

A large tree, 30 m (100 ft) or more in height, with a clear bole 10 m (33 ft) or more in length and 0·6 m (2 ft) in diameter. Burma and Indo-China.

A pale reddish-brown timber, turning darker on exposure and having some resemblance to American mahogany. Sapwood and heartwood not clearly defined. Grain interlocked, producing a narrow regular stripe or roe figure on quarter-sawn surfaces. Texture fine and even. Weight about 640–700 kg/m³ (40–44 lb/ft³), seasoned. Strength somewhat lower than European beech.

Dries slowly but well, with little tendency to distortion or checking. Working properties satisfactory, but in planing a cutting angle of 20° is necessary for good results owing to interlocked grain. Takes stains and polishes very well. The heartwood is durable and reported to be moderately resistant to termites in India. Resistant to preservative treatment.

A good quality timber for decorative purposes. It is employed in the East for high-class furniture, boat-building, walking-sticks and mathematical instruments, and is suitable for shopfitting, doors, panelling, etc.

TUPELO
Nyssa spp.

other names: cotton gum, tupelo gum, black gum.

The two principal species are *Nyssa aquatica* and *N. sylvatica*. The two timbers are similar in general character; *N. sylvatica*, also known as black gum, is on average somewhat heavier and stronger than *N. aquatica*.

THE TREE

Grows to a height of about 30 m (100 ft). Bole straight with greatly enlarged tapering base, diameter above base up to 1·2 m (4 ft). Southern and eastern United States.

THE TIMBER
properties

A yellowish or brownish timber, somewhat streaked. The wide greyish-white sapwood is not clearly defined from the heartwood. Grain typically irregular and interlocked. Texture fine and uniform. Average weight: *N. aquatica* about 510 kg/m³ (32 lb/ft³), *N. sylvatica* about 560 kg/m³ (35 lb/ft³), seasoned. Strength somewhat higher than obeche.

processing

Dries fairly readily without undue splitting. Some material may distort very severely; it is advisable to weight the load when kiln drying. Kiln Schedule E. Moderately hard to work. In planing a cutting angle of 20° is recommended on account of interlocked grain. Satisfactory in nailing but gluing properties are variable. It is employed in plywood manufacture, mainly for cores. Satisfactory in staining and polishing.

uses

A timber of medium weight and fine texture used in the United States for boxes, crates, furniture and vehicle parts. Suitable for flooring subjected to heavy wear, e.g. factory floors, bridge flooring, etc., but requires preservative treatment if used in situations where there is risk of decay.

ULMO
Eucryphia cordifolia

Reaches a maximum height of 40 m (130 ft), diameter 0·6 m (2 ft). Southern Chile.

Heartwood reddish- or greyish-brown, sometimes variegated. Sapwood paler in colour, not sharply demarcated from heartwood. The timber is generally straight grained, with a fine, uniform texture. Average weight about 610 kg/m³ (38 lb/ft³), seasoned. The timber is strong and moderately hard. It is stated that it dries reasonably well without checking, but is liable to twist badly on account of tension wood. Kiln Schedule C. Not difficult to work. Can be used for plywood manufacture. Heartwood non-durable. Used in Chile mainly for flooring but also for furniture, joinery and vehicle parts.

UTILE
Entandrophragma utile

other names: sipo (France and Ivory Coast); assié (Cameroon).

THE TREE Grows to a height of 45–60 m (150–200 ft). Narrow buttresses extend about 4·5 m (15 ft) up the stem. Bole straight and cylindrical, clear of branches for 21–24 m (70–80 ft), diameter above buttresses up to 2·5 m (8 ft). West and Central Africa.

THE TIMBER
properties

colour: Fairly uniform reddish- or purplish-brown.

sapwood: Light brown, distinct from heartwood.

grain: Interlocked and rather irregular. The stripe figure of quartered stock is less marked than in sapele. Texture more open than in sapele owing to the larger size of the pores.

weight: Ranges from 550 to 750 kg/m³ (34–47 lb/ft³), average about 660 kg/m³ (41 lb/ft³), seasoned.

strength: Comparable to European beech.

Moisture Content	Bending Strength		Modulus of Elasticity		Compression parallel to grain	
	N/mm²	lbf/in²	N/mm²	1000 lbf/in²	N/mm²	lbf/in²
Green	79	11 400	9600	1390	38·2	5540
12 per cent	103	15 000	10 800	1560	60·4	8760

movement: Medium.
Moisture content in 90 per cent relative humidity 20·5 per cent
Moisture content in 60 per cent relative humidity 13·5 per cent
Corresponding tangential movement 1·6 per cent ($\frac{3}{16}$ in/ft)
Corresponding radial movement 1·4 per cent ($\frac{11}{64}$ in/ft)

processing

drying: Dries at a moderate rate. Tendency for original shakes to extend and for distortion in the form of twist to develop, but distortion is not generally severe.
Kiln Schedule A.
Shrinkage: Green to 12 per cent moisture content:
Tangential about 3·5 per cent ($\frac{7}{16}$ in/ft)
Radial about 3·0 per cent ($\frac{3}{8}$ in/ft)

working properties:
Blunting: Moderate.
Sawing: Rip-sawing – Saw type HR 54.
Cross-cutting – Saw types 1, 2 and 3 satisfactory.
Narrow bandsawing – Satisfactory.
Wide bandsawing – Saw type B.
Machining: Planing – A cutting angle of 15° required to avoid tearing of interlocked grain.
Boring – Twist-fluted drills most suitable. A tendency to charring. Support required at drill exit.
Mortising – Chips do not clear well from hollow square chisel mortiser.
Nailing: Satisfactory.
Gluing: Good.

wood bending: Buckles severely when bent to any appreciable extent.
Classification – Very poor.
Limiting radius for 3·2 mm ($\frac{1}{8}$ in) laminae (unsteamed): 210 mm ($8\frac{1}{4}$ in).

plywood manufacture:
Employed in plywood and usually available in the United Kingdom. No information on movement of plywood or surface splitting on exposure to weather.

staining and polishing: Satisfactory when filled.

durability and preservation

insect attack: Sapwood liable to attack by powder-post beetles. Reported to be moderately resistant to termites in West Africa.

durability of heartwood: Durable.

preservative treatment: Extremely resistant.

uses

A reddish-brown timber, similar in general appearance and properties to sapele, but less liable to distort during drying and in subsequent use. Suitable for furniture and cabinet work, and interior and exterior joinery, and construction work.

VIROLA
Virola spp. and Dialyanthera spp.

The timber is derived from several species of *Virola* and *Dialyanthera* growing in northern South America.

properties

Woods of the various species are similar in character. Heartwood pale pinkish-brown, grain generally straight. Some variation in texture and weight according to species. Timber from Colombia, produced by species of *Dialyanthera*, is somewhat lighter in weight (about 400 kg/m³, 25 lb/ft³) than Brazilian timber, produced by species of *Virola* (about 530 kg/m³, 33 lb/ft³). Brazilian timber is also rather finer textured. The timbers are low in strength, except for stiffness which is somewhat above average for their density.

processing

Care required in drying if excessive degrade is to be avoided. Marked tendency to check and split, and distortion may be appreciable and is sometimes accompanied by collapse. Shrinkage on drying is high. Working properties generally satisfactory. The timbers nail well and glue satisfactorily.

durability and preservation

Logs are subject to pinhole-borer attack and rapid extraction, conversion and treatment are required to prevent serious degrade. Sapwood liable to attack by powder-post and furniture beetles. Reported to be susceptible to termite attack in South America. Heartwood is non-durable and permeable to preservatives.

uses

The timber is soft and light in weight. When produced in clean condition and dried satisfactorily it is suitable for general utility purposes, lightweight joinery, mouldings, and as core stock.

WAIKA CHEWSTICK
Symphonia globulifera

other names: manni (Guyana); chewstick (British Honduras); boar wood, hog gum (British West Indies); yellow mangue (Trinidad); matakki (Surinam); ossol (Gaboon).

Grows to an average height of 30 m (100 ft), or occasionally 40 m (130 ft). Long clear bole, slightly buttressed, diameter usually 0·5–0·8 m ($1\frac{1}{2}$–$2\frac{1}{2}$ ft). Tropical America and west tropical Africa.

Heartwood pale yellowish-brown to pale reddish-brown, generally of plain appearance. Sapwood paler in colour, 38–75 mm ($1\frac{1}{2}$–3 in) wide, sharply defined from heartwood. Grain typically straight, texture rather coarse. Weight about 640–780 kg/m³ (40–49 lb/ft³), seasoned. Strength properties somewhat higher than European beech.

Said to air dry under cover fairly rapidly with a tendency to check and split. Kiln Schedule C. Has a moderate blunting effect on cutting edges. Sawing and machining satisfactory. Tends to split when nailed. Employed in plywood manufacture but seldom seen in the United Kingdom.

Sapwood liable to attack by powder-post beetles. Reported to be non-resistant to termites in West Indies. Heartwood durable. Extremely resistant to preservative treatment. The sapwood is resistant.

A general utility timber used locally for building purposes, carpentry, crates and boxes, etc., but not always readily available in the United Kingdom.

WALLABA
Eperua falcata
and E. grandiflora

other names: *Eperua falcata:* soft wallaba (Guyana); *E. grandiflora:* ituri wallaba (Guyana).

THE TREE Grows to an average height of about 25 m (80 ft), maximum 30 m (100 ft). Basal swelling or buttresses up to 2·5 m (8 ft). Bole above buttresses straight and cylindrical for about 12–18 m (40–60 ft), diameter usually 0·5–0·6 m (1½–2 ft), but may sometimes reach 0·9 m (3 ft). Guyana.

THE TIMBER properties

colour: Dull reddish-brown, with characteristic dark gummy streaks which tend to spread over the surface.

sapwood: Pale in colour, sharply defined from heartwood, 25–50 mm (1–2 in) wide, showing dark gum streaks.

grain: Typically straight. Texture rather coarse.

odour: Green timber has an unpleasant rancid odour, not generally noticeable after drying.

weight: Generally in the range 850–950 kg/m³ (53–59 lb/ft³), average about 900 kg/m³ (56 lb/ft³), seasoned.

strength: About half-way between European beech and greenheart.

Moisture Content	Bending Strength		Modulus of Elasticity		Compression parallel to grain	
	N/mm²	lbf/in²	N/mm²	1000 lbf/in²	N/mm²	lbf/in²
Green	104	15 100	15 000	2180	57·8	8380
12 per cent	139	20 200	14 700	2130	77·3	11 210

movement: No information.

processing

drying: Dries very slowly with a marked tendency to check, split and distort. Honeycombing may develop in thicker material.
Kiln Schedule B.
(For material over 38 mm (1½ in) in thickness, humidity should be 10 per cent higher at each stage.)
Shrinkage: Green to 12 per cent moisture content:
        Tangential   about 8·5 per cent (1 1/16 in/ft)
        Radial   about 4·0 per cent (½ in/ft)

working properties:

Has a high content of gum which builds up on saw teeth and cutters, reducing clearance. Gum may exude on to machined surfaces for some time after machining.

Blunting: Moderate.

Sawing: Rip-sawing — Saw type HR 40.
Cross-cutting — Saw types 1 and 2.
Narrow bandsawing — 118 teeth/m (36 teeth/ft) recommended.
Wide bandsawing — Saw type C.

Machining: Generally satisfactory, but tends to char in boring.

Nailing: Pre-boring necessary.

Gluing: No information.

wood bending:

Not tolerant of pin knots. Resin exudation accompanies steaming.
Classification — Moderate.
Ratio radius/thickness for solid bends (steamed):
Supported: 16   Unsupported: 33
Limiting radius for 3·2 mm ($\frac{1}{8}$ in) laminae (unsteamed): 280 mm (11 in).

plywood manufacture: Unsuitable because of its weight.

staining and polishing:

Satisfactory, but finish is spoilt later by gum exudation.

durability and preservation

insect attack: Sapwood liable to attack by powder-post beetles. Reported to be attacked by dry-wood termites in West Indies.

durability of heartwood: Very durable.

preservative treatment: Extremely resistant.

uses

A hard and heavy timber, liable to gum exudation and therefore unsuitable for decorative purposes. Owing to its good durability should be useful for heavy wharf or other decking, etc. Has been used in the United Kingdom untreated, after removal of sapwood, for transmission poles, and in Guyana for fence posts, railway sleepers, etc. Suitable for general utility and industrial types of floor.

WALNUT, AMERICAN
Juglans nigra

other name: black walnut (Great Britain and United States).

THE TREE

Reaches a maximum height of 45 m (150 ft), more usually about 30 m (100 ft). Bole straight and free of branches for 15–18 m (50–60 ft), diameter 1·2–1·8 m (4–6 ft). Eastern United States and Canada.

THE TIMBER
properties

colour: Rich dark brown, deepening with age. Tends to be more uniform in colour than European walnut.

sapwood: Narrow, pale in colour and sharply defined from heartwood.

grain: Usually straight, sometimes wavy or curly. Texture rather coarse.

weight: Average about 640 kg/m³ (40 lb/ft³), seasoned.

strength: Comparable to, or slightly lower than, European beech.

Moisture Content	Bending Strength		Modulus of Elasticity		Compression parallel to grain	
	N/mm²	lbf/in²	N/mm²	1000 lbf/in²	N/mm²	lbf/in²
Green	69	10 000	9200	1330	30·8	4460
12 per cent	106	15 400	10 800	1570	54·3	7870

(Data from US Department of Agriculture)

movement: No information.

<table>
<tr><td>**processing**</td><td>**drying:**</td><td colspan="2">Dries rather slowly with a tendency to honeycomb.
Kiln Schedule E.</td></tr>
</table>

processing　　**drying:** Dries rather slowly with a tendency to honeycomb.
Kiln Schedule E.
Shrinkage: Green to 12 per cent moisture content:
　　　　　　Tangential　about 3·5 per cent ($\frac{7}{16}$ in/ft)
　　　　　　Radial　　　about 2·5 per cent ($\frac{5}{16}$ in/ft)

working properties:
Saws and machines without difficulty, giving a good finish.
Gluing:　　Satisfactory.

wood bending: No information.

plywood manufacture:
Employed in plywood but seldom seen in the United Kingdom. Widely used as a decorative veneer.
No information on movement of plywood or surface splitting on exposure.

staining and polishing: Good.

durability and preservation

insect attack: Sapwood liable to attack by powder-post beetles.

durability of heartwood: Very durable.

preservative treatment: No information.

uses　Used for high-quality furniture and widely employed as a decorative veneer. Also a standard timber for good rifle and gun stocks.

WALNUT, EUROPEAN
Juglans regia

other names: English walnut, French walnut, Italian walnut, etc., according to origin (Great Britain).

THE TREE　Grows to a height of 25–30 m (80–100 ft). Clear bole rarely exceeds 6 m (20 ft) in length, diameter usually 0·6–0·9 m (2–3 ft), but occasionally 1·5 m (5 ft). Native of south-eastern Europe and western and central Asia. Naturalised in other European countries including Great Britain. Timber-producing trees grown in commercial quantities mainly in Turkey, Italy, France and Yugoslavia.

THE TIMBER
properties　　**colour:**　Variable in colour, with a greyish-brown background marked with irregular dark streaks. The figure is sometimes accentuated by a natural wavy grain. Figured wood often forms a central core, more or less sharply defined from the outer zone of normal heartwood.

sapwood:　Pale in colour, distinct from heartwood.

grain:　Straight or sometimes wavy. Texture rather coarse.

weight:　Average about 640 kg/m³ (40 lb/ft³), seasoned.

chemical staining:
Blue-black stains liable to appear on the wood if it comes into contact with iron or iron compounds under damp conditions.

strength:　No data available, but probably comparable to European beech.

movement: Medium.
Moisture content in 90 per cent relative humidity　18·5 per cent
Moisture content in 60 per cent relative humidity　11·5 per cent
Corresponding tangential movement　2·0 per cent ($\frac{1}{4}$ in/ft)
Corresponding radial movement　　　1·6 per cent ($\frac{3}{16}$ in/ft)

processing　**drying:**　Dries well though rather slowly, with a tendency for honeycombing to develop in thick material.
Kiln Schedule E.
Shrinkage: Green to 12 per cent moisture content:
　　　　　　Tangential　about 5·5 per cent ($\frac{11}{16}$ in/ft)
　　　　　　Radial　　　about 3·0 per cent ($\frac{3}{8}$ in/ft)

working properties:

Blunting: Moderate.

Sawing: Rip-sawing – Saw type HR 54.

Cross-cutting – Satisfactory.

Narrow bandsawing – Satisfactory.

Wide bandsawing – Saw type B.

Machining: Satisfactory. Finishes well in most operations.

Gluing: Satisfactory.

wood bending:

Possesses very good bending properties, which are little affected by such defects as irregular grain and small knots.

Classification – Very good.

Ratio radius/thickness for solid bends (steamed) (English walnut):

Supported: 1·0 Unsupported: 11·0

Limiting radius for 3·2 mm ($\frac{1}{8}$ in) laminae (unsteamed): 91 mm ($3\frac{3}{8}$ in).

plywood manufacture:

Employed in plywood; also widely used as a decorative veneer.

No information on movement of plywood or surface splitting on exposure.

staining and polishing: Excellent.

durability and preservation

insect attack: Logs liable to attack by forest longhorn or Buprestid beetles. Sapwood liable to attack by powder-post beetles and by the common furniture beetle.

durability of heartwood: Moderately durable.

preservative treatment: Resistant. Sapwood permeable.

uses

Employed both as veneer and in the solid for high-quality furniture. The highly figured veneers employed in cabinet making and decorative panelling are obtained from the stumps, burrs and crotches of a relatively small proportion of trees. Also used for turned fancy goods, such as fruit bowls, and is the preferred timber for gun and rifle stocks.

Home-grown supplies are very scarce. Imported walnut is generally named according to the country of origin; the main supplies come from France and Italy, either in planks or veneers. The product of any one locality is variable in colour, figure and texture, but generally shows certain typical characteristics.

WAMARA
Swartzia leiocalycina

THE TREE

Reaches an average height of about 30 m (100 ft). Bole has low buttresses above which it is flanged. Clear bole about 20 m (65 ft) long, diameter 0·5 m ($1\frac{1}{2}$ ft). Guyana, Central America and West Indies.

THE TIMBER
properties

Heartwood chocolate- to purplish-brown with darker purple streaks, giving the wood an attractive appearance. Wide sapwood, pale in colour, sharply demarcated from heartwood. Grain interlocked and somewhat irregular. Texture moderately fine. A very heavy timber, varying in weight from 1000 to 1200 kg/m³ (62–73 lb/ft³), seasoned. Comparable in strength properties to greenheart. Medium movement.

processing

Dries slowly with appreciable surface checking and end-splitting. Original shakes tend to increase, but distortion is not generally serious. Kiln Schedule B. Has a severe blunting effect on cutting edges. In sawing there is a tendency for teeth to vibrate and saws to overheat. In bandsawing a reduced tooth pitch is recommended. Machining is difficult owing to the hardness of the timber. In planing a reduction of cutting angle to 20° is recommended because of interlocked grain. Heartwood is suitable for bends of moderate radius of curvature if well supported with a metal strap, but sapwood is unsuitable for bending. Stains will not penetrate the timber but it can be polished satisfactorily.

| durability and preservation | Reported to be non-resistant to termites in Central America. Heartwood durable and extremely resistant to preservative treatment. The sapwood is permeable. |

| uses | A heavy, strong and durable timber of attractive appearance, but difficult to work. Too heavy for furniture and joinery but should be very suitable for special purposes where hardness and high polish are required. Should also be suitable for heavy-duty flooring in factories or warehouses or as a decorative floor. |

WATTLE
Acacia spp., principally
Acacia mollissima

Native to Australia, but extensively planted in South Africa for the tannin industry which utilises the bark.

| THE TIMBER properties | Heartwood pale brown with a pinkish tinge. Grain commonly interlocked. Texture moderately fine and even. Average weight about 740 kg/m³ (46 lb/ft³), seasoned. Limited tests on wattle grown in South Africa indicate that its strength properties are slightly higher than those of European beech. Large movement. |

| processing | Dries readily but with pronounced distortion, particularly in the form of cupping. Shakes tend to increase and knots to split slightly. Kiln Schedule A. Shown by tests in Australia to be suitable for plywood manufacture. |

| durability and preservation | Sapwood liable to attack by powder-post beetles. Heartwood non-durable, and moderately resistant to preservative treatment. |

| uses | Grown primarily for its bark, for extraction of tannin. The timber remaining after stripping the bark is used locally for mining props, flooring blocks and strips, and for manufacture of hardboard. |

WHITEWOOD, AMERICAN
Liriodendron tulipifera

other names: Canary wood (Great Britain); tulip tree (Great Britain and United States); poplar, yellow poplar (United States). Not to be confused with whitewood (European spruce).

| THE TREE | Reaches a maximum height of about 50 m (150 ft), but generally less than 30 m (100 ft). Long, clear, cylindrical bole, diameter up to 1·8–2·5 m (6–8 ft). Eastern United States. |

| THE TIMBER properties | Heartwood yellowish-brown or pale olive brown. Sapwood almost white, narrow in old trees, but very wide in second-growth trees. Grain straight, texture fine and fairly even. Weight about 450–510 kg/m³ (28–32 lb/ft³), seasoned. Somewhat higher in strength than obeche, and appreciably higher in stiffness. |

| processing | Easy to work and finishes smoothly. Good nailing properties. Employed in plywood in the United States. Stains and polishes satisfactorily. |

| durability and preservation | Sapwood liable to attack by the common furniture beetle. Heartwood non-durable. |

| uses | A soft, light and easily worked timber. Suitable for interior parts of furniture, interior joinery and many purposes where a lightweight utility timber is required. Used extensively for plywood manufacture in the United States. |

WILLOW
Salix spp.

other names: white willow, common willow (*Salix alba*)
crack willow (*Salix fragilis*)
cricket-bat willow, close-bark willow (*Salix alba* var. *coerulea*).

THE TREE

Willows grow to a maximum height of about 21–27 m (70–90 ft), diameter 0·9–1·2 m (3–4 ft). If allowed to grow naturally they branch at about 5–8 m (15–25 ft) from the ground, but they are frequently pollarded at about 2–2·5 m (7 or 8 ft) height or grown as coppice. Trees grown for manufacture of cricket bats are felled when they have reached a diameter of about 0·5 m (1½ ft). Europe including British Isles, Western Asia and North Africa.

THE TIMBER
properties

Quality of timber is considerably affected by the general habit of the tree. Cricket-bat willow is characterised by extremely rapid growth and a shapely habit. These two factors combine to produce straight-grained lightweight wood which is without equal for cricket bats, but under less favourable conditions timber of inferior quality may be produced.

colour: Heartwood pinkish.

sapwood: Nearly white. Width of sapwood varies according to species and growth conditions, being particularly wide in fast-grown cricket-bat willow and white willow.

grain: Straight. Texture fine and even.

weight: Average about 450 kg/m³ (28 lb/ft³), seasoned. High-quality cricket-bat willow is rather lighter in weight (340–420 kg/m³, 21–26 lb/ft³).

strength: Comparable to obeche.

Species	Moisture Content	Bending Strength		Modulus of Elasticity		Compression parallel to grain	
		N/mm²	lbf/in²	N/mm²	1000 lbf/in²	N/mm²	lbf/in²
White willow	Green	36	5200	4800	690	14·7	2130
	12 per cent	63	9100	5800	840	28·4	4120
Crack willow	Green	35	5100	5600	810	14·8	2150
	12 per cent	66	9500	7000	1010	28·1	4080
Cricket-bat	Green	31	4500	5600	810	13·6	1970
willow	12 per cent	62	9000	6600	960	27·3	3960

movement: (Cricket-bat willow) Small.
Moisture content in 90 per cent relative humidity   22 per cent
Moisture content in 60 per cent relative humidity   13·5 per cent
Corresponding tangential movement   No data
Corresponding radial movement   0·5 per cent ($\frac{1}{16}$ in/ft)

processing

drying: Dries well and fairly rapidly, but local pockets of moisture are apt to remain in the timber. Special care is required in testing the moisture content to ensure that reasonable uniformity is achieved.
Kiln Schedule D for cricket-bat willow; Kiln Schedule H for other species.

working properties:
        Blunting: Slight.
        Sawing: Rip-sawing – Saw type HR 54.
                    Cross-cutting – Satisfactory.
                    Narrow bandsawing – Satisfactory.
                    Wide bandsawing – Saw type A.
        Machining: Machines easily to a good finish, provided that sharp cutters with a reduced sharpness angle are used.
        Gluing: Good.

wood bending: Has poor bending properties but no precise data available.
        Classification – Poor.

plywood manufacture: Shown by tests to be suitable for plywood.
        Movement: 1·5 mm plywood from 30 per cent to 90 per cent relative humidity – 0·31 per cent.
        Surface splitting on exposure to weather – Grade I.

staining and polishing: Satisfactory.

durability and preservation

insect attack: Trees and logs liable to attack by forest longhorn or Buprestid beetles. Sapwood liable to attack by powder-post beetles and the common furniture beetle and by *Ptilinus pectinicornis*. Wickerwork is especially susceptible to attack by the common furniture beetle. Death-watch beetle sometimes found in dead parts of trees.

durability of heartwood: Perishable.

preservative treatment: Resistant. Sapwood permeable.

uses

Only the best boles of cricket-bat willow are used for cricket bats. Other material of this species and timber of other willows is used for a variety of purposes requiring a lightweight, easily worked timber. Uses include artificial limbs, toys, chip baskets and other basket work.

Appendix 1

Properties of Hardwoods

This Appendix summarises in general terms the more important properties of the timbers described in detail in the text. It is a table of properties intended for quick reference, and to achieve this purpose it has been necessary to arrange the properties in a relatively small number of groups. Where there are two timbers that are very similar in a certain property, one may appear towards the top of a group whilst the other may come at the bottom of the next higher group. This is unavoidable, and is unimportant in a table intended only as a general guide. For more detailed information the full descriptions of the timbers should be consulted.

The classification adopted for the various properties is as follows:

Density

	Weight, kg/m³ at 12 per cent moisture content
Exceptionally light	Under 300
Light	300–450
Medium	450–650
Heavy	650–800
Very heavy	800–1000
Exceptionally heavy	Over 1000

Mechanical Properties (at 12 per cent moisture content)

	Bending Strength	Stiffness	Maximum crushing strength	Shock Resistance
	N/mm²	kN/mm²	N/mm²	m
Very low	Under 50	Under 10	Under 20	Under 0·6
Low	50–85	10–12	20–35	0·6–0·9
Medium	85–120	12–15	35–55	0·9–1·2
High	120–175	15–20	55–85	1·2–1·6
Very high	Over 175	Over 20	Over 85	Over 1·6

Drying Properties

Drying rate: The approximate time required to dry 25 mm (1 in) boards in a kiln from green to 12 per cent moisture content is taken as a measure of the drying rate and is classed as follows:

Rapid	Up to 1½ weeks
Fairly rapid	1½ to 2½ weeks
Rather slow	2½ to 4 weeks
Slow ⎫ Very slow ⎭	4 weeks or over

Movement: The movement in service is based on the sum of the tangential and radial movements corresponding to a change in humidity conditions from 90 per cent to 60 per cent relative humidity.

Small movement	Under 3 per cent
Medium movement	3·0–4·5 per cent
Large movement	Over 4·5 per cent

Woodworking Properties

Resistance in cutting: The classification is
Very low
Low
Medium
High
Very high

Blunting effect: The classification is
   Slight
   Moderate
   Fairly severe
   Severe

Durability and Preservation

Natural durability: This refers to heartwood only (see p. 8).

	Approximate life in contact with the ground (years)
Perishable	Less than 5
Non-durable	5–10
Moderately durable	10–15
Durable	15–25
Very durable	More than 25

Resistance to impregnation:

This classification refers to heartwood only (see p. 9), and is as follows. Timbers containing a large amount of permeable sapwood are marked with an asterisk.

Permeable
Moderate
Resistant
Extreme

Appendix 1
Properties of Hardwoods

Name	Density	Mechanical Properties			
		Bending strength	Stiffness	Crushing strength	Shock resistance
Abura	Medium	Low	Very low	Medium	Low
Afara or Limba	Medium	Low	Low	Medium	–
African walnut	Medium	Low	Very low	Medium	Low
Afrormosia	Heavy	High	Medium	High	Medium
Afzelia, West African	Very heavy	High	Medium	High	Low
Agba	Medium	Low	Very low	Medium	Low
Albizia, West African	Heavy	Medium	Low	High	Very low
Alder	Medium	Low	Very low	Medium	Low
Alstonia	Light	Low	Very low	Medium	Very low
Aningeria	Medium	–	–	–	–
Antiaris	Light	Low	Very low	Medium	Very low
Apple	Heavy	–	–	–	–
Ash, European	Heavy	Medium	Low	Medium	Medium
Aspen, Canadian	Light	Low	Low	Medium	–
Avodiré	Medium	Medium	Very low	Medium	Low
Ayan	Heavy	Medium	Low	High	Low
Balsa	Except. light	Very low	Very low	Very low	–
Banak	Medium	Low	Low	Medium	Very low
Basralocus	Heavy	High	High	High	–
Basswood	Light	Low	Very low	Low	Low
Beech, European	Heavy	Medium	Medium	High	Medium
Berlinia	Heavy	Medium	Low	Medium	Low
Binuang	Light	Low	Very low	Low	Very low
Birch, European	Heavy	High	Medium	High	Medium
Birch, Paper	Medium	Medium	Low	Medium	Medium
Birch, Yellow	Heavy	High	Medium	High	High
Black bean	Heavy	–	–	–	–
Blackbutt	Very heavy	High	High	High	–
Blackwood, African	Except. heavy	–	–	–	–
Blackwood, Australian	Heavy	Medium	Medium	High	–
Bombax	Light	–	–	–	–
Bombway, White	Medium	Low	Low	Medium	Low
Boxwoods	Very heavy	–	–	–	–
Brush box	Very heavy	High	Medium	High	–
Camphorwood, East African	Medium	Medium	Low	Medium	Low
Canarium, African	Medium	Low	Very low	Medium	Very low
Canarium, Indian	Light	Low	Very low	Medium	Very low
Ceiba	Light	–	–	–	–
Celtis	Heavy	High	High	High	High
'Central American cedar'	Medium	–	–	–	–
Cherry	Medium	Medium	Low	Medium	Medium
Chestnut, Sweet	Medium	Low	Very low	Medium	Very low
Chickrassy	Medium	Medium	Low	Medium	Medium
'Chilean laurel'	Medium	Medium	Very low	Medium	Very low

Natural Durability	Resistance to Impregnation	Drying Properties		Woodworking Properties	
		Drying rate	Movement in service	Resistance in cutting	Blunting effect
Perishable	Moderate*	Rapid	Small	Medium	Moderate (occ. severe)
Non-durable	Moderate	Rapid	Small	Medium	Slight
Moderate	Extreme	Fairly rapid	Small	Medium	Slight
Very durable	Extreme	Rather slow	Small	Medium	Moderate
Very durable	Extreme	Very slow	Small	High	Moderate
Durable	Resistant	Fairly rapid	Small	Medium	Slight
Very durable	Extreme	Slow	Small	Medium (variable)	Moderate
Perishable	Permeable	Fairly rapid	–	Low	Slight
Perishable	Permeable	Rapid	Small	Low	Slight
Perishable†	Permeable	–	–	–	Moderate (variable)
Perishable	Permeable	Fairly rapid	Small	Low	Slight
Non-durable	–	Slow	–	High	Moderate
Perishable	Moderate	Fairly rapid	Medium	Medium	Moderate
Non-durable	Extreme	–	–	Low	Slight
Non-durable	Extreme	Fairly rapid	Small	Medium	Slight
Moderate	Resistant	–	Small	Medium (variable)	Moderate (occ. severe)
Perishable	Resistant*	Rapid	Small	Very low	Slight
Perishable	Permeable	Rather slow	–	Low	Slight
Very durable	Extreme	–	–	–	Severe
Non-durable	Permeable	–	–	Low	Slight
Perishable	Permeable	Fairly rapid	Large	Medium (variable)	Moderate (variable)
Non-durable	Resistant	Rather slow	Medium	Medium (variable)	Moderate
Perishable	Moderate*	Slow	Small	Low	Slight
Perishable	Permeable	Fairly rapid	–	Medium	Moderate
Non-durable	Moderate	–	–	Medium	Moderate
Perishable	Moderate	Rather slow	Large	Medium	Moderate
Durable	Extreme	Very slow	Medium	High	Moderate
Very durable	Extreme	–	–	High	Moderate
Very durable†	–	–	Small	Very high	Fairly severe
Durable†	Extreme	–	–	Medium	Moderate
Perishable	Permeable	–	–	Low	Slight
Non-durable	Moderate	Fairly rapid	Small	Medium	Moderate
Durable†	–	Very slow	–	High	Moderate
Moderate	Extreme	–	–	High	Severe
Very durable	Extreme	Slow	Small	Medium	Slight
Non-durable	Extreme	Rather slow	Medium	Low	Severe
Perishable	Extreme	–	–	Low	Slight
Perishable	Permeable	Rapid	–	Low	Slight
Perishable	Moderate	Fairly rapid	Medium	High	Moderate
Durable	Extreme	Fairly rapid	Small	Low	Slight
Moderate	–	Fairly rapid	Medium	Medium	Moderate
Durable	Extreme	Slow	Small	Medium	Slight
Non-durable†	Extreme	–	Small	Medium	Moderate
–	Moderate	Fairly rapid	Large	Low	Slight

† Provisional.
* These timbers often contain large amounts of permeable sapwood.

Appendix 1
Properties of Hardwoods—*continued*

Name	Density	Mechanical Properties			
		Bending strength	Stiffness	Crushing strength	Shock resistance
Chuglam, White	Heavy	Medium	Medium	Medium	–
Coigue	Medium	Medium	Low	High	–
Cordia, West African	Light	Low	Very low	Medium	Very low
Crabwood	Medium	Medium	Medium	High	Low
Cramantee	Medium	Medium	Very low	Medium	–
Curupay	Except. heavy	Very high	High	Very high	Very high
Dahoma	Heavy	Medium	Low	High	Low
Danta	Heavy	High	Low	High	Medium
Degame	Very heavy	High	Medium	High	–
Dogwood	Very heavy	Medium	Low	High	–
Ebony, African	Very heavy	Very high	High	Very high	High
Ebony, East Indian	Except. heavy	–	–	–	–
Ekki	Except. heavy	Very high	High	Very high	High
Elm, English and Dutch	Medium	Low	Very low	Low	Very low
Elm, Rock	Heavy	Medium	Very low	Medium	Very high
Elm, White	Medium	Medium	Very low	Medium	High
Elm, Wych	Heavy	Medium	Low	Medium	Medium
Eng	Very heavy	High	High	High	High
Esia	Very heavy	High	Medium	High	Medium
Freijo	Medium	Medium	Medium	Medium	–
Gaboon	Light	Low	Very low	Medium	–
Gedu nohor	Medium	Low	Very low	Medium	Low
Greenheart	Except. heavy	Very high	Very high	Very high	High
Grevillea	Medium	–	–	–	–
Guarea (*cedrata*)	Medium	Medium	Very low	Medium	Low
Guarea (*thompsonii*)	Medium	Medium	Low	High	Low
Gum, Spotted	Very heavy	–	–	–	–
Gurjun – see Keruing					
Hickory	Very heavy	High	High	High	Very high
Holly	Heavy	–	–	–	–
Hornbeam	Heavy	Medium	Medium	High	Medium
Horse-chestnut	Medium	Low	Very low	Medium	–
Idigbo	Medium	Low	Very low	Medium	Very low
Ilomba	Medium	–	–	–	–
'Indian laurel'	Very heavy	Medium	Medium	High	–
Iroko	Medium	Medium	Very low	Medium	Very low
Ironbark	Except. heavy	Very high	Very high	Very high	–
Jacareuba (Santa Maria)	Medium	Medium	Low	High	Medium
Jarrah	Very heavy	Medium	Medium	High	–
Jelutong	Medium	Low	Very low	Low	–
Jequitiba	Medium	Medium	Very low	Medium	Low

Natural Durability	Resistance to Impregnation	Drying Properties		Woodworking Properties	
		Drying rate	Movement in service	Resistance in cutting	Blunting effect
Moderate†	Moderate to extreme	–	–	Medium	Moderate
Moderate†	Resistant*	–	–	Medium	Moderate
Non-durable to very durable	Resistant	Fairly rapid	Small	Low	Slight
Moderate	–	Rather slow	Small	Medium	Moderate
Moderate	Extreme	Rather slow	Small	Medium	Moderate
Very durable	Extreme	Very slow	Medium	Very high	Severe
Durable	Resistant	Slow	Medium	Medium	Moderate (variable)
Durable	Resistant	Rather slow	Medium	Medium	Moderate
–	–	–	–	–	–
–	–	Rather slow	Large	High	–
Very durable†	Extreme	–	–	Very high	Severe
Very durable	–	–	–	Very high	Severe
Very durable	Extreme	Very slow	Large	Very high	Severe
Non-durable	Moderate	Fairly rapid	Medium	Medium	Moderate
Non-durable	Resistant	–	–	Medium	Moderate
Non-durable	Moderate	–	–	Medium	Moderate
Non-durable	Resistant	Fairly rapid	–	Medium	Moderate
Moderate	–	Slow	–	High	Moderate (occ. severe)
Durable	Extreme	Slow	Large	High	Moderate
Durable	–	–	–	Low	Moderate
Non-durable	Resistant	Rapid	–	–	Moderate to severe
Moderate†	Extreme	Fairly rapid	Small	Medium	Moderate
Very durable	Extreme	Very slow	Medium	High	Moderate
Moderate†	Moderate	Slow	Medium	Low	–
Very durable	Extreme	Fairly rapid	Small	Medium	Moderate
Very durable	Extreme	Fairly rapid	Small	Medium	Slight
Moderate	Extreme	–	–	Medium	Moderate
Non-durable	Moderate	–	–	High (variable)	Moderate to severe
Perishable	–	–	Large	High	Moderate
Perishable	Permeable	Fairly rapid	Large	High	Moderate
Perishable	Permeable	–	Small	Low	Slight
Durable	Extreme	Rapid	Small	Medium (variable)	Slight
Perishable	Permeable	–	–	–	–
Moderate†	Resistant	Slow	–	High	Moderate
Very durable	Extreme	Fairly rapid	Small	Medium	Moderate (occ. severe)
Very durable	Extreme	–	–	Very high	Severe
Durable	Extreme	Slow	Medium	Medium	Moderate (occ. severe)
Very durable	Extreme	–	Medium	High	Moderate
Non-durable	Permeable	Fairly rapid	Small	Low	Slight
Durable	Extreme	–	–	Low	Slight

† Provisional.
* These timbers often contain large amounts of permeable sapwood.

P

Appendix 1
Properties of Hardwoods—*continued*

Name	Density	Mechanical Properties			
		Bending strength	Stiffness	Crushing strength	Shock resistance
Kapur	Heavy	High	Medium	High	Medium
Karri	Very heavy	High	High	High	–
Kempas	Very heavy	High	High	High	–
Keruing	Heavy	High	High	High	Medium
Kokko	Medium	–	–	–	–
Krabak (Mersawa)	Medium	–	–	–	–
Kurokai	Heavy	Medium	Medium	High	Medium
Lancewood	Very heavy	–	–	–	–
Lignum vitae	Except. heavy	–	–	Very high	–
Lime, European	Medium	Medium	Low	Medium	Low
Lingue	Medium	Medium	Low	Medium	–
Louro, Red	Medium	Low	Low	Medium	–
Mahogany, African	Medium	Low	Very low	Medium	Very low
Mahogany, African (*Khaya grandifoliola*)	Heavy	Medium	Low	High	Low
Mahogany, American	Medium	Low	Very low	Medium	Very low
Mahogany, Mozambique	Medium	Low	Very low	Medium	Very low
Makoré	Medium	Medium	Low	Medium	Low
Mansonia	Medium	High	Low	High	Medium
Maple, Rock	Heavy	High	Medium	High	High
Maple, Soft	Medium	Medium	Low	Medium	Low
Mengkulang	Heavy	Medium	Medium	High	Medium
Meranti, Dark red	Heavy	Medium	Low	Medium	Low
Meranti, Light red	Medium	Medium	Low	Medium	Low
Meranti, White	Heavy	–	–	–	–
Meranti, Yellow	Medium	Low	Very low	Medium	Low
Merbau	Heavy	–	–	–	–
Missanda	Very heavy	High	High	Very high	Low
Mora	Except. heavy	High	High	Very high	High
Mtambara	Medium	–	–	–	–
Mubura	Heavy	High	Medium	High	Low
Muhimbi	Very heavy	High	Medium	High	Medium
Muhuhu	Very heavy	Medium	Low	High	Very low
Muninga	Medium	Medium	Very low	High	Low
Musizi	Medium	Low	Very low	Medium	Very low
Mutenye	Very heavy	–	–	–	–
Niangon	Medium	Medium	Very low	Medium	Very low
Nyatoh	Heavy	–	–	–	–
Oak, American red	Heavy	Medium	Medium	High	–

Natural Durability	Resistance to Impregnation	Drying Properties		Woodworking Properties	
		Drying rate	Movement in service	Resistance in cutting	Blunting effect
Very durable	Extreme	Rather slow	Medium	Medium (variable)	Moderate (occ. severe)
Durable	Extreme	–	Large	High	Moderate to severe
Durable	Resistant	–	–	High	Moderate to severe
Moderate	Moderate to resistant	Slow	Medium to large	Medium (variable)	Moderate to severe
Moderate	Moderate	–	–	Medium	Moderate
Moderate	Moderate	Very slow	Medium	Medium	Severe
Non-durable	Extreme	Fairly rapid	Medium	Medium	Moderate
Non-durable	–	–	–	High	Moderate
Very durable	Extreme	Very slow	Medium	Very high	Moderate
Perishable	Permeable	Fairly rapid	Medium	Low	Slight
Moderate	–	–	–	Medium	Slight
Durable	Extreme	Rather slow	–	Medium	Slight
Moderate	Extreme	Fairly rapid	Small	Medium (variable)	Moderate
Durable†	Extreme	Rather slow	Small	High	Moderate
Durable	Extreme	Fairly rapid	Small	Medium	Slight
Moderate†	Extreme	Fairly rapid	Small	Medium	Moderate
Very durable	Extreme	Fairly rapid	Small	Medium	Severe
Very durable	Extreme	Fairly rapid	Medium	Medium	Moderate
Non-durable	Resistant	Slow	Medium	High	Moderate
Non-durable	Moderate	–	–	Medium	Moderate
Non-durable	Resistant	Rapid	Small	Medium	Severe
Moderate to durable	Resistant to extreme	Rather slow	Small	Medium (variable)	Slight
Non-durable to moderate	Resistant to extreme	Fairly rapid	Small	Low	Slight
Moderate†	Resistant to extreme	–	–	Medium	Severe
Moderate	Extreme	Slow	Small	Low	Moderate
Durable	–	Fairly rapid	Small	High	Moderate
Very durable	–	Slow	Small	High	–
Durable	Extreme	Very slow	Large	High	Moderate to severe
Perishable	Moderate	Rapid	Medium	Medium	–
Non-durable	Moderate	Very slow	Large	High	Severe
Durable	Moderate to resistant	Slow	Medium	High	Severe
Very durable	Extreme	Fairly rapid	Small	Medium	Moderate
Very durable	Resistant	Rather slow	Small	Medium	Moderate
Non-durable	Permeable	Fairly rapid	Small	Low	Slight
Moderate†	–	–	Medium	–	–
Durable	Extreme	Fairly rapid	Medium	Medium	Moderate
Non-durable to moderate	Extreme	Rather slow	–	–	Moderate to severe
Non-durable	Moderate	Slow	Medium	Medium	–

† Provisional.
* These timbers often contain large amounts of permeable sapwood.

P*

Name	Density	Mechanical Properties			
		Bending strength	Stiffness	Crushing strength	Shock resistance
Oak, American white	Heavy	Medium	Low	Medium	–
Oak, European	Heavy	Medium	Low	Medium	Low
Oak, Japanese	Heavy	–	–	–	–
Oak, Turkey	Very heavy	High	Low	High	High
Obeche	Light	Low	Very low	Low	Very low
Odoko	Medium	Medium	Medium	High	–
Ogea	Medium	Low	Very low	Medium	Very low
Okan	Very heavy	High	High	Very high	–
Okwen	Medium to heavy	Medium	Low	Medium	Low
Olive, East African	Very heavy	High	High	High	High
Omu	Medium	Medium	Low	High	–
Opepe	Heavy	Medium	Medium	High	Low
Padauk, African	Heavy	–	–	–	–
Padauk, Andaman	Heavy	Medium	Low	High	Low
Padauk, Burma	Very heavy	High	Medium	High	Medium
Panga panga	Very heavy	–	–	–	–
Pear	Heavy	–	–	–	–
Peroba Rosa	Heavy	Medium	Low	High	–
Peroba, White	Heavy	Medium	Low	High	Low
Persimmon	Very heavy	High	Medium	High	–
Plane, European	Medium	–	–	Medium	Medium
Poplar, Black Italian	Light	Low	Very low	Medium	Very low
Poplar, Grey	Medium	Low	Very low	Medium	Low
Pterygota	Medium	Medium	Low	Medium	Medium to low
Purpleheart	Very heavy	High	High	High	Medium
Pyinkado	Very heavy	High	High	High	Medium
Quaruba	Medium	Medium	Low	Medium	–
'Queensland walnut'	Heavy	Medium	Low	High	–
Ramin	Heavy	High	Medium	High	Low
Rauli	Medium	Medium	Very low	Medium	Low
'Rhodesian teak'	Very heavy	–	–	–	–
Robinia	Heavy	–	–	–	–
Rosewood, Honduras	Very heavy	–	–	–	–
Rosewood, Indian	Very heavy	High	Low	High	–
Sapele	Medium	Medium	Low	High	Medium
Satinwood, Ceylon	Very heavy	High	Medium	High	Low
Satinwood, West Indian	Very heavy	–	–	–	–
Sepetir, Swamp	Heavy	High	Medium	High	Low
Seraya – see Meranti					
Seraya, White	Medium	Low	Low	Medium	Low
Serrette	Heavy	High	Medium	High	Low
Silky oak, Australian	Medium	–	–	–	–

Natural Durability	Resistance to Impregnation	Drying Properties		Woodworking Properties	
		Drying rate	Movement in service	Resistance in cutting	Blunting effect
Durable	Extreme	Slow	Medium	Medium (variable)	Moderate
Durable	Extreme	Very slow	Medium	Medium (variable)	Moderate (variable)
Durable†	—	—	Medium	Medium	—
Moderate	Extreme	Very slow	Large	High	Moderate
Non-durable	Resistant*	Rapid	Small	Very low	Slight
Non-durable	Permeable	Fairly rapid	Medium	Medium	Moderate
Perishable	Moderate*	Fairly rapid	Medium	Low	Slight (variable)
Very durable	Extreme	Slow	—	Very high	Fairly severe
Moderate	Extreme	Slow	Medium	Medium	Moderate
Moderate	Moderate	Slow	Large	High	Moderate
Moderate	—	Rather slow	Medium	Medium	Moderate
Very durable	Moderate	Rather slow	Small	Medium	Moderate
Very durable	Moderate	Fairly rapid	Small	Medium	—
Very durable	Moderate	—	—	High	Moderate
Very durable	Extreme	Rather slow	—	High	Moderate
Durable†	Extreme	Very slow	Small	—	Moderate
Non-durable	—	Slow	—	High	Moderate
Durable	Extreme	—	—	Medium	Moderate
Very durable	Resistant	—	—	—	—
—	—	Fairly rapid	Large	—	—
Perishable	—	Fairly rapid	—	Medium	Moderate
Perishable	Moderate*	Fairly rapid	Medium	Medium	Slight
Perishable	Resistant*	Fairly rapid	Medium	Medium	Slight
Perishable	Permeable	Fairly rapid	Medium	Medium	Moderate
Very durable	Extreme	Fairly rapid	Small	High	Moderate to severe
Very durable	Extreme	—	Medium	Very high	Severe
Non-durable to moderate	Permeable to resistant	Fairly rapid	—	Low	Moderate
Non-durable	—	Fairly rapid	—	Medium	Severe
Perishable	Permeable	Fairly rapid	Large	Medium	Moderate
Durable†	Moderate	Rather slow	Small	Medium	Slight
Very durable	Extreme	Slow	Small	High	Severe
Durable	Extreme	Slow	—	Medium (variable)	Moderate
Very durable	—	Slow	—	High	Moderate
Very durable	—	Rather slow	Small	High	Moderate
Moderate	Resistant	Fairly rapid	Medium	Medium	Moderate
Durable†	Extreme	—	—	High	Moderate
Non-durable	—	—	—	High	Moderate
Durable	Extreme	Rather slow	Small	Medium	Moderate
Non-durable to moderate	Extreme	Fairly rapid	Small	Medium	Slight
Moderate	Moderate	Rather slow	Medium	Medium	Moderate
Moderate†	—	—	—	Low	Slight

† Provisional.
* These timbers often contain large amounts of permeable sapwood.

Appendix 1
Properties of Hardwoods—*continued*

Name	Density	Mechanical Properties			
		Bending strength	**Stiffness**	**Crushing strength**	**Shock resistance**
'Silver beech'	Medium	Medium	Low	Medium	Low
Sterculia, Brown	Very heavy	High	Medium	High	High
Sterculia, Yellow	Heavy	High	Medium	High	Medium
Sucupira	Very heavy	High	–	High	–
Sycamore	Medium	Medium	Very low	Medium	Low
Tallowwood	Very heavy	Very high	Very high	Very high	–
'Tasmanian myrtle'	Heavy	Medium	Medium	High	–
'Tasmanian oak'	Medium to heavy	Medium	Medium	High	–
Tawa	Medium	–	–	–	–
Teak	Medium	Medium	Low	High	Low
Tchitola	Medium	Medium	Low	High	Medium
Thingan	Heavy	–	–	–	–
Thitka	Heavy	–	–	–	–
Tupelo	Medium	Low	Very low	Medium	–
Utile	Heavy	Medium	Low	High	Low
Wallaba	Very heavy	High	Medium	High	Low
Walnut, American	Medium	Medium	Low	Medium	–
Walnut, European	Medium	–	–	–	–
Wamara	Except. heavy	Very high	Very high	Very high	Very high
Wattle	Heavy	High	Medium	High	High
Whitewood, American	Medium	Low	Low	Medium	–
Willow	Light	Low	Very low	Low	Low

Natural Durability	Resistance to Impregnation	Drying Properties		Woodworking Properties	
		Drying rate	Movement in service	Resistance in cutting	Blunting effect
Non-durable	Extreme	–	–	Medium	Moderate
Moderate	Extreme	Very slow	Large	Medium (variable)	Moderate
Non-durable	Extreme*	Slow	Medium	Medium	Moderate
Very durable†	–	–	–	High	–
Perishable	Permeable	Fairly rapid	Medium	Medium	Moderate
Very durable	Extreme	Very slow	Medium	High	Severe
Non-durable	–	–	–	Medium	Moderate
Moderate	Resistant	Fairly rapid	Medium	Medium (variable)	Moderate
Non-durable	–	Fairly rapid	–	Medium	Moderate
Very durable	Extreme	Rather slow	Small	Medium	Severe (variable)
Non-durable†	Permeable	–	–	–	–
–	–	Slow	Small	High	–
Durable	Resistant	Slow	–	Medium	Moderate
–	–	–	–	Medium	–
Durable	Extreme	Fairly rapid	Medium	Medium	Moderate
Very durable	Extreme	Very slow	–	Medium	Moderate
Very durable	–	Rather slow	–	Medium	–
Moderate	Resistant	Rather slow	Medium	Medium	Moderate
Durable	Extreme	Slow	Medium	High	Severe
Non-durable	Moderate	–	Large	–	–
Non-durable	–	–	–	Low	–
Perishable	Resistant*	Fairly rapid	Small	Low	Slight

† Provisional.
* These timbers often contain large amounts of permeable sapwood.

Appendix II

Types of Saws

Rip-sawing

Table 1
660 mm diameter circular saws used for ripping tests on seasoned timber

Type	No. of teeth	Pitch (mm)	Depth of gullet (mm)	Hook (deg.)	Clearance (deg.)	Top bevel (deg.)
A10				10		
A15	40	50·8	17·5	15	15	10
A20				20		
B10				10		
B15	46	44·4	17·5	15	15	10
B20				20		
B25				25		
C10				10		
C15	54	38·1	17·5	15	15	10
C20				20		
C25				25		
D10				10		
D15	66	31·7	12·7	15	15	10
D20				20		
D25				25		
E10	80	25·4	11·1	10	15	10
E15				15		

Saws are 2·77 mm thick, with set per side 0·4 mm.

Table 2

Specifications for circular saws recommended for rip-sawing the timbers described in the Handbook

Saw type	Density range (kg/m³, seasoned) Timber type	Hook angle	No. of teeth	Sharpness angle	Top bevel angle
SR 48	Less than 560 The lighter hardwoods	30°	48	45°	15°
HR 54	Over 560 to 800 Hardwoods in the medium and heavy density range	20°	54	55°	15°
HR 60	Over 800 to 1040 Hardwoods in the very heavy range	15°	60	60°	10°
HR 80	Over 1040 Exceptionally dense hardwoods	10°	80	65°	5°
HR 40	Abrasive and fibrous hardwoods	20°	40	55°	10°
Tungsten carbide-tipped saw	Abrasive hardwoods	15°	54	65°	0°

Notes: (1) The saw types SR 48, HR 54, HR 60 and HR 80 are the same as those defined in B.S. 411:1969, Table 5.

(2) For plate saws —
(a) the gullet depth to pitch ratio is 0·4 to 0·5;
(b) the set per side is 0·4 mm;
(c) the clearance angle is 15°.

Cross-cutting

Table 3

Specifications for 458 mm diameter circular saws used for cross-cutting tests on seasoned timber

Saw type	Type of cross-cutting operation	Type of tooth	Hook angle	No. of teeth	Clearance angle	Sharpness angle	Gullet depth/ pitch ratio	Set/ side mm
1	General	Topped	5° Neg.	64	25°	70°	0·7	0·25
2	Dimension	Peg	10° Neg.	110	40°	60°	—	0·25
3	Dimension	Peg and raker	10° Neg. and 15°	96 and 24	—	—	—	0 (hollow ground)

Bandsawing

Table 4
Narrow bandsaw used for cutting tests on seasoned timber

Blade width	Blade gauge	Pitch of teeth	Hook angle	Gullet angle	Set/side
20 mm	1·07 mm	6·3 mm	5°	60°	0·25 mm

Table 5
Wide bandsaws recommended for re-sawing

Type of wide bandsaw blade	Density range of timber kg/m³	lb/ft³	Hook angle	Clearance angle
A	under 560	under 35	25°–30°	15°
B	560 to 800	35 to 50	20°	15°
C	over 800	over 50	15°	10°

Note: This table is the same as Table 6 of B.S. 4411:1969.

Appendix III

Kiln Schedules*

It must be emphasised that these schedules are subject to revision and are approximate only, representing conditions suitable for average qualities of timber intended for normal use. Experience with timber from particular sources of supply destined for specific purposes will generally enable modifications to be made. Such modifications may indeed often prove to be very pronounced since, for example, good quality small dimension material will almost always be found to tolerate appreciably severer drying conditions than through and through planks from poorer quality wood of the same species.

These schedules are designed for use with timbers up to about 38 mm (1½ in) thick, dried in a forced draught kiln. Thicker dimensions require somewhat higher humidities to prevent severe moisture gradients from developing. When drying timber between 38 mm (1½ in) and 75 mm (3 in) thick, the relative humidity should be 5 per cent higher at each stage of the appropriate schedule, and 10 per cent higher with wood 75 mm (3 in) or more in thickness.

Kiln Schedule A

Suitable for timbers which must not darken in drying and for those which have a pronounced tendency to warp but are not particularly liable to check.

Moisture content (%) of the wettest timber on the air-inlet side at which changes are to be made	Temperature (Dry bulb)		Temperature (Wet bulb)		Relative humidity % (approx.)
	°C	°F	°C	°F	
Green	35	95	30·5	87	70
60	35	95	28·5	83	60
40	38	100	29	84	50
30	43·5	110	31·5	88	40
20	48·5	120	34	92	35
15	60	140	40·5	105	30

Kiln Schedule B

Suitable for timbers that are very prone to check.

Moisture content (%) of the wettest timber on the air-inlet side at which changes are to be made	Temperature (Dry bulb)		Temperature (Wet bulb)		Relative humidity % (approx.)
	°C	°F	°C	°F	
Green	40·5	105	38	101	85
40	40·5	105	37·5	99	80
30	43·5	110	39	102	75
25	46	115	40·5	105	70
20	54·5	130	46	115	60
15	60	140	47·5	118	50

*For full information on Kiln Drying see Kiln Operators Handbook – *A Guide to the Kiln Drying of Timber* by W C Stevens and G H Pratt (1961). HM Stationery Office, £1·38 (by post, £1·42).

Kiln Schedule C

Moisture content (%) of the wettest timber on the air-inlet side at which changes are to be made	Temperature (Dry bulb)		Temperature (Wet bulb)		Relative humidity % (approx.)
	°C	°F	°C	°F	
Green	40·5	105	38	101	85
60	40·5	105	37·5	99	80
40	43·5	110	39	102	75
35	43·5	110	38	100	70
30	46	115	39·5	103	65
25	51·5	125	43	109	60
20	60	140	47·5	118	50
15	65·5	150	49	121	40

Kiln Schedule D

Moisture content (%) of the wettest timber on the air-inlet side at which changes are to be made	Temperature (Dry bulb)		Temperature (Wet bulb)		Relative humidity % (approx.)
	°C	°F	°C	°F	
Green	40·5	105	38	101	85
60	40·5	105	37·5	99	80
40	40·5	105	35·5	96	70
35	43·5	110	36	97	60
30	46	115	36	97	50
25	51·5	125	38	101	40
20	60	140	40·5	105	30
15	65·5	150	44·5	112	30

Kiln Schedule E

Moisture content (%) of the wettest timber on the air-inlet side at which changes are to be made	Temperature (Dry bulb)		Temperature (Wet bulb)		Relative humidity % (approx.)
	°C	°F	°C	°F	
Green	48·5	120	46	115	85
60	48·5	120	45	113	80
40	51·5	125	46·5	116	75
30	54·5	130	47	117	65
25	60	140	49	120	55
20	68	155	53	127	45
15	76·5	170	58	136	40

Kiln Schedule F

Moisture content (%) of the wettest timber on the air-inlet side at which changes are to be made	Temperature (Dry bulb)		Temperature (Wet bulb)		Relative humidity % (approx.)
	°C	°F	°C	°F	
Green	48·5	120	44	111	75
60	48·5	120	43	109	70
40	51·5	125	43	109	60
30	54·5	130	43	109	50
25	60	140	46	115	45
20	68	155	51	124	40
15	76·5	170	58	136	40

Kiln Schedule G

Suitable for timbers which dry very slowly, but which are not particularly prone to warp.

Moisture content (%) of the wettest timber on the air-inlet side at which changes are to be made	Temperature (Dry bulb)		Temperature (Wet bulb)		Relative humidity % (approx.)
	°C	°F	°C	°F	
Green	48·5	120	46	115	85
60	48·5	120	45	113	80
40	54·5	130	50·5	123	80
30	60	140	55	131	75
25	71	160	63·5	146	70
20	76·5	170	64	147	55
15	82	180	62·5	144	40

Kiln Schedule H

Moisture content (%) of the wettest timber on the air-inlet side at which changes are to be made	Temperature (Dry bulb)		Temperature (Wet bulb)		Relative humidity % (approx.)
	°C	°F	°C	°F	
Green	57	135	53	127	80
50	57	135	52	126	75
40	60	140	52	126	65
30	65·5	150	54	129	55
20	76·5	170	58	136	40

R

Kiln Schedule J

Moisture content (%) of the wettest timber on the air-inlet side at which changes are to be made	Temperature (Dry bulb)		Temperature (Wet bulb)		Relative humidity % (approx.)
	°C	°F	°C	°F	
Green	57	135	50·5	123	70
50	57	135	48	119	60
40	60	140	47·5	118	50
30	65·5	150	49	121	40
20	76·5	170	53	127	30

Kiln Schedule K

Moisture content (%) of the wettest timber on the air-inlet side at which changes are to be made	Temperature (Dry bulb)		Temperature (Wet bulb)		Relative humidity % (approx.)
	°C	°F	°C	°F	
Green	71	160	66	151	80
50	76·5	170	68·5	156	70
30	82	180	70·5	159	60
20	88	190	67·5	153	40

Kiln Schedule L

Moisture content (%) of the wettest timber on the air-inlet side at which changes are to be made	Temperature (Dry bulb)		Temperature (Wet bulb)		Relative humidity % (approx.)
	°C	°F	°C	°F	
Green	82	180	74	165	70
40	93·5	200	72	162	40

Kiln Schedule M

Moisture content (%) of the wettest timber on the air-inlet side at which changes are to be made	Temperature (Dry bulb)		Temperature (Wet bulb)		Relative humidity % (approx.)
	°C	°F	°C	°F	
Green	93·5	200	84·5	184	70
50	99	210	81·5	179	50

Index of Botanical Names

A

	Page
Acacia melanoxylon	43
mollissima	211
spp.	211
Acer campestre	196
mono	123
nigrum	123
platanoides	196
pseudoplatanus	196
rubrum	124
saccharinum	124
saccharum	123
spp.	123
Adina cordifolia	86
Aesculus hippocastanum	90
Afrormosia elata	15
Afzelia africana	16
bipindensis	16
pachyloba	16
quanzensis	16
Albizia adianthifolia	19
ferruginea	19
lebbeck	108
zygia	19
Alnus glutinosa	21
incana	21
Alstonia boonei	22
congensis	22
Amblygonocarpus obtusangulus	33
Anadenanthera macrocarpa	63
Androstachys johnsonii	125
Aningeria adolfi-friederici	22
altissima	22
psuedo-racemosa	22
robusta	22
spp.	22
Anisoptera spp.	109
Antiaris africana	23
welwitschii	23
Aspidosperma peroba	165
spp.	165
Aucoumea klaineana	78

B

	Page
Baikiaea plurijuga	180
Beilschmiedia tawa	200
Berlinia confusa	37
grandiflora	37
occidentalis	37
spp.	37
Betula alleghaniensis	40
lenta	41
lutea	40
papyrifera	40
pubescens	39
verrucosa	39
Bombax buonopozense	44
spp.	44

	Page
Bowdichia spp.	196
nitida	196
Brachylaena hutchinsii	136
Brachystegia eurycoma	156
kennedyi	156
leonensis	156
nigerica	156
spp.	156
Brya ebenus	60
Bulnesia arborea	111
Burkea africana	119
Buxus macowani	45, 46
sempervirens	45
Byrsonima coriacea var. *spicata*	191

C

	Page
Calophyllum brasiliense	97
brasiliense var. *rekoi*	97
Calycophyllum candidissimum	66
Canarium euphyllum	51
schweinfurthii	49
Carapa guianensis	62
Cardwellia sublimis	191
Cariniana spp.	101
Carpinus betulus	88
Carya glabra	86
laciniosa	86
ovata	86
tomentosa	86
spp.	86
Castanea sativa	55
Castanospermum australe	42
Cedrela fissilis	53
odorata	53
spp.	53
Ceiba pentandra	51
Celtis adolfi-friderici	52
mildbraedii	52
zenkeri	52
spp.	52
Cephalosphaera usambarensis	134
Chlorophora excelsa	95
regia	95
Chloroxylon swietenia	186
Chukrasia tabularis	57
Cistanthera papaverifera	65
Combretodendron macrocarpum	77
africanum	77
Cordia goeldiana	77
millenii	61
platythyrsa	61
Cornus florida	67
Croton megalocarpus	139
Cylicodiscus gabunensis	155
Cynometra alexandri	135

D

	Page
Dalbergia latifolia	184

	Page
Dalbergia melanoxylon	43
nigra	183
stevensonii	183
Daniellia ogea	153
thurifera	153
Dialyanthera spp.	206
Dicorynia guianensis	33
paraensis	33
Diospyros celebica	68
crassiflora	67
ebenum	68
marmorata	68
melanoxylon	68
virginiana	167
spp.	67, 68
Diplotropis spp.	196
Dipterocarpus spp.	106
Distemonanthus benthamianus	29
Dracontomelum dao	164
Dryobalanops aromatica	102
beccarii	102
lanceolata	102
oblongifolia	102
spp.	102
Dyera costulata	100
lowii	100

E

Endiandra palmerstonii	176
Entandrophragma angolense	79
candollei	159
cylindricum	185
utile	205
Eperua falcata	207
grandiflora	207
Erythrophleum guineense	133
ivorense	133
Eucalyptus crebra	97
delegatensis	199
diversicolor	104
drepanophylla	97
maculata	85
marginata	99
microcorys	198
obliqua	199
paniculata	97
pilularis	42
regnans	199
saligna	85
siderophloia	97
sideroxylon	97
spp.	97
Eucryphia cordifolia	204

F

Fagara flava	187
Fagus crenata	36
sylvatica	35
spp.	36
Ferreirea spectabilis	196

	Page
Flindersia brayleyana	176
pimenteliana	176
Fraxinus americana	24
excelsior	25
mandschurica	27
nigra	24
pennsylvanica	24
spp.	24

G

Gonioma kamassi	45, 46
Gonystylus bancanus	177
spp.	177
Gossweilerodendron balsamiferum	18
Gossypiospermum praecox	45, 47
Goupia glabra	102
Grevillea robusta	82
Guarea cedrata	82
excelsa	62
thompsonii	82
Guaiacum officinale	111
spp.	111
Guibourtia arnoldiana	141
coleosperma	180
demeusii	48
ehie	161
pellegriniana	48
tessmannii	48

H

Heritiera simplicifolia	126
spp.	126
Hopea odorata	203
Hura crepitans	91

I

Ilex aquifolium	87
Intsia bijuga	132
palembanica	132

J

Juglans nigra	208
regia	209

K

Khaya anthotheca	114
grandifoliola	116
ivorensis	114
nyasica	119
senegalensis	118
Koompassia malaccensis	105

L

Laurelia aromatica	57
Liquidambar styraciflua	85
Liriodendron tulipifera	211
Lophira alata	69
Lovoa trichilioides	13

M	Page
Maesopsis eminii	140
Malus sylvestris	24
Mansonia altissima	121
Millettia stuhlmannii	164
laurentii	164
Mimusops heckelii	119
Mitragyna ciliata	11
Mora excelsa	133
gonggrijpii	133

N	
Nauclea diderrichii	160
Nesogordonia papaverifera	65
Nothofagus cunninghamii	198
dombeyi	60
fusca	192
menziesii	192
procera	179
truncata	192
Nyssa aquatica	204
sylvatica	204
spp.	204

O	
Ochroma lagopus	30
Ocotea barcellensis	113
rodiaei	80
rubra	113
usambarensis	48
Octomeles sumatrana	38
Olea hochstetteri	158
Oxandra lanceolata	110
Oxystigma oxyphyllum	201

P	
Palaquium spp.	143
Parashorea spp.	189
Paratecoma peroba	166
Parinari excelsa	134
Payena spp.	143
Peltogyne spp.	173
Pentace burmanica	204
Pentacme spp.	127
Pericopsis elata	15
Persea lingue	113
Phoebe porosa	94
Phyllostylon brasiliensis	45, 47
Piptadenia africana	63
macrocarpa	63
Piptadeniastrum africanum	63
Platanus acerifolia	168
hybrida	168
Populus alba	169
canadensis var. *serotina*	169
canescens	169
nigra	169
robusta	170
serotina	170
tremula	169

	Page
tremuloides	27
spp.	169
Protium decandrum	110
Prunus avium	54
Pseudosindora palustris	188
Pterocarpus angolensis	138
dalbergioides	162
macrocarpus	163
soyauxii	162
Pterygopodium oxyphyllum	201
Pterygota bequaertii	171
macrocarpa	171
Pycnanthus angolensis	93
Pygeum africanum	135
Pyrus communis	165

Q	
Quercus alba	144
castaneaefolia	149
cerris	149
falcata var. *falcata*	143
ilex	148
lyrata	144
michauxii	144
mongolica var. *grosseserrata*	149
pedunculata	146
petraea	146
prinus	144
robur	146
rubra	143
sessiliflora	146
spp.	143, 144, 149

R	
Robinia pseudoacacia	182

S	
Salix alba	212
alba var. *coerulea*	212
fragilis	212
spp.	212
Sarcocephalus diderrichii	160
Scottellia coriacea	152
Shorea pauciflora	127
spp.	127, 129, 130, 131
Sindora spp.	187
Sterculia oblonga	194
rhinopetala	193
Swartzia leiocalycina	210
Swietenia macrophylla	117
mahagoni	117
spp.	117
Symphonia globulifera	206

T	
Tarrietia cochinchinensis	60
simplicifolia	126
utilis	141
Tectona grandis	201

235

	Page		Page
Terminalia alata	94	*glabra*	75
bialata	58	*hollandica* var. *hollandica*	71
coriacea	94	*laciniata*	73
crenulata	94	*procera*	71
ivorensis	91	*thomasii*	73
superba	12	spp.	73
Tetramerista glabra	172		
Tieghemella heckelii	119		
Tilia americana	34	*Virola koschnyi*	32
vulgaris	112	spp.	206
spp.	112	*Vochysia guianensis*	175
Triplochiton scleroxylon	151	*hondurensis*	175
Tristania conferta	47	*surinamensis*	175
Turraeanthus africanus	28	*tetraphylla*	175
		spp.	175

U

X

Ulmus americana	74	*Xylia dolabriformis*	174
davidiana var. *japonica*	73	*xylocarpa*	174

Index of Trade and Local Names

Names in bold type are standard names under which the timbers appear in the text

A	Page		Page
Abel	49	**European**	25
Abem	37	French	25
Aboudikro	185	Green	24
Abura	11	**Japanese**	27
Acajou blanc	114	White	24
d'Afrique	114	Asna	94
Afara	12	**Aspen**	27
Black	91	**Canadian**	27
Dark	12	European	169
Light	12	Trembling	27
White	12	Assacu	91
African silky-oak	82	Assié	205
'African walnut'	13	Asta	110
Afrormosia	15	'Australian maple'	176
Afzelia	16	'Australian oak'	199
Agba	18	**Avodiré**	28
Agboin	63	Awari	171
Aguano	117	Awun	22
Ahun	22	**Ayan**	29
Aiélé	49	Ayanran	29
Akom	12	Aye	193
Akomu	93	Ayinre	19
Albizia, W. African	19	Ayous	151
Alder	21	Azobé	69
Black	21		
Grey	21	B	
Aligna	16		
Alona wood	13	Badam	45
'Alpine ash'	199	Badi	160
Alstonia	22	Bagtikan	189
Amaranth	173	Bahia	11
Amazakoue	161	Baitoa	47
Amazoue	161	Baku	119
Ambila	138	**Balsa**	30
Andaman marble-wood	68	**Banak**	32
Andaman redwood	162	**Banga wanga**	33
Andiroba	62	Barwood	162
Anegré	22	**Basralocus**	33
Angélique	33	**Basswood**	34
Aningeria	22	**Beech,** Carpathian	35
Angico preto	63	Danish	35
Anokye	161	English	35
Antiaris	23	**European**	35
Apa	16	French	35
Apitong	106	**Japanese**	36
Apopo	13	Rumanian	35
Apple	24	Yugoslavian	35
Aprono	121	Benge	141
Arere	151	'Benin walnut'	13
Ash, American	24	Benin wood	114
Belgian	25	Benuang	38
Black	24	**Berlinia**	37
Brown	24	Bété	121
English	25	Betula wood	40
		Bibolo	13

Some timbers, although called walnut, teak, etc., are botanically unrelated to these woods. They are commonly described by their geographical origin, e.g.. 'Venezuelan boxwood' and are indexed under the initial letter of the combined name.

	Page		Page
Bilinga	160	Cebil	63
Bilsted	85	Cebil colorado	63
Binuang	38	'Cedar'	53
Birch, American	40	Cedro	53
American white	40	**Ceiba**	51
Canadian white	40	**Celtis, African**	52
Canadian yellow	40	**'Central American cedar'**	53
English	39	Chanfuta	16
European	39	Chenchen	23
Birch, Finnish	39	Chêne-limbo	12
Hard	40	**Cherry, European**	54
Paper	40	Wild	54
Quebec	40	**Chestnut,** European	55
Red	40	Spanish	55
Silver	39	**Sweet**	55
Swedish	39	Chewstick	206
Sweet	41	**Chickrassy**	57
White	39, 40	**'Chilean laurel'**	57
Yellow	40	Chittagong wood	57
Bisselon	118	**Chuglam, White**	58
Bissilongo	118	**Chumprak**	60, 126
Bitis	143	'Cigar box cedar'	53
Black bean	42	'Clinker beech'	192
Blackbutt	42	Cocus	60
Black gum	204	**Cocus wood**	60
Black locust	182	**Coigue**	60
Blackwood, African	43	Coihue	60
Australian	43	Congowood	13
Boar wood	206	**Cordia, W. African**	61
Bombax	44	Cornel	67
Bombay blackwood	184	Cotton gum	204
Bombway, White	45	**Crabwood**	62
Bongele	194	**Cramantee**	62
Bongossi	69	Cupiuba	102
Bonsamdua	29	**Curupay**	63
'Borneo camphorwood'	102		
Bossé	82		
Box	45	**D**	
Boxwood, Abassian	45		
East London	46	Dabema	63
European	45	**Dahoma**	63
Iranian	45	**Danta**	65
Persian	45	Daniellia	153
Turkey	45	Dao	164
Brachystegia	156	**Degame**	66
'Brazilian cedar'	53	Denya	155
'British Honduras cedar'	53	Determa	113
Brush box	47	Dhup	51
Bubinga	48	Dhup, White	51
Buna	36	Dibetou	13
		Distemonanthus	29
C		**Dogwood**	67
		Douka	119
Camphorwood, E. African	48	Doussié	16
Camwood	162	Duku	22
Canarium, African	49		
Indian	51		
Canary wood	211	**E**	
Caoba	117		
Cape box	46	Eba	69
		Ebiara	37

Some timbers, although called walnut, teak, etc., are botanically unrelated to these woods. They are commonly described by their geographical origin, e.g., 'Venezuelan boxwood' and are indexed under the initial letter of the combined name.

	Page
Ebony, African	67
Cameroon	67
Ceylon	68
E. Indian	68
Gaboon	67
Indian	68
Kribi	67
Macassar	68
Nigerian	67
Edinam	79
Ehie	161
Ekhimi	63
Ekki	69
Ekonge	194
Ekpogoi	37
Elemi	49
Elm, American	74
Canadian rock	73
Cork	73
Cork bark	73
Dutch	71
English	71
Hickory	73
Japanese	73
Mountain	75
Nave	71
Red	71
Rock	73
Scotch	75
Soft	74
White	74
Wych	75
Embuia	94
Emeri	91
Emien	22
Eng	106
Erima	38
Erun	133
Esa	52
Esia	77

F

	Page
False acacia	182
Fara	153
Figueroa	62
Fraké	12
Framiré	91
Freijo	77
Fromager	51
Fuma	51

G

	Page
Gaboon	78
Gean	54
Gedu lohor	79
Gedu noha	79
Gedu nohor	79
'Ghana walnut'	13

	Page
Goupi	102
Greenheart	80
Grevillea	82
Gu	187
Guarea	82
Black	82
Scented	82
White	82
Gum, American red	85
Blue	85
Maculata	85
Red	85
Saligna	85
Sap	85
Spotted	85
Sweet	85
Sydney blue	85
Gurjun	106
'Guyana cedar'	53

H

	Page
Haldu	86
'Hard beech'	192
Hickory	86
Mockernut	86
Pignut	86
Red	86
Shagbark	86
Shellbark	86
Hnaw	86
Hog gum	206
Holly, European	87
'Honduras cedar'	53
Hornbeam	88
Horse-chestnut, European	90
Hura	91
Hyeduanini	161

I

	Page
Idigbo	91
Ilimo	38
Ilomba	93
Imbuya	94
In	106
Incenso	153
'Indian laurel'	94
Indian silver-grey wood	58
Insenso	153
Intule	95
Ipé peroba	166
Iroko	95
Ironbark	97
Ita	52
Iteballi	175

J

	Page
Jacareuba	97
Jarrah	99

Some timbers, although called walnut, teak, etc., are botanically unrelated to these woods. They are commonly described by their geographical origin, e.g., 'Venezuelan boxwood' and are indexed under the initial letter of the combined name.

	Page		Page
Jelutong	100	**Limba**	12
Jequitiba	101	bariolé	12
rosa	101	blanc	12
		clair	12
K		Dark	12
		Light	12
Kabukalli	102	noir	12
Kajat	138	Limbo	12
Kajatenhout	138	**Lime,** American	34
Kaku	69	**European**	112
Kali	22	**Lingue**	113
Kalungi	79	Lolagbola	201
Kamassi	46	**Louro** inamuhy	113
'Kamassi boxwood'	46	**inamui**	113
Kambala	95	**Red**	113
Kapoer	102	vermelho	113
Kapor	102	Lovoa wood	13
Kapur	102	Lumbayan	126
Karri	104		
Kashit	204	**M**	
Kassa	133		
Kaunghmu	109	Macula	85
Kembang	126	**Mahogany, African**	114, 116
Kempas	105	**American**	117
Keruing	106	Benin	114
Kevazingo	48	Brazilian	117
Khaya	114	British Honduras	117
Kiatt	138	Costa Rica	117
Kirundu	23	Cuban	117
Kitola	201	Degema	114
'Knysna boxwood'	45, 46	**Dry-zone**	118
Kokko	108	Ghana	114
Kokrodua	15	Grand Bassam	114
Kopie	102	Guatemala	117
Korina	12	Honduras	117
Kotibé	65	Ivory Coast	114
Koto	171	Lagos	114
Krabak	109	Mexican	117
Krala	114	**Mozambique**	119
Krappa	62	Nicaraguan	117
Kurokai	110	Nigerian	114
Kusia	160	Nyasaland	119
Kusiaba	160	Peruvian	117
Kwao	86	Spanish	117
Kwari	175	Takoradi	114
Kwow	86	Mai pradoo	163
Kyenkyen	23	**Makarati**	119
		Makata	187
L		**Makoré**	119
		Manni	206
Lacewood	168	**Mansonia**	121
Lagos wood	114	**Maple,** Black	123
Lancewood	110	Field	196
Landosan	22	Great	196
Lauan, Red	127	Hard	123
White	129	**Japanese**	123
Lemonwood	66	Norway	196
Libengi	141	Red	124
Lignum vitae	111	**Rock**	123
Maracaibo	111	silkwood	176

Some timbers, although called walnut, teak, etc., are botanically unrelated to these woods. They are commonly described by their geographical origin, e.g., 'Venezuelan boxwood' and are indexed under the initial letter of the combined name.

	Page		Page
Silver	124	Naga	156
Soft	124	Nemesu	127
Sugar	123	Ngollon	114
White	123	**Niangon**	141
'Maracaibo boxwood'	45, 47	'Nicaraguan cedar'	53
Matakki	206	'Nigerian golden walnut'	13
Mazzard	54	'Nigerian walnut'	13
Mbaua	119	Nire	73
Mbembakofi	16	Noyer d'Afrique	13
Meblo	156	Noyer de Gabon	13
Mecrusse	125	Noyer du Mayombe	12
Melapi	130	Nyankom	141
Melawis	177	**Nyatoh**	143
Mengkulang	126	batu	143
Meranti, Dark red	127		
Light red	129	**O**	
White	130	**Oak, American red**	143
Yellow	131	**American white**	144
Merbau	132	Chestnut	144
Mersawa	109	Durmast	146
Messmate stringybark	199	English	146
'Mexican cedar'	53	**European**	146
Minzu	77	Evergreen	148
Missanda	133	French	146
Mkora	16	**Holm**	148
Mninga	138	**Japanese**	149
Mora	133	Northern red	143
Morabukea	133	Overcup	144
Moreira	95	Pedunculate	146
'Mountain ash'	199	**Persian**	149
Movingui	29	Sessile	146
Mtambara	134	Southern red	143
Muave	133	Swamp chestnut	144
Mubura	134	**Turkey**	149
Mueri	135	White	144
Muhimbi	135	Yugoslavian	146
Muhindi	135	Oba suluk	127
Muhuhu	136	**Obeche**	151
Mujwa	22	Obobo	82
Mukali	22	Obobonekwi	82
Mukangu	22	Obobonufua	82
Mukarati	119	**Odoko**	152
Mukushi	180	Odum	95
Mukusi	180	Ofram	12
Mukwa	138	Ofun	121
Muna	22	**Ogea**	153
Muninga	138	Ogiovu	23
Musharagi	158	Ogwango	114
Musibi	180	Ohia	52
Musine	139	**Okan**	155
Musizi	140	Okoko	194
Mussacossa	16	Okoumé	78
Mussive	180	Okuro	19
Mutenye	141	**Okwen**	156
Mutti	94	**Olive, E. African**	158
Muxibe	180	Omo	61
Mvule	95	**Omu**	159
Mwafu	49	**Opepe**	160
'Myrtle beech'	198	Orhamwood	74

Some timbers, although called walnut, teak, etc., are botanically unrelated to these woods. They are commonly described by their geographical origin. e.g., 'Venezuelan boxwood' and are indexed under the initial letter of the combined name.

	Page
Oriental wood	176
Oro	23
Osan	22
Ossol	206
Otie	93
Otutu	65
Ovangkol	161
Owewe	77
Oziya	153

P

Padauk	162
African	162
Andaman	162
Burma	163
Paldao	164
Palo de sangre	32
Palosapis	109
Panga panga	164
Patternwood	22
Pear	165
Peroba amarella	166
branca	166
de campos	166
Ipé	166
Red	165
rosa	165
White	166
Persimmon	167
'Peruvian cedar'	53
Petir	187, 188
Plane	196
Plane, English	168
European	168
French	168
London	168
Poplar	169, 211
Black	169
Black Italian	169
Grey	169
White	169
Yellow	211
Poroposo	171
Possentrie	91
Potrodom	133
Pradoo	163
Pterygota	171
Punah	172
Purpleheart	173
Pycnanthus	93
Pyinkado	174

Q

Quaruba	175
branca	175
vermelha	175
'Queensland maple'	176
'Queensland walnut'	176

R

	Page
Ramin	177
Rauli	179
'Red beech'	192
Rhodesian copalwood	180
'Rhodesian teak'	180
Robinia	182
Rosewood, Bahia	183
Brazilian	183
East Indian	184
Honduras	183
Indian	184
Rio	183

S

Sain	94
Samba	151
'San Domingo boxwood'	45, 47
Sand box	91
Sangre palo	32
Santa Maria	97
Sapele	185
Sapelli	185
Sapupira	196
Sasswood	133
Satinwood, Ceylon	186
E. Indian	186
Jamaican	187
San Domingan	187
W. Indian	187
Sepetir	187, 188
paya	188
Swamp	188
Seraya, Dark red	127
Light red	129
White	189
Yellow	131
Serrette	191
Sida	13
Silkwood	176
Silky-oak	82
African	82
Australian	191
Silky wood, Canadian	40
'Silver Beech'	192
Sindru	22
Sipo	205
Siris	108
'South American cedar'	53
'Southland beech'	192
'Spanish cedar'	53
Sterculia, Brown	193
Yellow	194
Stoolwood	22
Subaha	11
Sucupira	196
Black	196
Yellow	196

Some timbers, although called walnut, teak, etc., are botanically unrelated to these woods. They are commonly described by their geographical origin, e.g., 'Venezuelan boxwood' and are indexed under the initial letter of the combined name.

	Page
Supa	187
Sycamore	196
plane	196

T

'Tabasco cedar'	53
Tali	133
Tallowwood	198
Tamo	27
Tangare	62
Tapsava	32
'Tasmanian beech'	198
'Tasmanian myrtle'	198
'Tasmanian oak'	199
Taukkyan	94
Tawa	200
Tchitola	201
Teak	201
Thingan	203
Thitka	204
Tiama	79
Tola	18
branca	18
chimfuta	201
mafuta	201
White	18
'Trinidad cedar'	53
Tsongutti	22
Tule	95
Tulip tree	211
Tupelo	204
gum	204

U

Ulmo	204
Umbaua	119
Umgusi	180
Utile	205

V

'Venezuelan boxwood'	47
Verawood	111

	Page
Vermilion wood	162
Virola	206

W

Waika chewstick	206
Walele	93
Wallaba	207
Ituri	207
Soft	207
Walnut, American	208
bean	176
Black	208
English	209
European	209
French	209
Italian	209
Wamara	210
Wane	113
Ware	171
Wattle	211
Wawa	151
Wawabima	193
Wengé	164
'West Indian boxwood'	47
'West Indian cedar'	53
Whitewood, American	211
Willow	212
Close-bark	212
Common	212
Crack	212
Cricket-bat	212
White	212

Y

Yang	106
Yellow mangue	206
Yemeri	175
Yinma	57

Z

Zambesi redwood	180
Zapatero	47
Zebra wood	68

Some timbers, although called walnut, teak, etc., are botanically unrelated to these woods. They are commonly described by their geographical origin. e.g., 'Venezuelan boxwood' and are indexed under the initial letter of the combined name.

Printed in England for Her Majesty's Stationery Office
by Ebenezer Baylis and Son Limited, The Trinity Press, Worcester, and London
Dd 289431 K32

THE VICTORS

THE VICTORS

EDITED BY BRIGADIER PETER YOUNG

HAMLYN
London · New York · Sidney · Toronto
A BISON BOOK

Published by
The Hamlyn Publishing Group Limited
London · New York · Sydney · Toronto
Astronaut House, Feltham
Middlesex, England

© Copyright Bison Books Limited 1981

Produced by
Bison Books Limited
4 Cromwell Place
London SW7

ISBN 0 600 34166 6

First published 1981

Printed in Hong Kong

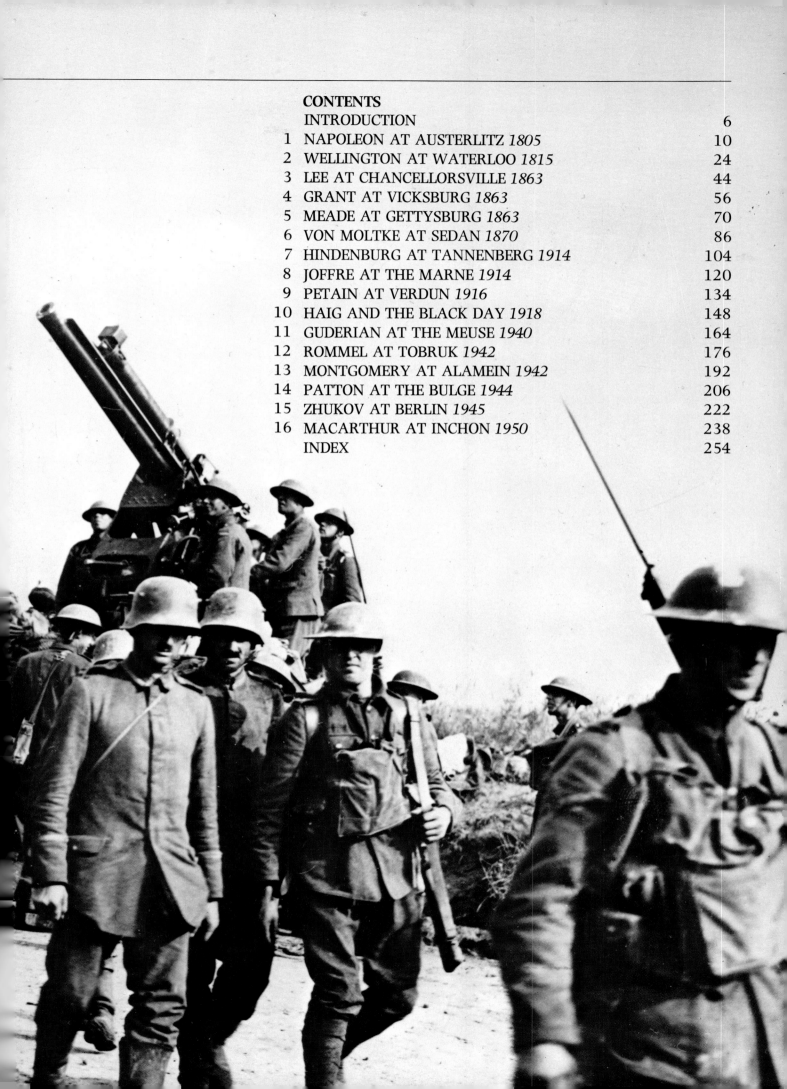

CONTENTS

INTRODUCTION 6
1 NAPOLEON AT AUSTERLITZ *1805* 10
2 WELLINGTON AT WATERLOO *1815* 24
3 LEE AT CHANCELLORSVILLE *1863* 44
4 GRANT AT VICKSBURG *1863* 56
5 MEADE AT GETTYSBURG *1863* 70
6 VON MOLTKE AT SEDAN *1870* 86
7 HINDENBURG AT TANNENBERG *1914* 104
8 JOFFRE AT THE MARNE *1914* 120
9 PETAIN AT VERDUN *1916* 134
10 HAIG AND THE BLACK DAY *1918* 148
11 GUDERIAN AT THE MEUSE *1940* 164
12 ROMMEL AT TOBRUK *1942* 176
13 MONTGOMERY AT ALAMEIN *1942* 192
14 PATTON AT THE BULGE *1944* 206
15 ZHUKOV AT BERLIN *1945* 222
16 MACARTHUR AT INCHON *1950* 238
INDEX 254

INTRODUCTION

After the disgraceful rout at Castlebar (27 August 1798) the Duke of Ormonde, the commander of the Kilkenny Militia, remarked that the British would have done better had there been no generals present. It was not the first nor the last occasion upon which such a thought occurred to a British colonel. No doubt that is true of other armies as well. How, for example, would the 7th Cavalry have got on had General Custer not been present at the Little Big Horn? (26 June 1876). Perhaps they would have stayed on their horses, and carved their way out. It was said of Inkerman (5 November 1854) that it was a soldiers' battle fought in a fog, the implication being that the generals, unable to see what was going on, could not influence the battle one way or the other. They could bring up reinforcements, but they could not tell what they were doing. Perhaps things had been much the same at Auerstadt (14 October 1806), when the Prussians, 63,000 strong, encountered a French corps (26,000) in a fog. They ought to have won, but unfortunately for them the French commander, Marshal Louis-Nicolas Davout was practically blind anyway: to him the 'fog of war' was no worse than usual and so he was able to sort things out very satisfactorily.

This 'fog of war' is a very real thing, and it is the general's job to pierce it, to answer the plaintive cry of his staff officers: 'What the hell is going on now?' Wellington had this gift, that is, the imagination to divine what was happening on 'the other side of the hill' as he put it, whether it was Soult, or Massena or Marmont, or Napoleon himself that he had to deal with.

The other thing which every general is up against is what Clausewitz called *friction de guerre*, and which may be translated as 'grit in the works.' It can never be altogether eradicated, there will always be valiant busybodies with a genius for misunderstanding their orders, even though accidents can to some extent be avoided by foresight and by training.

Good generals come in all different shapes and sizes. Some reach high rank as the result of long training, and carefully thought out selection procedures. Others are wafted upward by the changes and chances of the period in which they live. Some fortunate beings exercise command when they are still in the prime of life. Others find themselves faced with appalling crises when they are no longer young or fit.

What qualities would you look for if it fell to your lot to sit on a selection board charged with the promotion of general officers? Had you been one of Barras' colleagues would you have sent the young Bonaparte to Italy in 1796? Had Abraham Lincoln sought your advice would you have spoken up for Meade, or Reynolds or who. (There was not that much talent in the Army of the Potomac!) Would you have advised Winston S Churchill to send Montgomery to try conclusions with Rommel? Even with hindsight it is hard to say whether good decisions were made in these three cases. Bonaparte was to supplant the benefactors, who had given him a fair wind, and was to plunge Europe into wars of unprecedented violence. Meade, for all his good qualities, hardly ranks with the great captains of all time — the infamous Sherman, who does not figure in these pages, was, and it gives me no pleasure to say so, a much better general. As for the Montgomery problem it is difficult to see how, given the same quantity of reinforcements, and weapons, Auchinleck, who was after all an excellent tactician, could have failed to destroy the Afrika Korps. He might even have followed up victory with a relentless pursuit.

The victors paraded in these pages are 16 in number and the victories chosen to illustrate their talents fall in the 145 year period from 1805 to 1950. Some of them lived to a great age, and some did not. It is said that 'All they that take the sword shall perish with the sword' (St Matthew's gospel). This lot did not. Patton was the victim of a traffic accident and Rommel was compelled, by Hitler's agents, to take poison. The rest, it seems, died in their beds.

Other things being equal it is thought that a good young general is more likely to be useful than a good old one. Who can say? When Napoleon and Wellington, both born in 1769, met at Waterloo the one was stout and sick, the other young and brisk. Of the rest only Rommel, one of the youngest of all those under discussion, was seriously hampered by ill health during his campaigning days.

Needless to say some of these generals were more successful than others. Zhukov, who was only 48 when he took Berlin, alone of the 16 began his military career as a private soldier; from a brave NCO in the Czarist Army he became 'the general who never lost a battle.' Wellington and Moltke could certainly claim the same distinction, while Napoleon and Lee, for all their talents could not. Indeed had Napoleon been the servant of the French government instead of the Emperor one may doubt whether he would have been employed after 1812. The Moscow campaign, Leipzig and Waterloo were disasters that few reputations could have survived — except that of a Head of State!

Wellington would stand high in the list of the world's great generals, if only because he did not allow himself to be hypnotized by Napoleon's reputation, but — with a rather indifferent army — meted out to the Emperor the same treatment he had so often meted out to his marshals. '*Ça peut être une bataille d'Espagne*' said General Reille before the fight began. How right he proved. Napoleon proved no better able than Massena or Soult to solve the problems set by the defensive tactics which Wellington had perfected in the Peninsula. Moltke, one feels, would have approved of Wellington's ideas for not only did they both excel in the fields of staff work and logistics, but Moltke was 'the first to realize the great defensive power of modern firearms, and . . . inferred from it that an enveloping attack had become more formidable than the attempt to pierce an enemy's front' (Henry Spencer Wilkinson). Moltke's work is represented here by Sedan, but Königgratz (1866) was no less a masterpiece. He was an extraordinary man — scholar, traveller, strategist and linguist. He was so taciturn that he was said to be 'silent in seven languages.' His tall, spare figure, his tanned old face, his marvellous discretion all contributed to the legend of the 'man of gold,' the ideal character whom all admired, and who had no enemies. It is not given to many generals to make such an impression upon their contemporaries!

Robert E Lee enjoyed a rather similar reputation in the Confederate States' Army. It is true that he did not have a record of unbroken success to compare with those of Wellington or Moltke. Lee's ill-equipped armies were always outnumbered. In 1861 he was offered the command of all federal forces, and only rejected it because his own state, Virginia, seemed about to secede. The war between the States might have been brief indeed had Robert E Lee been Mr Lincoln's general. It is an interesting speculation.

Meade was a good officer, loyal and strict, and as the victor of Gettysburg deserves respect but he can scarcely be awarded a place among the great captains of all time. Nor can Ulysses S Grant, who, though he picked up his trade as he went along, was never more than a 'good, plain cook.' Hindenburg, too, has no serious claim to be considered one of the great captains. In fact, he merely supplied the backbone and personality for his chief of staff. Ludendorff had the brains and Hindenburg had the nerve, and at Tannenberg Hoffman provided the plan.

Joffre, though scarcely a thunderbolt of war, must take much of the credit for his one great victory, the Marne. He displayed throughout a monumental *phlegme*. He strained every nerve to see that the line did not break. Saddled with a fool for an ally — Sir John French — he managed by sheer force of personality to shame him into cooperation at the vital moment. Above all he proved ruthless in sacking the numerous generals who failed. He had determined and intelligent support from Messimy, the Minister for War, and from General Galliéni, the Governor of Paris, and one could say that the victory of the Marne was won by a triumvirate. Still let us not rob this phlegmatic Frenchman of his one great triumph. To his British allies, Joffre, with his monumental calm seemed very far from their idea of the volatile, unpredictable Gallic warrior. Somehow the conditions of 1914 demanded a Joffre. Who else of the French generals of 1914 could have survived the stresses and strains of the Battle of the Marne? It is not easy to supply a name. By 1940 his staff officer, Gamelin, had attained his rank. The little man was not the one to wield his master's sword. He had forgotten that an army must be prepared to take the offensive!

Haig is a very much underrated officer. His career commands but little interest these days. But is this fair? He had distinguished himself in the Sudan and in the Boer War and was a thoroughly well-trained officer, who had held appointments of the first importance. It is damning him with faint praise to say that he was a much better general than French, whom he succeeded, and in fact the worst things one can say of him is that — in an infantry/artillery war — he was a cavalryman, brought up to a very different style of fighting, and that he was incapable of talking to the soldiers in the 'matey' sort of way that Montgomery used to do in World War II — Haig had been properly brought up. The Somme and Passchendaele were at best Pyrrhic victories, but the same cannot be said of the great series of

battles in which, between 8 August and 11 November 1918, his armies pressed forward and broke the Hindenburg Line. Haig was the first of the Allied leaders to sense that the war could be won in 1918. In an order issued to his armies in the third week of August, he said that the situation had changed decisively, and that the time had come to press the enemy everywhere with the utmost energy. At this time Foch, no pessimist, was hoping for victory in 1919, while the British government was planning to make a final effort in 1920. Those who think of Haig as a butcher will do well to consider how many thousands of lives were saved by bringing the war to a swifter conclusion. After the war Haig devoted himself tirelessly to the cause of his old soldiers, uniting the various organizations for ex-service men into the British Legion, the largest benevolent organization ever created in Great Britain. His early death at 67 was evidently due to overwork. Although he did not appeal to the men in the way that Marlborough, Roberts, Wellington or Montgomery did, his veterans turned out in their thousands to line the route of his funeral procession.

MacArthur and Patton, Montgomery, Rommel and Zhukov all distinguished themselves in World War I. MacArthur collected a truly astonishing collection of medals in 1918, but, given the choice, the young Rommel, whose *Infanterie Greift An* (*Infantry Attacks!*) is still a classic, was the man one would choose to serve with! His flair for minor tactics was simply phenomenal. Rommel's personality was such that he was admired by the Eighth Army almost as much as by the Afrika Korps. In 1940 Guderian, with his 'get-a-move-on' attitude had shown the same flair for leadership. It came naturally. MacArthur, Patton and Montgomery all built themselves up as 'characters' in a way that was quite deliberate. Distasteful though it may seem it was no doubt a sensible exercise in Public Relations. Omar Bradley and 'Lightning Joe' Collins were more fortunate. The Press Corps took to them and no doubt to their surprise and amusement they found the title 'GI's general' conferred upon them. Nobody worked harder on his own image than Viscount Montgomery of Alamein — so much so that it is difficult to tell the substance from the shadow. He was well-served by his Chief of Staff, de Guingand, and by Sir Miles Dempsey, the commander, from Normandy to the Baltic, of the British Second Army — a man who was not in search of laurels and left no memoirs! If many an officer, British as well as Allied, found Montgomery abrasive and egocentric, he did not lack for a following of those who welcomed his firm hand on the wheel and were prepared to take him at his own valuation. The fact is that he *did* have good ideas in the field of planning, training and morale. The present writer commanded a unit in three of his campaigns. When Montgomery was in command the British soldier simply did not believe that he could be beaten. If that is not a pearl of great pride I know nothing. Success is what counts in war. Men who must stake their lives in battle do not like to think that they may be throwing them away in vain. After all, with the exception of Arnhem, Montgomery's plans worked pretty well.

It is not easy to become a successful general, if only because many of the best officers get themselves killed as subalterns. You may spring from an old military family, beloved of the throne. Your father may be a general and hold the Congressional Medal of Honor, but your path to high rank will still be beset by pitfalls. Your eccentricities, foolish and amusing pranks to your men, may be court-martial offenses to your more serious-minded superiors. Then again you may find yourself stuck in some distant garrison while the companions of your youth are winning medals with your country's main field army. A friend of mine, who had distinguished himself in Palestine, spent most of World War II acting as a staff officer in the Falkland Islands! Ordinary civilian life is full of hazards but they are as nothing compared with those encountered by those who seek their fortune — sword in hand — or as Patton did with a couple of pearl-handled pistols in his belt!

Some of the generals who figure in this book were sabreurs as well as sages. Rommel and Patton are obvious examples. Many of them thought deeply about the Art of War, and a few of them have left us their thoughts on the subject. Still it is evident that you do not have to be a great thinker to build up a fine fighting record. Prince Blücher's memoirs, had he left us any, might perhaps have confused rather than enlightened us! How, one wonders, would he have dealt with the period when he thought himself pregnant with an elephant? At the same time some of the acknowledged authorities on the art of war have left no great record as fighting men. Jomini and Clausewitz did all their campaigning with the staff. John Fuller, though he rose to be a Major General never commanded any formation in the field; and not did Sir Basil Liddell-Hart, though he spent some months in the trenches during the fairly quiet period leading up to the Battle of the Somme, before becoming a casualty of that unimaginatively planned battle. The point is that the best warriors are not necessarily the best military commentators — which is not to say that men like Fuller or Jomini cannot hold the highest staff appointments with distinction. The nature of armies is such that the same man is not always both the figurehead and the brains. Cleverness and valor do not necessarily go hand in hand. Maurice de Saxe, who had some respectable victories to his credit — Fontenoy, Lauffeld, Roucoux — and was one of only four Marshals of France to attain the rank of Maréchal-Général, wrote his *Thoughts on War*, in which he tells us 'War is a science shrouded in darkness, in the middle of which we do not move with an assured step; routine and prejudice are its basis, a natural consequence of ignorance. Other sciences have principles, war as yet has none; the great captains who have written do not give us any . . . Thus there are nothing but usages, the principles of which are unknown to us.' More than 170 years were to pass before Fuller made his attempt to remedy this state of affairs in an article published in 1916. Since that time the principles which govern the Conduct of Operations have been codified and enshrined in Field Service Regulations. How useful they are to a Montgomery or a Gort in the throes of planning a complicated operation one would hesitate to opine. A properly educated general should perhaps be above the need for so elementary a guide, but how many generals are properly educated? History offers us several examples of generals, who gave their men a generous ration of leadership, while their chief of staff provided the brains and the education. Blücher and Gneisenau; Hindenburg and Ludendorff; Ney and Jomini

are cases that spring to mind. Perhaps we should add Haig and Lawrence, even, dare one say it?, Montgomery and de Guingand.

The case of Ney and Jomini is not without interest. Ney, '*Le Brave des Braves*,' really was a man of quite exceptional courage, and heroes like Oudinot and Grouchy, the one with 26 wounds about his person, the other, who received no less than 14 on one day are simply not to be mentioned in the same breath. Yet Ney was of lowly birth, the son of an Alsation barrel-cooper, and had received no more education than you would expect of one whose only school had been the Régiment Colonel-Général des Hussards. On the battlefield Ney was superb, but in between times he was no great commander, and in particular he was far from being the strict disciplinarian that a nomadic Napoleonic Corps needed. Though it cannot be said that he was a military genius, he had a fairly impressionable mind. For example he had seen the balloon which 'spotted' for the French artillery at Fleurus in 1794, and some years later he was willing to put his hand in his pocket to help some inventor, who was trying to get a similar project off the ground.

While Ney was winning his laurels on the Rhine, Jomini, who was 10 years younger was analyzing the campaigns of Frederick the Great and Bonaparte, and writing his *Treatise on Grand Military Operations*, the *magnum opus* which was to enable him to achieve his life's ambition, which was nothing less than to teach others the Art of War. Armed with letters of introduction he approached several French generals, including Joachim Murat, who was not the brainiest of men. They all turned him down, and almost in despair Jomini presented his manuscript to Marshal Ney, who — no doubt to the young Swiss' astonishment — read it with great interest, lent him the money to publish it, and took him off with him on the Ulm campaign as a volunteer ADC. It was not long before Jomini had his chance to demonstrate his understanding of Napoleonic strategy. The Emperor placed Ney's VI Corps on the north bank of the Danube so as to prevent the Austrians, cut off in Ulm, from escaping in that direction, or striking at the communications of the Grand Army. At the same time Napoleon put Ney under Murat, who was in command of the French right wing. Murat, in his wisdom, decided to move Ney to the south bank despite the latter's objections. Jomini declined to pen the movement order, saying, with the effrontery of youth:

'Your highness will pardon me if I do not write, there are so many secretaries on the staff of Marshal Ney that there is no necessity for my taking part in a maneuver which I believe to be in direct opposition to the intentions of the Emperor.'

'Ah! Marshal Ney,' said Murat nastily, 'do you permit your officers to argue in this manner?'

'Pardon me, your highness,' replied Jomini, 'I am a Swiss officer, and serve here as a volunteer. Marshal Ney . . . sometimes permits me to discuss military operations with him. That is what I have just taken the liberty of doing.'

Murat, headstrong Gascon that he was, insisted on the move he had ordered, but as it happened 30,000 Austrians fell upon Ney's rearguard division before it had quit the north bank. Marshal Ney did not need much prodding from

Major Jomini to march to the sound of the guns. In a brilliant day's fighting he repulsed the Austrians and won the title of *Duc d'Elchingen* from a grateful Emperor.

After Austerlitz Jomini ventured to bring his book to Napoleon's notice, and an ADC read certain passages to the Emperor.

'They say the age does not advance!' exclaimed Napoleon, 'why here is a young major, a Swiss at that, who teaches us what my professors never taught me, and what very few generals understand!' The ADC read on, while his master became more and more excited.

'Why did the Minister of War allow such a book to be published?' he demanded. 'It teaches my whole system of war to my enemies. The book must be seized, and its circulation prevented.'

All was not lost. Fortunately for Jomini, after a little reflection Napoleon delivered himself of a characteristically cynical opinion: 'But I attach too much importance to this publication. The old generals who command against me will never read it, and young men who will read it do not command.'

Napoleon promoted Jomini and in 1806 attached him to his own staff. The new colonel asked permission to return to the VI Corps to collect his horses and baggage, and to see that Ney had a competent senior ADC.

'If your Majesty will grant me four days' leave, I can rejoin at Bamberg.'

'And who told you that I was going to Bamberg?' the Emperor demanded in an exasperated tone.

'The map of Germany, Sire.'

'How the map? There are a hundred roads on that map besides the Bamberg road.'

'Yes, Sir, but it is probable that your Majesty will perform the same maneuver against the left of the Prussians as you did against Mack's right at Ulm, and as by the St Bernard against Melas' right in the Marengo campaign; and that can only be done via Bamberg on Gera.'

'Very well,' replied the Emperor, 'be at Bamberg in four days; but do not say a word about it – not even to Berthier; no one must know that I go to Bamberg.'

Years after at St Helena the exiled Emperor recalled this remarkable episode. Napoleon said after the Russian campaign that had Jomini not then suffered a protracted illness he would have made him a Marshal of France. For all his talents the Swiss would never have made a corps commander. His was not the temperament that allows a man to think clearest with the *son du canon* in his ears. Still, he might have made an excellent Chief of Staff – better perhaps than the war-weary Berthier, who, though an excellent Chief Clerk was no strategist. Berthier, who detested Jomini with a hatred worthy of a jealous spouse, contrived his disgrace – for not sending in returns punctually! – so like a good Swiss mercenary, Jomini joined the Russian Army as a lieutenant general in 1813, and served the Muscovites faithfully for the next 56 years, during which he thoroughly overhauled their inchoate hordes on approved Napoleonic lines, all neatly organized in army corps.

Jomini could perceive in the great campaigns, which he analyzed, a system of war, which he described in detail. Clausewitz, who unfortunately died of cholera before he could revise – and dare one say cut – his great work, had a tremendous influence on the old Prussian Army of Moltke and Schlieffen, but his writing is seldom easy to follow or to comprehend. Every now and then he says something of limpid clarity such as: 'In war everything is very simple; and even the simple is very difficult.'

This book begins in a period when war, thanks to the relative inefficiency of weapons, was still tolerable. Indeed the connoisseur of tactics and war games could hardly wish for a more interesting series of battles than those of the French Revolutionary and Napoleonic Wars. Even the uniforms were not unhandsome, though for my part I prefer those of the late 18th century.

The increase in the range of firearms begins, with the Crimea, to lay a deadening hand on the Art. Casualties mount, frontal attacks become almost impossible. Then, clearly heralded by the Russo-Japanese War of 1904, comes what Fuller called the War of 'cattle-wire and spade.' This gave way to the fast-moving mechanized war for which he had always tried to prepare his fellow countrymen. One can see now, looking back, that World War II produced a far more talented collection of generals than World War I, men who were not prepared to be shackled by the trench, the machine gun and barbed wire.

By 1934 it was not too difficult to discern the shadow of another War on the horizon — yet there were many who would not see it. One wonders now, in the autumn of 1980, whether perhaps we are sitting in a similar shadow — albeit mushroom shaped. If war comes what will the old-style generals do? Shall we still want leaders like the Victors whose exploits are described in these pages? Will the principles of war be found to have changed entirely?

We live, we are often told, in an age when the young — those of military age— are devoted to drink, drugs, vandalism and violence. From their ranks may rise an alcoholic hooligan with just sufficient talent to steer one or other of the great powers to the conquest of the world: a world so ravaged as to be a prize that no sensible person would wish to win. World War II has had its critics, but I am glad that it fell to my lot to go through that one, rather than the nuclear holocaust we are promised for next time. Who will be the Victors then?

Peter Young.

CHAPTER 1

NAPOLEON AT AUSTERLITZ 1805

Marshal Louis-Alexandre Berthier, Napoleon's Chief of Staff, described Austerlitz as 'the Emperor Napoleon's most famous victory.' It was certainly one of the Emperor's most impressive achievements. On 2 December 1805, deep inside hostile territory between the Moravian towns of Brünn and Olmütz, he inflicted a catastrophic defeat on a numerically superior Allied army. It was certainly not his first victory, nor was it to be his last. The campaign and battle of Austerlitz must, however, rank among the most remarkable episodes of a dazzling career.

The 2 December was a fitting date for victory. Only a year before Austerlitz, on 2 December 1804, Napoleon had crowned himself Emperor of the French in the Cathedral of Notre Dame. This in itself was no mean feat, representing as it did the culmination of a rise from obscure origins to undisputed eminence within the state. Born at Ajaccio in Corsica in 1769, Napoleon Bonaparte was commissioned into the French artillery in 1785. The turmoils which followed the revolution, and his distinguished performance at the siege of Toulon in 1793, facilitated his speedy promotion, and 1796 saw him set the seal on his rising military reputation in the brilliant Italian campaign. His subsequent adventures in Egypt brought him few laurels, but his failure in the East did not prevent him from returning to France and replacing the Directory with the Consulate. The aspiring general became First Consul and, in the summer of 1802, with fresh victories over the Austrians to his credit, he was voted Consul for life.

Napoleon spent 1801–03 in the reorganization and revival of France. Fruitful though these efforts were, they brought him domestic but not international stability. War between Britain and France flared up once more in May 1803; the British blockaded French ports and began to seize French colonies, while Napoleon responded by closing European ports to British trade and assembling a formidable army of invasion on the channel coast. Britain naturally sought continental allies, and attracted Russia and Austria, followed by Sweden and Naples, into an anti-French league, the Third Coalition. Prussia remained neutral, but Spain, Bavaria and Württemberg adhered to France. The Coalition's war plan was complex. The Austrian Archduke Charles commanded a powerful Austro-Russian force in northern Italy, and a smaller Austrian army under the Archduke John was stationed in the Tyrol. Further north, the Archduke Ferdinand, under the tutelage of his Chief of Staff, General Mack, threatened Bavaria. From the northeast came the doughty warriors of Holy Russia. One army under General Levin Bennigsen approached Prussia, while another, led by the plump and avuncular General Mikhail Ilarionovich Kutuzov, marched toward the Upper Danube to join with Ferdinand. Between these two great columns tramped a third under General Buxhöwden, who was to support Bennigsen before moving south to join Kutuzov. Away to the north an improbable association of Russians and Swedes was, with British assistance, to advance from Stralsund to Hanover, and on the southern flank an Anglo-Russian force was to appear in northern Italy.

An ambitious plan like this, filling the map of Europe with converging forces, would have been difficult enough to execute if slick staff work and wholehearted cooperation had

been features of the alliance. As things were, the scheme was flawed from its inception by convoluted command arrangements and serious inter-Allied misunderstandings. Among the latter was the failure to appreciate that the Austrian and Russian calendars differed by 10 days, an oversight which was to have unfortunate consequences.

While the Allied projects took shape, the bulk of the French army was concentrated around Boulogne on the channel coast in preparation for a descent on England. By mid-August 1805 this operation seemed less practicable than ever, and Napoleon realized that the inability of the French and Spanish fleets to keep the British out of the channel rendered a crossing impossible. Moreover, the situation in Central Europe grew more ominous day by day. The Emperor therefore decided to shelve his plans for an invasion of England, and to march eastward with all possible speed, falling upon Ferdinand and Mack before going on to deal with the advancing Russians.

Napoleon's plan was nothing if not bold. Speed and secrecy were its prerequisites, but it was no easy task to move an army of over 200,000 men across the face of Europe. That the scheme was at all feasible was a tribute to Napoleon's achievements during his army's stay at Boulogne. The stormy years which followed the Revolution had witnessed the expansion of the French army and the removal of social barriers to promotion. However, until 1803–04 there had been a marked tendency for soldiers to identify more strongly with their division, or with the particular army in which they served, than with the French army as a whole. The Boulogne camp, whatever its real strategic purpose – and there were those who suspected that Napoleon was never entirely serious about the British invasion plan – played a vitally important role in training and unifying the army. It also aided the Emperor's own efforts to consolidate his hold on the army, and to participate in lavish ceremonies which seemed to turn even staunch republicans into devoted supporters of the Empire.

The Austerlitz campaign was to involve about 200,000 French soldiers, just under half the total strength of the *Grande Armée*. The force comprised the Imperial Guard, the Reserve Cavalry and seven corps. Each of the latter consisted of two or more infantry divisions with attached cavalry, artillery and engineers. There was, though, no standard corps organization, and these formations varied in size between Nicolas Jean de Dieu Soult's huge IV Corps with its 41,000 men and the tiny VII Corps of Pierre François Charles Augereau with a mere 15,000. Corps were normally commanded by marshals, a rank reintroduced by Napoleon in 1804. Most of the marshals were astonishingly young; Charles Augereau and Jean-Baptiste Jules Bernadotte however were over 40 years of age. The corps was an essential ingredient of the 1805 plan and, indeed, an indispensable instrument of Napoleonic strategy generally. A corps was self-contained and flexible, and yet used to operating as a cohesive body. Providing that its staff work was sound, an army composed of corps could march divided – easing the pressure on roads and provisioning facilities – but could unite on the battlefield.

If the hard training carried out at the camp of Boulogne had imparted a high gloss to the fighting units and formations of the *Grande Armée*, the same period had witnessed

the final development of Imperial Headquarters into the form which, with minor alterations, it was to retain for the next decade. The most important of its three branches was Napoleon's own *Maison*, 'the cabinet of genius,' in which he formulated his detailed plans. Marshal Louis Alexandre Berthier, Napoleon's Chief of Staff until 1814, headed the General Staff of the *Grande Armée*, which dealt with more routine matters of administration. Finally, the General Commissary of Army Stores supervised the acquisition and distribution of stores and supplies of all sorts. The pivot of this entire apparatus was the Emperor himself. His boundless energy and retentive brain dominated Imperial Headquarters, and through it the army as a whole. Although the simple green undress uniform of a colonel of the *Chasseurs à Cheval* of his Guard which the Emperor habitually wore could not conceal his growing corpulence, he was capable of spending hours in the saddle and remaining fresh throughout a 14-hour working day. He lived simply when in the field, spending the night in whatever quarters were available or in his well-equipped coach. His elevation to Imperial honors had little effect upon his campaigning habits; he remained a field commander of almost republican simplicity.

On 27 August Napoleon unleashed the *Grande Armée*. The seven corps, Reserve Cavalry and Imperial Guard scythed through France and the Rhineland, each corps marching on its own axis to finish up on a 70-mile front between Ulm and Ingolstadt on the Danube. Ferdinand and Mack, around Ulm, wrangled over strategy until it was almost too late. The Archduke got away with some cavalry on the night of 14–15 October, but Mack capitulated, with about 24,000 men, on the 20th.

Napoleon's sudden irruption into the Upper Danube brought the Allied war machine grinding to a halt. Kutuzov had been making for Ulm and had reached Braunau on the

Above: General Prince Mikhail Ilarionovich Golenishev-Kutuzov was the formal commander of Austro-Russian armies in 1805. He was regarded as the grand old man of the Russian forces.

Right: Napoleon enters Vienna after the Austrians declared the city open on 12 November 1805.

Far right top: Nicolas Jean-de-Dieu Soult, who was appointed Marshal of the Empire on 19 May 1804.

Far right: Joachim Murat, another of Napoleon's Marshals. His prompt action secured the Vienna bridge without loss of life.

Inn by 23 October. His junction with Mack had been impeded by the rigors of the march as well as the calendar problem; now it would never take place. The French advance was resumed on the 25th, but Kutuzov avoided encirclement and made a successful fighting retreat, crossing the Danube at Krems on 8 November. Four days later the French bluffed their way across the Danube and entered Vienna. On 13 November Joachim Murat, leading the pursuit with his cavalry, was duped into accepting an armistice, and by the time this had been repudiated by an enraged Napoleon Kutuzov had slipped away, although his rearguards fought a brisk action at Shöngraben on the 16th. Three days later the old general and his exhausted force joined Czar Alexander I and the main Russian army near Olmütz. The French reached Brünn on the same day, but the *Grande Armée*, a gruelling campaign of marching and fighting behind it, was so footsore that Napoleon was forced to suspend active operations four days later.

The first phase of Napoleon's daring and dramatic campaign had gone remarkably well. He had marched undetected to the Danube, and gobbled up the luckless Mack. The second phase, his pursuit of Kutuzov, had been rather less satisfactory, although it had brought him the prestigious plum of Vienna. Now opened the third phase, which seemed fraught with peril for the French. In the south, the Archdukes John and Charles were concentrated with a powerful army around Marburg, watched by part of Auguste Frédéric Louis Viesse Marmont's corps. The main Austro-Russian army was at Olmütz, growing stronger by the day as reinforcements tramped in from the East. A force under Ferdinand was approaching Iglau, northwest of Brünn, and threatened the French rear. The Allies might have lost Vienna, but they enjoyed unobstructed communications with Russia and Silesia, and their growing forces threatened the French from three sides. Napoleon, on the other hand, was deep in hostile territory, and at the end of a tenuous line of communications, faced by powerful adversaries eager to revenge the setbacks of the earlier part of the campaign.

Napoleon's response to this seemingly depressing strategic situation was as characteristic as it was brilliant. He shunned the safest course of action, retreat, and decided instead to shatter the Allied threat by defeating its strongest element, the army concentrated around Olmütz. The execution of this plan depended upon an elaborate deception scheme. While Marmont, with instructions to avoid a pitched battle at all costs, stood off the Archdukes in the south, the Guard, Soult, Jean Lannes and three cavalry divisions, concentrated at Brünn, with a hussar brigade forward at Wischau on the Olmütz road. Once the Allies had been lured forward to snap up the forces at Brünn, Louis Nicolas Davout would be summoned from Vienna and Jean-Baptiste Bernadotte from Iglau. This would give Napoleon some 75,000 men as opposed to the 88,000 of the Allied central force. The moves of Bernadotte and Davout had to be accomplished in total secrecy, for if the Allies knew of their arrival around Brünn there was every chance that they would decline to attack — and an Allied attack on his supposedly weak forces was exactly what Napoleon hoped to induce. He expected the Allies to attempt to outflank him from the south; when they did this he planned to launch his own counterattack, stabbing, in the

classic pattern of double envelopment, at both flanks of the Allied force.

The plan was perilously finely balanced. It depended upon the Allied commanders taking the bait, and relied upon the Emperor's ability to concentrate rapidly before his enemies realized what was afoot. If the Allies attacked too early — or, indeed, declined to attack at all — then the plan would fail and the ominous strategic situation could only worsen as winter progressed. However, as events were to show, Napoleon had gauged the mood of his opponents precisely. Friction between the Austrians and Russians had worsened; the Russians now professed contempt for the battered remnants of the Austrian army, blaming the reverses of the autumn upon Austrian cowardice or stupidity. Kutuzov remained in nominal command of the Allied army, but real power was in the hands of the young Czar and his coterie of elegant and ambitious advisers. There was general support for a counteroffensive which would teach the upstart Bonaparte a lesson he would never forget, a policy which commended itself not only to the Russians but also to the Emperor Francis of Austria, a

fugitive from his own capital, and to his Chief of Staff, the able but pedantic Major General Weyrother. Wiser heads in the Allied councils recommended caution, but Alexander, lured on by extravagent reports of French demoralization, was eager to bring Napoleon to battle.

On 27–28 November the Allies moved westward from Olmütz, driving in the French outpost line. Napoleon, informed of the move on the afternoon of the 28th, rode forward to see the state of affairs for himself. He now began the delicate process of concentrating his forces east of Brünn, while at the same time giving the Allies the impression that he was afraid of giving battle. General Anne Jean Marie René Savary had been sent on a diplomatic mission to Allied headquarters on the 28th, ostensibly to negotiate but in reality to measure Allied feelings and to convey the impression that Napoleon wished to avoid battle. On the following day Napoleon received a visit from Count Dolgoruki and, although the latter was eventually sent off with a flea in his ear, the incident simply confirmed the Allied leaders in their conviction that they were dealing with a beaten man.

On 29 November both armies slid into position between Brünn and Olmütz. The ground taken up by the French was already familiar to Napoleon. Eight days before he had pointed to a stubby hill, the Santon, and informed his entourage that a battle would take place in that area. Indeed, his engineers and gunners had begun preparing the region for defense almost immediately. Just south of the Santon ran the straight, tree-lined road linking Olmütz with Brünn, where it turned south to Vienna. The so-called Mountains of Moravia, heavily-wooded hills, lay to the northwest, while about five miles southeast of the Santon stood the village of Austerlitz, a cluster of houses around the chateau of Prince Kaunitz. Two miles west of this the ground rose onto the Pratzen plateau. The village of Pratzen lay between the high points of the Pratzeberg and the Staré Vinohrady, in the neck of a small valley on the plateau's western edge. A mile or so west of Pratzen the Goldbach

Top left: Louis-Nicolas Davout was in command of III Corps at Austerlitz.

Left: A Napoleonic War cartoon showing Napoleon playing leapfrog with the nations of Europe.

Right: Napoleon painted by the official court painter, Jacques Louis David.

stream wended its muddy way through the villages of Puntowitz, Kobelnitz, Zokolnitz and Telnitz. At Puntowitz the Goldbach was joined by the Bosenitz stream, which ran through Jirschikowitz before going under the Brünn–Olmütz road a few hundred yards from the Santon. The valley of the Goldbach was boggy, and south of Telnitz, where it joined the Littawa, were a number of shallow fish ponds, covered by a thick layer of ice.

Several details of the ground had brought themselves forcefully to Napoleon's attention. Firstly, the Santon, rising sharply above the surrounding countryside, formed a solid pivot for the French left. It dominated the road, and was of equal value as a defensive buttress or as a springboard for an advance. Secondly, the soggy valley of the Goldbach, with its miserable villages, had some potential as a temporary defensive line, and at its northern end it formed a useful concentration area, for it was 'dead ground' to observers on the Pratzen. Finally, the Pratzen itself dominated the center of the position. It was not as steep as is sometimes supposed; its western and northeastern slopes were easily passable to troops of all arms, but its southeastern edge, falling away to the valley of the Littawa, was much more precipitous.

On the afternoon of 1 December, Napoleon, his familiar figure muffled up in a singed gray riding coat, established himself on the Zurlan hill, south of the main road, between the Goldbach and Bosenitz streams. Much of his army had already moved into position. Lannes' V Corps, containing the infantry divisions of Suchet and Caffarelli as well as a light cavalry division, held the Santon sector, with a formidable battery on the hill itself. Behind the V Corps were the Imperial Guard and Oudinot's Grenadier Division, in the area of the Zurlan. Bernadotte's I Corps, fresh – if that is quite the right word – from its march from Iglau, stood slightly west of the Zurlan, screened from observation.

The valley of the Goldbach itself was thinly held by the widely extended IV Corps of Soult, whose left was linked to the concentration in the northern sector by Murat's cavalry in Schlapanitz. The villages of Telnitz and Zokolnitz, the latter with its castle and walled game park, were garrisoned by Legrand's division of the IV Corps. Soult's overextension was, to a degree, intentional, being an element of Napoleon's plan for inducing the Allies to attack his right. In part, however, it reflected the disturbing fact that Davout's III Corps, on the march from Vienna, had not yet arrived. Its leading division did not stumble into Gros-Raigern, south of Telnitz, until after dark, after an exhausting march of 76 miles in 46 hours.

The late arrival of Davout led to a modification of the original plan. Napoleon had hoped to persuade the Allies to attack his weak right-center, in the area of Zokolnitz and Kobelnitz, and then to strike them with his concentrated forces in the north, along the axis of the main road, and with Davout's corps in the south. By nightfall on the 1st it had become clear to the Emperor that this would not do. Davout's men were simply too tired to participate in a vigorous offensive, and in any case it appeared that the ponderous Allied army was itself edging to the south, extending itself over the ground across which Davout was to have attacked. The two-pronged assault was therefore replaced by an oblique attack, spearheaded by the divisions of Vandamme and Saint-Hilaire (IV Corps), which were to thrust eastward onto the Pratzen once the Allied mass had ground its way onto the plateau's southwestern edge. In the north, Lannes was to hold the Santon, while Bernadotte deployed between Puntowitz and Jirschikowitz in preparation for an assault on Bläswitz. On the southern flank, Legrand's men were to do their best to hold the line of the Goldbach until Davout came up in support. The Guard and Oudinot's Grenadiers were to be held in reserve.

Napoleon's orders went out at about 2000 hours, and the Emperor then dined on his favorite dish of potatoes and onions fried in oil, in the relative comfort of an improvised hut on the Zurlan. He chatted with a small group of senior officers, and snatched some sleep before General Savary awoke him at midnight with the news that the Allies had just entered Telnitz, and were clearly present in great strength on the southern end of the Pratzen. Napoleon

Left: Napoleon and his marshals before the battle on 2 December 1805, from a painting by Carle Vernet.

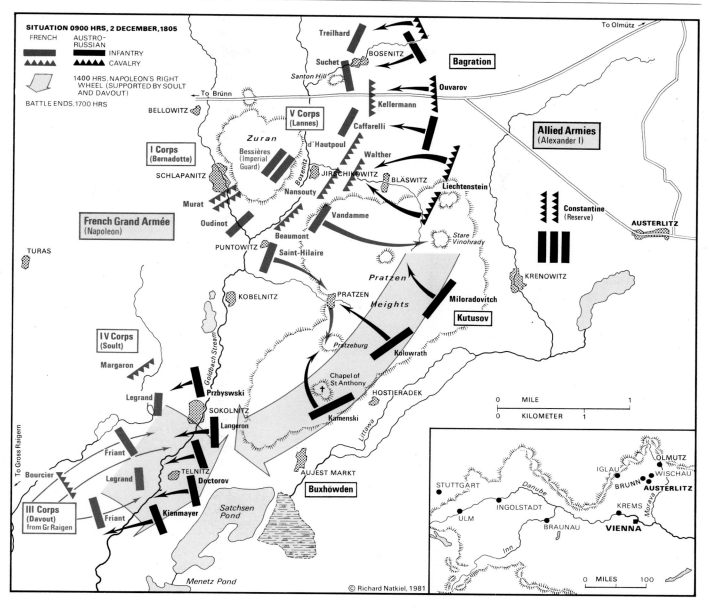

SITUATION 0900 HRS, 2 DECEMBER, 1805
FRENCH / AUSTRO-RUSSIAN
INFANTRY
CAVALRY
1400 HRS, NAPOLEON'S RIGHT WHEEL (SUPPORTED BY SOULT AND DAVOUT)
BATTLE ENDS, 1700 HRS

© Richard Natkiel, 1981

rode south to see for himself and, after a muddy scramble with a Cossack patrol in the valley of the Goldbach, he established that the Allied advance seemed to have stopped. As he walked back to the Zurlan through the bivouacs of the Guard he was recognized by a grenadier who twisted a wisp of straw into a torch to light the Emperor's way. The man's comrades followed suit, and Napoleon returned to his humble quarters in a blaze of light. Back on the Zurlan, he made some final modifications to his plan to allow for the fact that the Allies had slipped further south than he had originally anticipated. Soult's attack was shifted slightly southward, and Bernadotte's I Corps, although still in reserve, was moved behind Soult's left-hand division, that of Vandamme. Murat's role now assumed greater importance, for the infantry strength of the northern sector had decreased, and Murat's cavalry would have to prevent the French left and center from drifting apart. As the final adjustments to his dispositions took place, in the freezing predawn of 2 December, Napoleon managed to get a little more sleep.

Slumber was the last thing in the minds of most of the Allied commanders. During the night the Allied army swung clumsily forward onto the Pratzen in a plan designed to pulverize the French right and bundle Napoleon back against the mountains of Moravia. This grand design was largely the work of Weyrother, who had expounded it, in somniferous detail, to senior commanders in Kutuzov's headquarters at Krenowitz at about midnight. The attack was to be carried out by a number of mixed columns. On the southern flank, the Austrian Lieutenant General Kienmayer was to screen the left of Lieutenant General Doctorov's 13,500 Russian infantrymen, who were to cross the Goldbach at Telnitz before swinging north. On Doctorov's right the columns of Generals Andrault Langeron and Przbyswski aimed for the country between Telnitz and Zokolnitz, and for the latter village itself. The three southerly columns were to operate under the far from eagle eye of the Russian General Buxhöwden, who seems to have spent much of the time trying to keep the cold at bay by consuming alcohol. North of Buxhöwden's force marched 24,000 infantry under Generals Kollowrath and Miloradovitch, making for the Goldbach around Kobelnitz. Prince John of Liechtenstein's cavalry were to deploy in the area of Bläswitz to link the Allied center to its

General Rapp brings news of the success of the main attack to Napoleon, during the Battle of Austerlitz.

right wing, the detachment of Prince Bagration. Bagration was under orders to cover the main road until the assault was seen to go in; he was then to assist by attacking up the road. The only formation left in reserve was the newly arrived Russian Imperial Guard, posted behind Bläswitz to support Bagration and Liechtenstein.

Weyrother's scheme had many attractions. Nevertheless, it contained an almost equal number of flaws. Firstly there was an unshakable assumption — partly the result of Napoleon's deception — that the French were morally paralyzed and would not interfere with the descent from the Pratzen and the crossing of the Goldbach. Secondly, the final details of the project were produced only after some of the columns had already started to move. Orders had to be translated and sent out in the darkness to a massive army on the march. The Allied columns bore no resemblance to French corps; they were extemporized formations, cobbled together in haste and often led by officers whose knowledge of the master plan was somewhat sketchy. It is small wonder that there was confusion in the Allied ranks long before the first shot echoed over the marshes of the Goldbach.

Napoleon had issued his customary prebattle proclamation to his troops on the previous night and, despite their chilly bivouacs, his men were in good heart as they stood to arms before first light. They might have been less confident had they known how heavily the numerical balance was tilted against them. Of the 73,000 or so French troops in the area only about 60,000 were fit to fight, while about 85,000 Austrians and Russians were already on the move, and further Allied reinforcements were expected.

Few battles start neatly, and Austerlitz was no exception. At about 0700 hours Kienmayer's force entered Telnitz, and fighting flared up around the village as both Doctorov and Davout fed in more troops. The Russians took the village, and Przbyswski's men stormed Zokolnitz, a mile to the north. Smoke, therefore, already lay thick in the valley of the Goldbach when Napoleon, munching a hasty breakfast, summoned his corps commanders to a final briefing on the Zurlan. The news from Telnitz confirmed that the Allied attack was developing as expected, and Napoleon sent his marshals back to their commands after only a few words. Soult, however, he kept the longest. 'I beg Your Majesty to hold me back no longer,' urged the anxious marshal. 'I have 20,000 men to set in motion.' 'How long will it take you to climb the Pratzen?' asked Napoleon. 'Less than 20 minutes, Sire,' replied Soult. 'In that case,' concluded the Emperor, 'we will wait for another quarter of an hour.'

Soult's men had been issued with a triple ration of brandy and, fortified by almost half a pint of raw spirit each, as well as by their commander's oratory, they were ready for anything. At about 0900 hours the divisions of Vandamme and Saint-Hilaire advanced, bursting from the mist into bright sunlight halfway through their ascent. Saint-Hilaire seized Pratzen and the Pratzeberg, and Vandamme, after a brush with an off-course Allied detachment, secured the Staré Vinohrady. It was now only 0930 hours, and the keys of the Pratzen were in French hands. On the southern end of the battlefield, in the smoky chaos of Zokolnitz and Telnitz, the Allied attack had lost momentum in the face of a staunch defense.

It was in the north that events were taking a crucial turn. The pugnacious Bagration, dissatisfied with the role allotted to him by the Weyrother plan, decided to move forward on his own initiative, and at about 1000 hours he collided with Lannes north of Bläswitz. The leading elements of Liechtenstein's cavalry, who had found themselves well to the south of their proper position and had ridden through several advancing columns to get there, gave some support to Bagration's left, while to Bagration's left rear the Russian Guard stamped its immaculate way toward Bläswitz.

Although by midmorning Napoleon had the satisfaction of knowing that his opponents' plan had gone badly awry, his own was not proceeding as he might have wished. Bagration's determined attack gave Lannes all he could cope with, and although the southern flank was holding up much better than had been expected, the Emperor was now becoming preoccupied with the fate of his center. Vandamme and Saint-Hilaire had indeed established themselves on the Pratzen, but they soon found themselves vying for the tenancy of this vital feature with the forces of Kollowrath and Miloradovitch, which had fallen victim to numerous delays and were consequently still on the Pratzen. After a period of very fierce fighting, Saint-Hilaire's men managed to win secure possession of the southern crest of the plateau, beating off spirited counterattacks from Kollowrath and from a brigade of Langeron's column which had swung north to help. Vandamme was more fortunate, and established himself around the Staré Vinohrady with relative ease.

Until the Pratzen was secured Napoleon remained at his command post on the Zurlan, receiving information from couriers, watching the progress of the fighting and sending off gallopers with orders. At about 1130 hours, with the Staré Vinohrady firmly in French hands, he rode forward onto the plateau, followed by his entire reserve — the Guard, Oudinot's Grenadiers and the I Corps. He seemed to have won the battle in the center, and the fighting along the

Above: The scene at the battle showing Napoleon's view over the valley, based on a drawing of an eye witness.

Left: The charge of the Mamelukes at Austerlitz. The Mamelukes were an elite cavalry force first raised in 1803, and part of the Imperial Guard Cavalry.

Far left top: General Prince Peter Bagration, who was in command of 13,000 troops positioned east of the Goldbach Heights during the battle.

Goldbach in the south crashed on in stalemate. In the north those dogged rivals, Lannes and Bagration, were locked in a bitter struggle which swayed to and fro between Bläswitz and Bosenitz. It was not until early afternoon that Bagration was at last pushed back, and even then he retired in his customary resolute style. The fighting in the north had been much fiercer than Napoleon had anticipated. Moreover, both Lannes and Murat had suffered numerous casualties and were in no condition to mount a vigorous attack.

However disappointed Napoleon might have been by the lack of progress on his left, he at least had the satisfaction of knowing that he retained the Guard, Bernadotte and Oudinot — the latter soon to be sent off to the south — in reserve under his hand. The only reserve formation now left in the hands of the Allied commanders was the Russian Imperial Guard, under the swarthy and savage Grand Duke Constantine. He had approached Bläswitz toward mid-morning, but discovered, to his surprise and chagrin, that the village was held by the French. In response to an appeal from his Imperial brother, wandering like an unhappy wraith among the wreckage of the Allied center, he dispatched a battalion to help Kollowrath and Milora-dovich on the Pratzen. He then took up position about a mile southeast of Bläswitz, and sent forward the Preo-brazhensky and Semenovsky Regiments, supported by the Guard Cuirassiers. This attack, delivered with commendable determination, broke two of Vandamme's regiments. Thus

Napoleon in defeat on the *Bellerophon* taking him to St Helena.

it was that no sooner had Napoleon ridden onto the Staré Vinohrady than a crowd of fugitives ran past him, rather sheepishly shouting *'Vive L'Empereur'* as they did so.

Napoleon at once ordered Marshal Bessières to support Vandamme with his own Guard. Bessières' first attack failed, and Constantine fed his remaining regiments into the fighting. Bessières renewed his efforts, and Bernadotte sent D'Erlon's division of the I Corps to assist. The Russian Guard fought manfully but was at last driven back, with heavy losses in men and guns. The defeat of the Russian Guard removed the threat to the French center, and extinguished the only Allied reserve. Nevertheless, although Napoleon was now in a dominant position, his oblique attack on the Allied right had not worked, and any French victory would be inconclusive. Napoleon was not content to accept this. He decided to carry out a great right wheel by his forces on the Pratzen, with the intention of cutting off Buxhöwden's force at the southern end of the battlefield.

The move was under way shortly after 1400 hours. The infantry of Vandamme and Saint-Hilaire, with powerful support, struck southwest from the Pratzen, and the Guard followed them. Friant chose just this moment to push his division of the III Corps into a counterattack, and the stolid Russian infantry around Žokolnitz were overwhelmed and their commander, Przbyswski, was captured. Kienmayer and Doctorov, further south, disengaged the remnants of their columns with very considerable skill, although many of their men were, inevitably, cut off. Buxhöwden and his

staff got away over the Littawa somewhat in advance of their retreating forces, but as Vandamme's men secured Aujest, escape for much of Buxhöwden's command became impossible. Some Allied soldiers tried to escape across the frozen ponds of Satschan and Menitz, and a number were drowned when the ice broke. However, subsequent French claims that 20,000 Allies perished in the ponds were wildly exaggerated.

Napoleon had again moved to a position from which he could observe the decisive sector of the battlefield. As Soult's men descended from the Pratzen he had established himself at the chapel of St Anthony, on the southern slopes overlooking Aujest. From this vantage point he watched the partial encirclement of the Allied left, and urged on his commanders, replacing one of them, General Boyé, whose dragoons had not produced satisfactory results.

The battle was over by 1700 hours. Most of the French spent the night on the battlefield, while the debris of the Allied armies lurched away to the east. Bagration, Liechtenstein and Constantine brought their survivors off in good order, but the remainder of the Allied force presented a sorry spectacle as it shuffled back through the sleet. Allied casualties totalled some 27,000 men — roughly one-third of those engaged. French losses were much smaller, something under 10,000 men, many of whom were wounded.

The Emperors of Austria and Russia, the latter dazed by a day of wandering about the field with only a few companions, met at Czeitsch on the road to Hungary at noon on 3 December. Although fresh troops were at hand, neither monarch had the stomach for more fighting, and the Russians resumed their retreat while Joseph made terms with the victors. He met Napoleon on the afternoon of the 4th. The latter had prepared for the occasion by donning a clean shirt — his first for a week. The meeting was a success; an armistice was agreed upon, and the Russians subsequently undertook to leave Austria as quickly as possible.

Contemporaries were well aware of the momentous nature of Austerlitz. News of the battle broke the heart of William Pitt, the British Prime Minister; he died soon afterward. The Coalition was shattered. Austria made a humiliating peace, losing great tracts of territory and over two million subjects. Napoleon was less successful with Russia, for the Czar repudiated a tentative agreement in September 1806. Prussia, cowed by the presence of the triumphant *Grande Armée* on its southern borders, concluded a unilateral treaty with France in return for the Kingdom of Hanover. Not only was Prussia thereby brought within the French sphere of influence, but she also became an object of contempt and suspicion to her former allies.

The slopes of the Pratzen still bore their grisly harvest of unburied dead when military experts began to dissect Napoleon's victory. The Russians and Austrians found it convenient to blame their column commanders for incompetence or worse, but many officers appreciated that their armies had attempted to execute a plan for which they were simply not prepared. The Weyrother project, admirable though it looked on paper, did not commend itself to execution by an Allied force whose staff work was decidedly patchy. However, whatever was wrong with the Allied army on 2 December was not the fault of its junior ranks. It would be hard to match the desperate courage of the

Napoleon meets Francis I, the Austrian emperor, after the battle.

men of the Galicia, Azov and Narva Regiments around Kobelnitz, or the staunch valor of the Croats and Hungarians of Kienmayer's detachment. The Allied army was not so much outfought as outgeneralled, and for this nobody was more responsible than Napoleon himself.

It is hard to cut through the myth and propaganda surrounding Napoleon's role at Austerlitz. His supporters, naturally enough, regarded the battle as a master stroke of sublime brilliance, whereas some historians have suggested that Napoleon was dangerously overextended, and was saved from catastrophe only by the blunders of the Allies. The truth lies somewhere between these extremes. It is, in any case, dangerous to consider the battle of Austerlitz outside the context of the campaign; Napoleon himself wrote that 'the victory of Austerlitz was only the natural outcome of the Moravian plan of campaign. In an art as difficult as that of war, the system of a campaign often reveals the plan of battle.' Napoleon's 1805 campaign was indeed bold to the point of rashness, but it represented one of the few viable strategic courses open to him. To have waited for the Allies to link up in Germany and threaten France, while the British blockade strangled French industry, was a recipe for disaster. It was far better to catch the Allies off guard and defeat them in the critical theater of operations. Moreover, Napoleon had to hand an instrument forged in the crucible of revolutionary France, tempered in the wars of the Revolution and polished at the Boulogne camp. He knew exactly what it was capable of, and was able to undertake a strategic deployment which would have confounded a less maneuverable force.

Once in Moravia, Napoleon succeeded in imposing his own will on his opponents. The fact that the Allies were disunited and overconfident does not detract from this achievement. Just as a fencer judges the ability of his opponent in the first few seconds of a bout, so Napoleon took the measure of the Allied commanders and correctly deduced that they could be persuaded to lunge at him if he seemed to drop his guard. His own riposte – the original double envelopment scheme – was planned on exactly this basis.

History is full of generals who have shown themselves adept at planning battles. Capable of bringing their armies onto the battlefield and setting them in motion according to a sound policy, they have nevertheless often failed to respond to the fluid circumstances of battle or to answer the challenge of imponderables. Napoleon's conduct at Austerlitz vividly demonstrates his ability to fight as well as to plan a battle. He made two modifications to his original plan, reshaping it to fit changing circumstances. Finally, when part of his plan went wrong, he was able to improvise and execute an alternative which rendered his victory conclusive. Although the battle did not follow the course he had anticipated, he remained firmly in control, meeting eventualities as they arose. Historians may quibble about which battle may justly be termed Napoleon's finest. But on 2 December 1805, in the aptly-named Battle of the Three Emperors, Napoleon established himself as the military colossus of his age. The Allied generals underestimated him on the eve of Austerlitz. It was a mistake nobody would make again.

CHAPTER 2

WELLINGTON AT WATERLOO 1815

The Duke of Wellington was rightly wary of historians. 'I recommend to you,' he wrote, 'to leave the battle of Waterloo as it is,' and later in life he remarked that he entertained no hope of ever reading a true account of the battle. It is likely that the Duke's customary perspicacity did not fail him in this instance, for few accounts of Waterloo can be deemed entirely satisfactory; several facts remain disputed, and major differences of interpretation still occur. As Wellington might have expected, his own reputation has not emerged unchallenged.

Waterloo was Wellington's last and most conclusive battle. It crowned a military career which had commenced in the twilight of the eighteenth century when, in 1787, young Arthur Wellesley had received his ensign's commission. Yet although the battle was, on the face of things, a remarkable triumph over the greatest soldier of the age, none of Wellington's actions have been the subject of such critical attention as the campaign and battle of Waterloo. The Duke's own assessment of the battle – that it had been 'a damned serious business' and 'the nearest run thing you ever saw in your life' – has lent fuel to the fires of criticism. It has been asserted that Waterloo was, at best, an accidental victory in which Wellington, after fighting a stolid and unenterprising defensive battle, was saved from well-deserved defeat only by the timely arrival of the Prussians. French historians have, understandably enough, found this interpretation particularly attractive, for it lends poignancy to a lost cause and permits the French army to stagger from the field of Waterloo with its reputation intact.

Wellington enjoyed substantial prestige even before Waterloo. His career had begun in earnest in India where, between 1799 and 1805, he had commanded a number of sieges and pitched battles, including a remarkably hard-fought contest at Assaye in September 1803. Wellington had emerged as a capable leader in the rough-and-tumble of Indian warfare – but reputations acquired in India lost their gloss beneath the paler skies of Europe, and it was all too easy for the critics of Sir Arthur Wellesley, as he then was, to write him off as just another Sepoy General. The belief that Wellington was essentially a man for sideshows was to linger on. Napoleon seems to have retained this conviction, with fatal consequences, until 1815.

After his return from India in 1805, Sir Arthur was employed in two fruitless military commands before being appointed Chief Secretary for Ireland. In the summer of 1808 the government's decision to send a force to Spain, recently invaded by the French, gave Wellesley his golden opportunity. He went out to Portugal in temporary command of an expeditionary force, and defeated the French at Roliça and Vimiero before being superseded by staid and senior officers who allowed the French to negotiate their escape from Portugal. Nevertheless, the experience of Vimiero was invaluable. Wellesley had decided that one of the reasons behind the French success was that many of their opponents were half beaten before the battle began. 'I, at least, will not be frightened beforehand,' he remarked, and at Vimiero he received the attack of French columns with his infantry in line behind the crest of a ridge. The columns, swept by close-range musketry to which they could make no adequate response, wilted and died. The lessons of Vimiero were to remain imprinted on Wellesley's mind.

Sir Arthur's return from Portugal was followed by another brief interlude at Dublin Castle before, in the spring of 1809, he again sailed for Portugal, this time in undisputed command. His initial success in driving Marshal Soult out of Portugal was the prelude to a more difficult period in which the aspiring general savored the questionable delights of cooperating with a proud but ineffective Spanish ally. In July 1809 he beat off a French attack at Talavera – a victory which raised him to the peerage as Lord Wellington – but in the autumn of 1810 the appearance of powerful new French forces persuaded him to fall back on a strong position covering Lisbon.

Marshal André Massena's unfortunate Frenchmen starved in front of the lines of Torres Vedras in the winter of 1810, and in early 1811 fell back into Spain. Wellington took Almeida after a scrambling victory at Fuentes de Oñoro, and in January 1812 stormed the vital fortress of Cuidad Rodrigo. Badajoz fell in April, and Wellington pressed on into Spain. In July he jockeyed for position with Marmont, and on the 22nd he exploited the latter's overextension to win a brilliant offensive battle at Salamanca. The winter of 1812–13 saw Wellington once more in the safety of Portugal, but he emerged the following spring, cleared Portugal for good and swept the French back against the Pyrenees, defeating another adversary, Marshal Jourdan, at Vitoria in June.

The winter of 1813–14 was spent fencing among the inhospitable Pyrenees, and in April 1814 Wellington, now firmly on French soil, dislodged his old opponent Soult from Toulouse. Scarcely had Wellington entered the town than an officer arrived with the news that Napoleon had abdicated. Wellington's incomparable army did not long survive the termination of hostilities. It was rapidly disbanded, and a number of its seasoned regiments were sent off to North America to take part in the war against the United States.

Wellington himself went to Paris, to take over the British Embassy. He had been there less than a week when he was sent on a special mission to Spain, in an effort to dissuade the newly restored Ferdinand VII from goading his subjects into civil war. He had some success with this task and, while in Madrid, received the news that he had been created a duke. He then returned to England, where he remained until 8 August, when he left for France to take up his ambassadorial duties. On his way, however, he spent three weeks examining the border defenses between Belgium – now incorporated with Holland – and France. He recommended the construction of a number of small fortresses, and noted a number of good positions for a defending army. One of these lay at the edge of the forest of Soignes, on the main road from Brussels to Charleroi, near the little village of Waterloo.

On 24 August, when Wellington presented his credentials to Louis XVIII, he seemed to have seen the last of the profession of arms. He could, it is true, look back with satisfaction upon his military career. To his exploits in India he had now added the laurels of the Peninsula. He had shown himself adept at siege warfare, and although his battles showed a predilection for the tactical defensive, his attack at Salamanca was a classic example of switching from the defensive to the offensive to take advantage of a

The Duke of Wellington before the battle.

fleeting opportunity. He had a good grasp of the problems of supply and transport, an attentive eye for discipline and morale, and was familiar with the perils of alliance warfare. Yet something was lacking. His exploits, remarkable and significant though they were, remained peripheral. The Spanish ulcer had debilitated Napoleon, but it was the fighting in Russia, Germany and Northern France that had broken him. Moreover, Wellington had rarely faced an opponent whose forte was independent command. Napoleon's marshals were, with a few exceptions, more capable as subordinates than as commanders in their own right. Had Europe remained at peace from 1814 onward, Wellington would probably be remembered as a competent and workmanlike general, whose capabilities were never fully tested against a first-rate opponent in a major battle.

Be this as it may, there can be no doubt that Wellington successfully made the transition from soldier to diplomat in 1814–15. The Paris post grew somewhat uncomfortable as Louis' subjects grew increasingly nostalgic for the exiled Emperor. Wellington was reluctant to relinquish the post lest it should seem that he was being driven from France by the threat of violence. He was also markedly unenthusiastic about the possibility of becoming commander in chief in America. On Christmas Eve 1814, however, peace was signed between Britain and the United States at Ghent, and with this the specter of the American command receded. A more congenial appointment fell vacant when Lord Castlereagh, British representative at the Congress of Vienna, was forced to return to England for the opening of Parliament. Wellington took his place, and joined the squabbling Allies in their efforts to redraw the map of Europe in the aftermath of the defeat of France.

The Duke was immersed in his labors at Vienna when, on 7 March, he received word that Napoleon had left the island of Elba for an unknown destination. The wily Prince Metternich astutely observed that he would make straight for Paris, and indeed this is what he did. Louis XVIII's government collapsed like a house of cards; most of his army went over to Napoleon without firing a shot. The representatives of the powers at Vienna, united at last by this threat to the very security they were striving to create, had already declared their steadfast opposition to Napoleon. Wellington was offered an ill-defined advisory post by the Czar, but declined, declaring later that he would prefer to carry a musket. He was instead given command of the right wing of the Allied armies – a force of British, German and Dutch-Belgian troops in the Low Countries. A Prussian army, under the veteran Marshal Blücher, with whom Wellington enjoyed friendly relations undiminished by mutual incomprehension, was on the Duke's left flank. An Austrian force was concentrating in the Black Forest, and more Austrians threatened the South of France while a Russian army marched ponderously toward the central Rhine.

Wellington arrived in Brussels early on the morning of 5 April, and took command of the Anglo-Dutch force, hitherto in the hands of the Prince of Orange. The problems which immediately confronted him placed no less of a strain on his abilities as a diplomat than had the negotiations in Vienna. The British portion of his force, initially some 14,000 men, was of uncertain quality, and included a number of inexperienced and poorly trained units. Yet if this caused Wellington dismay, the state of his Dutch-Belgian allies was even more alarming. Many Dutch-Belgian officers – including the Minister of War – had fought for Napoleon and were pro-French. Political considerations forced Wellington to employ the Prince of Orange and the young Prince Frederick of the Netherlands in senior command appointments, despite their lack of experience. This in turn offended some of Wellington's British subordinates, notably the valiant and profane Sir Thomas Picton. The training, equipment and staff work of the Dutch-Belgians left much to be desired, and the frontier fortresses had scarcely improved since Wellington's visit the previous summer.

The Duke at once set to work reorganizing this heterogeneous force into an army. He made strenuous attempts to obtain more British troops, particularly veteran infantry, and many Peninsular battalions were on their way back from America. He reduced the number of companies in his King's German Legion battalions and used the surplus officers and noncommissioned officers to strengthen Hanoverian militia battalions. Wellington realized that his army was of decidedly patchy quality, and strove to weld it together by ensuring that, wherever possible, good units were intermingled with bad or inexperienced ones. His I Corps, under the Prince of Orange, contained two British and two Dutch-Belgian divisions, and Lord Hill's II Corps was composed of two British divisions, a Dutch-Belgian division and a Dutch-Belgian brigade. The latter two formations were, nominally at least, commanded by Prince Frederick of the Netherlands. Two British divisions, the Brunswick Corps and a Nassau contingent, formed the Duke's reserve. The Earl of Uxbridge commanded the British and Hanoverian cavalry, eight brigades in all, to which were added three Dutch-Belgian brigades. Wellington mixed Englishman and foreigner, veteran and recruit in all the British infantry divisions, except the 1st Division. Sir Charles Alten's 3rd Division, for example, contained an inexperienced British brigade, a veteran brigade of the King's German Legion, and a regular Hanoverian brigade.

Wellington's energetic efforts to create order out of chaos were, ironically, mirrored by those of his opponent. Although war had not yet been formally declared, Napoleon was well aware that if his regime was to survive he must defeat the Allied armies in detail, before their combined offensive was ready. He therefore turned his fierce energy onto the problem of raising men and providing them with arms and equipment. A defensive strategy was out of the question. Napoleon had to attack, and the Low Countries and the multinational force they contained were the most obvious target. Not only were the armies of Wellington and Blücher the most immediate threat to France, but the Low Countries also contained a strong Bonapartist faction. If Napoleon struck fast, he had every chance of jamming the complex Allied war machine and exploiting the military and political divisions which bedeviled Wellington in the early summer of 1815.

Wellington and Blücher had every hope of launching their own offensive without French interference, but were nevertheless aware that Napoleon might try to preempt their attack. A conference at Tirlemont on 3 May estab-

Marshal Michel Ney, Prince of the Moskowa.

lished an interarmy boundary and laid down certain general principles: liaison staffs were exchanged and arrangements made for the passage of intelligence between the armies. The two commanders seem to have agreed to take the Roman Road from Bavay to Maastricht as the line of demarcation, and to concentrate the bulk of their forces on the front from Ghent to Namur. Each army would support the other if attacked, and tentative arrangements were made for a Prussian concentration at Sombreffe and an Anglo-Dutch concentration at Nivelles. Vague though many details of the conference are, the meeting was nevertheless of prime importance: it cemented the excellent personal understanding between Wellington and Blücher, and established the principle of mutual assistance upon which the campaign was to turn.

By the end of May the Prussian army was deployed with two of its corps forward at Charleroi and Ciney, a third close at hand just north of Namur and the fourth in reserve near Liège. Just across the Roman Road from the Prussian I Corps at Charleroi was the Anglo-Dutch I Corps, deployed in the area Mons-Genappe-Enghien. The II Corps lay to its west, in the area Leuze-Alost-Renaix, while the Reserve Corps was billeted around Brussels. Uxbridge's cavalry occupied a narrow pocket between the corps, in the valley of the Dender. Both armies were fairly widely dispersed in order to make the best use of the billeting facilities available.

By early June it was apparent that something was happening behind the French frontier, though it was far from easy to say what it was. Napoleon, aided by French sympathizers in the Netherlands, had a reasonably clear picture of his enemy's dispositions while Wellington had little reliable intelligence. Napoleon still hoped to defeat the Allies in detail; he planned to throw his main weight onto the boundary between the two armies, hoping that each would fall back on its own lines of communication.

As he partook of his customary early breakfast in Brussels on the morning of 15 June, Wellington can have entertained few worries for the immediate future. He had every confidence in Blücher, and could concentrate his entire force on either flank within 48 hours. Much less time than that would be needed to deploy the bulk of his army to meet any threat within his area of responsibility. If the French attacked Blücher, the Duke would concentrate at Nivelles; if, on the other hand, the blow fell on Wellington, the Prussians would march on Sombreffe. As the Duke enjoyed his tea and toast, the possibility of a French offensive was probably not uppermost in his mind. All being well, he hoped that Blücher and he would mount their own attack before the French were ready.

This satisfactory state of affairs did not survive the day. In midafternoon Wellington heard that the Prussian I Corps had been sharply attacked early that morning. This disturbing news was therefore rather old by the time it reached Brussels, indicating that the flow of intelligence between the armies was by no means as smooth as it should have been. Furthermore, the information was vague, giving the Duke no idea of the strength or extent of the French attack. Wellington, in default of more precise intelligence, did what he could on the facts available, and late in the afternoon orders went out for divisions to concentrate at their assembly points. The Duke's first fear seems to have been for his right, and he was thinking in terms of a concentration at Nivelles to meet a French thrust through Mons.

On the night of the 15th Wellington attended the Duchess of Richmond's ball. This was by no means as irresponsible as it might appear. Cancelling the ball would have encouraged the pro-French element in Brussels, and it was no bad thing for the Duke to have a number of his senior officers about him at a crucial time. More news — and news of a decidedly gloomy nature — arrived while Wellington was actually at the ball. He first heard that the Prussian advance guard had been driven back by the French at Fleurus — only a few miles from Blücher's planned concentration point at Sombreffe. Then, during supper, he heard from the Prince of Orange that French scouts had reached Quatre Bras, an important junction on the main route between the armies. The Dutch-Belgian commander at Quatre Bras had, on his own initiative, decided to hold the crossroads rather than march to his concentration point at Nivelles. Nevertheless, the situation was delicate: the main weight of the French offensive seemed likely to fall on the interarmy boundary, rather than on the Anglo-Dutch right where Wellington had expected it. 'Napoleon has humbugged me, by God!' said Wellington to his host. 'He has gained 24 hours' march on me.' 'What do you intend doing?' the Duke of Richmond asked. 'I have ordered the army to concentrate at Quatre Bras,' replied Wellington, 'but we shall not stop him there, and so I must fight him *here*,' and a thumb-nail jabbed at the map, indicating a ridge across the Brussels-Charleroi road, just south of the village of Waterloo.

The Duke's reserve left Brussels at dawn on the 16th, with orders to march to the road junction at Mont Saint Jean. Wellington himself rode forward later, and by midmorning had arrived at Quatre Bras. So long as this crossroads remained in his hands, he enjoyed good lateral communications with the Prussian concentration point at Sombreffe, eight miles to the east along the Nivelles-Namur road. Wellington reached the junction to find it securely held by General Perponcher's Dutch-Belgian division. He ordered his reserve forward from Mont Saint Jean, dispatched officers to bring up those units ordered to concentrate at Nivelles, and sent a message to Blücher with details — some of them unintentionally false — of his army's dispositions.

These steps taken, Wellington crossed the Roman Road, and met Blücher at a windmill near the village of Brye. The Prussian told him that three of his corps were now concentrated in front of Sombreffe, holding a forward slope dotted with small villages. This position was certainly not

to the Duke's taste; he feared that the Prussians would be 'damnably mauled' if they fought there. Wellington then outlined his own plan for a concentration at Quatre Bras, and added that he thought it quite likely that the French would attack him there. He declined a suggestion made by General Graf von Gneisenau, Blücher's Chief of Staff, that he should reinforce the Prussians, but eventually agreed to move all available troops from Quatre Bras to Sombreffe, always provided that he was not attacked in the meantime.

By the time the Duke returned to Quatre Bras, the situation there had taken a turn for the worse. The French were clearly about to launch an attack, and when it came many of the Dutch-Belgians recalled an urgent appointment elsewhere and slid off to the rear. Fortunately for the Duke, his

Left: Marshal Soult restrains Napoleon from making a last desperate attack during the battle.

Below: The French cuirassiers charge the British squares by Felix Philippoteaux, painted in 1874.

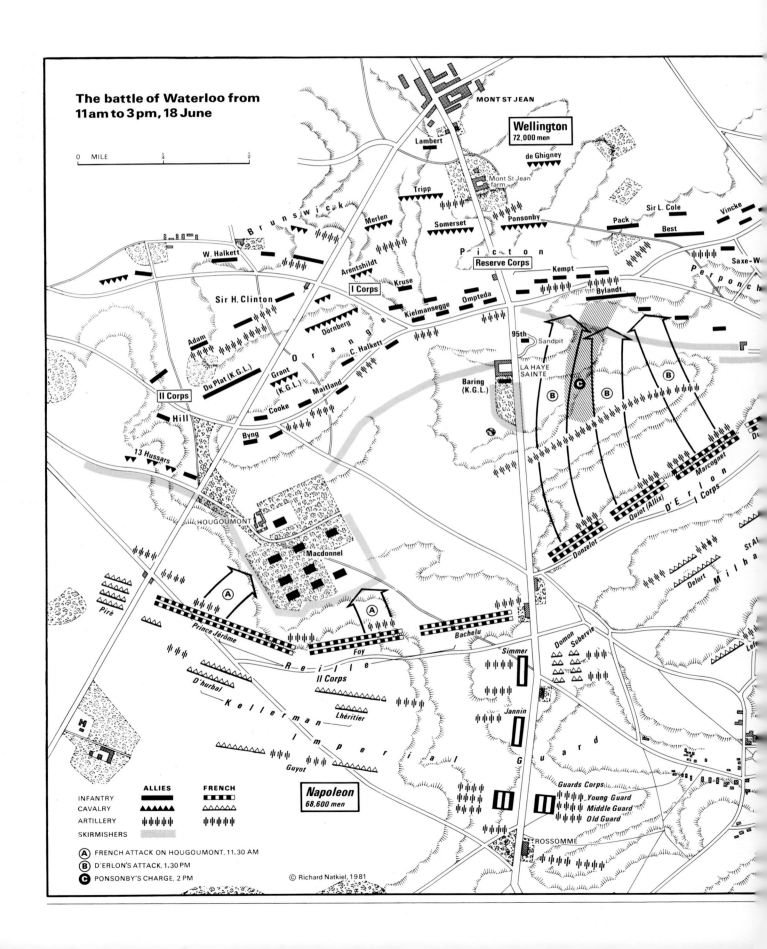

The battle of Waterloo from 11 am to 3 pm, 18 June

MONT ST JEAN

Wellington 72,000 men

0 MILE ¼ ½

Lambert

de Ghigney

Mont St Jean farm

Tripp

Somerset

Ponsonby

Sir L. Cole

Vincke

Merlen

P i c t o n

Pack

Best

Brunswick

W. Halkett

Arentshildt

Kruse

Reserve Corps

Kemp t

Bylandt

Saxe-W

Perponch

I Corps

Kielmansegge

Ompteda

Sir H. Clinton

O r a n g e

Dörnberg

C. Halkett

95th

Sandpit

LA HAYE SAINTE

Adam

Grant (K.G.L.)

Maitland

Baring (K.G.L.)

B

C

B

Du Plat (K.G.L.)

Cooke

B

II Corps

Hill

Marcognet

Byng

Ouiot (Allix)

D'Erlon I Corps

13 Hussars

Dontelot

St A

HOUGOUMONT

Milha

Macdonnel

Delort

A

A

Bachelu

Domon

Subervie

St Al

Piré

Prince Jérôme

Foy

Simmer

Lefe

R e i l l e

II Corps

D'hurbal

Jannin

Lhéritier

K e l l e r m a n

I m p e r i a l

G u a r d

Guards Corps

Young Guard

Guyot

Middle Guard

Napoleon 68,600 men

Old Guard

ROSSOMME

	ALLIES	FRENCH
INFANTRY	▬▬▬	▦▦▦
CAVALRY	▲▲▲▲▲	△△△△△
ARTILLERY	�┼┼┼┼	┼┼┼┼┼
SKIRMISHERS		

Ⓐ FRENCH ATTACK ON HOUGOUMONT, 11.30 AM

Ⓑ D'ERLON'S ATTACK, 1.30 PM

Ⓒ PONSONBY'S CHARGE, 2 PM

© Richard Natkiel, 1981

Right: The charge of the Life Guards ended the second phase of the battle. It repelled d'Erlon's attack but it also laid low 2500 Allied horsemen.

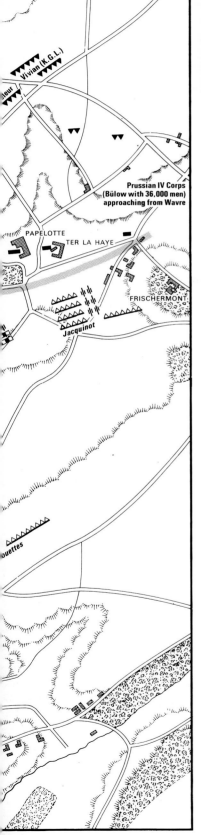

Prussian IV Corps (Bülow with 36,000 men) approaching from Wavre

PAPELOTTE

TER LA HAYE

FRISCHERMONT

Jacquinot

ouettes

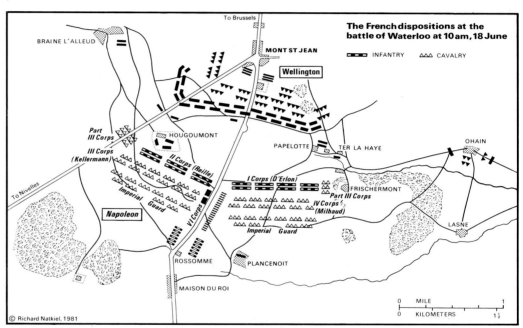

The French dispositions at the battle of Waterloo at 10am, 18 June

◼▭◼ INFANTRY   △△△ CAVALRY

To Brussels

BRAINE L'ALLEUD

MONT ST JEAN

Wellington

Part III Corps

III Corps (Kellermann)

HOUGOUMONT

II Corps (Reille)

PAPELOTTE   TER LA HAYE   OHAIN

Imperial Guard

Napoleon

VI Corps

I Corps (D'Erlon)

FRISCHERMONT

IV Corps (Milhaud)

LASNE

Imperial Guard

To Nivelles

ROSSOMME   PLANCENOIT

MAISON DU ROI

© Richard Natkiel, 1981

0   MILE   1
0   KILOMETERS   1½

Flanking march of the Prussian Army on 18 June

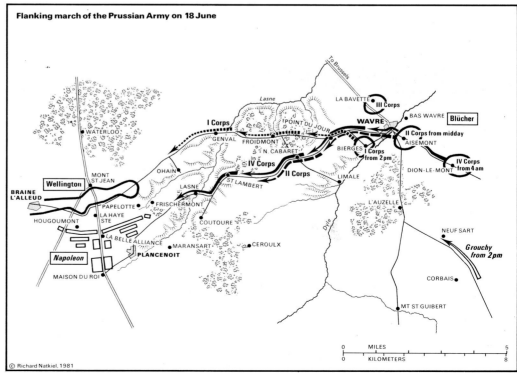

To Brussels

Lasne   LA BAVETTE   III Corps

I Corps

POINT DU JOUR   WAVRE   BAS WAVRE   Blücher

WATERLOO   GENVAL   FROIDMONT   N. CABARET   II Corps from midday

BIERGES   AISEMONT

IV Corps   I Corps from 2pm

OHAIN   IV Corps   II Corps   LIMALE   DION-LE-MONT   IV Corps from 4am

MONT ST JEAN   LASNE   ST LAMBERT

Wellington   L'AUZELLE

BRAINE L'ALLEUD   PAPELOTTE   FRISCHERMONT

HOUGOUMONT   LA HAYE STE   COUTOURE   NEUF SART

LA BELLE ALLIANCE   Dyle   Grouchy from 2pm

Napoleon   MARANSART   CEROULX

PLANCENOIT

MAISON DU ROI   CORBAIS

MT ST GUIBERT

© Richard Natkiel, 1981

0   MILES   5
0   KILOMETERS   8

opponent at Quatre Bras was Marshal Michel Ney, who had uncomfortable memories of Wellingtonian positions in Spain, and feared that the slight ridge which screened part of the position might conceal the full strength of the Allied army. Reinforcements, notably Picton's 1st Division, arrived as the day wore on, but it was an anxious time for Wellington, whose position was far more exposed than he would have wished. Nevertheless, he held Quatre Bras throughout a desperate June afternoon, partly by bluff and partly because of the sheer fighting ability of the indomitable 1st Division.

Things had gone less well with Blücher, who had been badly mauled at Ligny, receiving — as the Duke had predicted — 'a damned good hiding.' Napoleon's *corps de réserve* had spent the day fruitlessly marching between the battlefields of Ligny and Quatre Bras, when its appearance on either field would probably have clinched matters. The Prussians were, however, forced to withdraw; the only question was in which direction this retirement would take place. Gneisenau, who believed that Wellington's service in India had made him as devious as any native ruler, felt that the Duke had failed in his promise to support his allies, and suggested that the army should fall back along the best roads available, which led eastward, toward the Prussian base. Blücher had been rolled on by his wounded horse while leading a cavalry charge during the battle, but he had dosed himself with a horrid brew of gin, rhubarb and garlic and was still full of fight. He remembered that Wellington's promise had been conditional upon his not being attacked himself, and he therefore determined to fall back on Wavre, from where he could still support Wellington, or withdraw to the east if necessary.

Wellington had no definite news of the Prussian with-drawal until about 0730 hours on the morning of the 17th. A little later he received a full account from a Prussian staff officer, and told Müffling, his Prussian liaison officer, that he would give battle on the Mont Saint Jean position if Blücher would support him with at least a corps. It was obviously essential for the Duke to conform to Blücher's northerly movement, and, having agreed with Müffling that the French were likely to make a slow start that morning, he ordered his men to breakfast before setting off. By midmorning all units ordered up from Nivelles had arrived, and Wellington began to disengage, feeding troops back down the Brussels road. At about midday he observed with satisfaction, 'Well, there is the last of the infantry gone, and I don't care now.' His forces were all but clear of Quatre Bras by 1400 hours when Napoleon himself appeared at the head of a mass of lancers. The British retirement was accompanied by a violent thunderstorm which added to the miseries of the combatants but also drenched the fields, confining movement to the only metalled road. This prevented French cavalry from out-flanking the retreating force, and made it easier for Wellington to accomplish that most difficult of maneuvers, a with-drawal in contact with the enemy. There was a brisk rearguard action in Genappe, but the pursuit had died away long before midnight, as the Duke's army took position in the rainy cornfields below Mont Saint Jean.

Napoleon's principal worry was that the Duke would slip away altogether, depriving him of the benefits of his initial move and the fighting on the 16th. In the early hours of the 18th the Emperor rode along his outpost line and

The view from La Haie Sainte during the last phase of the battle.

saw, with relief, the glow of bivouac fires along the ridge to the north, showing that Wellington intended to offer battle. Napoleon desired nothing more: 'I have them now, these English,' he remarked with pleasure.

Wellington's position lay along a low ridge south of the hamlet of Mont Saint Jean. Although it was by no means as good as many of the reverse slope positions he had used in the Peninsula, it was not without numerous advantages. Most of his force was posted just over the crest line, and could avoid the worst effects of French artillery fire by lying down. The ridge also shielded movements between the flanks of the position, facilitating the deployment of reserves. A strong outpost line garnished the forward slope, and the farms of Papellotte, La Haie Sainte and the Chateau of Hougoumont had been prepared for defense and garrisoned. The Brussels road ran squarely through the position, and it was just to the west of this route that Wellington had posted the bulk of his force. He recognized that the right flank of his position was in the air, and suspected that Napoleon might try to roll him up from the west. To obstruct a really wide outflanking move, strong detachments were posted at Hal and Tubize, about eight miles to the west. This force took no part in the battle, but it offered useful insurance to the Duke's weak right flank.

Satisfactory though the Mont Saint Jean position seemed, Wellington was only prepared to fight in it if he had guaranteed Prussian support. He had some 68,000 men under his hand, and was well aware that Napoleon had over 100,000 troops, who could be employed against either the Anglo-Dutch or the Prussians. In the small hours of the 18th he received word from Blücher that two Prussian corps would move to his support at dawn. Wellington expected them to be making their presence felt by midday, and he was confi-

dent of holding Napoleon till then; with the arrival of this message, therefore, his decision to stand and fight became irrevocable.

However great the Duke's apparent confidence at dawn on the 18th, his impassive profile must have concealed certain misgivings. Quatre Bras had confirmed the unreliability of the Dutch-Belgians, who could be expected to perform no better in a major battle. The Duke believed that his position was, ultimately, tenable only with Prussian assistance, but his experiences in Spain, coupled with more recent events, gave ample evidence of the fallibility of allies. Moreover, this was the first time that Wellington had met Napoleon in a pitched battle. Wellington had a poor opinion of the Emperor as a man, calling him 'low and ungentleman-like.' Despite this he felt that Napoleon's presence on a battlefield was worth 40,000 men, and he was by no means unaware of his formidable military record. The Duke had faced French attacks often enough in the Peninsula to know what to expect if Napoleon launched a frontal assault, and probably felt that this was a far less frightening eventuality than a fluid battle of maneuver in which the more supple French army would be likely to excel. Finally, despite the presence of the force at Tubize and Hal, the ugly prospect of a flanking attack cannot have been far from the Duke's mind. Whatever his doubts, Wellington kept most of them to himself, although he did send a discreet letter to a lady in Brussels suggesting that she might care to visit Antwerp. He appears to have made no plans for retirement in the event of a French victory, and when his second in command, Lord Uxbridge,

Wellington encourages the British infantry squares.

asked him what his plans were, the Duke replied that 'Bonaparte has not given me any idea of his projects, and as my plans will depend upon his, how can you expect me to tell you what mine are?' Then he softened for a moment, put his hand on Uxbridge's shoulder, and said: 'There is one thing certain, Uxbridge; that is, that whatever happens you and I will do our duty.'

Napoleon's intentions, founded as they were upon sublime assurance, were simple enough. He proposed to bombard the Anglo-Dutch position, and to use cavalry to force Wellington to disclose his troops; he would then administer the knock-out blow with an infantry assault. Some of those on his staff, memories of the Peninsula fresh in their minds, protested that the task might prove rather more difficult than this. 'Just because you have been beaten by Wellington,' snapped the Emperor, 'you think he's a good general. I tell you, Wellington is a bad general, the English are bad troops, and this affair is nothing more than eating breakfast.' It was to prove a singularly indigestible meal.

The Emperor's conviction that the simple recipe of frontal assault would beat the Anglo-Dutch, coupled with his belief that the Prussians had been so badly hammered at Ligny that they would be intent on escaping and not aiding Wellington, led him to make two important decisions. Firstly, he postponed the main attack from 0900 until 1300 hours, to allow the ground to dry out, permitting his artillery to move more easily off the roads and enhancing the ricochet effect of its fire. Secondly, his orders to Marshal Grouchy, who was following the Prussians, failed to emphasize the importance of preventing a junction between Blücher and Wellington. These decisions were to prove crucial. Time was of the essence, and Napoleon's only hope of victory lay in crushing the Anglo-Belgians before the Prussians could intervene.

Bad Prussian staff work went some way toward compensating for the inaccuracy of Napoleon's appreciation of the condition of Blücher's army. Bülow's corps, which had not been engaged on the 16th, was ordered to lead the advance, although it lay furthest from Wellington's army. Its line of march lay across that of the other two corps — Blücher's remaining corps, under Thielmann, faced Grouchy beyond Wavre — and the resulting confusion meant that there was no possibility of Bülow arriving on the battlefield before 1600 hours. Wellington would therefore have to wait for Prussian help for at least four hours longer than he expected.

The Duke spent the morning putting the finishing touches to his dispositions, and watching the French army as it deployed. He wore a blue frock-coat and a low black cocked hat and, eminently practical as he was, from time to time donned a short cloak to keep off the showers. He rode his favorite charger, Copenhagen, a fact which caused some dismay to the infantry in whose squares he later took refuge, for the horse was a notorious kicker. By midmorning the French were drawn up in parade array, and it was clear that Napoleon had over 70,000 men in the field. The corps of Reille and d'Erlon stood on either side of the Brussels road, with a great battery of 84 guns just to the west of the road. Behind this were Lobau's VI Corps and the Imperial Guard. The cavalry corps of Kellermann and Milhaud, a formidable mass of mounted men, were behind Reille and

General Blücher marching to join Wellington's forces.

d'Erlon ready for the attack.

A fanciful depiction of the British Guards attacking.

The battle began at about 1130 hours with a French attack on Hougoumont. Napoleon, possibly recognizing Wellington's anxiety for his right, intended this as a feint, but its initial repulse so enraged the divisional commander concerned that he renewed the assault, assisted by some troops from a neighboring division. Wellington, giving an early demonstration of his ability to be always on hand at the critical point, sent a battalion forward to counterattack, and supported it with another battalion soon afterward. The fighting at Hougoumont grew from a flicker into a blaze, steadily drawing in more and more troops – about 3500 of Wellington's men and 14,000 French. Not only was the attack a failure as a diversion, for it did not induce Wellington to weaken his center, but the Chateau also remained in British hands all day, providing a serious obstacle to French attacks west of the Brussels road.

The great bombardment began just after 1300 hours. Shortly before his guns opened fire, Napoleon had received the unwelcome news that the leading elements of Bülow's corps had been seen to the east, beyond the Bois de Paris. Undeterred by this, the Emperor assured Soult, his Chief of Staff, that the odds were still in his favor. The bombardment, deprived as it was of a solid target, did relatively little damage, and when d'Erlon's men advanced at about 1330 hours they were warmly received by the Allied artillery and the infantry of Picton's division. Picton was killed leading his men into the French, and his soldiers were soon hard pressed to sustain the weight of d'Erlon's attack. Moreover, on d'Erlon's left a *cuirassier* brigade had already swamped a Hanoverian battalion and now threatened Wellington's

center. The Duke, his eye on the trouble spot, ordered Uxbridge to take forward Somerset's Household Brigade and Ponsonby's Union Brigade. The British heavy cavalry moved past the German infantry battalions in square west of the Brussels road, swept through the *cuirassiers*, whose horses were blown by their uphill charge, and crashed into d'Erlon's flank. Three of d'Erlon's four divisions were routed, and over 3000 men were captured. However, indiscipline proved the undoing of the British cavalry; elated by success, they surged on to the enemy gun line, where they were sharply counterattacked by fresh cavalry. Barely half of them returned.

The crippling of two fine cavalry brigades was a serious disappointment to Wellington, but it was a small enough price to pay for the laceration of a French corps and the protection of his center. The fighting had also consumed valuable time, for it was not until 1530 hours that Ney ordered forward the next attack. He was repulsed from La Haie Sainte but, seeing transport apparently retreating down the Brussels road, he brought a *cuirassier* brigade to accelerate what he believed to be a withdrawal. Two cavalry divisions followed, more or less on their own initiative, and at about 1600 hours Ney led forward some 5000 horsemen on a 700-yard front between Hougoumont and La Haie Sainte. Receiving a cavalry charge in square was a trying experience for infantry, but as long as they stood firm there was little to fear. Wellington had instructed his gunners to remain at their places until the last possible

General Blücher sends his cavalry into battle in the nick of time.

moment, firing from point-blank range before running to the shelter of the squares behind them.

Ney's charge, awesome though it was, achieved little of practical value. Its impact was broken by rolling volleys of musketry from the squares, and although the French rallied and came on again, they could not shake that imperturbable infantry. 'I never saw British infantry behave so well,' wrote Wellington afterward. Napoleon, by now seriously worried by the growing Prussian pressure on his right, ordered forward more cavalry, and at about 1700 hours another great charge swirled up the slopes between Hougoumont and La Haie Sainte. Again and again the tearing musketry of the squares checked the French horsemen, but Ney, once again demonstrating the accuracy of his nickname 'bravest of the brave,' persisted in leading charge after charge. Wellington, his finger firmly on the pulse of the battle, rode from square to square encouraging his battered infantry, ordering up reserves or unleashing his own cavalry against any French who had penetrated the infantry line.

At about 1800 hours Ney, his cavalry exhausted, led forward a division and a half of Reille's corps which had escaped involvement in the cauldron of Hougoumont. It fared no better than his previous attacks. He was, though, more fortunate shortly afterward when, on Napoleon's direct orders, he wrested La Haie Sainte from its exhausted garrison. The fall of La Haie Sainte enabled Ney to push artillery and *tirailleurs* forward against Wellington's center. The Duke himself was above Hougoumont, doing what he could to make the fire of the *tirailleurs* less galling to his squares, when he heard of the desperate state of his line

north of La Haie Sainte. He received the bad news coolly, ordered up some fresh Brunswickers, sent for every available gun and put the light cavalry brigades of Vivian and Vandeleur into line behind the infantry. The Brunswickers, arriving on the battlefield at this desperate juncture, turned and fled, but were checked by Vivian's cavalry to their rear before being rallied and led forward by the Duke himself. This stabilized the situation in the center, and the threat posed by Ney's artillery and *tirailleurs* diminished.

The fall of La Haie Sainte and the erosion of Wellington's center was the decisive point of the battle. Ney, in spite of his rather limited perception, seems to have realized this, and appealed to his master for the Imperial Guard, an instrument kept in reserve for use at just such a time as this. The Emperor, preoccupied by the appearance of the Prussians in Plancenoit, used part of the Guard to restore the situation there and declined to give the rest of it to Ney. It was a fatal miscalculation. Wellington, as we have seen, recovered his balance, and by the time that Napoleon decided to commit the Guard against the Anglo-Dutch center, at about 1900 hours, the opportunity had gone for ever. Bülow's men were by now heavily engaged on the edge of Plancenoit, and Ziethen's corps was coming up to reinforce Wellington's left.

It was just after 1900 hours when Napoleon led his Guard past his command post at La Belle Alliance. The Duke had put the last half hour to good use, and sorely tried though his army was, it presented an unbroken front, with

General Hill calls upon the Imperial Guard to surrender. General Cambronne's reply was defiant and brief.

The battle scene on the morning after. Casualties were 25,000 French and 22,000 Allied.

some depth in its center, to Napoleon's last gamble. As nine battalions of the Guard, now with the dishevelled figure of Ney at their head, marched up the trampled slope, Wellington positioned himself behind Maitland's Guards Brigade, behind the crest at the point where the threat seemed most serious. The French gunners gave what support they could to the Guard, and the surviving infantry of Reille and d'Erlon prepared for one last effort.

The advancing guardsmen pressed on through a torrent of artillery fire. They now formed two columns, the largest of which breasted the rise only 40 yards from Maitland's men. The Duke's voice rang out: 'Now, Maitland! Now is your time!' After a short pause, he ordered: 'Stand up Guards! Make ready! Fire!' With this, Maitland's men, whose front overlapped the column on either side, poured volleys into the dense French formation. Adjacent units joined in, and after a brief and intense fire fight the column was checked and driven down the slope. The second column, striking the Anglo-Dutch line further east, fared no better.

With the repulse of the Guard, French morale trembled in the balance. Wellington rode over to a mound near the Brussels road and surveyed the battlefield. The French were badly shaken but not yet broken, but Wellington sensed that they had reached the end of their tether. A shaft of sunlight broke through the cloud to illuminate the Duke's trim figure as he motioned with his cocked hat – the army was to advance and complete the victory. He galloped forward to one of his most capable battalion commanders, Sir John Colborne of the 52nd. 'Go on, Colborne! Go on!' he shouted. 'They won't stand. Don't give them a chance to rally!' Wellington had gauged the mood of the French army precisely; as his own stumbled forward and the Prussians pressed in from the east, Napoleon's last army

disintegrated in flight.

Wellington and Blücher met in the twilight south of La Belle Alliance. They embraced on horseback, the Duke getting the full benefit of Blücher's unconventional medicine, and made quick decisions concerning the pursuit, which was entrusted to the Prussians. Wellington then rode back to the little inn at Waterloo where he had spent the previous night. He ate supper at a table laid for all his staff, but there were many empty places. He wept when he read the casualty returns. 'I don't know what it is to lose a battle,' he said, 'but certainly nothing can be more painful than to gain one with the loss of so many of one's friends.'

Although large numbers of French troops remained in Northern France after Waterloo, the battle was decisive. Napoleon's political support withered away, and on 22 June he abdicated. Paris surrendered on 4 July, and the Allied armies entered it a few days later. Wellington stayed on, after the conclusion of peace in November, as Allied Commander in Chief. He carried out his duties with his customary skill and diplomacy, and ended the occupation a year early, in 1819.

Wellington considered that Waterloo had been a close-run affair, and said, with characteristic frankness, 'By God! I don't think it would have been done if I had not been there.' The evidence amply supports this immodest verdict. Wellington's personal contribution to victory was decisive, in the prebattle phase of the campaign as well as during the battle itself. The fact that the Anglo-Dutch army remained a cohesive whole, despite the poor performance of some of its units, bears witness to his organizational ability. His personal relationship with Blücher was of vital importance, otherwise it seems certain that the Allied armies would have drifted apart on 16–17 June. The Duke's decision to fight on the 18th was dependent upon the promise of Prussian support. It is, therefore, fruitless to criticize Wellington for having been rescued only by the timely arrival of the Prussians. He counted upon Prussian assistance, and had every reason to expect it sooner than it came.

Wellington's selection of the Waterloo position has also been the topic of considerable debate. It was certainly not the classic ridge line of the Peninsula. A general, however, must take the ground in a theater of war as he finds it, and the ridge of Mont Saint Jean was one of the few satisfactory positions blocking the Charleroi-Brussels road. During the battle itself, Wellington never lost his grip, in spite of frequent fluctuations in his fortunes. He showed that rare ability to be always at the right place at the right time; it was his personal intervention which shored up the damaged center after the fall of La Haie Sainte, and his own order which initiated the destruction of the Imperial Guard. The Duke's constant appearance at points of crisis did much for the morale of his troops; he hazarded himself as much as the most lowly private soldier, and the heavy casualties among his staff bear solemn testimony to the risks he ran. His conduct of the campaign was, nevertheless, certainly not flawless. He was, perhaps, a little too complacent shortly before the French offensive. A more serious weakness was his unwillingness to trust his senior subordinates with the secrets of his mind. Had the Duke become a casualty, as well he might, on the morning of the 18th, the Allied army would have lost its brain as well as its soul.

Historians continue to ask themselves questions about Waterloo. Did Napoleon's mistakes, rather than Wellington's skill, determine the course of the battle? Should the Duke have detached the Hal-Tubize force? How would he have fared against the Emperor at his best? These questions, interesting though they may be, must remain mere speculation – while the facts speak for themselves. On 18 July Wellington, at the head of a diverse Anglo-Dutch force, fought to a standstill a confident and enthusiastic army led by a man whose military prowess had dazzled Europe. It was no supine and passive defensive battle, but a conflict in which the defender capitalized upon his enemy's weaknesses as well as his own strengths. Great though the part played by the private soldier was, Waterloo was no mere soldiers' battle, in which victory was extemporized by individual valor. Without Wellington's guiding hand such valor would not have denied the Guard its triumphal march through Brussels. At Waterloo, Wellington's presence was everything. His grateful countrymen nicknamed him the 'Iron Duke,' but they did him less than justice. Wellington's military character was, rather, of the finest spring steel, and never was its strength and resilience better shown than at Waterloo.

Wellington and Blücher meet as victors at Belle Alliance after the battle. Blücher greeted Wellington with the words 'I stink don't I' because he had taken a garlic cure during the battle.

CHAPTER 3

LEE AT CHANCELLORS-VILLE 1863

Great battles and flamboyant generals have long been topics for study. In the evaluation of battles of the Civil War one battle in particular is set apart from the others in this regard. Combined in one engagement are the elements of bold strategy, classic maneuvering, brilliant generals and an army fighting desperately for survival against overwhelmingly superior odds. The details of General Robert E Lee and the Army of Northern Virginia at Chancellorsville are those which are found in best-selling novels or on the silver screen. Yet those were in fact the prime ingredients at Chancellorsville. What emerges in analysis however, is a confrontation of one general prepared to gamble everything in desperately bold aggressiveness against another general floundering in overconfident misinformation and indecisiveness.

Of all Civil War generals none can compare with the mystique and admiration which has surrounded Robert E Lee. Beyond commanding an army, Lee held the love and respect of his soldiers, his fellow countrymen and even his enemies. He was in every respect a gentleman-soldier; a gentleman caught in a struggle of personal beliefs and a strong commitment to his countrymen, Northern and Southern; a soldier sworn to his duty to the Confederacy and his rank.

Robert Edward Lee was born in 1807, son of a Revolutionary War hero who had become governor of Virginia. The surname Lee is associated with others of the highest distinction and is considered one of the foremost family names in Virginia. To complete his education young Lee attended the Military Academy at West Point, graduating second in his class. As a young officer he gained a reputation for resourcefulness and daring during the Mexican War. With the passing of time he continued to climb within the military ranks and even served as the Superintendent of West Point from 1852–55.

When the secession of southern states made war an imminent possibility, Lee was offered command over all Federal troops. However, he refused to commit himself until he learned if his beloved Virginia would also secede and join the Confederate States of America. In later writings Lee referred to this period of decision as the most difficult of his lifetime. While he did not believe in secession, neither did he believe in slavery. Lee was convinced that war was a terrible mistake, but he felt honor bound as a true Virginian to offer his services to the Confederate cause.

During the early months of the Civil War General Lee held a local command in his home state, acting as a military advisor to the Confederate President, Jefferson Davis. In May of 1862, Lee was offered command of the Army of Northern Virginia. Thus began the creation of a legend and powerful mystique, the likes of which have rarely been seen. Lee and his army hold a fascination for students of military history with which perhaps only Rommel and his Afrika Korps can compare.

In spite of this, one grain of doubt remains when considering the reputation Lee acquired through the exploits of his army. Ever present in the shadow of Lee's overwhelming popularity was the influence of another great Confederate commander, General Thomas Jonathan 'Stonewall' Jackson. Robert E Lee has been cited as a brilliant commanding general, yet 'Stonewall' Jackson's role in bringing about that

acclaim cannot be denied and was clearly evident in and essential to the victory at Chancellorsville.

During the months of 1862 after his assumption of command over the Army of Northern Virginia Lee campaigned through his home state, denying the Union hope of capturing the Confederate capital, Richmond. The strategical and tactical abilities of Lee's staff achieved a level which frustrated not only the Union commanders but President Lincoln as well. It caused the Union to doubt its ability to win the war and caused many to fear that the United States

General Robert E Lee, the Commander in Chief of the Confederate Army.

would indeed remain divided.

Lee and the Army of Northern Virginia continued to accumulate victories as Lincoln sifted the ranks in an attempt to find a commander who would be a match for Lee's obvious talents. After General Burnside's fiasco at Fredericksburg in December 1862, when Union forces were thrown against the breastworks of the Confederate defense to be senselessly slaughtered, Lincoln began to search once again for a competent commanding general. The decision was made which brought forward the main Union character in the forthcoming Battle of Chancellorsville, Major General Joseph 'Fighting Joe' Hooker, to assume command of the Army of the Potomac in January 1863.

General Hooker had also gained a reputation in the Mexican War, and in the early days of the Civil War had shown himself to be a hard working, able divisional commander. Unfortunately for himself and the Union army, 'Fighting Joe' had one fatal flaw — overconfidence in himself as a man and as a commander. This led him to believe that his personal evaluation of a given situation was always accurate and never needed revision. His demeanor was considered pompous by his fellow officers, and the troops sniggered at his overindulgences of drink and women. His excesses in the latter category led the troops to give the name 'Hooker' to the prostitutes who followed the army.

The obvious flaws in his character could not completely override the abilities which Hooker did possess. The basis for his campaign at Chancellorsville was extremely creative and based on sound assessments. Had the plan been executed by a competent Union commander of different

General George McClellan by Matthew Brady.

temperament the battle may have provided the Union with its first decisive victory in the Eastern Campaign. Hooker also set forward policies aimed at raising the morale, confidence and readiness of the troops. He instituted a furlough policy and commenced training programs which included mock battles within the brigades. He set about rebuilding the Army of the Potomac, reorganizing the troops into large fighting corps capable of sustaining themselves in battle. He also restructured the Union cavalry in attempts to bring it to a level on a par with its Southern counterpart. Finally, Hooker created the first independent intelligence service for the Army of the Potomac. This service was composed of a network of riders, scouts and spies under Colonel Sharp's command, who gathered information about the enemy and in some cases planted false information to deceive and confuse the Confederate command. Hooker's predecessors, McClellan and Burnside, had always relied on the Pinkerton Agency to supply information. Unfortunately such information was often erroneous or exaggerated. Within weeks of its formation Hooker's intelligence network had given him an accurate knowledge of not only Lee's position but Confederate troop strength and condition. Although Hooker's future as a commander held disaster for himself and the Army of the Potomac, he had instituted changes and procedures which would prove valuable to the long-term welfare of that army.

An introduction to the Battle of Chancellorsville would not be complete without some discussion of the man who had acted for some time as General Lee's right hand, General 'Stonewall' Jackson. In character and bearing Jackson was the epitome of the Southern gentleman and officer. He was a religious man and a devoted husband and father. Jackson began his military career with West Point training, later to become an instructor at the Virginia Military Institute. He was extremely proud of this private academy and believed in the school motto that those who occupied its halls were 'yet to be heard from.' On the day of the Battle of Chancellorsville, in the midst of the assault against the Union right flank, 'Stonewall' Jackson is said to have remarked to one of his staff, 'The Yankees will hear from VMI today!'

As a soldier Jackson had the unique ability to take any plausible plan of action and implement it successfully on the tactical battlefield. With reference to the Battle of Chancellorsville it will be easily recognized that while Lee did indeed conceive the strategy and act as the restraint in the execution of the plan at its most critical moment, Jackson ensured the success of the maneuver which led to ultimate victory.

Those two men, Lee and Jackson, so completely complemented each other that it is impossible to do justice to either without discussing both. Both Lee and Jackson were unusual in the respect, admiration and concern each felt for the simple soldiers under their command. Unlike many generals of that time, neither believed in nor condoned the wasting of precious lives, sent into battle as cannon fodder. The obvious concern Lee and Jackson demonstrated for the individuals under their command was the primary reason that their troops could and would perform miraculous feats on the field of battle. Earlier in the war, during the Valley Campaign, Jackson's troops maneuvered so rapidly on the

General Thomas 'Stonewall' Jackson.

battlefield that they earned the nickname 'foot cavalry.' Such maneuvering was only possible because the men of his command would go without food or rest in obedience to his orders.

There is an incident which bears telling at this point, for not only does it demonstrate the mutual respect between Jackson and his men, but it also holds a consequence which would prove fatal in the Chancellorsville Campaign. Hearing of an advance on the night before the Confederate attack on the Union right flank, Jackson sped to the front leaving his personal belongings and wagons far behind in the rear. In the bitter cold of the night one of Jackson's young aides-de-camp, Lieutenant J P Smith, discovered the general huddled in sleep with neither blanket nor cloak to keep him warm. The lieutenant was appalled by Jackson's condition and insisted that the general take his cape. Jackson refused the offer, but when it seemed that the young lieutenant's feelings had been extremely hurt he accepted the offer, not wishing to appear ungrateful. Jackson dozed again and Lieutenant Smith huddled near the fire and also fell asleep. Sometime during the night Jackson must have awakened to see the lieutenant shivering on the cold ground, for he rose up and without waking the young officer, replaced his cape. In the early morning hours Lieutenant Smith woke to find the general as he had the night before, without coverings, and his cape gently spread over his own form. The irony of the incident lies in the fact that it was on that night that Jackson apparently caught the cold which soon afterward contributed to his death.

For all his sterling qualities, Jackson too had one principal flaw as a commander. Too often he was inclined to pursue the enemy without concern for the total scope of the engagement. As exemplified by the Valley Campaign, he could win the majority of the battles yet lose sight of the true and ultimate objectives. Once again this demonstrates how perfectly well suited were Jackson and Lee, and how their natures achieved a delicate balance of strategy and tactics. Lee understood the total scope of an encounter, while Jackson implemented these objectives.

Lee's Army of Northern Virginia held the field with approximately 60,000 veteran Confederate soldiers. In opposition to his forces, Hooker's Army of the Potomac was some 200,000 strong. The overwhelming difference in manpower was of course crucial to Lee's thinking prior to, during and after the campaign.

Lee had positioned his troops outside Fredericksburg at approximately the same locations where, during the Battle of Fredericksburg, he had dealt General Burnside a staggeringly costly defeat. That defeat had almost completely destroyed the morale of the Union army, causing the soldiers and even President Lincoln to lose confidence in its generals. Although he therefore faced an army of deteriorating morale, Lee also suffered from many handicaps. Most obvious was the outnumbering of his own forces by more than three to one. Perhaps more crucial, however, was the army's poor state of readiness to do battle once again. Much in line with the old quotation 'the spirit was willing but the flesh was weak,' Lee's troops were near the point of exhaustion. Few troops, if any, had had a leave of absence within easy memory. Food and supplies were dwindling, and the splendid uniforms in which many had marched to war had become tattered rags. Both arms and equipment were in short supply, and the troops had to rely heavily on what could be scavenged from successful skirmishes to re-equip themselves. More than a few Confederate officers believed that if the Southern troops were as well supplied and equipped as their Northern counterparts the war would have ended the previous year.

Compounding those disadvantages, Lee's army had been forced to fight one defensive action after another. The initiative had swung into Union hands. Lee was convinced that if he were to have any major success in the future and thus entice foreign support to the Confederate cause, he would have to regain that initiative. In the days prior to the Battle of Chancellorsville Lee and his officers struggled to resolve that dilemma without arriving at any sound course of action.

Jackson and Major General J E B Stuart were in favor of instituting an all-out offensive against the Union forces, but Lee recognized their eagerness as voiced pride in their men and army. It would take more than high morale, pride and confidence to bring about the changes which were so desperately needed. Much to Lee's amazement it was the commander of the Army of the Potomac who gave the necessary aid to the Confederate Army. General Hooker brought the battle to Lee and then handed him the reins of initiative through indecision.

That circumstance was certainly not a part of Hooker's battle plans. After rebuilding Union morale and reorganizing the troops, Hooker's intention was to trap Lee into maintaining his position on the heights above Fredericksburg while the majority of the Army of the Potomac swung in a broad movement which would surround Lee's forces.

To that end, three corps would encircle the heights, two would attack above Fredericksburg, and another two corps would wait in reserve to deliver the final, crushing blow.

Hooker's confidence in his superior numbers and the planned maneuver was so strong that he gave no thought to any outcome other than the destruction of the Confederate forces wherever they chose to give battle, or Lee's full retreat along the road to Richmond. Hooker was convinced that Lee would turn tail and run rather than face the superior odds of the Army of the Potomac. It was a sound and operable plan, if only Hooker had not been so rigid in his predictions of Lee's reactions.

On 27 April General Hooker began his flanking maneuver and the Battle of Chancellorsville was underway. What occurred from that point on must be seen from two perspectives: the daring of Lee and Jackson in attacking the attacker; and Hooker's overconfident bumbling.

The 27–28 April saw the three Union corps swinging wide around the Confederate's left northernmost flank. Confederate cavalry units on that flank reported the enemy movements to General Stuart and by the early morning of 29 April Lee had been informed of the unusual Union activity. The information was somewhat fragmented and Lee pondered over the significance of the maneuver as he watched Major General John Sedgwick cross the Rappahannock River with the Union XI Corps. In spite of the obvious advance of the Union forces, and the threatening posture in which they positioned themselves along his front, Lee could not believe that the main assault would be aimed in that area. Therefore, on the morning of the 29th he dispatched General R H Anderson with part of his division to cover the left flank.

During those morning hours Lee received constant and insistant proddings from his generals, particularly Jackson, to sweep down from the heights and destroy Sedgwick's corps while it attempted to take formation below them. Jackson argued that his units alone could defeat Sedgwick and dispose of any imminent threat to the Confederate Army's position. Lee's wisdom, sense and consideration for the scope of planning came into full display at that time.

His generals felt certain that Hooker's flanking maneuver was merely a ploy to draw defenses away from a major frontal assault but Lee believed otherwise. Lee remembered these savage days of December 1862 when General Burnside threw Union regiments, brigades, divisions and corps against the heights in wholesale slaughter. That fiasco had so sickened and appalled Lee that nothing could convince him that the Union generals intended another such assault. He therefore determined that the forces circling to his north constituted the major threat. Lee held his generals firmly in check while he awaited further information from his

The burning of Baton Rouge, Louisiana in December 1862 was another example of the Union's tactics of total war.

Right: General Hooker's
headquarters at Chancellorsville.

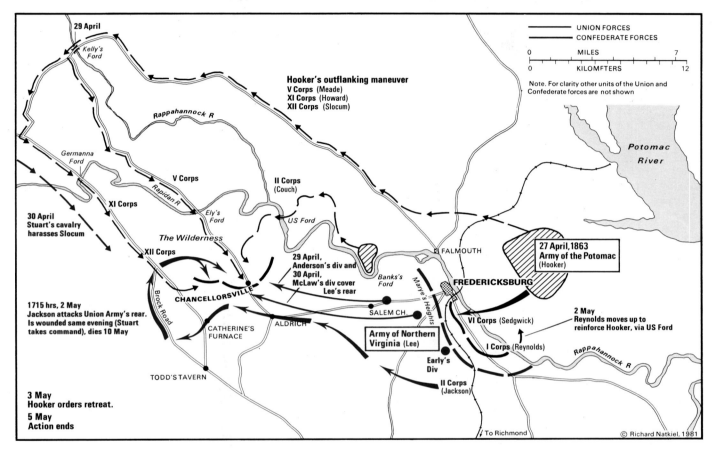

cavalry. Only after discerning the type of maneuver Hooker actually intended would Lee commit himself to action.

By the morning of 30 April Anderson's forces had begun to entrench east of Chancellorsville, to Lee's rear. Reports filtered back to Lee throughout the day that the three Union corps had approached Chancellorsville and were poised to strike against the Confederate rear. Later in the day another Union corps was located and one lone Confederate division was all that stood between the combined strength of four Union corps and the rear of the Confederate army.

On the same day General Hooker arrived at the scene and assumed personal command of his forces. His confidence in his strategy was higher than ever as he received reports of constant rail movements from Fredericksburg along the lines which led to Richmond. Hooker began to believe that Lee had indeed chosen the lesser of two evils and had begun a retreat.

Lee was by then fully aware of Hooker's intent. He realized that if Hooker could pin and encircle the Confederate forces the four Union corps, who alone outnumbered Lee's army, would crush the Army of Northern Virginia between them.

Lee was exploring the feasibility of a retreat, realizing

that while losses would be high they would be acceptable if the army could indeed escape, when Hooker suddenly handed Lee a reprieve. Hooker's self-assuredness in victory had become so strong that he halted his troops for the night. Lee found that fact unbelievable, yet reports continued to arrive confirming the halt. He acted immediately, deciding that if Hooker were willing to give up the initiative he, Lee, was more than happy to accept it.

On the night of 30 April Lee dispatched Major General Lafayette McLaw's division to reinforce General Anderson and made the decision to attack Hooker's forces in the rear. He left one Confederate division to hold the heights and check any assault which Sedgwick's forces might attempt. He sent Jackson with three divisions on a sweep around the Union's right, northern flank, hoping to throw the Union forces of balance with his attack and forcing them on the defensive or into a retreat.

This crucial part of the plan could only succeed because of one man, 'Stonewall' Jackson, for although Lee knew what must be done only Jackson could do it. At that moment of decision Lee shifted the focus of command to Jackson as he shifted the campaign from one of strategy to one of tactics.

Upon reaching Anderson's and McLaw's forces, Jackson quickly assessed the situation and initiated the plans for the counterflanking maneuver. As his troops set out Lee arrived to confer with Jackson. The exact details of that conference remain a mystery, as the generals spoke in strict privacy. Apparently, however, Lee approved of Jackson's plans, yet he questioned the time factor and the ability of Confederate troops to maneuver into a position which would allow for both the surprise and numbers necessary for success. Neither man could forget that the course on which they had set themselves meant the pitting of a small corps of less than six divisions against a force four times that number. The fact remained that they had captured the initiative and although the risks were high each general had supreme confidence in the other's instincts and abilities.

Thus the final battle plan was agreed upon. Lee, with some 20,000 troops, would hold the center, keeping Hooker's attention focused on himself through minor skirmish actions. In that way he could buy the time Jackson needed for the full day's march across the enemy's front. If Hooker recognized the ploy then the results could prove fatal for the Army of Northern Virginia.

The 2 May was a day of anxious anticipation and apprehension; a day which would hold great victory and crushing loss. Early that morning Jackson's phase in the plan of action was implemented. Two factors were essential to the success of his mission. First he had to traverse the area between the fronts of the opposing armies, swing his corps

Major General John Reynolds marches to join Union forces at the Battle of Chancellorsville.

round and position his troops at the weakest point of the Union right flank, all without the Union commander realizing what was happening. Jackson's second major concern, about which he could do little but hope, was that Hooker would refrain from implementing his own offensive against any sector of the Confederate army. The key factors in the complete Confederate plan were so delicately

balanced that only with speed, control and a fair measure of luck could it hope to succeed.

Jackson realized that at best it would be midafternoon or early evening before his troops were in a position to carry out Lee's strategy. With General R E Rodes' division in the lead and Fitzhugh Lee's cavalry covering his flank, Jackson set his forces to their task. After a 15-mile march across the front, Jackson found himself between Catherine's Furnace and the Brock Road. It was there that he turned his force to accomplish an outflanking move. It was also just outside Catherine's Furnace that Union troops, noticing the large concentration of Confederate soldiers moving across a clearing, began harassing the ranks with rifle and cannon fire. As Confederate cavalry exposed itself to screen those troops Jackson felt sure that the element of surprise was lost. Union officers did begin sending messages to their corps commanders stating that a Confederate maneuver had been sighted which appeared to be setting upon the Union right flank. Those reports were soon before General Hooker with requests for orders concerning the new developments.

After assessing his position and the information brought to him, Hooker drew two disastrous conclusions. First, he decided that his position and troop strength were so formidable that he would pay little attention to a maneuver which he believed was merely a weak Confederate ploy to draw the focus from Lee's presumed retreat. His second fatal error was a result of the lateness of the evening. He decided not to make any move at that time, but to hold his position and wait for what morning would bring. Little did he realize that it would only bring disaster.

At approximately 1700 hours on 2 May Jackson, believing that his luck could not hold, began positioning his troops in a line along the Union right flank in an area known as 'the Wilderness.' Periodically Union fire power had been brought to bear on his troops around Catherine's Furnace and they had also skirmished with Union pickets in that area. Although communications had been received many of the Union officers were convinced that more was afoot than an effort to shield Lee's retreat. The junior officers

General Hooker's Army marches to Chancellorsville, where they were to adopt a defensive position and dig entrenchments near a road junction outside the city.

in some of those units were immigrants who had fought in European armies who realized the need to maintain contact with the enemy to avoid being taken by surprise. Although they sent messages to their commander demanding some response to the Confederate maneuvers, their demands fell on deaf ears. In all but a few isolated pockets along the lines, Union troops were in a formation which more closely resembled a bivouac than an army facing a formidable enemy.

By 1715 hours a sufficient number of Confederate troops had completed the maneuver to allow Jackson to give the order to attack. Much to his surprise the Union forces were totally unprepared, and the XI Corps was caught completely off guard. General Oliver O Howard, Union commander in that area, had considered 'the Wilderness' of trees, brush, and brambles which lay along his right flank an adequate protection from any enemy assault. However, as the XI Corps was preparing for its evening meal, Jackson's troops attacked. With bugles blaring Confederate troops surged through 'the Wilderness' with such speed and force that few Union units had time to form a resistance to the onslaught.

In the ferocity of the attack, and with complete disregard for their own safety, Confederate troops fell upon the enemy, many literally having their clothing ripped from their bodies as they pushed through the thorny undergrowth. Both Union and Confederate officers were amazed at the soldiers' determination in pushing the assault to its limits, destroying any hope the Union officers might have had in rallying their own forces. One unknown Union officer, after his capture, declared that in their unbelievable charge through 'the Wilderness' Confederate troops descended like demons on the Union camp.

As Southern troops rolled up the Union right flank, darkness and chaos descended on the battlefield. In some isolated areas Union officers managed to form defensive lines, only to have them shattered by routed friendly forces moving as a sea which dragged everything in its path. Some sources go so far as to note that Union officers and noncommissioned officers attempted to keep their men from fleeing by shooting any soldier who broke ranks. This seemed to be the last straw for the Union troops, and the already brittle state of morale was shattered.

The Confederate charge slowly began to lose impetus, not through lack of spirit but through a lack of cohesiveness between the units and their commanders. Although 'the Wilderness' had been the source of surprise it had also caused disruption to the Confederate line all across the front. Jackson and his divisional commanders were faced with the situation of having their troops scattered along an erratic front, confused not only by the events and the increasing darkness, but also through separation from their commanders.

In the pandemonium Confederate units found themselves mixed with other regiments. Both sides fired not on known enemy positions, for there were none, but on the sound of other gunfire. Many incidents are recorded of units accidentally firing on friendly troops out of confusion and fear. It was expressly this type of confusion which would later lead to the greatest Confederate disaster of all.

Throughout the entire day, and even as the events on

the right flank were taking place, Lee's troops skirmished in an attempt to occupy Hooker's attention, particularly with Anderson's division. There are no recorded conversations or written evidence which give any conception of Lee's thoughts during Jackson's maneuver, but it can be speculated that it was with fear and anxiety that he awaited the results.

With the impetus of the Confederate attack slowing down in the deepening twilight, Jackson became impatient, feeling that with slightly more drive he could gain possession of the United States Ford Road, which led directly to the Union army's rear area. The majority of his officers had lost control of their troops, and each felt that some type of consolidation was necessary before pressing on. Rumors were rife among the ranks. Union cavalry had been caught up in the attack and had broken up into small groups attempting to fight their way back to their own lines. These cavalry troops contacted and skirmished briefly with Confederate infantry.

Infantry caught in the open feared the cavalrymen, who could attack without warning. These brief skirmishes and the confusion of the moment only added to the increasing fears of the Confederate soldiers. Jackson himself had become aware of a rumor that Union forces were massing in preparation for a charge down the road from Chancellorsville.

It was at that time that Jackson's impatience overrode his better judgement. Taking his staff and several couriers he set off down the road toward the area where the attack was said to be massing. Having decided that his personal reconnaissance was necessary, he would not heed the warnings of his fellow officers to send a younger, less indis-

General Hooker's forces fight off a Confederate attack on Sunday 3 May 1863. The Union forces were taken by surprise and fought a losing battle for four days.

pensable, person than himself survey the situation. Jackson believed that he could only command properly if he knew exactly what was transpiring.

After travelling a short distance along the road, Jackson's contingent was fired upon by a group of Union soldiers who were trying desperately to rejoin their regiment. The Union troops had also heard rumors and feared that Confederate cavalry, which was supposedly moving along the front, was bearing down on them. Several members of Jackson's group were killed or wounded by the volley and it is believed that it was then that Jackson received a wound in his right hand. The contingent turned and fled back toward Confederate lines.

As the small cavalcade approached the Confederate lines a cry of 'Yankee cavalry' went up from the ranks. The bewildered Confederate soldiers opened fire. Two more of Jackson's party fell dead and the general himself was shot twice more, in the left wrist and between the left elbow and shoulder. Wounded in both arms Jackson was unable to control his mount which bolted into the thick undergrowth. As his horse careered through the brush, Jackson was struck by a low hanging branch which caused him to reel in the saddle. Captain Wilbourn of the Signal Corps, in a valiant effort to save the general from further harm, caught Jackson as he fell to the ground.

Almost immediately General A P Hill and his staff came upon the scene. They sent for the corps surgeon and an ambulance wagon for the general. As Jackson was lifted to a litter, disaster struck once again. Union artillery, guessing at the location of the enemy from the sound of gunshot, opened fire with shell and grape shot. Several more members of both generals' staffs were killed and General Hill himself received a minor wound. Even as Jackson passed his command to Hill, that general was also removed from action, leaving General Rodes the next officer in line.

General Rodes however was not the man to manage such a maneuver nor consolidate the troops, and command passed to General 'Jeb' Stuart who like Jackson was a very daring and resourceful figure.

The assault was now in complete chaos. The Union forces were in uncontrolled retreat and the Confederate troops were attempting to reorganize and establish a front in the darkness. These efforts were difficult enough, yet were compounded by the rumors which circulated about the Union forces and the incident with Jackson and Hill. Stuart did what lay in his power to calm the army, but with little apparent success.

By the end of the first day's fighting both armies were shrouded in gloom. The Union despaired over an attack which had completely routed the right flank and the Confederates were disconsolate over the circumstances which caused the removal of their beloved Jackson. Although Jackson's wounds were not severe, he did lose his left arm. Eight days later, as a result of his weakened condition, the cold which he had caught on the night before the attack worsened into pneumonia, causing his untimely death. The Confederate success was indeed paid for dearly.

With order somewhat restored the Confederate forces continued their attack on 3 May. Many had gone for more than 30 hours without adequate food or rest. They dis-

covered that although the right flank had been routed, for the most part the Union Army was still intact and offering stout resistance. The Union troops were displaying a staunch attitude but the same could not be said for their commanding general. Hooker's spirit and confidence had been totally destroyed on the previous night. He could not believe that the Confederate attack had inflicted such a loss, nor even that such an extraordinary maneuver had been mounted against him. Hooker suddenly doubted his intelligence reports, his army, his plans and, most importantly, his own infallibility. To make matters worse, Hooker's headquarters was hit by a Confederate artillery round early in the day, which knocked the general temporarily unconscious. For the remainder of the day he appeared to be in a daze. His mind apparently clouded, he ordered a full retreat of the army.

The order for retreat came in spite of the fact that nowhere along the line of attack were Confederate forces making any substantial headway. Lee had taken complete command of the army once again, and although he tried all day to cut off the Union forces as they retreated from the battlefield he could not accomplish any tactical advantage. Union forces therefore retreated safely across the Rappahannock River. Although Lee and Stuart made every attempt, they simply were not able to accomplish what each felt Jackson could easily have done.

On 4 May, as the retreat continued, Hooker ordered Sedgwick to attack Major-General Jubal Early's lone division at its position on Marye's Heights. By noon Sedgwick had cracked the Confederate defenses and was headed toward Chancellorsville. With the news of this Union success, and with hopes dwindling of inflicting further losses against Hooker in the main frontal area, Lee turned his attention to the latest assault. To meet the threat Lee shifted his entire army against Sedgwick. In a succession

of unsuccessful attacks and skirmishes Lee managed only to hold Sedgwick's advance, inflicting very few casualties, before Sedgwick turned and escaped across the river. Lee changed direction once again, marching back toward Chancellorsville, only to find that Hooker's forces were too far away for the exhausted Confederate troops to give battle.

By 5 May the Chancellorsville Campaign was officially over. Jackson's leadership had resulted in Hooker fleeing the field. Lee was credited with defeating the new commander of the Army of the Potomac, though he did not defeat the army itself. What in fact did he accomplish?

First and foremost, Lee defeated Hooker, throwing the Union command once again into chaos and undermining the morale and confidence of the Union troops in their leaders. He turned back a Union Army three times the troop strength of his own Confederate force and with the aid of a competent staff, primarily Jackson, extinguished Union hopes of a decisive victory at that time. Through continued success the myth of Lee's invincibility was once more confirmed in the minds of the Union rank and file and started to spread to the citizens of the North. Worse yet, President Lincoln and his staff began to believe the legend.

Lee once again proved that he had an uncanny knack for understanding his enemy's strategies. He established

Confederate prisoners marching north after Chancellorsville.

his true mastery over given situations. His ability to command was aided by the great respect paid him by his officers and the trust with which they awaited his judgment of the situation, rather than striking out on their own and acting independently. A key example of that ability was demonstrated in the manner in which he held his generals in check, particularly Jackson, when they wanted desperately to attack Sedgwick's forces in the early days of the battle. In spite of their own individual qualities as commanders and generals, Lee's staff responded to their commanding general, awaiting his orders. The quality of leadership was truly Lee's.

In spite of the fact that Lee saw through Hooker's plan he held off for the right moment and situation and came up with an aggressively daring strategy which would wrest victory from his enemy. This triumph is overshadowed by two definite factors. First is the previously mentioned point that he was a strategist and not a tactician. Second is his relationship with Jackson. In short, Lee was the 'thinker' and Jackson was the 'doer.' This is clearly evident in the manner in which Jackson accomplished the flanking maneuver, and Lee's inability to deliver the final blow to the retreating Union army once Jackson's influence had been removed.

From 3 May onward Lee's forces were lacking in fighting strength and direction. Lee marched to and fro on the battlefield between fronts attempting to trap or defeat some part of the Union forces. Had Jackson been present this would surely have been a success or at least greater damage would have been inflicted upon the Union forces.

Another bitter fact of the engagement was the number of casualties. After the battle Union casualties numbered some 17,000, while Confederate losses were listed at 13,000. Most of the Confederate casualties occurred after Jackson's removal from the field, while Lee was in command. While numerical losses were similar, the percentages of overall troop strength make the figures staggeringly disproportionate.

In combining the losses which Lee would ill afford, the indecisive maneuvering from 3–5 May, the inability to inflict a crucial defeat and the loss of 'Stonewall' Jackson to the Confederate army, Chancellorsville was a very hollow victory. If it were not for Hooker's loss of nerve, Lee would probably have found himself facing defeat.

Yet these points do not indicate that Lee should be denied his title as a great general. He was after all only a man and not the demigod which many believed him to be. It can perhaps even be said that Lee's heart went out of the war with the loss of his trusted aide and confidante, Jackson. Indeed, Jackson's death caused a great cry of mourning to echo throughout the South. A lunatic in a Baltimore asylum is credited with Jackson's most telling epitaph when he said, 'a terrible battle must be raging in heaven when God and his archangels need the services of "Stonewall" Jackson.' There would be no replacement for Jackson in the service of Lee's Army of Northern Virginia.

In the final analysis of General Robert E Lee during his years as the commander of the Army of Northern Virginia it is perhaps sufficient to say, 'Lee was competent, Lee and Jackson were invincible; Lee was a general, Lee and Jackson were a legend.'

CHAPTER 4

GRANT
AT VICKSBURG 1863

In the campaign staged by Ulysses S Grant at Vicksburg, a multitude of influences, strategies and events combined to produce a battle unlike any other of the Civil War. In essence the importance of the campaign lay not so much in the defeat of an enemy or the turning of a tide, but in the battle's effects on future events.

With the exception of Robert E Lee, Ulysses Simpson Grant is perhaps the most renowned Civil War general. His success was partly a result of his unusual attitude to war, which is best described by a colleague in the Vicksburg Campaign. A man of kindred spirit, General William Tecumseh Sherman voiced the definition well when he quite simply stated, 'War is hell.'

Grant saw war as a necessity which comes and goes as societies demand. Once the inevitable act of war was undertaken, warfare itself became a job which had to be done. Grant, unlike most of his contemporaries, was not a general in the Civil War for glory and love of country, nor to defeat slavery or uphold the Union. He was a soldier by profession and war was the business and commodity he knew best.

It is with that perspective that the key to Grant's success can be best understood. The objective of war is victory for oneself and defeat for the enemy. To that end one must strive to completely destroy the enemy's ability to fight. Not only must the enemy be initially defeated, but his power to continue the conflict must be eliminated. It was that concept which Grant implemented in his strategies.

For many months the Union searched for a man capable of defeating his Confederate counterparts and bringing victory to the Union. By March 1864 such a man had risen to supreme command of the Union armies and had become known as 'Mr Lincoln's General.' Yet the man, who would become not only a legendary figure of the Civil War but the eighteenth President of the United States, at one time seemed likely to end his career in ignominy.

Ulysses Simpson Grant was born in 1822. He attended the Academy at West Point during the same time period as others who would figure prominently in the Civil War. Unlike Lee, Grant was not academically inclined and he eventually graduated in the bottom half of his class at West Point. He received his initiation into active service during the Mexican War where he was decorated on several occasions for his heroism. Grant also displayed a keen grasp of tactics which earmarked him for a promising military career. Unfortunately, after the war the boredom of the frontier, combined with his inability to cope with alcohol led Grant to resign his commission rather than face court-martial.

In civilian life Grant drifted from one misadventure to another and by the time the Civil War broke out he was completely destitute. With little hope of success in a civilian position, Grant viewed the war as a means of extricating himself from his difficulties. He secured a position with the 21st Illinois Volunteers only with difficulty, as his civilian reputation of failure and drunkenness preceeded him. It was with great reluctance and through sheer necessity that he was given the command.

His primary duty was not only to command but also to drill the new volunteers. In spite of the various problems which beset him, Grant quickly rose to the rank of Brigadier General of the Illinois Military District. In that capacity he became involved in the war in the West along the Mississippi River, generally considered of secondary importance compared to the conflict raging in the East around the two capitals.

Although in his efforts during the later months of 1861 Grant showed tendencies toward indecisiveness, he received national acclaim in February 1862 with the brilliant seizure of Fort Henry and Fort Donelson on the upper tributaries of the Mississippi. In April of the same year Confederate forces made a surprise attack on his position and Grant converted possible disaster into victory in what has become known as the Battle of Shiloh. The following year Grant became an integral force in the campaign which proved to be not only the turning point of the Civil War but the turning point of Grant's career — the Battle of Vicksburg.

It seems curious that the Battle of Vicksburg, and not Gettysburg which has become more legendary, was the true turning point. It is perhaps because Vicksburg was a true military campaign rather than the simple confrontation of two armies. However it was Vicksburg's location which gave it its major importance. Located on the eastern bank of the Mississippi, on a sharp bend in the river, Vicksburg was a natural fortification and control point. No ship could travel the narrow passage without passing under the guns of the city.

Being one of the largest rivers in America, and owing to its rail links and its wide-ranging tributaries, the Mississippi River was a natural avenue for transporting goods from the midwest of the United States to New Orleans and

General Grant at Cold Harbor, Virginia in June 1864.

other national and international markets. In the opening days of the Civil War few people considered that the war would be of extended duration, thus little Northern attention was focused on the importance of the Southern control of the Mississippi. However, as the war progressed the disruption of river traffic and the deviation from the river's traditional use bred discontent in the western states. Pressure increased in those states for the acceptance and recognition of the Confederacy, even to the idea of secession themselves, in order to resume free commerce on the great river.

In Washington, Lincoln recognized the growing discord and realized that the dissent would continue to grow unless the Mississippi was reopened. His strategy was simple, capture the river and open it for trade. The execution of this strategy was another matter entirely.

During the second half of 1862 several attempts were made to overcome Vicksburg as it was the primary fortification blocking the Mississippi. One venture involved the combined efforts of river fleets attacking from both north and south. Admiral David Glasgow Farragut was responsible for the capture of New Orleans during that action, and although some degree of success was initially achieved the problems of supply, cooperation between the services and the ample Confederate defenses along the river proved overpowering. It should be pointed out that although the action was not successfully completed, this was the first instance where a concerted effort was made to combine the army and the navy in one operation.

The many difficulties encountered through the use of flotilla tactics led to the use of basic assault tactics. Another

Vicksburg before the siege.

attempt was made to capture Vicksburg through the use of overwhelming numbers of troops. This, however, proved equally ineffectual against the Confederate defenses. With the Confederate forces in possession of the high ground and the most strategic locations, the concept of landing an army and supporting it with fleet artillery simply did not function. Grant determined to devise a successful strategy.

His first plan was to bypass Vicksburg by cutting a canal across the narrow finger of land which formed the bend in the river on the western side, opposite Vicksburg. He received support for his plan from President Lincoln and Sherman's XV Corps was assigned the task of digging the canal. Once completed the canal would allow the free passage of ships outside the range of Vicksburg firepower. The canal would also allow for the future encirclement and isolation of Vicksburg in the Union scheme.

Several obvious flaws in the project became apparent as construction began. First and foremost was the fact that the Confederates could view the Union activity from their vantage point on the heights of Vicksburg. Their immediate retaliation to the impending canal was to reposition several batteries at Warrentown, bringing the southern mouth of the canal under range of their guns. The second major problem occurred in March 1863 when the river rose, breaking through the upper dam and flooding the canal and the surrounding lands. This natural disaster destroyed Union stores and equipment and completely disrupted Sherman's corps.

At General Grant's headquarters alternative plans were already being devised. In February 1863 a navy flotilla was sent down a tributary of the Yazoo River north of Vicksburg with the support of infantry columns. The troops were to capture Confederate dockyards at Yazoo City and secure all positions of high ground at Snyder's Bluff. This plan was thwarted by harassment from Confederate forces, who blocked the waterways and hindered the navy at every opportunity. The Confederates were assisted in their efforts by the natural silting of the narrow river passageways. The Union forces were forced to move at a snail's pace when speed and rapid deployment were vital.

General William T Sherman, photographed by Matthew Brady.

By April the navy was forced to withdraw. Confederate troops had been successful enough in their delaying tactics to be able to reinforce their position, creating a formidable enough front to dissuade the Union forces from further attacks in that area. A second expedition set out in mid-March, with General Sherman taking part. This was equally unsuccessful. The fact that movement was slow, whether by water or land, meant that Confederate forces were always fully aware of Union intentions and able to act accordingly. The failure of these attempts to attack the city from the north caused Grant to again reassess his ideas and reevaluate his strategy.

However, despite the failure of those missions Grant learned a valuable lesson which would stand him in good

The Union forces tried to cut a canal across a loop of the Mississippi in order to approach Vicksburg. The canal was either flooded or silted and proved a wasted effort.

stead during the coming Vicksburg campaign. Forced to abandon their assault by the fact that Confederate forces had severed the Union lines of communication and supply, Grant discovered that his force could subsist quite well by living off the land during their retreat. He would implement this knowledge in the near future.

Grant developed a strategy which was not only daring but once set in motion was undoubtedly irrevocable. Two issues had troubled Grant throughout the campaign. First was the fact that despite all efforts his forces were forever in an adverse position when attempting to bring pressure against the Vicksburg fortifications. Second was the fact that the garrison was reinforced and resupplied with consistent ease. In his new strategy Grant decided to eliminate the support before concentrating on the garrison itself.

In mid-April Grant sent seven gunboats, a ram and several cargo steamers down the river past Vicksburg. This maneuver was accomplished under cover of darkness to minimize the effectiveness of the guns. From 20 April Grant began to move his army down the Louisiana bank of the river, constructing an effective roadway through the bogs and forests toward the river city of New Carthage. After great success in that region Grant's forces continued south along the river banks, yet by-passing Vicksburg, on a course for Bruinsburg. Simultaneously ships continued to run the gauntlet at Vicksburg, creating a sizable force south of the city. These vessels not only carried supplies for the army but would figure as major components in the next stage of Grant's strategy.

By the end of April Grant's forces had entered Bruinsburg and on 30 April he began to ferry his troops from the western to the eastern banks of the Mississippi. That maneuver took the Confederates completely by surprise. Thus far all Union attacks on Vicksburg had been from the north. Grant exploited the opportunity for surprise and now was threatening by land from the south, a contingency for which the Confederate command was not prepared.

The details of Grant's scheme were yet to be realized. As Confederate commanders considered the possibilities of Grant's next move against Vicksburg, Grant was concentrating on his prime objective – the isolating of Vicksburg. His first course of action would have to be a move inland to sever the Confederate lines of communication. This would mean that his own lines of supply would cease to exist. As the area south of Vicksburg had seen little land combat, Grant felt certain that he would be able to forage all necessary supplies.

On 1 May Grant sent part of Major General John A McClernand's XIII Corps to quell any Confederate resistance at Port Gibson. With Grand Gulf threatened on the landward side by the Union's XVII Corps, the Confederates decided it would be wise to evacuate those troops and join forces with troops sent from Vicksburg, holding a line along the Big Black River.

On 7 May Sherman's XV Corps took advantage of the Confederate evacuation, marching from Milliken's Bend and crossing the river at Grand Gulf. With that maneuver Grant shifted most of his army to the Vicksburg side of the river, and with Sherman moving rapidly to join him, Grant set about executing the next phase of his plan.

Rather than marching straight on Vicksburg from the

Above: The head of the canal opposite Vicksburg. This shows the swampy nature of the countryside.

Right: Skirmishing in the woods before Vicksburg.

south, Grant struck further inland, converging his forces on the Mississippi state capital, Jackson. General Joseph Eggleston Johnston, Confederate commander in the West, had at that time only two brigades at his disposal. Realizing that he would have to attempt to buy time, Johnston retreated northward to link up with reinforcements he expected from the East. Within a few days he hoped to be able to field an army of 30,000 men. Johnston also sent word to Lieutenant General John Clifford Pemberton, Commander of the Mississippi Military District, to take charge of combining the garrison forces of Port Hudson and Vicksburg. Johnston's field army and the combined force should have brought his command to more than 50,000.

Possessing what he thought to be double the number of men under Grant, Johnston felt certain that he could not only defeat Grant but maintain a strong Confederate dominance in the war in the West. However, Johnston was not aware that Pemberton had previous orders. Jefferson Davis had personally given the order that neither Port Hudson nor Vicksburg was to be left ungarrisoned. In following that order Pemberton left behind 10,000 of his

troops to man the defenses at Vicksburg and, rather than move to the immediate support of Johnston's army, decided to take his smaller force and sever Grant's lines of communication and supply.

As Pemberton ignored Johnston's orders he came to realize that Grant had disregarded military convention and had no lines of communication or supply. Instead of disrupting Grant's rear area Pemberton followed the movement of Grant's troops toward Jackson, Mississippi. On 14 May Grant entered Jackson, destroying the Confederate stores and severing the Jackson–Vicksburg railroad. Grant was well aware that if he could defeat any of the Confederate armies in an open field battle Vicksburg would be temporarily isolated. On 16 May, after five hours of seesaw engagements Pemberton's forces were defeated at the Battle of Champion's Hill.

More significant than the defeat was the fact that Pemberton had made a grave tactical error. Instead of retreating toward Johnston to combine their strengths and attempt to pin Grant's forces between the field army and Vicksburg, Pemberton retreated to the mock safety of the Vicksburg defenses. In succumbing to the temptation of a safe haven Pemberton again divided his troops to cover his retreat. Over the next 24 hours that decision would cost him more than 1700 men and the loss of some 18 heavy guns as the Union troops swept away the rearguard at the Big Black River.

On 19 May, a little more than one month after Grant had set his new strategy in motion, his army was poised just outside Vicksburg preparing for the attack. Grant confidently believed that victory was at hand. He felt certain that his enemy was so demoralized by the past weeks' events and the recent defeat that panic would surely be racing through the ranks, and that an immediate

assault on Vicksburg would probably overpower the defenses of the city.

In all the past month's displays of brilliant maneuvering and strategy this was Grant's first mistake. His error in judgment was based primarily on a lack of intelligence information concerning the garrison and an overconfidence attributable to easy victories. Although the defenders' morale had indeed been severely shaken, Grant failed to compensate for the determination that the Confederate forces gained in the realization that they must either make a stand for victory or surrender in defeat. They had no alternative with the Union forces encircling the city. This determination, coupled with the fact that 10,000 fresh troops had remained in the garrison, gave new heart to the Confederate troops who had already faced Grant's army. Thus the first Union assault was soundly repulsed and Union forces withdrew from the attack greatly bewildered by the staunch resistance which they had encountered.

On 22 May Grant ordered a second assault. He was concerned that Johnston's force was by then only some 50 miles east of the garrison. In spite of the fact that Johnston had a much smaller force Grant did not wish to allow him any time to put together an army which could arrive to relieve the city. At 1000 hours Grant ordered a heavy bombardment of the fortification along his army's entire front. He was confident that his second assault was better organized, with all three Corps concentrating their attacks simultaneously. Sherman's XV Corps attacked from the north, McPherson's XVII Corps from the east, and McClernand's XIII Corps from the southeast. Union forces pressed home a fierce attack against the defenders, but nowhere along the line was an actual breech in the defenses made. McClernand's XIII Corps did however manage to overrun several entrenchments, causing him to request reinforcements that he might carry his position. Grant, unable to have an overall view of the battle, had committed his reinforcements to renewed assaults by Sherman and McPherson. That commitment resulted only in increased casualties on the Union side without the gain of an extra foot of ground. Without reinforcements the XIII Corps could not hold their ground and foot by foot were driven back from the entrenchments.

The second assault had failed with three times the casualties suffered in the first assault. Grant hastily reviewed the situation. The first question he undoubtedly asked himself was whether or not the troops had lost heart. His answer, based on their performance was an emphatic 'no.' During the second assault each corps had managed to plant its flag on top on the entrenchments and had fought stubbornly to gain a victory. If the troops were not to blame it had to be the tactics. Two frontal assaults had gained nothing but casualties and meant that a third would also be unsuccessful. If assaults were unsuccessful then the only alternative left was siege.

Yet how successful could such a siege be? Grant had completely cut off the city of Vicksburg from any avenue of resupply or reinforcement. The surrounding land was in the complete control of Union forces and the Mississippi River had been blocked by Union war vessels. His campaign had until that point been successful, and he had managed

to deliver the South a severe blow even if he had failed to take the fortifications at Vicksburg. But Vicksburg was the key and Grant would not be dissuaded.

In 20 days Grant had marched some 180 miles through enemy territory, had accumulated vast quantities of military supplies, captured a state capital and fought five major engagements, all on five days' rations. His forces had managed to capture more than 6000 enemy soldiers and 89 field pieces, all at the loss of less than 2000 killed and 7500 wounded of his 43,000-strong army. As Sherman is said to have pointed out, they had achieved a great victory even if Vicksburg held out. Not only had they encircled and isolated the fortification, but Grant was aware that Johnston's army was not in a position to offer any assistance to the defenders. The reserves that Johnston had counted on so desperately had been diverted to the East to support Lee in his Northern Campaign into Pennsylvania.

Grant was therefore faced with the problem of seizing Vicksburg as quickly as possible, denying the Confederate forces time to recover from his previous assaults. The mounting of a full siege was a prospect which Grant did not relish. He knew that siege warfare was a very specialized art and from his studies at West Point realized that whole armies could be wasted on a mismanaged siege. Grant had

On 22 May 1863 Grant ordered another attack which was preceded by heavy bombardment. After a day's heavy fighting the attack was seen to have failed.

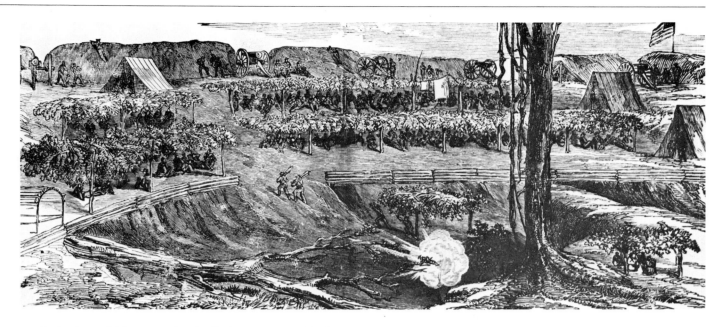

no desire to end his campaign in that manner, yet to be so close to the prize and then to lose it would be more than he could bear.

As the troops began to dig themselves in Grant reviewed his command and found that he was very short of trained military engineers. He had a good number of men in his volunteer units who had built bridges, constructed roads and with Sherman's XV Corps had built a canal. However, very few of these men understood the business of con-

Following the failure of the 22 May attack Union forces built fortifications around the Confederate lines to prevent them from breaking out. The picture shows Sherman's Corps.

structing parallel trenches, mining and countermining, and lines of circumvallation and countervallation, all of which are an integral part of siege warfare. Grant was not to be found wanting. As a young cadet he had received training in the field of military engineering, as did all cadets no matter what branch of the armed service they would eventually enter. As he had no true Engineer Corps Grant decided to create one. He put a request to all units that all officers who were West Point graduates report to one of the four regular engineering officers to serve in the capacity of assistant engineers. With a man of ample capability to guide them, in the person of Captain F E Prime and his successor Captain Cyrus B Comstock, Grant felt certain that he could mold his officers and western volunteers into an engineering force which could rival any regular engineering unit.

Grant was also granted some breathing space in that while he assembled his engineering unit and his troops were digging themselves in for the siege, a truce had been called on two separate days so that the Union forces might recover and bury their dead. With Confederate troops assisting their Union counterparts in the recovery, Colonel H S Lockett, commander of the Confederate engineers, decided that he had a prime opportunity to examine the Union siege works. Lockett was quite sure that Grant was no fool. It would be in Grant's best interest to assign his engineering officers to command the recovery details, thus allowing them a close look at the Confederate entrenchments and fortifications. Lockett decided that he would perform the same role for the Confederates. Unfortunately, Lockett was consistantly hindered from seeing many aspects of the Union works by General Sherman, who kept Lockett engaged in courteous conversation throughout the day. This gave the Union engineers a certain advantage over the Confederate forces.

Grant's next most important problem lay in the fact that he had no true siege guns, except for his six 32-pounders. For the most part he would have to rely on his

field guns to soften the fortifications. Admiral Porter did
manage to transfer a battery of large caliber naval guns
to the siege troops. Although greatly needed they were
still far too few to make a great difference.

At that time Grant made a very wise decision. He turned
the siege operation over to his engineers, maintaining
control over the army as a whole yet allowing the engineers
to formulate the detailed plans of attack on the various
sections. In doing this Grant displayed two things. First and
foremost he demonstrated that he was a capable and confident commander who was not afraid to allow subordi-
nants to take charge and formulate the best avenues of
attack according to their specialities. Secondly, it showed
that Grant was well aware of the great magnitude of the
task, realizing that it was too large and intricate an operation for one man and his staff to command efficiently. His
best possible strategy was to apply a loose command over
the army, giving it direction and purpose, while allowing
the details of the plan to be handled at the lower levels of
command.

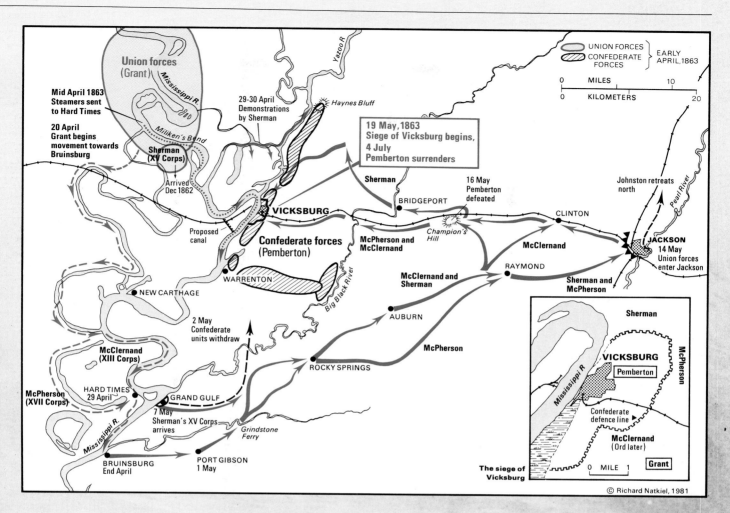

UNION FORCES (Grant)
CONFEDERATE FORCES
} EARLY APRIL, 1863

MILES 10
KILOMETERS 20

Yazoo R.

Union forces (Grant)
Mississippi R.

Mid April 1863 Steamers sent to Hard Times

29-30 April Demonstrations by Sherman

Haynes Bluff

20 April Grant begins movement towards Bruinsburg

Miliken's Bend

Sherman (XV Corps)

Arrived Dec 1862

Proposed canal

VICKSBURG

Confederate forces (Pemberton)

WARRENTON

NEW CARTHAGE

2 May Confederate units withdraw

McClernand (XIII Corps)

HARD TIMES 29 April

McPherson (XVII Corps)

Mississippi R.

GRAND GULF

7 May Sherman's XV Corps arrives

Grindstone Ferry

BRUINSBURG End April

PORT GIBSON 1 May

19 May, 1863 Siege of Vicksburg begins, 4 July Pemberton surrenders

Sherman

BRIDGEPORT

16 May Pemberton defeated

CLINTON

Johnston retreats north

Pearl River

McPherson and McClernand

Champion's Hill

Big Black River

McClernand and Sherman

AUBURN

McPherson

ROCKY SPRINGS

RAYMOND

McClernand

Sherman and McPherson

JACKSON

14 May Union forces enter Jackson

The siege of Vicksburg

Sherman

VICKSBURG

Pemberton

McPherson

Mississippi R.

Confederate defence line

McClernand (Ord later)

Grant

0 MILE 1

© Richard Natkiel, 1981

Although the actual basic mechanisms of the siege did not necessarily reflect on Grant as an able military commander, the siege was highly important for several reasons. The concept of siege was not a new one, but the manner in which the siege at Vicksburg was executed gave a preview of the type of trench warfare which would become prevalent in World War I. The idea of two armies dug in, entrenched only yards from one another, with mining and countermining a daily routine, would be the blueprints of trench warfare in the future. The two initial assaults on Vicksburg serve to exemplify the devastating results of attacking troops in entrenched and prepared positions. The losses sustained by Grant's forces were only a preview to the number of soldiers who would be lost in 'no man's land' on the battlefields of France.

The whole concept of war was changing. Less and less frequently would armies meet in a short, decisive battle. More often battles would degenerate into a war of attrition. Had Vicksburg been in a position to resupply, and if the fortifications had been woven with more interlocking points on a grand scale of defense and support, it would have given a staunch resistance, more clearly displaying what was forthcoming in the art of warfare.

Another point concerning the siege merely confirmed that no matter how formidable or well conceived a fortification may be, once isolated and unable to be resupplied or relieved, that fortification is doomed. Grant's strategy of encircling his objective as well as removing any chance for support or relief was responsible for ending the siege of Vicksburg so quickly. His tactic of first removing hope

before setting about the basic details of the engagement would be used as the guidelines for many siege engagements of the future, most notably at Dien Bien Phu by General Giap.

Those strategies and tactics would echo in the future, as would the life led by Grant's soldiers fighting the siege. Life became a set routine of digging trenches at night, defending the newly built earthworks during the day, and trying to survive the snipers' bullets and the harassment of the enemy artillery. By 30 June they had produced miles of trenches encircling the fortification on all landward sides and had created positions for some 200 pieces of artillery. When it rained the trenches became muddy troughs, and in the heat of the southern summer sun became baked dust pits. The life of the Union soldier at Vicksburg was to sweat and dig all night and exchange a few shots during the day.

An unusual phenomena occurred to break the monotony for the Union troops in early July. Northerners, elated by the news of the successes at Vicksburg, flocked to the siege site. Friends, family, volunteers, politicians and others arrived to give aid and comfort or simply to view the proceedings. A carnival atmosphere prevailed before the troops returned to the inescapable task at hand.

It was the Confederate soldier who had to face the worst the siege could offer. The first problem confronting the defenders was lack of food. The fact that Vicksburg had a

The fight in the crater at Fort Hill after the explosion of the mine, 25 June 1863.

civilian population as well as a military one meant that rations dwindled rapidly. By mid-June the defenders' diets consisted mainly of mule stew, cane shoots and roast rat. Early in the siege yams and blackberry leaves were substituted for coffee which ran out quickly. Flour was eked out with cornmeal or ground peas, while they lasted. All food was strictly guarded but supplies were very limited.

The lack of food began to show its effects as Vicksburg defenders became so weak that their maximum efforts and abilities were consumed by the mere manning of their positions. There were stories that in the early days of the siege Union and Confederate soldiers, whose trenches were so close that they could throw hand grenades, managed to communicate long enough to toss hardtack in exchange for Confederate tobacco. As the siege dragged on such exchanges rapidly disappeared.

The Confederate's second problem dealt with another much needed commodity, ammunition. Most personal accounts written by Confederate soldiers at Vicksburg remark of the Union soldiers firing constantly at the Confederate lines with both artillery and muskets, while the Confederate soldiers had to horde their ammunition to be used at its most advantageous time.

Finally, and of grave importance to the morale of the Confederate troops, was the realization that relief was not forthcoming. Union forces were virtually in full control of the Mississippi River. Johnston's army, which was supposed to come to the aid of Vicksburg, could barely muster sufficient troops for an attack when it was most needed. As the defenders watched Grant's forces grow day by day, watched their rations dwindle, and watched for relief which they knew would not arrive, their spirit and will to fight slowly died.

On the night of 2 July General Pemberton called together his staff and most of his officers and informed them that the situation was desperate. He discussed whether or not they should surrender the garrison. Although several officers wanted to hold out as long as possible, in order to cause more casualties to the Union forces and to keep tied down the main Union army in the West, the majority of officers felt that it was only a matter of a few days before ammunition became so scarce that even this object could not be accomplished. Pemberton accepted the majority opinion and decided to surrender.

At about 1000 hours on the morning of 3 July flags of truce were flown over the garrison. Pemberton sent his aide-de-camp, Colonel Montgomery, and General Bowen, a divisional commander, with a message of armistice. Both sides took heart upon seeing the white flags. For the Confederates, even though it meant defeat, the bloodshed of a hopeless situation would soon end. Once surrender terms were properly ironed out medical aid and food for the beleaguered troops would be available within a matter of hours. For the Union soldiers it meant an end to a hard fought campaign and victory over what the Confederate president had termed an 'impregnable, unconquerable fortification.' Victory would indeed be sweet.

However, the terms indicated in Pemberton's letter were not in any way to Grant's liking. Pemberton offered a period of armistice and proposed that the two commanders set about laying the terms which would be required before

Fighting with hand grenades on the last day of the siege of Vicksburg before the terms of the surrender were agreed.

the Confederates would relinquish their position. Grant's reply was returned in no uncertain terms itself, and is recorded in the *Personal Memoirs of U S Grant*. He replied: 'Your note of this date is just received, proposing an armistice for several hours, for the purpose of arranging terms of capitulation through commissioners, to be appointed, etc. The useless effusion of blood you propose stopping by this course can be ended at any time you may choose, by the unconditional surrender of the city and garrison. Men who have shown so much endurance and courage as those now in Vicksburg, will always challenge the respect on an adversary, and I can assure you will be treated with all the respect due to prisoners of war. I do not favor the proposition of appointing commissioners to arrange the terms of capitulation, because I have no terms other than those indicated above.'

Unconditional surrender was the key to Grant's terms. At 1500 hours Grant and Pemberton met on a hillside a few hundred yards from the Confederate lines. The two generals, who had served together in the same division during the Mexican War, greeted one another as old acquaintances. Perhaps it was a result of past friendship, or out of respect for the valiant manner of the garrison troops, but more than that it was a reflection of Grant's military character which dictated the terms he was to see to his unconditional surrender.

Although the surrender was not finalized until 2000 hours on 3 July the ceasefire began immediately. Whereas Pemberton balked at the word 'unconditional' his troops seemed prepared to accept the fate which awaited them. Many soldiers from both camps had been recruited from

A rare photograph of Vicksburg under Union fire during the early part of 1863.

Missouri, and the Vicksburg campaign was one in which brother had literally fought brother. The troops took the opportunity to begin searching for friends and relatives in the opposing camp.

Finally, after 2000 hours Grant had sent his proposal for surrender to General Pemberton. The unconditional surrender stood but several modifications and provisions were introduced. First, the city of Vicksburg would surrender and the garrison would lay down its arms. The garrison force would be released on parole, with the stipulation that they would participate in no further action in the war unless properly exchanged for Union prisoners of war. Confederate officers would be allowed to retain their side arms and swords, along with their mounts. The army would be allowed to keep any personal effects such as food, rations or clothing, but nothing else. Civilian rights and disciplines were to be strictly maintained but all slaves would be forfeited and set free as called for in Mr Lincoln's Emancipation Proclamation.

It was obvious by General Pemberton's relief at the terms that he had underestimated his adversary. Grant was not one, once the battle had been won, to add humiliation or subjugation to the defeat of his foe. On the morning of 4 July Pemberton surrendered the city and garrison of Vicksburg and Grant's campaign was victorious.

The surrender meant that the Union forces acquired more than 170 pieces of artillery and some 60,000 small arms. The small arms quota especially pleased the Union command as many of the rifles were of more recent date than those supplied to the Union troops. Grant issued orders that the new Enfield Rifles be issued to his troops, replacing the antiquated US or Belgian rifles then in use.

Without a doubt, however, the main prize in the victory was the fall of Vicksburg as the main block on the Mississippi River. The fall of Port Hudson five days later opened the river completely and within one week Union shipping could leave the docks of St Louis and travel unimpeded to New Orleans.

The success of the plan which Grant formulated for the defeat of Vicksburg sets him in the ranks of the greatest generals in history. His policy could be said to have been somewhat ahead of its time. Like all generals, Grant suffered from outside interference. His superiors, politicians and even the President hindered his success in the early days of his second campaign, and could be said to have cost him a victory in the first campaign almost a year earlier. There comes a point in the command of every great military leader when he must abandon what is considered acceptable and, as in Grant's Vicksburg campaign, take the 'bull by the horns' and set out to conquer rather than simply remove an obstacle.

Grant's early campaigns convinced him that any further attempts to take Vicksburg from the north or northeast would be a senseless waste of both time and lives. He set out to destroy the enemy's means and will to fight by staging a successful campaign against the surrounding key areas, such as the capital, Jackson. His daring and success set him apart as a master among the great minds of strategical and tactical brilliance. But the brilliance of one man's strategy is often overlooked when comparing his role to that of others of his era. Jackson's magnificent flanking maneuver at Chancellorsville and Meade's rapid rise to glory at Gettysburg are fine examples of well-executed, courageously fought engagements. Grant's campaign was a different type of warfare entirely.

Not only did Grant have to consider his objective, but each and every encounter which led up to that final engagement. In the space of time which was allotted to him he had to maintain the impetus of his army at its maximum, never losing the initiative, while keeping his enemy separated, isolated and off balance until his goals were accomplished. Gettysburg and Chancellorsville were crucial engagements, with ranging effects, yet the generals concerned were faced with dealing with a basic strategy, the execution of which would determine the success or failure of that one encounter. Grant's strategy was such that he had to execute each battle to facilitate the final execution of his plan at Vicksburg. There is often a great deal of chance and a certain degree of luck involved in any given battle, but in a campaign nothing can be left to chance. Planning, execution and sound strategy are the factors which dictate success or failure.

In an era of attack-counterattack, move-countermove,

General Grant meets the defeated General Pemberton at the State Hous in Vicksburg on the morning of 4 July. The loss of Vicksburg was a terrible blow to Southern morale.

Grant worked out a deliberate long-range plan for the defeat of Vicksburg, with the actual assault on the city as his final goal. He set about to accomplish three major goals before the assault on Vicksburg. First, to destroy Confederate supplies and ammunition stored in Jackson. Second, to cut the Confederate railroad lines of communication and movement. Third, to defeat any forces in the area which could reinforce or assist the garrison. The final act in Grant's strategy was the outright defeat of Vicksburg. With no food, no ammunition, and no hope for relief he expected the victory to be more easily obtained than it actually was. The first two direct assaults on the city were valiantly driven back. This however was more a result of the fact that 10,000 fresh troops were garrisoned there and the undeniable influence of the defense being a last-gasp effort. The point remained, despite the initial set backs in the assaults, that Vicksburg was virtually an island. Grant had removed all chance of outside support, save Johnston's distant force, and while he may have misjudged the Confederate defenders' initial state of morale, the outcome was obvious. Sooner or later Vicksburg would fall.

An interesting point in the campaign lay in the development of another as yet little explored military technique.

The Vicksburg campaigns, both unsuccessful and successful, employed the use and cooperation of the naval forces of the Mississippi in their strategies. While the slow moving ironclads were often made to appear foolish in their efforts to navigate Confederate patrolled tributaries, the armored vessels were often used to carry troops and supplies to key areas along the river banks. The guns on the decks of the river craft were well appreciated by land-based troops for the support they gave. Grant demonstrated considerable forethought in his application of the naval forces in the overall picture of his campaign, another preview of tactics of the future in the art of warfare.

Grant and his Vicksburg campaign are not merely examples of a great leader and a monumental battle, but are also examples of innovative strategy and tactics which can be considered ageless. Grant's concepts form the foundation for much modern strategy. Grant deserves to be remembered not only as a victor but as one of the founding fathers of modern warfare.

MEADE
AT GETTYSBURG 1863

When considering crucial battles, famous generals and the success or failure of campaigns, seldom does one encounter a battle such as Gettysburg. In this campaign the victorious general is quite often referred to as little more than one of the many commanders in the fray. In most historical accounts, Major General George Gordon Meade has been a victim of severe criticism and injustice in a refusal to give credit to his leadership and judgment, as shown not only during the battle of Gettysburg but throughout his career as an officer in the Union Army. Too often reviews of the battle concentrate on the mystique of Lee and the invulnerability of the Army of Northern Virginia, overshadowing Meade's finest qualities. The fact remains that in a time span of less than one week General Meade did more to turn the tide in the Eastern Campaign than any other individual.

Although born at Cadiz in Spain in 1815, George Gordon Meade was brought up in Pennsylvania and eventually completed his education by attending the Military Academy at West Point. During his service in the Union Army Meade gained a reputation as 'the most professional looking soldier in the North.' Although this has often been misinterpreted as an insult to his abilities, Meade was indeed an extremely professional military man. A grave, stern man, often impatient with his subordinates, Meade was a dedicated commander who demanded perfection in himself and those who served with him. He was methodical in his approach to any course of action, and had the ability to quickly assess the essentials of a situation. Meade's self-discipline meant that he could be counted upon to keep a clear head in the most volatile of situations, particularly when he could operate from a defensive posture.

Critics often dwell on the fact that Meade's methodical approach made him unable to control a fast moving battlefield and therefore unsuitable to command the Army of the Potomac, particularly at that point in time. Meade himself realized his shortcomings, considering his own cautious nature and attention to detail fine attributes for an administrator, staff officer or corps commander, but not those required of an army field commander. However, knowing one's own shortcomings is often the greatest of attributes.

Primarily Meade must be accepted for exactly what he was — a simple soldier fighting for his country. This cannot be said of his predecessors, particularly Hooker, who appeared to visualize himself as the 'savior of the Republic.' Perhaps the greatest tribute paid to Meade came from the lips of General Robert E Lee. When news of Meade's appointment reached the Confederate headquarters, Lee is reported to have referred to Meade as 'a soldier who would not commit a blunder, but who would make haste to take advantage if I should commit one.' This comment demonstrated that the Confederate Staff realized that Lincoln had at last chosen a commander who knew his profession and whose abilities could not be taken as lightly as those before him.

The second of Meade's basic qualifications lay in the fact that he was not afraid to act. He was well aware that the task of his army was to check and ultimately defeat its adversary. As an administrator and staff officer Meade

General Ulysses S Grant was besieging Vicksburg at the time of the Battle of Gettysburg.

knew how to move an army; as an efficient general, he knew why. Taking command a mere three days before the Battle of Gettysburg, Meade marched his army from the border of Virginia, through Maryland, to the hills of southern Pennsylvania to meet his enemy. So positive an effort displays capabilities which were lacking in his predecessors.

The manner in which Meade assumed command of the Army of the Potomac is another reflection of the man. The Army of the Potomac, having taken the war into the South, failed to convert advantages into victory. In every major encounter Union forces bungled their opportunities and courted disaster. Morale was at an unprecedented low. In many units the confidence of the troops had deteriorated to the point where soldiers believed that the defeat of the Army of Northern Virginia was an impossible dream in the minds of those who sat in Washington.

Lee's invasion of Pennsylvania had brought the chaotic situation of command to its climax. In Washington Lincoln considered his options and personal preferences in appointing a commander who would set the Army of the Potomac back on its feet. The first possible choice as permanent commander of the Army was Major General Joseph Hooker, the acting commander. The second possibility was Major General John F Reynolds, considered a capable and aggressive leader. Finally, though perhaps not the strongest candidate, was Major General Meade.

In considering Hooker for the post Lincoln decided that the general's past record, particularly his indecisiveness, proved that he was not the man for the position. In fact,

neither Lincoln nor his advisors had been proponents of the original decision to give him the command. The President, supported by General Halleck and backed by the Secretary of War, considered Reynolds the most appropriate man to assume command. However, Reynolds had stated that unless he was given full command and complete freedom of action he did not want the command. Lincoln's office declared him Commander in Chief of the Army, and he was not prepared to let the reins slip from his grasp. Unable to accept Reynolds' terms, Lincoln and his Secretary of War, Edwin Stanton, were left with the dilemma of continuing to tolerate Hooker's ineptitude or taking the risk of promoting Meade.

While these decisions were being made, Hooker, Reynolds and Meade were encamped, carrying out their separate duties during a lull in the fighting. Rumors had been circulating and most officers were aware that something of major importance was afoot. It was generally believed, indeed even Meade was convinced, that the foremost candidates for the permanent command position were Hooker and Reynolds. Meade so firmly believed that Reynolds could be the only choice that when one of his junior officers approached him with the suggestion that Meade was also being considered, Meade dismissed it as 'another ridiculous camp rumor.'

In the early morning hours of 28 June 1863, General James J Hardie, Chief of Staff to the Secretary of War,

entered Meade's camp dressed unobtrusively in civilian clothing. Finding Meade asleep in his tent, Hardie woke him and began to explain that as acting representative to the President, Hardie was commissioned to oversee Meade's immediate assumption of Command of the Army of the Potomac. Meade was unable to digest the information. Hardie slowly and with determined simplicity repeated his assignment to a stunned and speechless Meade. Upon regaining his composure, Meade is said to have become extremely agitated. He insisted that if anyone should succeed Hooker it should rightfully be Reynolds, and condemned the President's decision as an injustice to all parties concerned.

After several trying hours and many appeals that Hardie return and have the orders amended, Meade realized that as a soldier his first duty was to obey orders. He and Hardie then set out for Hooker's headquarters. Meade demanded that Hardie deliver the news to the general in private, believing that his presence would only cause Hooker more personal discomfort in an already delicate situation. Shortly thereafter Meade assumed command as quickly and unostentatiously as possible.

In a later interview Meade exposed himself by expressing his shock and lack of preparedness for the promotion. One of Meade's anxieties was removed when his long standing friend General Reynolds arrived to pay respects to his new commander. Meade and Reynolds discussed the twist in

Above: General Meade consults his generals at a Council of War on the night of 2 July in his farmhouse headquarters on Cemetery Ridge. Meade and his generals decided that in spite of the heavy casualties the Union forces should hold their ground.

© Richard Natkiel, 1981

events and Meade again expressed his wish that Reynolds had been given the command. Reynolds assured Meade that he was happy for his friend and would support 'with all earnest' his new Commanding Officer. The two generals then spent some time exchanging ideas concerning Lee's movements into Pennsylvania, parting only when other duties required their attention.

In the afternoon hours of 28 June Generals Meade and Hooker drew up the official change of command and General George G Meade embarked on his career as Commander of the Army of the Potomac.

Two important points reflecting Meade's character were apparent in his acceptance of the command. Firstly, Meade displayed an unselfish concern not only for the army but for the sensibilities of his fellow officers. Secondly, he showed a willingness, despite surprise and doubt, to accept the responsibility and attempt the seemingly impossible in true military fashion. Meade realized that he was no 'Prima Donna,' no 'savior of the Republic.' He was commander of an army with a responsibility which could not

be treated lightly nor with self-indulgence. The Army of the Potomac had at last acquired a leader who would serve it with pride and concern.

Meade's place in history was, of course, caused by his involvement in one of the most famous battles in modern history. On 28 June when Meade took command of the Army of the Potomac, the main force was located in and around Frederick, Maryland, and Lee had already moved the Army of Northern Virginia into Pennsylvania. Moreover, Lee's march north had been almost completely uneventful, with little or no resistance or pressure, except from local militia groups. Meade decided to begin his command with action. Unlike Hooker, who sent Union cavalry units to probe the direction and strength of the Confederate forces, Meade considered that the situation was too grave to allow the two armies to be separated by such a distance. On the day his command began Meade adopted the plan of moving directly north to stay in contact with and counter Lee's army, setting the Army of the Potomac on a definite and direct course of action against its opposition.

The effect of this decision on the Union troops cannot be overemphasized. He demonstrated to the troops that he was willing to take affirmative action. In keeping with the true soldier he was, Meade intended to turn his army into an efficient fighting force, not soldiers en masse fumbling desperately to block a Confederate assault on Washington. It is often asserted that the protection of Washington was Meade's only concern. In fact, his primary concern was the defeat of the enemy, rather than the protection of a city whose garrison and fortifications would have proved a powerful obstacle to the number of troops at Lee's disposal.

Another factor which gave a much needed boost to the morale of the Union troops was their return to an area which was sympathetic to them and their cause. Since the beginning of the war the Army of the Potomac had spent the majority of its time faced with hostile civilian populations, Lee's seemingly invincible army, declining morale due to failures in leadership and the attrition of war. Morale soared as the Union soldiers crossed the Mason-Dixon line to be greeted by cheering crowds and friendly faces. Food and drink were placed into the hands of the troops as they marched along the roads. Pennsylvania was proud of its native sons, whose numbers swelled the ranks of the army. Little did they realize that in a few short days a battle would be fought on the soil of their beloved Pennsylvania, led by a general who was himself an adopted Pennsylvanian. The new confidence within the army which resulted from the affirmative action and the move in a definite path toward the enemy would stand Meade in good stead in the days ahead.

By 30 June neither Meade nor Lee was aware of the exact location of their respective enemies, nor could they determine which army had the most advantageous position. The lack of tangible evidence with regard to the enemy's position caused Meade to worry over the possibility of Confederate raiders blocking the roads to Washington and Baltimore. However, his main concern lay in determining whether or not he could maneuver Lee into a confrontation on a battlefield which would be advantageous to the Union force. Lee on the other hand struggled with the problem of having to rely heavily on information gained through the questionable source of his spies and delayed intelligence gleaned from captured local newspapers. Stuart's cavalry, Lee's primary intelligence gathering source, had vanished into the Maryland and Pennsylvania wilderness. The majority of the remaining Confederate cavalry were occupied in guarding the passes of the Blue Ridge Mountains to keep the avenues of communication, supply and possible retreat open.

On the morning of 30 June Meade sent engineers to scout areas along the Big Pipe Creek, hoping to find a location which would prove suitable and advantageous to a defensively positioned Union force. He sent his III Corps to the town of Emmitsburg, just south of the Pennsylvania border. The I and II Corps were sent north to take position around the sleepy college town of Gettysburg. Meade and his headquarters staff were a day's ride behind the most forward troops near Taneytown. This gap was not a major concern as Meade believed it would be several days before

General James Longstreet was called 'Lee's old war horse' and was accused of procrastinating during the Battle of Gettysburg.

he or Lee would be prepared to commit themselves to the inevitable battle.

Lee's headquarters at that time was only some 10–15 miles northeast of Gettysburg. On that same morning he decided to detach a portion of his troops to Gettysburg in a bid to capture equipment and clothing said to be in storage there.

At approximately 0800 hours on 1 July Union cavalry under the command of Brigadier General John Buford, scouting along the road between Gettysburg and Cashtown where Meade believed Lee's forces were concentrated, encountered the Confederate troops which Lee had sent on a supply search. This chance encounter developed, through circumstance, from a mere clash between pickets into the battle itself.

Before conducting his reconnaissance mission along the Chambersburg Pike, Buford had received assurances from Reynolds that the I Corps would support Buford should contact with the enemy be made. The Confederate troops, under Major General Henry Heth, had been given the same verbal support by their III Corps commander, Lieutenant General Ambrose Powell Hill.

Upon encountering the enemy, Buford's cavalry dismounted. For more than two hours, with a single artillery battery for support, they held back the Confederate advance. By 0930 Reynolds' I Corps was fully aware of the confrontation and was on the march in Buford's support. What had begun as a minor skirmish was rapidly escalating into a major engagement. Neither Meade nor Lee made the decision to fight a major battle in the hills surrounding Gettysburg. It developed from the circumstances of combined support assurances and the aggressiveness with which Reynolds entered the fray.

Before 1100 hours the Union 1st Division of the I Corps was reinforcing Buford's cavalry, attacking Heth's column as it moved ever closer to Gettysburg. The strain on Buford's forces was becoming apparent. In spite of Reynolds' Divisional support, Buford's cavalry was forced to abandon its position on Seminary Ridge in the face of Heth's overwhelming superiority of numbers. Locked in the struggle, both factions were totally committed to battle.

As noon approached a Confederate sharpshooter, marking a Union general in his sights, shot and killed General John Reynolds as he rallied troops on the battlefield. By order of rank, command fell on the less able shoulders of Major General Abner Doubleday who, in direct contrast to Reynolds' forcefully aggressive nature, was a man of indecision. Groping to decide what his next move should be, Doubleday caused the Union forces to lose momentum along the entire front. In spite of the fact that Heth had slowed his attack in order to consolidate his forces and await further reinforcements, the Union impetus at Gettysburg was in danger of crumbling.

Major General Oliver Otis Howard, commander of the Union XI Corps, which had been advancing to join Reynolds' force, received word that Doubleday had assumed command of the I Corps. Unable to accept the thought of Doubleday in command, Howard immediately relinquished his own command to General Schurz and rode forward to relieve Doubleday of his burden.

At about the same time Meade received reports of the developing engagement. Although stunned by the news that his friend Reynolds was dead, Meade began to assert his role as commander of the army. He commenced with an action which caused much controversy yet gives an insight into his powers for the proper delegation of authority. Of the opinion that he had no one near the front whom he could entrust with the impending responsibilities, Meade sent Major General Winfield Scott Hancock to assume principal command of the three Corps which were engaged in combat. Hancock arrived at the front and informed General Howard of Meade's decision, whereby Howard challenged his right to assume command. It spoke well of Hancock's demeanor that he again quite simply informed Howard that he had his orders and would immediately undertake the command responsibility, at which point Howard withdrew as gracefully as possible.

General Winfield Scott Hancock was the commander of the II Corps and was sent by Meade to Gettysburg to take charge after Reynold's death.

As an officer of the Army of the Potomac, Hancock was a well-known figure, considered a courageous and sensible officer, though of less seniority than Major General Howard. Meade's decision to place Hancock in command was a bold act unprecedented in the rank-conscious Union Army. Meade had elected his course of action on the basic principle that Hancock was the most capable man for assessing the situation and keeping Meade well informed and abreast of the battle situation. Meade's conduct in this instance is one of several decisive moves which show him to have been a responsible commander. Hancock's arrival on the field of battle restored confidence, established order and held the Union troops in check. He was primarily responsible for converting the early, disastrous conflict into a well ordered defense around Cemetery Hill, causing Lee to postpone any thoughts of an assault at that time.

Meade's most important decision during that first day of fighting was his acceptance and continuance of the battle at Gettysburg despite the fact that his closest reinforcements were some distance from the battlefield. He had evidence that Lee's troop concentrations were much higher than had been expected. The decision to fight Lee's army with the Union I, XI and III Corps on the battlefield demonstrated that Meade was not afraid to challenge Lee's mystique. Chance had forced Meade's hand, but he would not tolerate the idea of retreat. In short, Meade was willing to take a calculated risk. Such is often the main ingredient needed to win great battles, and establish great generals. His conviction and aggressiveness in the initial stages of the battle established his control over the army and paved the way for the Union victory.

Meade finally arrived at the battle scene near midnight of 1 July, and personally assumed command. Surveying the situation and troop positions with his staff, Meade was again faced with a critical decision. He could continue with the engagement, in spite of the fact that he was aware that only three-quarters of his army could possibly reach the field by the morning of 2 July. Further time would be wasted in positioning the newly arrived troops, shuffling the units already in position, and reinforcing the weak sectors of the front. The time element was not particularly favorable. His alternative was a strategic withdrawal and regroupment in preparation for a battle more of his own choosing.

However, Meade considered that his army's position was tolerable. He was convinced that if his forces could defend their ground through 2 July, his reinforcements could strengthen him sufficiently to give him an advantage.

Predicting the enemy's movements is another quality inherent in a man of command capabilities. Meade rode over the battlefield at 0100, and after careful deliberation and a conference with Major General Howard, decided that the Confederates would be most likely to renew their attack with a major assault against Cemetery Hill. He concluded that their main objective would be an attempt at a sweeping maneuver to engulf the Union right flank.

The morning of 2 July saw the beginning of one of the greatest controversies of the battle of Gettysburg, and perhaps of Meade's entire command. Early on that morning Meade gave orders for Major General Daniel E Sickles to move his III Corps to a position of high ground on the

army's left flank at a location known as the Little Round Top. Sickles was to hold that position and to place his artillery where it would have command over the entire battlefield in that sector. The controversy arose over Sickles' misinterpretation of his commander's orders. Whether deliberately or through simple misunderstanding, Sickles chose to elaborate on Meade's orders.

Comparing his position with an area of high ground approximately one mile beyond his position, Sickles decided to advance and occupy that alternate location. Unknown to Sickles as he began his relocation, Confederate forces commanded by Lieutenant General James Longstreet were advancing. Although the Confederates did not have complete superiority, they were maneuvering into a position which could spell disaster for Sickles' exposed Corps.

Officers watching the shift of Sickles' Corps from other sectors of the front were amazed to see the abandonment of the Little Round Top commence. Knowing of Sickles' ambitious nature and viewing the move as another of his personal ploys, General Hunt was goaded into openly referring to Sickles as 'a politician . . . a man after his own ends.'

General Meade was not a commander to sit in stunned surprise shaking a bewildered head at the scene unfolding along his left flank. Without delay or thought for his personal safety, Meade rushed to Sickles' position, his anger mounting as he sped to the exposed troops.

Above: Close fighting on the last day of Gettysburg, 3 July 1863.

Above and below: The fighting at
Gettysburg lasted three days and
was the decisive battle of the Civil
War. Both sides fought desperately
for every inch of territory.

General George Pickett.

Morale was perhaps the most major consideration as Meade weighed the possibilities open to him on that second day of battle. The location of the engagement was not of Meade's choosing, and with circumstances going awry so early in the day despite Meade's planning, he could have begun a withdrawal of his forces at the earliest opportunity. He could not forget however that the Army of the Potomac had become infamous in its apparent 'destiny with disaster.' Ill fate seemed ever present and ever ready to snatch victory from the Union army. A defeat of the III Corps would probably have resulted in a massive Union retreat. A defeat of the Army of the Potomac on northern soil would have compounded the legend of Lee's Army of Northern Virginia and meant a complete collapse in the morale of the Union troops.

Upon reaching Sickles, Meade lashed out a reprimand and demanded an explanation. Sickles defended himself by claiming that Meade had led him to believe that he was authorized to decide where best to station his Corps. Sickles felt that he was in complete command of his Corps and could best decide the most advantageous position. Meade stated quite firmly that when he positioned Sickles, Sickles *was* in command — in command to *maintain* that position. In the face of Meade's obvious wrath, Sickles agreed to return to the Little Round Top. Surveying the unmistakable approach of the Confederate forces, Meade replied, 'I wish to God you could, but the enemy won't let you!'

This early test of Meade's mettle as commander of the Army of the Potomac was an all-important factor in the outcome of the Battle of Gettysburg. The results of his success as a commanding general were bought and paid for in those moments of decision as he stood surveying Sickles' Corps stretched along the Emmitsburg Road. The most obvious problem was blatantly simple. Meade could not afford the losses which would surely result if Sickles attempted to reestablish his original position. Neither could he leave the corps unsupported in its position.

While the gravity of the possibilities weighed on him, Meade could not forget that on the battlefield one must take action with swift deliberateness as circumstances demanded, not when or where one might wish. Sickles' maneuver courted disaster for the army, but at all cost the Confederate move against his army's left flank had to be repelled. Meade sped away to find troops with which to reinforce the III Corps. He personally rode along the flank, positioning troops, giving encouragement and driving the reinforcements forward.

The manner in which General Meade handled the situation which Sickles had created was one of many actions which earned him the rightful title 'victor.' He was delivered what appeared to be a lost cause, acted promptly and with forethought, employed his reinforcements, and ensured that victory would not be lost due to Sickles' blunder.

Meade's obvious options were few. It required outstanding command abilities to assess the situation as rapidly as he did, personally relocate reinforcements and still maintain an understanding of the battle as a whole. His efforts converted disaster to his advantage, breaking the spell of the Army of the Potomac's destiny with disaster.

This is not to say that the Union forces went unscathed in their efforts to maintain a solid front. Losses to the left flank were a staggering 50 percent, which caused a considerable dilemma in the strength and morale of the entire army, as Meade had stripped troops from various sectors of the front. However, the determination with which they repulsed the Confederate assault helped to temper the Union forces and instilled a confidence in themselves and their new commander. Meade's decisions, actions, and

presence on the battlefield demonstrated courage and dedication which for once need not be in vain.

By the evening of 2 July Meade was again faced with a decision which could decide the fate of the army. The left flank had held, but casualties continued to remain enormous. The army's center and the right flank, which had seen the majority of the day's action, had repulsed the attacks but casualties in those areas had reached a point which prompted Meade's staff to suggest that the army disengage and seek a better position. This move would have given the army time to recover and reengage at a more favorable time and location. While the staff was

Pickett's charge at Gettysburg was the climax of the battle. Lee hoped that Pickett would break the Union center.

shrouded in gloom the spirit of the army itself was high. Corps commanders reported a strict confidence in the ability of their troops to fight and win the battle along the existing lines. Meade himself questioned the ability of his troops to withstand such heavy losses and retain the morale necessary for victory.

In the bleak hours of early morning on 2 July General Meade had directed staff officer General Butterfield to draw up plans for a withdrawal of the army. Later that night the corps and division commanders met collectively to express their conviction that victory would surely be theirs if they could but rob the Confederates of any substantial success along the front. After listening to and evaluating their opinions, Meade solemnly replied, 'Well Gentlemen, the question is settled. We will remain here. But I want to say that I consider this no fit place to fight a battle.'

Thus, Meade chose to act, despite reservations, on the faith he had in his frontline commanders, whom he considered to have the most direct perception of the army's ability to continue the battle. General Henry Slocum, commander of the XI Corps, is said to have been the man in whom Meade had the most faith and on whom he based the decision to remain. Slocum was later to write that though many criticized Meade's thought of retreat, it was not due to military incompetence nor 'a timidness for battle,' but as a result of the serious sense of responsibility Meade felt for the army and the lives of the men under his command.

It was also during that night's meeting that Meade and his staff decided that the Confederate army's main assault on the morrow would most probably be directed against the center of the Union line. Lee had tested the flanks and had been unable to breech them.

The first offensive on the morning of 3 July was Ewell's attack on the Union right on Culp's Hill. Devastated by Union crossfire, the Confederates were forced to withdraw after three hours. The expected attack on the Union center began after a two-hour artillery barrage, as 15,000 Confederate led by Major General George Pickett charged up Cemetery Ridge. Outnumbered and outgunned the attack collapsed with a loss of almost two-thirds of Pickett's force. With a number of his generals seriously wounded, Meade did not order a counterattack against the heavy mass of Confederate cannon.

The primary questions and criticisms which surface with regard to that final day of battle deal not so much with the actual actions fought as with Meade's conduct following the defeat of the Confederate Army. Meade has come under consistently severe criticism for his failure to follow-up the engagement with an offensive aimed at the destruction of the Confederate forces as they retreated south. There are those who voice the opinion that Meade helped to extend the war by at least one year. Armchair strategists who criticize Meade because he did not sweep down from his position to annihilate his enemy apparently fail to consider the true situation.

Both armies had fought one of the most costly battles which they had been or would be engaged in. Casualties and losses were staggering and great fatigue was felt by both sides. Accusations concerning Meade's lack of response were unfounded. Early on the morning of 4 July

he ordered Union forces under Commander French in Frederick, Maryland, to cut off the Confederate retreat to the best of their ability. Meade himself began to move his army against Lee, but he was thwarted on two accounts. First, the morning of 4 July brought a severe storm which turned the battlefield into a sea of mud. Second, and perhaps more responsible for the lack of eagerness with which the army viewed continued action, was the task of caring for the wounded and burying the dead. It may be claimed that such tasks are best left until the battle is truly won, but these concerns seemed to have more priority in the minds of the men of the Army of the Potomac than engaging an enemy which was already defeated. Knowing

General Hancock (on horse, left center) directed the defense of the Union center. Moments later he was badly wounded.

Meade for the 'fine old soldier' that he was, Lee was certain that he would not allow the wounded Confederate troops left behind to go unattended. Although Meade sent patrols to reconnoitre the enemy position, the tending of wounded and prisoners seemed more appropriate that day than a vainglorious counterattack. By 5–6 July the Army of the Potomac was once again on the move, but the chances of contacting the retreating Confederates were nil.

Those who have experienced war, particularly actions of long duration or suffering heavy losses, realize that comforting thoughts stem not from who has ultimately won or lost but the relief that the battle is over. It takes time to mentally and emotionally prepare for final victory or defeat, and as the Battle of Gettysburg was not the last battle to be faced, time was needed to prepare for what lay ahead. Meade did what lay in his power to conduct himself and the army in the most competent military manner. The Battle of Gettysburg had been too costly and too wearisome to execute the counterattack which Meade's critics think he should have attempted.

Meade's place as a great commanding general, with Gettysburg the pinnacle of his career, has long been over-shadowed by Lee and the mystique which surrounds the

The fighting was so intense and the bombardment of lines so massive that visibility was poor. The flower of Lee's army, some 15,000 men, was lost at Gettysburg.

Army of Northern Virginia. However, without a doubt Meade conducted himself and his army with great distinction and honor in what can be considered one of the great battles of American history. The conditions and accomplishments of Meade's command have been exceptionally well documented by Francis A Walker, Brevet Brigadier General of the US Volunteers. It is difficult to elaborate on his description of Meade's handling of the situation at Gettysburg, as he follows precisely the main points of Meade's success.

First and foremost, Meade was in command of the entire Army of the Potomac for only three days before the Battle of Gettysburg began. The command he assumed was not one over an army of efficient, aggressive troops. Thrust upon him was the task of bringing cohesion to an army which had long suffered from a lack of competent leadership and deteriorating morale.

Meade was in fact nowhere near the battlefield when the battle began and was actually considering the possibility of giving battle in another location entirely. It was Reynolds who forced his hand and committed Meade to battle at Gettysburg. In effect, Meade inherited a battle, as he inherited the army, rather than executing an engagement according to his own strategy. Fortunately he had the necessary level-headed adaptability to circumstances which the commanders before him, particularly Hooker, had lacked.

Meade's immediate decision upon assuming command was to move the army northward in pursuit. This was

perhaps the most influential factor in the success he would achieve. After many years the Army of the Potomac was again fighting on and for northern soil. A large number of the men, including Meade, Reynolds and other officers, were Pennsylvanians that were not merely headed toward another battle, but a battle for home and state, a battle of Pennsylvanians for Pennsylvania.

Meade had an ability to understand not only the situation but also his commanders in the field. His ability to select the right man for the job was emphasized with his dispatching of Hancock to the front to take full command and do what was necessary to save the day. Meade chose Hancock not only for his abilities but as a man who would report to and carry out the express wishes of his supreme commander. Also, the fact that Hancock was not the senior-ranking officer demonstrated that Meade was concerned with effectiveness and capability, not with tradition and rank distinction. Meade continued to display a keen evaluation of, and regard for, his field commanders in his decision to remain at Gettysburg on the strength of their convictions. Too often a disregard for the opinions of such officers leads to a slanted perspective on the part of the commanding general, as amply seen during Hooker's reign over the army at Chancellorsville.

Meade's determination to maintain a defensive strategy also contributed to his eventual success. A defensive posture was more appropriate considering the terrain on which the Union forces took position and the lack of positive knowledge of the deployment of enemy troops.

Above: After the battle: a Confederate soldier lies in a crude breastwork in one of the niches in Devil's Den.

Above left: Cameras could not take action shots but after the battle the dead were commemorated in a series of historic photographs.

Meade's personality and his military concepts also had much to do with the decision to remain defensive. He believed that when truly in doubt, if the opportunity favorably presented itself, defend. Meade did not lose the initiative to Lee, he forced it upon him, causing Lee to gamble all in assaults against the Union lines. With only three corps in position by the end of the first day, Meade took the best possible option. Defense was not the more glorious option, but Meade was concerned with results, not glory.

By 2 July, in spite of Sickles, Meade had positioned the Union Army in a solid convex line facing Lee's forces. This caused the battle to separate into multiple fronts and engagements and made it easier to shift troops from one area to another. He could reinforce endangered areas and combat the Confederates in piecemeal fashion rather than face one massive line assault.

Many high ranking Union officers were killed or wounded during the days at Gettysburg, including Reynolds. The heroism and accomplishments of those who lost their lives have been stressed in many accounts. However, as with Sickles' blunder at the Little Round Top, it was Meade who entered into direct contact with the troops to rally and deploy them. The heavy loss of troops and officers is often overstated while Meade's involvement is understated.

The two principal criticisms levelled at Meade as Commander of the Army of the Potomac during the summer of 1863 are he mismanaged the episode with Sickles and charges that he allowed the Army of Northern Virginia to escape destruction. Meade's reaction to criticism was indicated by the manner in which he pursued his command in the best interest of the men and army as a whole. He believed that individual actions and results were rudimentary in the winning of battles and ultimately the war. Perhaps he lacked the flamboyance of other officers of the day, but Meade was no glory seeker. He refused to be manipulated by critics and detractors, content to apply his skills to the best of his ability for the good of his troops, his army and his country.

General George Gordon Meade was, in the classic sense, a great general and a worthy victor. One word is the key in the evaluation of Meade's character. That word is 'soldier,' as defined in the *Random House Dictionary*, 'a man of military skill or experience.' In the true context of the term soldier, Meade applied his skills to the best of his ability in the brief period before and during the Battle of Gettysburg. He was bound by his sense of honor, duty and service to the position he held.

Meade will never compete with other flamboyant figures of his era, yet he was without a doubt every inch the soldier, the general, necessary to accomplish the feats required during that decisive time. He was truly one of the great, though little appreciated, generals of Civil War.

CHAPTER 6

VON MOLTKE AT SEDAN 1870

t is strange to realize that a man whose name, perhaps more than any other, is associated with Prussian militarism was himself not even a German but a Dane. Helmuth Carl Bernhard, Baron von Moltke, born at Parchim in Northern Denmark, was the son of a Danish general and began his own military career in his own country's forces.

It was at the age of 22 that he transferred to the army he was to serve with such distinction. The joint reformer of the Prussian army with von Roon, he later became the strategic architect of victories over his own native country in 1864 and over Austria two years later. Yet it was his victories during the late summer and early autumn of 1870, when he was himself as old as the century, which were the most resonant. France, long regarded as the greatest among the military powers, found herself deposed by the nation which was regarded at that time as being among the least.

Moltke's contemporaries and colleagues included not only von Roon, the Prussian War Minister, but also Otto Eduard Leopold von Bismarck, who if not the sole creator of German unity was certainly the man who contributed most to its realization. Following the defeat of Austria, hitherto unquestioned leader of the German-speaking peoples, 20 states north of the River Main joined Bismarck's North German Confederation with his own Prussia as its head. He realized, however, that only some external intervention would persuade the states of the Catholic South to overcome their religious aversion to uniting with the Protestant North.

One such intervention could be a war between Prussia and France. If this occurred the French armies would be compelled to strike through South Germany. In coming to

Count Otto von Bismarck.

Napoleon III, Emperor of France.

its aid, Prussia would also assert her own physical presence there and in fact Bismarck had gone so far in persuading the Southern States of the menace they faced from France that they had united with Prussia in a series of mutual defense treaties.

At the same time, he had looked to the condition of the Prussian army which he had himself helped to reform in 1862, while Moltke was to draw up the plans for encompassing France's defeat. What he finally proposed involved crossing the Rhine with the large forces available to him through the Confederation, driving a wedge between the principal forward bases of Metz and Strasbourg and thus dividing the enemy's forces.

Although quite prepared to incite the French to action, Bismarck had reason to hope they might themselves oblige by providing an excuse for war. The French had been growing increasingly ambitious. The spirit behind this new mood was Charles Louis Napoleon, nephew of the great Napoleon, who, elected President of the Republic in 1848, had proclaimed himself emperor under the title of Napoleon III, four years later.

His desire to add further luster to the family name had already led him to rebuild a sizeable chunk of Paris and, with the assistance of his elegant, Spanish-born consort, Eugénie, to make it the center of fashion and style it has been ever since. It had also led him into a number of activities less innocuous such as his futile attempt to install the Hapsburg princeling, Ferdinand-Maximilian, as client-emperor of Mexico.

So long as the German Confederation was restricted to the Northern States, Napoleon was prepared to cast a benevolent eye upon it. Matters would be different, however, as Bismarck realized, once a strong and united Germany pressed on France's borders.

William I of Prussia acceded to the Prussian throne in 1861 and became the first German emperor.

A potential challenge occurred in 1870 when Napoleon's ally, Queen Isabella of Spain, was deposed. In the July, Bismarck, knowing they would acquiesce, suggested to the Spanish Parliament, that they give the vacant throne to a member of the Sigmaringen-Hohenzollern family. A cadet branch of Prussia's own Hohenzollern dynasty and having some claim to the Spanish crown, Bismarck rightly judged that the French would not tolerate a potentially hostile power across the Pyrenées.

He was right. Angry speeches in the Chamber of Deputies were matched by angrier leaders in the Paris press. The capital had only just stopped laughing at Offenbach's *The Grand Duchess* in which figures recognizable as Bismarck and the Prussian General Staff were parodied. In such a mood, prudence flew out of the window. Instead of discreetly indicating to Spain that she was running the risk of incurring its displeasure, the French government at once revealed the Prussian machinations and was thereby brought into direct confrontation.

Nonetheless, French persuasions at first achieved their objective when the Prussian monarchs, Wilhelm I and Augusta, who disliked Bismarck, prevailed on the Hohenzollern candidate to withdraw. There the matter could and should have rested. The French press and the more right-wing politicians, however, were still baying and would only be satisfied by an abject confession of Prussian guilt. The French ambassador, Benedetti, was therefore instructed to obtain the King's public acknowledgment that there would be no repetition of the Spanish claim; this was tantamount to an admission of Prussian involvement. The furthest Wilhelm would go was a statement that the renunciation had had his approval. The king was both old and unwell. The discussions with Benedetti took place while he was undergoing a cure at Ems. Pressed by the importunate ambassador, he told him with exasperation that he should not be troubled with matters which were his ministers' responsibility. That evening he telegraphed Bismarck a summarized version of the interview, no doubt to prepare him for a possible encounter.

When his message arrived Bismarck was entertaining Moltke and von Roon to dinner. He saw it as just what he needed. Without making any textual changes, but by skillful editing he managed to convey the impression that the King had virtually broken off diplomatic relations. 'It will be like a red rag to the Gallic bull,' he predicted, before releasing it for publication. If in nothing else, Bismarck showed a matador's finesse in judging the reactions of his Gallic bull. Amid the angry bellows of the French Assembly war credits were voted eagerly.

He could now have his war under ideal conditions. The world at large which thought — and continued to think — that France was being paranoically touchy in detecting Bismarck's hand in every turn of events, saw it as the frivolous villain, especially after Bismarck revealed an earlier French plan to divide Belgium between them.

The French expected a brief and glorious war with a rapid advance across the Rhine to neutralize the South German States. Even Britain though openly favoring Prussia, took a similar view. The French army was admittedly smaller than the Prussian numbering about 300,000 men. Moltke expected to field 330,000 men if the Southern States remained neutral, 360,000 if they did not — as proved the case.

France's advantage lay with its vaunted 'professionalism.' It was an imperial power and required a standing army ready to be dispatched to any outpost of empire at short notice. To achieve this with a conscript army meant a long period of service and if this was not to drain the nation of its manhood, it had to be selective. Only 100,000 men, chosen by lottery, were conscripted every year, but until 1868 they had to serve seven years with special bonuses for reenlistment.

As a consequence the French regarded the Prussian army where every single able-bodied man was called up and served a mere three years, as little more than a *levée en masse*. This, of course, ignored the essential differences. For Prussia was a nation state, albeit a recently founded one, and required an army which could be mobilized quickly to guard the integrity of its own frontiers; one with a large pool of reserves, in other words. These the French army lacked, as men who had served with the colors for seven years and often much more, were unwilling to face recall.

There were other, deeper deficiencies. Splendid as they were on the parade ground, the French troops were drawn from the lowest and poorest elements of the peasantry, those unable to buy their way out of the call-up. The officers, men of the bourgeoisie, no less brave than their troops, lacked initiative, while the General Staff was frankly incompetent and openly relied on '*le Système D*,' D standing for '*débrouillage*' or 'muddling through.'

It could scarcely compete with Moltke's almost inhumanly efficient Prussian General Staff and this began to affect French conduct from the moment the war began. The nation's greatest advantage lay in the fact that with its standing army it should have been able to field the greater force initially. It was widely assumed, at home and abroad, that the declaration of war would be followed by an immediate invasion of Germany. This was prevented by '*le Système D.*' From the start there were commissariat failures which left units without cooking equipment and sometimes without food. Men were pushed hither and thither all over France and sometimes shipped off to Algiers and brought back.

Meanwhile Moltke could hardly believe his luck as he made his final preparations, unhindered by enemy action. In fact, the French had no real plans for an invasion of South Germany, indeed no plans of any kind save for those drawn up virtually in his spare time by General Frossard, the Emperor's military tutor. In character these were purely defensive. An Army of the Rhine would stand guard over Strasbourg and an Army of the Moselle over Metz. They would therefore be exactly where Moltke wanted them. There were, however, to be two reserve forces, an Army of Châlons and an Army of Paris. The backbone of the latter would be the *Garde Mobile*, proposals for which had been formulated in 1868 as part of General Niel's army reforms. It was to be an elite corps of the National Guard, ready to be sent anywhere, and lists of eligible men had been drawn up. The men involved had been rounded up when mobilization began. Though untrained civilians, they were at once dispatched to various units. Sullen at finding themselves snatched from civil life and left with no means of filling their time, they quickly became a drunken rabble, burdening every commander they were sent to and endangering general discipline.

Then, at the very last moment, what plans the French did have were scrapped. The Rhine, Moselle and Reserve Armies were merged into one cumbrous body assigned to hold some 200 miles of frontier. At the same time, Napoleon appointed himself Commander in Chief with Leboeuf, the Minister of War as his Chief of Staff. To watch over affairs of state in his absence, he appointed the Empress as Regent.

Napoleon's love of the army, certainly as great as his august uncle's, was not equalled by his talent, let alone experience. Charles Louis had never held command, could only read a map with difficulty and, now at 62, suffered from a chronic bladder weakness which required the constant presence of a surgeon to relieve it.

On 2 August, a French force under Frossard defiantly crossed the border and occupied Saarbrücken. This gave Moltke a nasty moment until it was clear it had been done solely to give the army employment and did not presage a concerted attack. On 4 August, Moltke himself moved. The Prussian Third Army, under the Crown Prince Frederick-Wilhelm, took the minor fortress town of Wissembourg, the French commander dying in the action. To meet this threat, the armies had to regroup and Saarbrücken was promptly abandoned, Frossard omitting, as he did so, to burn the town's bridges. It was now obvious to Napoleon that while his armies would not be in any position to invade for months, the unthinkable was happening and the Prussians were marching into his own country. With this realization he seems to have lost all interest in the campaign.

Further blows fell before there was any chance of a French regroupment. On 6 August Frossard was forced to withdraw from a line he had taken up from Spicheren to Forbach. The loss of the latter was particularly serious as it was the main supply depot for the French offensive. At almost the same time, the Prussian Third Army, followed up the capture of Wissembourg by that of Froeschwiller ridge. The enemy was visibly interposing himself between Metz and Strasbourg.

Military logic dictated only one course: a rapid retreat to positions which would allow for reconcentration and the formation of a continuous line. Logic was not paramount. The regime of Napoleon III had from the outset faced opposition from republican and liberal opinion, while his main support had come from the bourgeoisie and the conservative peasantry. This support had been steadily eroded largely as a result of the ill-advised adventures which were regarded as breaking his promise to restore peace to the country. To make matters worse, the Paris press had hailed the occupation of Saarbrücken as a major victory. When this was followed not merely by retirement but by three defeats for French arms, it looked as though fury would be boundless. To forestall this, on 12 August, the emperor gave up command of the armies to Marshal Bazaine.

Bazaine's prescription was indeed withdrawal and he successfully eluded Moltke's pursuit. Then, for reasons of political expediency, he was instructed to stand and fight. He took up a position at Borny where the astonished Prussians, believing their enemy in retreat, blundered into them. The daylong battle ended with the elated French holding their positions, but the truth was they still had to retreat and had lost time by taking a stand. Furthermore, the French troops had so spent themselves that when, the following day, they began to stream into the fortress of Metz they could not be dislodged from its refuge. On 16 August the Prussians reached them. From behind their fortifications the French rained down slaughter, but Bazaine was

Marshal Bazaine was the Commander of the III Corps at the start of the Franco-Prussian War. Napoleon gave him command of the Army of the Rhine.

Left: Strasbourg under fire. After the disastrous battles of 6 August the French armies retreated leaving Alsace and Lorraine.

Right: Marshal MacMahon was the Commander of the French I Corps. He had gained his military experience fighting in Algeria and was in command of Algerian troops.

Below right: At the Battle of Beaumont Failly's V Corps was taken by surprise and suffered 7000 casualties. He was replaced by Wimpffen after the engagement.

Below: A Cavalry action at Buzancy on 27 August 1870.

concerned that he might be besieged and tried to extend his line leftward toward the rest of the army. The result was a defeat which had just the effect he had tried to prevent.

The situation was not irretrievable as there was still the Army of Châlons. If this could be drawn back to form a protective cordon in front of Paris, survival was still possible. Even the *Gardes Mobiles* might be persuaded to fight since they were largely drawn from the capital. Militating against this course was not only the political situation, but also the ambitions of the Empress-Regent. Her husband had slipped out of Metz and was now at Châlons, a forlorn wreck, frequently in pain and hardly the sort of leader to inspire confidence in the ragbag of soldiery the Army of Châlons had become as stragglers from the previous debacles were absorbed into it. Nonetheless, both the Empress and the War Minister, Palikao, urged Napoleon to participate in a great relief expedition on Metz. It was this or the final collapse of the regime, they insisted.

The commander of the Châlons Army was Marshal MacMahon, a soldier experienced enough to realize that he was being asked to consign to hopeless battle an army which should have been kept intact at all costs. However, he could hesitate no longer when news came that Bazaine was attempting a breakout. Decency required that the Army of Châlons advance and stretch forth a helping hand. In fact, there is no evidence Bazaine had any intention of breaking out.

MacMahon suffered a further reverse. The pro-government press in Paris, to answer accusations of inertia against government, dynasty and army, saluted the departure of the relief force with a fanfare of headlines which had the unhelpful effect of telling Moltke, previously mystified, what his enemy was up to. Twist, turn, double-back on themselves as the French might, they never lost their pursuers who were often at their destination before them. At Beaumont,

the V Corps, under Failly, was surprised, showed little stomach for the fight and retired. This was the last blow to MacMahon's tortured and weakening nerves. He pulled back to the little town of Sedan, up the Meuse and behind the shelter of the Belgian border. There, too, Moltke had not only divined his intentions, but was preparing to thwart them.

Sedan lies in the valley of the Meuse. On its south side is a marshy plain, criss-crossed by streams, while wooded slopes surround its other sides. A main road passes through the town, roughly following the meandering line of the river. Somewhat less than a mile from it, however, at the hamlet of Floing, is a crossroads with a second road going eastward and away from the river to a junction at Olly Farm. Here the road sweeps back southwest to meet the Sedan road once more at Bazeilles. In other words, it forms a kind of large triangle with Bazeilles, Floing and Olly Farm as its three extremities and Sedan midway along its base line.

It was within the limits of this triangle that MacMahon disposed his forces. The VII Corps, under Douay, took up a position along the ridge from Floing to the Calvaire d'Illy which lay just below the point of the triangle at Olly Farm. Then the I Corps, under Ducrot, occupied a line running from Olly Farm halfway toward Bazeilles. Lebrun's XII Corps continued the line down to and including Bazeilles. The V Corps, battered at Beaumont, was kept as a reserve in the center.

This last unit was under the command of Failly who, because of his dismal performance at that battle, was about to be replaced. His successor was to be Wimpffen, formerly a colonial governor in Algeria, who had returned hotfoot on the declaration of war in search of glory in the field. Besides succeeding Failly, however, he was also instructed by Paris to take over as Commander in Chief should the misfortunes of war overtake MacMahon.

How MacMahon intended to fight the battle has never been entirely clear. He was adamant, in his briefing of the corps' commanders, that he had no intention of being bottled up in Sedan as Bazaine had been at Metz. In any event, supplies and ammunition at Metz were plainly inadequate for a prolonged struggle. Rather he seems to have envisaged a kind of rearguard action, pulling back along the banks of the Meuse toward Mézières, a move taking him away from Metz, the fortress, which as the Empress was quick to point out, he was supposed to be relieving.

If there was on the French side that uncertainty which had marked their conduct of the entire war, there was none on Moltke's. His plans for what he appreciated was the last and, thanks to the ill-judged retirement on Sedan, decisive phase of the war were complete before MacMahon began making his dispositions. The total area available to the French forces was a narrow wedge of land. On one side was the River Meuse and on the other the Belgian frontier. Bismarck had already dispatched a warning to the Belgian government to see that no French unit was allowed to

Right above and below: The I Bavarian Corps in action at Bazeilles on the morning of 1 September. Led by von der Tann the German unit attacked prematurely at 0400 hours following up some heavy bombardment of the town. After fierce fighting there was a stalemate.

escape by that route and to make assurance doubly sure the Third Army was ordered to pursue into Belgium if necessary. This unit was also allotted the task of holding the strip of land between frontier and river. On 19 August, the commander of the Second Army, Prince Frederick-Charles, had complained that his force, consisting of six corps and two independent cavalry divisions was grossly unwieldy. Three of its corps and the cavalry divisions had accordingly been taken from it to make another army, the Army of the Meuse, under the command of the resourceful Crown Prince of Saxony. This was to advance along the river's bank, cutting off escape across it.

On the night of 31 August, Moltke ordered Blumenthal, commanding the XI Corps, to cross the Meuse and close the remaining opening, that to the west. 'Now we have them in a mousetrap,' he told the king. The selfsame belief was expressed by Ducrot, commanding the French I Corps, in somewhat earthier language: '*Nous sommes dans un pot de chambre et nous y serons emmerdés.*' As night fell, one after another of the Prussian watchfires punctuated the darkness, staking out the limits of the French perimeter.

MacMahon talked airily of maneuvering under the nose of the enemy as he prepared for the withdrawal of the morrow. 'Tomorrow,' Douay, commander of the VII Corps told him, 'the enemy will not give you time' — a statement which coming events would soon confirm.

Throughout the campaign, the *mitrailleuse*, a multi-barrelled, quick-firing forerunner of the machine gun, dubbed by one Prussian commander 'that damned coffee-mill' had taken severe toil, while the French *chassepot* rifle had proved superior to the Prussian needle-firing gun. Unfortunately from the French point of view, both required that the enemy come within range. This he was often able to avoid doing as the Krupps' steel, breech-loading guns of the Prussian artillery far outranged the French bronze, muzzle-loading weapons. Firing salvo after salvo from positions well out of reach the former could tear gaping holes in trench and earthwork, frequently reducing the defenders to cowering impotence. On the night of 31 August, the Bavarian I Corps, under von der Tann, began just such a bombardment on the French position at Bazeilles as the preliminary to attack.

All the same, there is reason to believe Moltke was in no hurry. He wanted his trap right in front of the mouse hole by moving his two wings round the enemy before he did. What frustrated him was the overeagerness of von der Tann to attack.

That the Prussian Army had become a sharper and more accurate weapon with each engagement was largely due to deep differences of philosophy between French and Prussian General Staffs. Ever since the days of the first Napoleon, the one had been designed to be the responsive instrument of a single, brilliant, dominant will. Field commanders were assigned their positions, took them up and defended them, if necessary, to the death. No one ever moved to the assistance of a comrade, however hard-pressed, unless ordered to do so and such were the jealousies the system engendered that it was not uncommon for one commander to watch, gloating, as his brother-in-arms suffered disaster. Under Moltke's tutelage, he and his head-quarters were merely a nerve center. The system of staff training he had inaugurated was based on two precepts, the first being that field commanders should develop an almost telepathic sensitivity to the needs of the campaign or battle as a whole, referring to higher authority as little as possible.

Obviously this had its defects and could result in individual officers taking too much upon themselves. One repeated offender was the commander of the First Army, Steinmetz, who in the early stages of the Battle of Spicheren, for instance, defied orders by pursuing an enemy he believed in full retreat (which he was not), thus risking a separation between himself and Second Army next to him.

Such manifestations were, of course, to be expected in as new an army as that of the North German Confederation where many of those now subservient to Moltke had formerly held command in state armies and were, besides, often members of its landed gentry, more accustomed to rule than to obey. Nor was it easy for him to dismiss them, since doing so might offend local susceptibilities. The problem was usually avoided because experience showed that close collusion and the adherence to strategic intentions produced the best results. This was not always the case.

At 0400 hours, through a cold and misty darkness, the V and XI Corps, forming Moltke's wings, began their crossing of the Meuse at Donchéry, the junction of all rail lines and roads running westward. There was no resistance and by 0730 hours they were astride the Sedan-Mézières road. Believing themselves the first unit to take the field they were surprised to hear firing from an easterly direction.

It came from von der Tann's Bavarians who had attacked prematurely, while the western exits were still open, and were now locked in a ferocious struggle with French Marines, among the finest troops in the erstwhile Army of Châlons. In a very short time Bazeilles, or such parts of it as were left unburning from the artillery bombardment, was ablaze from end to end. As daylight came the positions were virtually deadlocked. Later, von der Tann was to

justify his action on the grounds that he was afraid the French were slipping away in the darkness.

However, as Lebrun, the XII Corps commander, could see for himself, there was no escape: the Prussians were covering his whole front. The next village to Bazeilles in the northeasterly direction was La Moncelle and beyond this Daigny, both along the road to Olly Farm, the apex of the defensive triangle. Soon enemy artillery was occupying the slopes of the hills on the far side of the Givonne valley. A column of Saxon troops took Moncelle, joined up with von der Tann's Bavarians on their left, and another column

Right: The capture of three *mitrailleuses* by the Saxon Infantry Regiment August-Frederick. The *mitrailleuse* was Napoleon's secret weapon but unfortunately few French poilus knew how to operate it.

Below: On 1 September at 0600 MacMahon was injured in the leg by a shell splinter as he rode to Bazeilles. He was evacuated and he chose Ducrot to replace him.

to the right was threatening Daigny. The defense here, by Zouaves, a French colonial unit, was so tenacious it took four hours to capture the village.

Already, the French had suffered another mishap which because of the confused state of staff organization had more far-reaching effects than was necessary. MacMahon riding out toward Bazeilles just after the Bavarian attack began, was struck by a shell splinter and had to be carried back to Sedan. This, of course, was the situation in which Wimpffen was to take over. He had not yet arrived, nor was there any warning of his coming or of the orders from Paris he bore.

In these circumstances, MacMahon's chosen successor, Ducrot naturally took over.

Appreciating just what was happening, at 0800 hours he gave orders for a withdrawl by the V and XII Corps as the prelude to a general disengagement and withdrawal westward. He had not yet learned that the Germans were occupying Donchéry, blocking the western exit.

The plan was never tested, for no sooner had orders gone out than Wimpffen arrived on the scene and, full of vigor and determination which made him an incongruous figure in the present company, declared himself to be Commander

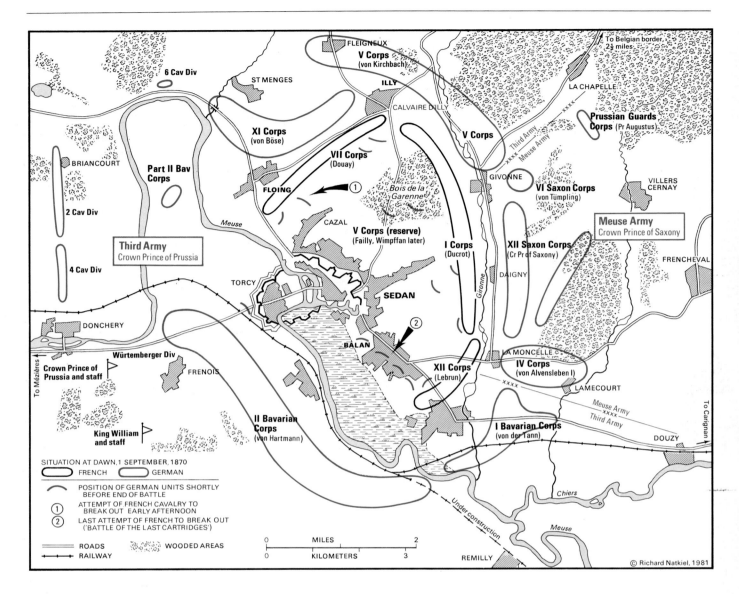

SITUATION AT DAWN, 1 SEPTEMBER, 1870

⬭ FRENCH   ◯ GERMAN

◜ POSITION OF GERMAN UNITS SHORTLY
   BEFORE END OF BATTLE

① ATTEMPT OF FRENCH CAVALRY TO
   BREAK OUT EARLY AFTERNOON

② LAST ATTEMPT OF FRENCH TO BREAK OUT
   ('BATTLE OF THE LAST CARTRIDGES')

═══ ROADS   ░░░ WOODED AREAS

╂╂╂ RAILWAY

© Richard Natkiel, 1981

Right: French troops scramble
across marshy ground.

Far right above: The 105th
Infantry Regiment at Daigny,
near Sedan.

in Chief. The Empress and the Paris politicians had impressed on him the need for victory and, accordingly, he announced there would be no withdrawal. Those units under extreme pressure would be reinforced and in due course they would turn to the offensive.

If Ducrot's answer typified the mood of the defenders, it also had the merit of realism. He told Wimpffen, 'You will be lucky, *mon général*, if by this evening you even have a retreat.' As commentators like Michael Howard have pointed out Wimpffen was actually ensuring the total destruction of the French army.

As it happened, the Prussian V and XI Corps, now advancing beyond Donchéry were experiencing some difficulty, the product of von der Tann's overeagerness. The second of Moltke's basic precepts was that of marching to the guns, which, in contrast to the French habit of waiting for orders, produced a build up of strength wherever a struggle was raging. Its defect was that it could also produce severe traffic jams. When the V and XI Corps heard the firing from the east, every available unit began marching toward it with the consequence that infantry, cavalry, waggons and guns converged on the roads and required an hour to disentangle themselves. Nonetheless, by 0900 hours, while the XI Corps was advancing on Floing, the V Corps were crossing open country in the direction of Fleigneux. The delay, however, had had the effect of leaving a gap and realizing this the French cavalry had been sent out to keep it open by checking the Prussian advance.

One of the lessons of the present war had been to demonstrate the futility of cavalry charges against infantry armed with the rapid-firing, longer range modern rifles. Men and horses were cut down long before their foe was within saber range, while at the same time the artillery was firing into them. A few managed to escape via Fleigneux and the Ardennes Forest, but, soon after, this gap too was closed when the Prussian Guard Corps marched up to take position facing the apex of the defensive triangle.

Moltke need have done no more than wait for a result, using probing attacks to find weak points in the enemy perimeter and exploiting them in his own time. The sun had come up on a glorious day and perhaps it was this, combined with the panoramic view of the battlefield which the German position afforded that led him to succumb to the temptation to turn this, the world's first truly modern battle, into an exhibition of his own virtuosity. As the morning air grew warmer a procession of notabilities began mounting the hills near Frenois. Among them were not only the Kings of Prussia, but also Bismarck himself with Moltke and von Roon and their staffs, various Foreign Office officials, the military attachés of Britain, Russia, the United States and a gaggle of German princes whose titles recalled those states which by the end of the day would have taken such an irrevocable step toward dissolution into the larger entity of Germany — Leopold of Bavaria, William of Württemburg, Frederick of Schleswig-Holstein, the Duke of Saxe-Coburg, the Grand Dukes of Weimar and Mecklenburg-Strelitz. Bringing up the rear were the newspaper correspondents, whom Moltke was delighted to have as witnesses and heralds to his triumph.

Lingering mist still hid the more distant parts of the

Left: The destruction of the French cavalry at the Battle of Sedan.

Right: Prussian infantry overrun French *mitrailleuse* positions at the height of the battle.

Below: The Prussian Guard artillery, under Prince Kraft zu Hohenlohe-Ingelfingen, deploying above the Fond de Givonne where it faced the French I Corps.

landscape. As it rose, the spectators saw that what lay below them far outshone their own splendor: a panoply of war such as was never again to be arrayed, men in their thousands, marching, wheeling, advancing cautiously but boldly over open country. Everywhere there was a brilliance of color. The Zouaves, in tasselled caps and embroidered jackets, the glinting helmets and breastplates of the *cuirassiers*, even the humble French infantry in dark blue *képi* and tunic to set off their bright red *pantalons*; the German regiments in Prussian blue with glinting *pickelhaube*. Behind them were horse-drawn ambulances, supply waggons and gun teams, dragging weapons and limbers into positions. Guns seemed to be aimed at every point of the defensive concentration, while a line of them ran almost the entire length of the south bank of the Meuse below the spectators' line of vision.

The contrast in mood on the two sides was stark. Napoleon was in agony from his bladder and without hope of relief since duty demanded he should be seen by his troops. To this end, his face rouged to cover the pallor of sickness, he mounted his horse and spent the morning riding from one unit to another. Though he constantly came into the view and range of the enemy, hoping for the bullet which might end pains as much moral as physical, by lunchtime when he returned to his quarters this anodyne had eluded him.

There was one small hope. Wimpffen realizing that, whatever else, he was not short of men, had decided he would break out, but eastward toward Douzy and Carignan rather than west and toward Mézières. His first inclination was to make the attempt right away, then he decided to wait until darkness. With his plans thus crystallized, he rode over to see how the men who would have to hold his exit door at Bazeilles, were faring. Their commander was not unhopeful: he would hold as long the corps to his right did so and Wimpffen promised to reinforce this with men from the I Corps. By the time his rounds were finished, even the Commander in Chief understood the truth of his position. If he was to break out, he must do it at once, and at 1300 hours he sent out messengers to begin mustering the troops. At the same time, Napoleon was invited to take his place at their head. Long convinced the war and so this particular battle was lost, he declined the honor.

Nor was there much sign of men or officers, though at Balan, between Sedan and Bazeilles, Wimpffen assembled enough to mount an attack which compelled the II Bavarian Corps to ask for reinforcement.

He had not yet abandoned the thought of the breakthrough when the line of the VI Corps broke and he found himself engulfed in a tide of disorderly withdrawal which was more like a rout and belied its commander's promise to hold. Douay blamed the reverse on the failure of his right, but the truth was that Wimpffen's reinforcements, if they had ever arrived, had not been able to take up their positions before they were swept up by a German infantry advance. Ducrot, who rode up to see for himself what was happening, came upon hosts of stragglers in the Bois de la Garenne which lay behind the position. An attempt was made to rally them and a line of sorts actually formed from a hotchpotch of regiments. It held for two hours, then at 1500 hours began to break up as men sought the shelter of the woods, still under fire from some 90 Prussian guns.

Another crisis developed behind Floing, where the defenders had staved off every attack from the village for six hours. It was almost exactly as Wimpffen was giving his orders for the breakout that their guns were silenced by the superior German ones, however, and soon after, the Prussians began reaching round the defenders' flank.

There were many acts of sacrificial, if vain, gallantry that day, but none was more conspicuous than that of the French Lancers. The Germans were now beginning to appear on the plateau above Floing and had to be repulsed at all costs. Not only this, but the French commanders were returning to the idea of a breakout westward once more. The cavalry would, therefore, act as the broom to sweep the way.

General Margueritte, their commander, left immediately to reconnoiter, while his squadrons formed up under shell-fire. As his horse picked its way down the slopes toward Floing Margueritte was hit by a bullet which smashed his jaw and ripped his tongue. With the lower part of his face a mass of blood, he rode up the hill, pointed with one arm toward the enemy, then collapsed.

A cry of 'Avenge' went up from his men and they spurred to rush the slopes, bursting in upon the German skirmishers who broke up before them. That was their sole achievement. The infantry stood their ground inflicting such slaughter with their rifle fire that the cavalry was forced to split up and go round it instead of overpowering it. They became more and more separated from each other as they tried to find their way through the enemy lines and back to their own lines while all the time the killing went on, so that for weeks after the battle the carcasses were found, often miles from the battlefield.

Some came back over the hill's crest, nevertheless, and Ducrot could only ask General de Gallifet, who was now commanding them, if they would try again. 'So long as there's one of us left,' the cavalry general told him. Twice more they were unleashed down the hill.

Watching from his imperial grandstand, Wilhelm was compelled to exclaim in admiration, '*Ah, les braves gens,*' a phrase recorded in the memorial to them which stands on the site. Their last attempt came at 1500 hours. As they went, Ducrot and his staff rode among the infantry urging them to follow, to seize such initiative as the Lancers' bravery had given, perhaps to break through. But it was in vain. They did not lack the courage; they were exhausted and utterly disspirited. As Gallifet and his last brave few reached the valley, the enemy ceased fire; officers raised their swords in salute and the *braves gens* rode away.

Perhaps it could be said that the Germans felt able to afford such gestures. All along the French lines were signs of disintegration. As it became increasingly obvious that the Germans were not to be stopped, men abandoned their positions. The charge of the Lancers was almost the last throw of the battle. However, Wimpffen, taking Lebrun with him, returned to Sedan and somehow scraped together a force some 1000 strong with a few guns, and advanced toward Balan. On the way they collected a few more unvanquished spirits and attacked the village with such ferocity that the II Bavarian Corps was driven from it and men and guns had to be rushed up to prevent exploitation of the attack. There was none since no one would follow Wimpffen and Lebrun beyond the village.

This – the Battle of the Last Cartridges – was truly the end. All round the fields silence fell in the brilliant coppery evening light. Men of the French II Corps who had taken shelter among the trees of the Bois de la Garenne, there to be ruthlessly shelled, came out, half-dazed bands of stragglers, to give themselves up.

Ever since the Emperor knew the day was lost, his one concern had been to end the carnage, and even before Wimpffen and Lebrun's shortlived rally, he had ordered that a white flag should be hoisted over Sedan. It was only at this point that Moltke spotted it and sent one of his officers, Bronsart von Schellendorf, to ascertain its meaning. He was taken to Napoleon who sent him back with a member of his own suite, bearing a letter to the Prussian

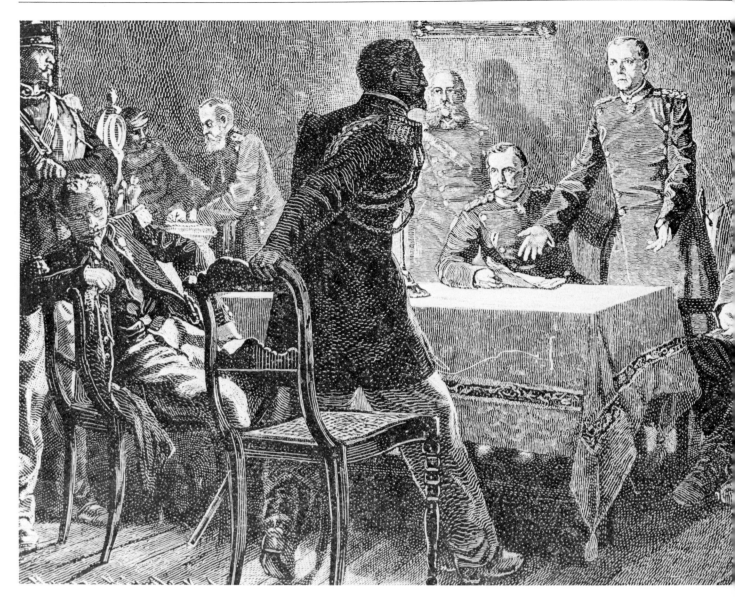

Above: The surrender ceremony after Sedan. From left to right: General Faure, Count Nostitz, General Wimpffen, von Clair, Polbielski, von Moltke, Oberst Verdy, Count von Bismarck, General Castelnau and Oberstleutnant Bronsart.

Above far right: A cartoon of the end at Sedan showing General Ducrot waving a white flag.

king. The two armies had just fought what was the first engagement in a new, yet more dreadful epoch of warfare, yet the air of a great royal tournament imparted to the occasion by the hilltop spectators was maintained in that missive. 'My Brother,' it ran, 'not having been granted death in the middle of my troops, it only remains for me to place my sword in Your Majesty's hands. I am Your Majesty's true brother, Napoleon.'

Wilhelm's answer, dictated by Bismarck, continued the strain: 'My Brother, in regretting the circumstances in which we meet, I accept Your Majesty's sword.' There followed a reminder that what had been at stake had been more than a piece of gingerbread encrusted with a gilded fleur-de-lis, the traditional reward of a knightly tourney. Napoleon was asked to send plenipoteniaries to negotiate

the capitulation of his army. On his side, Wilhelm designated Moltke.

It was an inevitable final humiliation for Wimpffen that he should have been sent by Napoleon. He tried to evade it, but got short shrift from his brothers-in-arms who reminded him that he had been happy enough to command an army he thought might bring him glory.

The following day, 3 September, the two sides met at a chateau on the Meuse. There was little to be negotiated. Wimpffen was simply passing his forces, such as had survived, into captivity. Including those taken during the battle they amounted to 105,000 men, to this could be added 1000 waggons, 6000 horses and 419 guns. Prussian losses amounted to 9000 officers and men.

Among those taken into custody that day was a young officer, Joseph-Simon Gallieni, who was to profit by and later to put to successful use what he learned of German military organization. Another, and perhaps the least reluctant to leave France, was Napoleon himself with all his entourage.

The fruits of Moltke's victory are well known. The fall of the empire; its replacement by the republic; the formation of the national army to try to prolong the struggle; the Siege of Paris; the Commune and the brutal massacre with which

it was repressed.

Following the victory of Sedan, the Prussian army became the world's model and the shouts of the Feldwebel echoed across the parade grounds of Japan, Turkey, Russia and Latin America, leaving as their permanent legacy the goosestep. Despite his Danish origins, Moltke was, like Clausewitz, a peculiarly German phenomenon, for it is a national characteristic to believe that everything – musical composition, human history, the universe, the fluidities of the battlefield – can be reduced to a single, all-embracing theory. This led Moltke to blunders in predicting enemy reactions. Believing they would comport themselves according to the exigencies of military logic, he overlooked the other, more human factors.

He was a gifted man rather than a military genius and his gifts were of two kinds. First, he recognized (as the French General Staff failed to) that the nature of war had changed. The battlefield was no longer to be ruled by the will of a single commander who, like Raglan at Balaclava, overlooked the entire action from his vantage point as if it were taking place on chessboard, sending out his aides with orders to this or that commander. The lengthening range of weapons increased the distance between the contenders.

If the Commander in Chief was to impose his will, he must do it before battle was joined. It had to be imprinted on the minds of subordinates.

His other gift was for sheer organization, for painstakingly meticulous preparation. The victory of Sedan was no strategic tour de force; there are those who claim the result could not have been otherwise. History shows they are wrong. One has only to think of the overweening confidence of the French chivalry before Agincourt or of the Russian before the Russo-Japanese War. Moltke did not throw away his advantages as they did. He did all that was required, making full use of his numerical superiority and the longer reach of his guns.

It might be added that he was the mind behind the entire campaign whose objective was to harry the enemy, driving him to inevitable, irreparable error. This the French made when they decided to make a stand with their whole army at a place where they could easily be surrounded. The French situation – an unpopular regime, embarked on an ill-considered and unpopular war and desperate for victory, and, when that eluded them, desperate to mask the magnitude of the catastrophe – made its great contribution to Moltke's victory.

HINDENBURG
AT TANNENBERG 1914

The reputation of Field Marshal Paul von Beneckendorff und von Hindenburg, which was to carry him to be Commander in Chief of the German Armies in the last two years of the First World War, was made by a single victory. When he died in 1934, by that time president of his country and having seen Hitler come to power, a grandiose mausoleum was erected on the battle's alleged site as a resting place for his remains.

Tannenberg was a significant victory, a blow to the enemy gut from which he was never to recover. The question is how much was all this due to Hindenburg?

In August 1914, Germany faced two enemies. To the west there was France. To the east there was Russia. The two were bound together by the 1892 treaty. The forces the latter could deploy were enormous: 6,800,000 men, divided into 114 infantry and 36 cavalry divisions. In comparison Germany's 87 infantry divisions, to cover both fronts, seemed paltry.

The answer to the predicament of a two-front war had been provided by General Count Alfried von Schlieffen, the German strategical genius and the Army Chief of Staff from 1891–1905. The Russian Army would take months to mobilize its forces because of its great size. Germany's plan must, therefore, be first to defeat the more readily mobilized western armies — and he provided the blueprint for doing just this — and then to turn in full strength on the eastern one.

If the Russians were able to mobilize quickly enough, there was only one place where they could mount an attack. This was East Prussia, which stuck out into Russian Poland. There was only one way they could attack it; by advancing one army in a westerly direction from Lithuania and a second in a northerly direction from Poland. For most of their approach march these two armies would be separated from one another by the Masurian Lakes in the southeast corner of East Prussia. The German defenders would simply strike with all available strength at the first Russian army that came within reach, incapacitate it, and having the advantage of superior internal communications, switch their forces to turn on the second.

When war came, the Germans felt they had greater reason for confidence than they had had in 1905 when von Schlieffen's plan was adopted. Now the southern part of the Russian front was guarded by their Austrian ally with its 49 infantry divisions. Their commander, Conrad von Hötzendorf had great offensive plans of his own, which should have kept the Russians busy.

This being so the High Command felt few qualms about entrusting the defense of the region to a single army, the Eighth. Numbering some 200,000 men, it consisted of three corps of regulars and one of militia or *Landswehr* drawn, in accordance with German practice, from men of the locality so that they had a vested interested in its defense.

However, there was one factor the Germans failed to take account of simply because they knew nothing about it. Under a secret protocol added to the original Franco-Russian Treaty of 1892 was a Russian undertaking that if war broke out they would field an army of 800,000 men by the fifteenth day of mobilization, while the French would respond in the west with a force of 1,300,000.

The Russians can only have made this commitment tongue in cheek. The size of the army was a reflection of the size of the country, across whose length and breadth its reserves were scattered and it was notoriously ill-served by railroads and roads, while the system of mobilization was of Byzantine complexity.

It was events in the west which compelled the Russians to keep their word. In Belgium and France, things had gone desperately awry. An attempted offensive toward Germany by the French armies of the right had failed, while the German forces threatening their left were nearly twice as great as their peacetime intelligence estimates had made them.

Through its ambassador in St Petersburg and its military mission at Russian headquarters in Baranovichi, the French demanded that the Russians fulfill their treaty obligations. On 6 August the Russian Commander in Chief, the Grand Duke Nicholas Nicholaievich, the Czar's uncle, ordered the armies of his Northwestern Front, facing East Prussia, to launch an attack to relieve their ally, though 'only when sufficient strength was available.' By 14 August, under pressure from increasingly plaintive French appeals, this caution was omitted and immediate moves were ordered.

The Northwest Front commander, General Yakov Grigorievich Zhilinski, had by chance been the man who had signed the secret protocol. Now it was to be his misfortune to have to fulfill its stipulations.

The Russian armies took their names from the towns in which they concentrated and Northwest Front had two at its disposal. The 'Vilna' or First Army, under Pavel K

Czar Nicholas II of Russia.

Rennenkampf and the 'Warsaw' or Second Army, under Alexander Samsonov. Rennenkampf was a dashing cavalryman with a roving eye, who had earned an international reputation in command of a force during the Chinese Boxer Rising. Samsonov was a solid, good-natured, competent, but unimaginative officer. The two loathed one another — to such an extent that during the Russo-Japanese War of 1904 they had come to blows in front of the assembled military observers of the neutral nations at Mukden railroad station, hardly a promising basis on which to embark on a battle which would require the closest possible cooperation.

The Russian plan was the same as the one von Schlieffen had foreseen. Rennenkampf's Vilna Army would advance in a westerly direction to engage the enemy frontally; Samsonov would advance on a northern axis to cut off its retreat behind the Vistula, and with luck envelop it. The only way to circumvent the Germans' execution of Schlieffen's response was by exact timing. The Second Army must arrive on the battlefield as the defenders were reeling back from the blow delivered by the First.

That this was unlikely to be achieved in practice was amply demonstrated in trial runs. Samsonov had the greater distance to traverse and this through terrain which, with its forests, marshes and lakes, lent itself to defense. while the Germans had improved on nature by building railroads and paved roads for the quick movement of troops.

The Russians were proceeding with such haste that some of their corps were not up to strength, while there was a shortage of ammunition for the big guns, themselves in short supply, and of such other necessities of modern war such as telegraph-wire for signalling. Not even the field-bakeries and kitchens on which they prided themselves were available in sufficient numbers to feed the forces they had to support.

Nonetheless, with the French growing daily more restive, Rennenkampf began crossing the East Prussian border on 17 August — within three days of the Grand Duke's order. This was not, in fact, his first incursion. Jealous to retain his reputation for dash he had sent Cossack raiding parties into the area in the first week of the war.

The results had been two-fold. The German High Command might be prepared to view with equanimity the sacrifice of a border village or two, perhaps even of a small town, in pursuance of its greater strategic objectives. However, this took no account of the feelings of the sacrificial victims. The Germans, and particularly those living near Russia, had a mortal terror of the 'Mongol hordes,' as if they still lived in the time of Genghis Khan. At the first sight of Rennenkampf's wild horsemen many of them fled. East Prussia was the Junker heartland. The first Hohenzollern, the dynasty currently ruling the country, had been crowned at Königsberg (now Kaliningrad) and East Prussia contained the hunting forests in which the Kaiser annually enjoyed himself, so that he was familiar with many of the leading local families. Some of them had reached Berlin and were begging for prompt and effective succor. Helmuth von Moltke, the German Commander in Chief, found himself besieged with demands that he move troops from the Western to the Eastern Theater at the very time when events in France were taking a turn for the worse.

The other result was one which worked to the detriment of the Russians. In the process of decamping, the population near the frontier took with them anything movable, including telegraph poles and railroad rolling stock which the invaders desperately needed. Rennenkampf found himself moving into a region totally deserted and offering little on which his troops could subsist. He advanced, regardless. His was thus the 'first Russian army to come within reach' according to Schlieffen's prescription.

Command of the German Eighth Army was exercised by General Max von Prittwitz und Gaffron, known for his corpulence. He owed his position less to proven military skill than to the stock of scurrilous stories about his brother-officers with which he kept the Kaiser entertained. However, he had on his staff the leading German expert on Russian affairs, Colonel (later General) Max Hoffmann. Scarcely less corpulent than his commander, he was also ambitious, intelligent and possessed of a penetrating wit which often made him enemies.

Prittwitz's response was to follow Schlieffen's textbook. He dispatched the greater part of his strength far forward to make a stand before the town of Gumbinnen (now Gusiev). There the XVII Corps, under August von Mackensen, would attack the enemy center, while the I Corps under Hermann von François, would reach round his flank. A single corps, the XX, under Scholtz, was sent south to hold or delay any enemy forces which advanced from Poland.

Both François and the troops he commanded were East Prussian. Inflamed by patriotic feeling, he decided of his own accord to push still further forward, to Stallüpönen, only five miles from the Russian border. There he engaged the First Army and, by the end of the day, sent it back across the frontier. Although his disobedience had had the result he intended, it had also upset Prittwitz's intention of a defense before Gumbinnen and could have led to the other units being dragged forward, too. He therefore ordered

François back and, by pure chance, as they had made no attempt at reconnaissance, the Russians struck on the next day.

The I Corps' flank attack was such a success that Rennenkampf contemplated going back over the border once more. He was saved by the failure of Mackensen's attack on his center. The troops of XVII Corps actually collided with the Russian artillery which fired salvo after salvo into them. The more determined units were torn to shreds; the less determined broke in panic and fled. The attack had to be called off, the Germans retired and the town of Gumbinnen fell.

Rennenkampf, though he did not appear to have realized it, had won a strategic victory. There was, however, much

Top: Hermann von François was a brilliant commander but was not very good at following orders.

Top far left: General Pavel Rennenkampf.

Far left: General Alexander Samsonov. Samsonov and Rennenkampf were supposed to act together at Tannenberg but their mutual antagonism totally jeopardized the Russian position.

Left: The mobilization of Russian troops at the outbreak of World War I.

Left: The Czar addressing his troops in 1914.

Right: Major General Erich Ludendorff, the strong man behind Hindenburg.

Below right: Russian dragoons cross Tilsit on their way through East Prussia.

Below: Supplies for troops in the Front at Warsaw Station.

crowing in St Petersburg and a parade of captured guns. Having court connexions he was anxious to maintain, he even sent a German machine gun to the Czar, who put it on display.

After this blow, Prittwitz received a second: another Russian force was advancing from the south. This, of course, was Samsonov's Warsaw Army. The summer of 1914 had been exceptionally dry and the terrain through which he was moving, which in winter could be a quagmire, was now dry, ashy sand into which the men's boots sank, while the guns and their limbers were continually bogged down. There was a shortage of food and forage because, as had happened on Rennenkampf's front, the first approach of the enemy had led to a mass exodus of

local population, taking everything they could with them. All the same Samsonov was keeping to schedule, though at punishing cost to his troops, and Scholtz's XX Corps was being pushed steadily back, leaving the deserted little country towns to the Russians.

Prittwitz was faced by the fact that while his action at Gumbinnen had failed, Samsonov's army was likely to fall on his flank at any moment. Before taking up his appointment, Moltke had told him that he must at all costs avoid being locked up in the fortress of Königsberg, but that he might in extremity fall back behind the Vistula. In the deepest depression, therefore, he telephoned High Command at Koblenz and told them that this was what he now proposed to do. However, he doubted his ability to hold

the river line as the summer's drought had reduced its waters to a mere trickle.

Moltke, already under pressure to reinforce in the east at the expense of the west, came off the telephone no less depressed. He had been opposed to Prittwitz's appointment in the first place and saw his misgivings justified. He first ordered one of his own aides to call the various Eighth Army corps' commanders and find out whether they shared Prittwitz's pessimism and then himself gave thought to the matter of a replacement, a man to save a critical situation.

His mind turned at once to a Major General Erich Ludendorff, commanding a brigade in Belgium, who when the advance was delayed by the resistance of the fortress of Liège, had captured it by the simple expedient of knocking

on the front door and demanding its surrender. The problem was that Ludendorff had no title and this might make for difficulties in his dealings with fellow officers and subordinates who had.

Then his mind turned to Hindenburg, on the retired list but still only 67. He was admittedly neither a particularly distinguished nor a particularly intelligent soldier, but if made nominal commander he could, at least, supply the 'von' while Ludendorff supplied the brain. He forthwith set about reaching Belgium.

In the meantime, a deep discussion had been going on at Eighth Army headquarters. While Prittwitz was on the telephone, Hoffmann suggested a solution to their current problem. After the losses at Stallüpönen and Gumbinnen, Rennenkampf would be in no position to pursue in strength, he believed. This being so, the defenders' response should be to leave only screening forces in front of Rennenkampf and, making use of their excellent communications, swing the weight of their force over to Samsonov's front, defeating him before he was past the barrier of the Masurian Lakes and could form a continuous line with the First Army.

An attack on Samsonov, however, might cause Rennenkampf to react at whatever cost to himself in order to save his brother-in-arms. But Hoffmann had been a witness of the famous fisticuffs on Mukden station and knew this would not happen.

The proposal was put to Prittwitz when he came back into the room. At first dubious, he was gradually won over to it, and orders amending those sent out earlier were dispatched. Unfortunately, no one thought of telling Koblenz of the changed arrangements. There, the telephone calls to Prittwitz's corps' commanders had been completed and the consensus view was that although serious, the situation was not yet desperate.

Ludendorff, meanwhile, had been contacted and was on his way to headquarters in a fast car. Hindenburg had also

been reached, but it was symptomatic of Moltke's attitude that he had not bothered to bring him to Koblenz for briefing.

After his interview with Moltke and another with his Supreme War Lord, Kaiser Wilhelm II, Ludendorff entrained for the journey east. At Hanover, he was joined by Hindenburg, the first ever meeting between the two men who were later to be so closely associated that they were to speak of their relationship as 'a happy marriage.'

Solely to undermine Prittwitz, Ludendorff's only action had been to telephone the Eighth Army commanders from Koblenz and tell them to act independently until his arrival at headquarters.

There Prittwitz was in a positively cheerful mood. With no sign of movement on Rennenkampf's part, the transfer southward to forestall and defeat Samsonov had begun. In fact, Hoffmann's analysis had been perfectly correct. Not only had the First Army suffered 20 percent casualties, but what was worse in view of its shortages, it had fired off the allocation of artillery shell for the entire campaign in the Gumbinnen battle. Unable to pursue, Rennenkampf, for reasons of his own, had omitted even to send out scouts and so had no idea where the enemy was making for.

In the early afternoon, Ludendorff and Hindenburg arrived at Marienburg where the Eighth Army headquarters had been set up. Prittwitz, who had no knowledge of his own replacement until that moment, shook hands with his fellow officers and gave up his command to the newcomers, in Hoffmann's words 'without any complaint at his treatment.'

Told of the new orders, Ludendorff simply confirmed them, though, according to the evidence not without a great deal of persuasion. By the following day, the 24th, he seems to have conquered his apprehensions. For one thing, he now had an insight into Russian intentions. Short of telegraph wire, Rennenkampf was forced to use radio to promulgate his orders and, though sent in code, these were of such a puerile nature, that the Eighth Army cryptographers had no difficulty in deciphering them. These confirmed that he was not going to move.

Instead of striking at only one of Samsonov's flanks and preventing the junction with Rennenkampf, as had been proposed earlier, Ludendorff decided to strike at both and the pace of the transfer from the First to the Second Army front was increased accordingly and the attack's scope widened.

Samsonov was now encountering increased opposition and duly reported this to Northwestern Front command. There Zhilinski, convinced that the greater German forces were retreating before the First Army, accused Samsonov of cowardice and urged him to hurry as his slowness was endangering Rennenkampf – a suggestion which contained, of course, not an iota of truth. Zhilinski continued in his mistaken belief about relative German strength on the two fronts even after a Russian flyer had reported a

massive movement toward the south in which every conceivable form of transport, including taxicabs, was being used.

On the night of 24th the Germans had a momentary shock when Rennenkampf again began moving. However, from the decoded radio messages it soon became clear that instead of advancing westward, he was veering north and toward Königsberg where he believed the enemy to have gone. This was taking him away from Samsonov and, by widening the gap, making the possibility of a link up more difficult.

Samsonov was facing a dilemma of his own. As the forces in front of him increased, the need to find Rennenkampf's left flank, which he believed would be stretching out to him to form a single line, became ever more urgent. At the same time, he was being impelled forward under Zhilinski's orders.

As his right was groping in the dark for Rennenkampf,

Left: Russian heavy artillery unit in 1914.

Below: German troops advancing through East Prussia to the Russian border to protect their homeland.

Above center: A German defensive position on the shores of a Masurian Lake.

Right: German infantry in a defensive position during the Battle of Tannenberg. They have built trenches in a farmhouse.

Far right: Russian troops pressing home an attack.

Map legend

- FRONT LINE, EVENING 25 AUG 1914
- RUSSIAN ATTACKS
- GERMAN MOVEMENTS
- GERMAN FORTIFIED POSITIONS HELD DURING RUSSIAN ADVANCE
- MAIN RAILWAYS
- OTHER RAILWAYS

MILES 0 — 50
KILOMETERS 0 — 80

Königsberg Garrison
KÖNIGSBERG
LABIAU
I Corps
TAPIAU
INSTERBURG
STALLUPÖNEN
XX Corps
1 Cav Corps
III Corps
18 Aug
GUMBINNEN
21 Aug
First Army (Rennenkampf)
Crossed border 17 Aug
IV Corps
1 Cav Div Remained here
ALLENBURG
XVII Corps
I Res Corps
NORDENBURG
GOLDAP
Angerapp
BRAUNSBERG
3 Res Div
ANGERBURG
1 Cav Div
I Corps to Usdau
BARTENSTEIN
SUWALKI
G E R M A N Y
Alle
Eighth Army (Prittwitz, then Hindenburg)
RASTENBURG
LÖTZEN
Masurian Lakes
E A S T   P R U S S I A
AUGUSTOW
WARTENBURG
SENS-BURG
BISCHOFSBURG
LYCK
ALLENSTEIN
3 Res Div
NIKOLAIKEN
II Corps
21 Aug, transferred from Second to First Army
OSTERODE
JOHANNISBURG
HOHENSTEIN
GRAJEWO
TANNENBERG
ORTELSBURG
OSOWIEC
R U S S I A
XX Corps
WILLENBERG
4 Cav Div
USDAU
NIEDENBURG
VI Corps
XIII Corps
SOLDAU
XV Corps
Second Army (Samsonov)
XXIII Corps
Crossed border 21/22 Aug
I Corps
MLAWA
OSTROLENKA
Narew
P O L A N D

FRONT LINE, EVENING 25 AUG 1914
" " " 29 AUG "
RUSSIAN ATTACKS
GERMAN COUNTERATTACKS
RUSSIAN RETREAT
GERMAN FORTIFIED POSITIONS
HELD DURING RUSSIAN ADVANCE

MILES
0   30
0   50
KILOMETERS

First Army
(Rennenkampf)

II Corps

IV Corps

NORDENBURG

ANGERBURG

Angerapp

Alle

BARTENSTEIN

1 Cav Div

RASTENBURG

BISCHOFSTEIN

LÖTZEN

Masurian Lakes

XVII Corps

G E R M A N Y

Eighth Army
(Hindenburg)

I Res Corps

BISCHOFSBURG

SENSBURG

NIKOLAIKEN

ALLENSTEIN

BARTELSDORF

VI Corps

E A S T

P R U S S I A

OSTERODE

III Res Corps

HOHENSTEIN

PASSENHEIM

4 Cav Div

JOHANNISBURG

ORTELSBURG

FRÖGENAU   XX Corps

KURKEN

JEDWABNO

TANNENBERG

XIII Corps
XV Corps
XXIII Corps

ORLAU

WILLENBERG

USDAU

NIEDENBURG

I Corps

LAUTENBERG

I Corps
SOLDAU

Second Army
(Samsonov)

R U S S I A

MLAWA

P O L A N D

Left: Russian casualties after Tannenberg.

Right: Russian troops surrender. Some 92,000 prisoners were taken by the Germans.

Below left: Hugo Vögel's painting of the German commanders at Tannenberg. General Hoffman is looking through a rangefinder and to the right stand Hindenburg and Ludendorff.

it was becoming further and further separated from the rest of the army. Concerned about the thinning in his line this was leading to, he therefore ordered the VII Corps, on his extreme right to close up. He then realized, after the order had gone out, that this would leave his right flank open, and he was forced to countermand so as to make it a flank guard.

His communications were in an even more parlous state than Rennenkampf's. Not only was he lacking in telegraphs, but each of his corps was using a different code and the First Army headquarters did not have all the requisite keys. He was compelled to send out his orders by radio and *en clair*. They were monitored without trouble by the enemy.

Soon after his VII Corps had begun retracing its steps, it ran into a German force. The only explanation of its presence the Russians could think of was that it must be a scattered unit in retreat from Rennenkampf. This could only mean the two Russian armies were about to meet.

It was nothing of the kind. It was part of Mackensen's XVII Corps which was beginning to reach Samsonov's front. As the battle flared up along the VII Corps' line, the corps to its left was embroiled. At the same time, Scholtz's XX Corps, which had turned from retreat to attack, was holding Samsonov's center, making it impossible for it to give assistance to the struggling units to its right. By nightfall, a breach six miles wide gaped between Second Army's center and the Masurian Lakes, and Mackensen's men were beginning to sidle round it to reach the Russian flank.

The Russian I Corps, under Artamanov, was sent to fill the gap but arrived too late and fell back across the frontier under withering German artillery fire. A third German corps was now reaching the south. This was François's I, which had the greatest distance to travel. Insubordination among German commanders was a malaise which had afflicted the army from the time it was formed under the united flag of the North German Confederation in the time of the first Moltke. François had already given a demonstration of it at Stallüpönen. Now, under orders to attack the Russian left wing at Usdau, he made a token gesture in that direction before wheeling to strike at their flank at Soldau, his excuse being that his artillery had not arrived and this made a frontal attack too hazardous and costly. His action was later to be justified since it was actually to seal the fate of the Second Army and of its commander.

The struggle had now been raging for two days. On the 27th, it reached its climax. The artillery of the I Corps arrived and François began an all-out assault at Usdau. The Russian troops, tired from their long, arduous marches and hungry from lack of supplies, fell back. A pursuit was ordered as far as Neidenburg, this placing the I Corps in the Russian rear, a move checked, though only temporarily, by a counterattack.

The brutal truth was that more than a counterattack was needed. The only way in which the Russians could extricate themselves was by a hasty withdrawal of their forces back to the border.

Samsonov did the opposite, either because he was still smarting from accusations that his own tardiness was endangering Rennenkampf or because while the First Army was advancing, as he still believed it to be, he did not want to be the one to instigate retreat. With his left folded back on itself and his right under threat he continued to push his center forward. It soon came into contact with Scholtz's XX Corps once more. It succeeded in pushing it back, but with severe loss and at the cost of further fatigue to the men.

Nonetheless, news of this repulse to the XX Corps so unnerved Ludendorff and Hindenburg that they considered abandonment of the entire plan and even sent François instructions to withdraw back to support Scholtz.

In fact their advance had led the Russians deeper into the trap, as the I Corps' commander was quick to point out.

It is true that at this stage Samsonov seemed to have realized that there was no alternative to retreat. His troops, however, were already taking the law into their own hands and first in small bands, then in ever growing numbers, were finding their way back across the frontier.

Their escape route was not open to them for long. Having reached Neidenburg, François' troops pushed further east to Willenburg, setting up a line of outposts across the Russian rear. The area was largely one of forests, totally unfamiliar to the invaders who, lost and confused among the trees, found themselves under machine-gun fire from the I Corps' picket line whenever they emerged from cover.

They had stood as much as was humanly possible and, on the whole, had fought with exemplary if futile courage.

Now, they began giving themselves up in their thousands.

Riding out to discover for himself the exact situation, Samsonov got caught up in the tide of retreat. Then, with his immediate staff, he got lost in the forest himself. In the confusion, he was able to slip away, unmissed, until a single shot was heard. His companions found their Commander in Chief had shot himself.

Yet even at that stage the Russians' position was not totally lost. Artamanov's I Corps which had retreated over the frontier, was now on its way back, reinforced. In any case, Mackensen's XVII Corps had not yet joined with François in the Russian rear, leaving an escape route still open, had the Russians been able to fight their way out. There was still Rennenkampf who, had he attacked vigor-

ously, would have found the way to the heart of East Prussia open to him.

None of these things happened. In consequence, Samsonov's army suffered an enormous defeat, one of the worst in military history. Of its five corps, two and a half were annihilated. Ninety thousand of its men were taken prisoner. Total Russian casualties in the first East Prussian campaign were 310,000 men and 650 guns, the equivalent of seven and a half corps.

At the height of the battle, the Eighth Army Staff had set

A Russian field battery after the battle. The three corps which were encircled during the battle included Martos' XV Corps and Kliouviev's XII Corps which were considered among the best in action.

Left: The ruins of the market place at Ortelsburg, in the heart of the Masurian Lakes area. The town was on the escape route of the VI Corps.

Right: Hindenburg's victory brought him the supreme command of the German Army in 1916. This pictures was taken shortly after his appointment and shows him with Kaiser Wilhelm and Ludendorff at the German headquarters at Spa.

Below left: A street in Hohenstein after the expulsion of the Germans.

up temporary headquarters in a farmhouse at Frogenau, a few miles from where the battle was being fought. Someone pointed out that not far away was the village of Tannenberg. Here in 1410, the Teutonic knights had suffered a great reverse. This, surely, was the revenge. Accordingly, the great defeat of Samsonov's army was called the Battle of Tannenberg.

By his inactivity while his old enemy was writhing, Rennenkampf had merely ensured his own ultimate defeat. For, with the threat of the Second Army removed, the Germans could, *à la* Schlieffen, turn on him in full strength. This they shortly did, pushing him back across the frontier from where his Cossack bands once more began their forays, though there was little enough for them to find.

Even before this, recriminations against the First Army Commander had begun. In a play on his German name, he was rechristened 'Rennohne-Kampf,' ran-without-fighting. There was even a suggestion of treason, which, of course, was totally without foundation. All the same, in contrast with the speed with which Prittwitz had been replaced, it was not until the early November that the Czar confirmed his dismissal as commander.

Yet, taking the war overall, the defeat of Tannenberg was not without effect. Two days after his appointment of Hindenburg and Ludendorff as saviors of East Prussia, Moltke was still wondering if he had done enough. Two corps on his Western Front — the XI and Guard Reserve — seemed underemployed, since both the French and British forces in front of them were in full retreat. He sent them east.

On 6 September the Franco-British forces, taking advantage of the German offer of their flank along the Marne, swung from the defensive to the offensive. The British

Expeditionary Force found itself advancing into a 30-mile gap. It had been left by the removal of the troops sent to Ludendorff. It was German apprehensions about this gap which in the end persuaded them to disengage and retreat behind the Aisne, removing the threat to Paris.

Just after Tannenberg the French liaison officer at Russian headquarters, who had done so much to spur them to adventure, commiserated with the Grand Duke who, with breathtaking chivalry, told him, 'We are happy to make such sacrifices for our allies.'

What of Hindenburg's part in the battle which is regarded by the German people as his special victory. All the moves toward it had, as we have seen, been taken before he even entered the scene and had been confirmed by Prittwitz. Both Hindenburg's account and that of Ludendorff make it very clear that they had considerable trepidation about the battle. Certainly, the latter was prepared to abandon the whole enterprise when a Russian threat appeared.

The Eighth Army staff were, naturally, perfectly well aware of the truth. Hoffmann best sums up their feelings. After the war, it would sometimes fall to his lot to show awed visitors over the house at Frogenau which had served as headquarters. Taking them into one room he would tell them, 'This is the room in which General Hindenburg slept before the Battle of Tannenberg.' And into another, he would say, 'And this is the room in which General Hindenburg slept after the Battle of Tannenberg.' And into a third, 'And this is the room in which General Hindenburg slept during the Battle of Tannenberg.'

Rarely can a general have slept so greatly to his country's and his own benefit.

JOFFRE
AT THE MARNE 1914

To the modern mind, accustomed to protracted wars ending only when one side has been bankrupted and physically devastated, the idea that a struggle between two great nations could be decided in a mere month or two seems hardly credible. Yet before 1914, the long war was thought to belong to a past growing ever more remote. Countries had their standing armies, usually backed by a large pool of reserves; their General Staffs drew up plans of campaign, periodically revised, which envisaged two or three battles culminating in a final and decisive one. In theory, that was to be that. In practice, what is more, it sometimes was.

The idea that the army and its first-line reserves might serve only to give time for the economy to be geared to war and for enormous armies, drawing on a nation's entire manhood, to be mustered had occurred to none. There was nothing outrageous, therefore, in the German plan to defeat France in the six weeks between 1 August and 15 September 1914. Indeed, they would be improving only slightly on their 1870 performance. A quick, decisive end was also a necessity, for they would be facing not one but two enemies. To their east was Russia, bound to France by the Treaty of 1892. The western foe had to be neutralized in the period of grace granted while the unwieldy Russian war machine creaked into action.

The plans for achieving this had been drawn up by General Count Alfried von Schlieffen, Chief of the German General Staff from 1891 to 1905. What he proposed had the merit of originality. Instead of the obvious route of an advance across the Rhine, as in Moltke's plans of 1870, he advocated the longer one through Luxembourg and Belgium. The armies would march almost to the coast before wheeling west to advance into France with the extreme right, as Schlieffen put it, 'brushing the Channel with their sleeves.' In this way, like the trunk of an enormous elephant, the German phalanxes would reach round the waist of the French defense, pick it up and, presumably, deposit it lock, stock and barrel in the German prison cages.

The French were, as a matter of fact, aware that the Germans might have some such intention, but, though they had hardly distinguished themselves in recent wars, felt they could face the next with confidence. For one thing, they had as allies not only massive Russia, but also Britain which besides its command of the seas, had an indubitably successful, if comparatively small army.

For another, the French had not merely a counterplan, but an entirely new military philosophy. The plan was the acme of simplicity: their army was simply going to march into Germany by the shortest route, advancing across the old battlegrounds of 1870, liberating the lost provinces of Alsace and Lorraine, and crossing the Rhine. In sum, while the Germans advanced their left, the French would hurl themselves on their right. The more they strengthened the one, the more they would weaken the other — *et voilà*, the road to Berlin would lie open.

What transformed this mere intention into something different was their new found philosophy. This, with equal simplicity, stated that the absolute key to military success was *l'attaque*. The next war would be won by *l'offensive à l'outrance*, offensive to the uttermost. The French army, resplendent in blue tunic and red trousers, inspired by their

General Joseph Galliéni, Military Governor of Paris.

sword-brandishing officers, would advance without consideration of losses, and would overcome by sheer élan.

As the French response to the Schlieffen Plan, it was given the sternly businesslike title of Plan 17 and in July 1911 its execution, should necessity arise, was entrusted to General Joseph Jacques Césaire Joffre, then 59. This paunchy figure with a thick white moustache, who looked like a healthy Norman peasant dressed in a general's uniform, had not been the first choice for the post. The man the government originally wanted was a very different man, Joseph Simon Galliéni, who had refused it on the grounds that as a colonial soldier his appointment might be resented by the Metropolitan army.

However, when necessity did arise, the French intentions suffered two terrifying reverses. First, all the enormous French courage of their *offensive à l'outrance* was sacrificed in a useless carnage which should have convinced all observers that in this new war the weapons of defense — machine gun and barbed wire, to say nothing of fire power — were more potent than the most indomitable will in *l'attaque*. Within a short time, the struggle resolved itself into one of attrition of a kind to become familiar along the whole front. Its effect on German intentions was virtually nil. By the end of the month, the French First and Second Armies, their flower destroyed, were holding a line north of the Moselle in what was rapidly becoming a secondary theater. Worse still, the advance of the German right which was to have been little more than a march into French custody took on an entirely different character. For one thing the invaders were nearly twice as numerous as French intelligence estimates had made them.

In a desperate effort both to succor their Belgian ally and to halt the rolling German masses, Joffre ordered the Fifth Army under Lanrezac to strike at the enemy flank as it moved toward the French frontiers. On 22 August Lanrezac was caught at Charleroi, defeated and thrown back. To maintain continuity of line, the two French armies next to him, the Fourth and Third also had to retreat. On 24 August a million German troops began crossing the borders, a line of reapers 75 miles in width consisting of four armies. In the way of their advance, like a clump of poppies in a harvest field, was the French capital, Paris.

Later to be credited with a policy of 'masterly inactivity,' on this occasion Joffre reacted with laudable promptness. On 25 August the armies facing the German advance were to undertake a slow retirement, seeking as they did so to harass the enemy and delay his advance. At the same time a Sixth Army would be formed from troops railed from Lorraine to take their place on the left, along the Somme, a position which should pose them on the flanks of the invaders. It was planned to achieve this on 2 September – Sedan Day – which would mark the start of a general offensive.

There was a snag. To the left of Lanrezac's Fifth Army were the four (later five) divisions of the British Expeditionary Force with only a thin screen of French cavalry and territorials between them and the English Channel. Liaison between the British and French armies was almost

General Alexander von Kluck, Commander of the First Army.

Sir John French, Commander in Chief of the British Expeditionary Force.

non-existent. The Commander of the BEF, Sir John French, deeply suspicious of his ally, unable to pronounce their place names, let alone speak their language, still had ringing in his ears the instructions given to him when he left Britain: he must not allow his army to be sacrificed.

When, on 23 August, the day after it had taken its place in the line, the BEF was surprised on the march by the Germans miles forward of where they were thought to be, the British held their own at the Battle of Mons but, totally outnumbered, were forced to retire. Even more alarming, Sir John discovered that instead of forming part of a solid line with the French Fifth, he was actually on his own and Lanrezac was in full retreat after Charleroi. If, as he suspected and perhaps half-hoped, 1914 was to be a repetition of 1870, it was now his overriding duty to save his forces. While it is untrue, as the French alleged and continued to allege, that he began forthwith to make for the Channel ports, he certainly decided to set as great a distance as possible between himself and the advancing foe.

He had, of course, totally misread the situation. What was now happening was the reverse of 1870. Then total defeat had largely been due to the fact that for political reasons the French armies had failed to retreat and maintain a line when military necessity dictated that they should have done so.

On the German side, General Alexander von Kluck, commanding their First Army on the far right, had come to a similar conclusion to Sir John, but in his estimation, not only the French, but also the British were worsted, and in this belief he pushed his men relentlessly forward.

They pursued with such speed that on 26 August they caught up with the BEF's II Corps, under Smith-Dorrien, at Le Cateau. Compelled to give battle, the corps mauled their tormentors, but only at the cost of 20 percent casualties and further retreat.

Since Kluck was able to continue his advance, the flow

of euphoric communiqués from his headquarters went on uninterrupted. These, however, were being received with increasing skepticism by the German Commander in Chief, Helmuth von Moltke, nephew of the hero of Sedan. As great a pessimist as Kluck was an optimist, Moltke was asking himself why it was, since the Franco-British forces were supposed to be retreating in disorder, there were so few prisoners and so little booty. He had an uneasy suspicion they might actually be escaping his trap and luring him into one of their own.

It must be remembered he had other causes for anxiety. There had already been delays and modifications to the original plans. First, the line of fortresses on the Belgian frontier had proved more intractable than had been supposed. Then, the Belgians themselves had failed to collapse before the Germans' might and there were rumors they were being reinforced by the British. Though grossly exaggerated, these caused Moltke to detach some of his forces to stand guard over Antwerp. This reduced the breadth of the intended swing on the English Channel, which now had to be carried out further inland than Schlieffen had stipulated. On top of all this, came stories of Russian troops reaching France in numbers which increased with each retelling. Pure legend as they were, until denied, they were enough to unsettle still further Moltke's

never very strong nerves.

And there was worse. The armies of the Czar which, it had been confidently estimated, would require three months to mobilize, actually began invading East Prussia on 17 August. Schlieffen had not ignored this eventuality and had catered for it. What he had not catered for was the panic into which the arrival of the Cossacks in German territory threw the local population, leading to a demand for massive reinforcement of the forces there.

These doubts at enemy headquarters would, naturally enough, have been balm to the souls of the harassed British and French commanders, had they known of them. As it was their horizons were dominated by what they saw before them — a remorseless advance toward the capital.

The question of the moment, therefore, was Paris and how it should be treated. We know that militarily Joffre viewed Paris as just another city 'comme les autres,' that he did not believe its loss betokened the end of the struggle. We also know, of course, that he was preparing the way for his own riposte to the German invasion and one in which Paris played no part. However, the sacrifice of the city was hardly likely to appeal either to France's popula-

The German armies swept through Belgium according to the plan laid down by Schlieffen. On 23 August 1914 they reached Mons and the British cavalry had to retreat.

tion or its leaders. Their misgivings were not merely emotional. The strategist might see Paris as *'comme les autres,'* but from the point of view of morale it was, after all, the capital. The effect of its loss would be devastating. There was one man who realized this.

When Joffre had been appointed Commander in Chief, his predecessor, Michel, had been offered as a sop the military governorship of Paris. The War Ministry had as little faith in him in his new post as in his previous one and on 24 August replaced him with Galliéni, Joffre's former rival to the post of *generalissimo*. Tall, austere, bespectacled Galliéni looked like an eccentric university professor among the French generals and because of his strangely variegated

uniform was regarded by the British as little more than 'a clown' – a judgment they, like others, were soon to revise. Now 65, recently bereaved of his wife, he was already suffering from the cancer which was to kill him within two years.

Though he believed that Paris could and should be defended, his acceptance of the post of governor had been conditional. To man the 100-mile perimeter round the city would require an army of three active corps, far more than its garrison could muster and this, in any case, was being milked by Joffre for his coming attack on the German flank. There followed a protracted struggle to persuade Joffre to part with some of the men he was husbanding for this, but the government had little recourse since it had voluntarily

Top: Prince Rupprecht of Bavaria was in command of the Sixth Army.

Top left: General Joseph Maunoury was put in command of the newly formed Sixth Army which was to protect Paris.

Left: German transports taking bridging equipment during the march to the Marne.

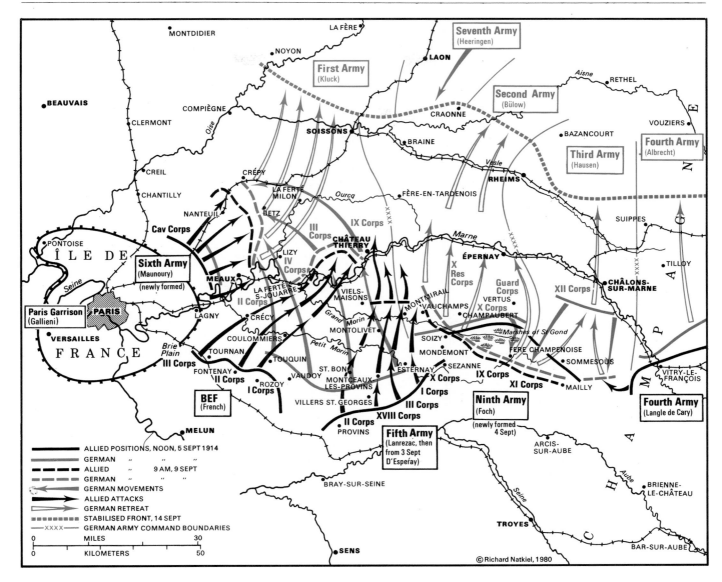

MONTDIDIER · NOYON · LA FÈRE · LAON
Seventh Army (Heeringen)
First Army (Kluck)
BEAUVAIS · COMPIÈGNE · CLERMONT · CRAONNE · RETHEL
Aisne
Second Army (Bülow)
SOISSONS · BRAINE · BAZANCOURT · VOUZIERS
CREIL · CHANTILLY · CRÉPY · **Third Army** (Hausen) · **Fourth Army** (Albrecht)
LA FERTÉ MILON · *Ourcq* · FÈRE-EN-TARDENOIS · RHEIMS · *Vesle*
NANTEUIL · BETZ · III Corps · IX Corps · SUIPPES
Cav Corps · CHÂTEAU THIERRY · *Marne* · TILLOY
PONTOISE · ÎLE DE · LIZY · ÉPERNAY · CHÂLONS-SUR-MARNE
Sixth Army (Maunoury) (newly formed) · IV Corps · X Res Corps · XII Corps
MEAUX · LA FERTÉ-S-JOUARRE · II Corps · VIELS-MAISONS · VERTUS · Guard Corps · X Corps · CHAMPAUBERT
Seine · LAGNY · CRÉCY · *Grand Morin* · MONTMIRAIL · VAUCHAMPS · *Marshes of St Gond*
Paris Garrison (Gallieni) · PARIS · MONTOLIVET · SOIZY · FÈRE CHAMPENOISE
VERSAILLES · *Petit Morin* · COULOMMIERS · MONDEMONT · SOMMESOUS
FRANCE · *Brie Plain* · TOURNAN · TOUQUIN · VAUDOY · ESTERNAY · SÉZANNE · VITRY-LE-FRANÇOIS
III Corps · FONTENAY · ST. BON · X Corps · IX Corps · MAILLY
II Corps · ROZOY · MONTCEAUX-LES-PROVINS · I Corps · XI Corps · **Fourth Army** (Langle de Cary)
BEF (French) · I Corps · VILLERS ST. GEORGES · III Corps · **Ninth Army** (Foch)
II Corps · XVIII Corps · (newly formed 4 Sept)
MELUN · PROVINS · **Fifth Army** (Lanrezac, then from 3 Sept D'Esperay) · ARCIS-SUR-AUBE · BRIENNE-LE-CHÂTEAU

ALLIED POSITIONS, NOON, 5 SEPT 1914
GERMAN "
ALLIED " 9 AM, 9 SEPT
GERMAN "
GERMAN MOVEMENTS
ALLIED ATTACKS
GERMAN RETREAT
STABILISED FRONT, 14 SEPT
XXXX GERMAN ARMY COMMAND BOUNDARIES

0 MILES 30
0 KILOMETERS 50

BRAY-SUR-SEINE · *Seine* · TROYES · *Aube* · BAR-SUR-AUBE
SENS · © Richard Natkiel, 1980

foregone much of its own power in military matters.

Among the lessons which the French believed they had learned from 1870 was that strategy and politics made poor bedfellows. Then, the armies had embarked on hopeless adventures simply because the politicians had overruled the commanders. To avoid a repetition of this, at the beginning of the present war France had been split in two. The Zone of the Interior was the province of the politicians; the Zone of the Armies was ruled by Joffre as by a dictator. The trouble was the latter was fluid. As the armies moved back so did their zone. By 28 August it encompassed Paris and within it the control of government and municipality had ceased. Thus Joffre refused Galliéni's requests on the grounds that a fight for the capital was no part of his plans.

The resolution of the dispute was due to the enemy. Kluck's army had been on the verge of victory over Smith-Dorrien's British II Corps when night fell. Under its cover, the British decamped with such efficiency the Germans were totally nonplussed. Aggrieved, as though he had been unfairly tricked, Kluck set off in pursuit. But where had they gone? Toward the Channel Ports, no doubt. It was, therefore, in this direction that he ordered his army to march. The British were not making for the ports at all; they were pulling back toward Paris. Far from finding his enemy, Kluck was moving away from them.

On the other hand, his route was taking him toward the French, for it was in this area that Joffre's Sixth Army, newly arrived from the northeast, was beginning its concentration. To avoid a head on collision, it too had to be moved toward Paris and so became the force Galliéni had been demanding.

Kluck's chase after British shadows was having another effect: it was widening the gap between himself and the army to his left, Bülow's Second. The Second Army had already run into trouble from Lanrezac's Fifth Army which had been ordered to turn and fight to reduce the pressure on the British. When Bülow's position became critical, Kluck wheeled to come to his aid. The move was made with the blessing of Moltke, who was worried about the fanning out of his forces as they had advanced deeper into France, and the thinning of the lines it had brought about. At the same time, he had himself contributed to this weakening as, in response to cries of help from East Prussia, he had sent two of the corps from his right to the east, although the attempted Russian 'march on Berlin,' never more than a hasty improvisation to aid France, had been stopped and one of their armies destroyed.

What he now realized was that these miscellaneous

Left: Taxis and cars were requisitioned in Paris to take troops to the Marne Front.

Right: French prisoners take a rest as they are marched from the front, 26 August 1914.

Below: Belgian cavalry and infantry evacuate Antwerp and fall back to join the BEF in early September 1914.

moves in the west added up to the entire abandonment of Schlieffen's grandiose plan. Instead of the broad sweep of the armies far to the north of Paris and its formidable defenses he was now forced to a compromise with something infinitely more modest. His Fourth, Fifth, Sixth and Seventh Armies would concentrate on that part of the French Army on the far side of Paris. The capital's garrison was to be held in check meanwhile by stationing Kluck's First and Bülow's Second Armies in line facing west.

The outstanding flaw in these arrangements was that Kluck was ahead of Bülow and advancing southward and so upon Paris. Considerable shuffling of enormous bodies of men would be necessary to comply with Moltke's orders and Kluck saw them as denying him the chance of decisive victory. There was, in any case, little time to make such complex redeployments.

In the meantime, Galliéni was preparing for what everyone in Paris and outside it expected: a second siege. In the days following his appointment he worked tirelessly to prepare, knowing he was short of men and that the capital's defenses were unready. Charges were laid on bridges, even those scheduled as works of art; every conceivable means of infiltrating the city was barricaded, including sewers and drains; ammunition was stockpiled as it came from the factories; cattle and sheep were brought in to graze in the Bois de Boulogne and in such other open spaces as the city offered.

Through this Joffre was single-mindedly refusing to allow himself to be diverted from his own plans by the city's needs. What he now wanted was to create a stable line and this, of necessity, required that the British cease their retreat. To persuade them to do so he paid a personal visit to Sir John French at the new headquarters he had established at Compiègne. His efforts were entirely wasted. The British Commander in Chief intended to retire to St Nazaire in the west for reinforcement and reequipping over the winter. While he did not make Joffre privy to these plans, he argued that his army was exhausted and reduced by its

losses and so would be unable to take its place in the line for some ten days. The losses, though small by later standards, were, it was true, considerable, but as to exhaustion, the British troops had had far more opportunity for rest than their French brothers-in-arms or, come to that, the Germans. During the time that Lanrezac's men had been struggling to relieve the pressure on their ally, the British Army had quite simply slept.

Nor did French's estimate of his condition tally with that of his field officers or the men themselves. Both believed they were at least a match for the Germans and were longing to show their mettle. If the consideration of any of these points caused Sir John any hesitation, he certainly did not allow them to weaken his resolve.

All the time the crisis was developing by the day. On 30 August Paris experienced the first of the daily air raids to which it was to become accustomed when a handful of bombs were dropped from the cockpit of a single-seater *Taube*, causing some casualties mainly round the Gare de l'Est area. Besides its more lethal spawn, it also deposited leaflets warning the Parisians that they were now besieged 'as in 1870' and had no alternative to surrender.

The message of the leaflets had very little effect, but it brought home – in view of the short range of the aircraft of those days – how near the enemy was. In fact, Kluck's advance guards were only some 20 miles from the city gates and this being so Galliéni warned the government it might be necessary for it to leave, though it was a further three days before the move was actually made.

Expectation of the coming battle for the city grew by the day and a kind of nervous tension, half-excited, half-apprehensive grasped its citizens or such as had not left. All realized it would affect fundamentally the course of the war, a view those who remembered events before and after the Siege of Paris could only confirm.

It was at precisely this moment, when France needed to draw together all her strength that relations with the BEF reached a crisis. In a dispatch to Lord Kitchener, the British War Minister, French was proposing nothing less than the abandonment of the struggle and a return home with his armies now some 150,000 strong. Horrified by the military and political consequences even the suggestion of such a move could have, Kitchener donned his field marshal's uniform and hastened through the night of 31 August–1 September toward Paris. The result of his meeting with Sir John French was to secure his undertaking to 'conform with the movement of the French Army' – and no more. In his interpretation, Sir John was given the widest latitude.

Infuriated by Kitchener's 'political' intervention, he actually interpreted it by continuing his retreat, though managing in doing so to keep some kind of tenuous contact with his French neighbors, at least initially.

So much for Joffre's efforts to secure a continuous and stabilized line as a springboard for his own offensive. In justification, it has to be said that French was convinced Paris would fall; that France would be defeated. Thereafter an angry – and victorious – Kaiser might well decide to turn on Britain. Then it would have reason to be grateful to him for saving the army.

The conviction that Paris was soon to fall was one shared not only by many of French's fellow-Britons, but also people in France. In a matter of 24 hours, Kluck's armies had halved the distance separating them from the city.

Then came a succession of indicators, at first hardly credited, which betokened a dramatic and unhoped for shift in German intentions. Between the day of Kitchener's visit and 3 September, it became increasingly plain that Kluck was, after all, going to bypass Paris. Instead of advancing in a southwesterly direction and so toward the city, he was veering southeastward. He was, of course, marching to take up his assigned position in Moltke's new plan.

On 3 September Maunoury, commander of the Sixth Army now part of the Paris garrison, reported that there were no Germans west of Paris along a line reaching out as far as Senlis. This was confirmed by British air reconnaissance.

That Kluck believed he could make such a turn under the noses of the enemy could only be accounted for by his conviction that they were incapable any longer of responding to the challenge his exposed flank offered them.

Galliéni, however, was not the man to miss such an opportunity and at once saw that, instead of merely defending the city, the men under his command were going to be in a position to strike a decisive blow. On 3 September he told Maunoury to begin cavalry and air reconnaissance and at 0900 hours on the 4th, ordered him to prepare for a move against the German flank. If this was to be more than a parry it required greater strength than he disposed,

Left: German officers take a break from the advance during September 1914.

Far left: BEF gunners attack.

Far left below: Shrapnel hits the 1st Middlesex First-Line Transport at Signy on 8 September 1914. Nine horses were killed and the man with hands raised was badly wounded in the incident.

Below: German Jäger troops use a mobile phone. World War I saw a revolution in military communications.

while, in any case, the large-scale offensive for which he thought the opportunity was presenting itself, would need the sanction of Joffre to whom he was subordinated.

He hurried to General Headquarters only to find the Commander in Chief in an intractable mood. Joffre was still considering his own plans for the blow to take place well to the northwest of Paris and the farthest he would go was to promise to think over Galliéni's proposals. Leaving his chief to this meditation, Galliéni next made for Compiègne and the British Headquarters where Sir John French was absent and he saw the even more cautious Chief of Staff, Murray. It was absolutely plain that nothing would shift the British from their planned withdrawal and they even went so far as to say that had they known of the condition of the French Army they would never have entered the conflict.

For Galliéni it had been a thoroughly depressing and largely wasted day. Or so it must have seemed.

Nonetheless, despite outward appearances, Joffre had not been entirely unmoved by the arguments presented to him. He went so far as to address a letter to Franchet d'Esperey, who had succeeded Lanrezac as commander of the Fifth Army, asking him, though without implying a sense of urgency, whether he thought his forces could join in a general attack with any hope of success. Franchet's answer was that they would be in no position to do so in less than 48 hours. In the meantime, they, like the British, were continuing a retreat which would take them even further from Paris.

Even though Joffre was coming round to Galliéni's view, there was a danger, increased with every passing hour, that he would be too late since the Sixth Army was advancing alone against a much stronger enemy. What was more, the armies on its flanks were actually marching away, leaving an ominous gap. The situation clearly gave the advantage to Kluck.

At this late stage Joffre began to move, as did, even if reluctantly, the British Commander in Chief. Galliéni arrived back at his own headquarters in Paris after his discouraging interviews with Joffre and Murray late on the 5th. He found a message from the former awaiting him. Permission for the counteroffensive was granted, but it was suggested

that it should be south, not north, of the Marne as Galliéni wanted. He rang Joffre and by sheer vehemence won him over. That night orders were issued for a general offensive to begin on the 6th. Joffre promised to obtain British co-operation in the attack.

The battle order on the eve of the Marne Battle ran, left to right: the French Sixth Army, whose sector began roughly on a line with Paris and extended southwest of the Ourcq to Lagny, then the BEF from Lagny along the River Grand Morin, then the French Fifth Army, north of the Seine and to the marshes of St Gond. The Fourth Army had originally stood next in line, but its tenacious defense of a sector south of the Ornain had caused it to draw upon its own left along the Marne to reinforce. To fill the gap thus created, Joffre had inserted a so-called Detachment Foch, named after its commander, General Ferdinand Foch, and this had become the Ninth Army, still commanded by Foch, whose sector ran for much of its distance southwest of the St Gond marshes.

Beyond the Fourth Army's sector, from the confluence of the Marne and Ornain rivers, that of the Third Army ran up to Verdun and then followed the French fortress line to Nancy.

With one notable absentee, all these armies from the Ourcq to the Meuse were to turn about and go over to the offensive next day. Along this extensive line, the crucial area was that on the French left up to the St Gond marshes where Kluck was exposing his flank. It was here that there was one army which apparently had no intention of participating in the attack. The BEF's front, though small, chanced to stride the point of junction between the German First and Second Armies. It had been from here, just because there was no sign of enemy activity, that Moltke had taken the two corps he had sent east a day or two earlier.

Aside from the opportunities which existed for them because of this, the possibility that the British would absent themselves was a dreadful one for Joffre to contemplate. Not only did it virtually divide the attack in half — with a hole in the middle — but it also allowed an opportunity for the enemy to outflank his attackers and convert the offensive into the most appalling of catastrophes. Yet for most of the 5th, his efforts to make the BEF commander

Left: General Pau was the Commander of the Army of Alsace during the Battle of the Frontiers.

Center left: General Franchet d'Espérey was given command of the Fifth Army at the height of the battle.

Far left: General de Castelnau was in command of the Second Army in 1914.

Right: 'The Last Gunner' at Mons commemorates the bravery of a soldier who held off the advancing Germans despite the fact that the other members of his crew had been killed.

Below right: General Joseph Galliéni, who must take the credit for saving Paris.

change his mind proved unavailing. Not even when he asked the French government, now evacuated to Bordeaux, to appeal directly to Britain's was he any more successful.

When Joffre's order to his own forces to begin the offensive was promulgated, a copy of it was delivered to Compiègne, in accordance with established practice. Its bearer reported back that the British seemed lukewarm to the whole enterprise. To depress French spirits still further news came that day that Sir John French was proposing to move his own headquarters further back, to Fontainebleau. In fact, at those headquarters one British officer, General Sir Henry Wilson, a loyal and unswerving friend to France, was just beginning to persuade his commander.

At 0930 hours Joffre could stand the suspense no longer. He had appointed as his personal chauffeur a French racing driver. With him at the wheel, he set out on the 150-mile journey to see Sir John. He arrived at 1400 hours and presented himself before his ally's Commander in Chief. Normally a laconic man, what now poured from Joffre's lips was a passionate and eloquent appeal. It ended with the words, 'Monsieur le Marechal, the honor of England is at stake.' At these words French is said to have blushed, struggled to say something in Joffre's own tongue as a tear welled up and coursed down his cheek. Despairing of his French, he burst out, 'Dammit, tell him we will do all we can.'

If the British were no longer to be absent from the battle, they still had to retrace their steps. So, to some extent, did Franchet d'Espérey's Fifth Army. Yet strangely, as Basil Liddell Hart has pointed out, the British absence at the beginning actually helped rather than hindered the successful outcome as it further convinced Kluck he was dealing with a beaten foe and so could afford risks he would not otherwise have taken.

Two armies were, however, in position on the 6th. These were Maunoury's Sixth and Foch's Ninth. The latter's performance is instructive since he was to end the war as Commander in Chief of the combined Anglo-French forces. Under orders to hold the southern outlets of the St Gond

marshes as the Fifth Army put pressure on Kluck's right, what he actually did was to mount a private offensive of his own. This very nearly ran into trouble from a determined German counterattack which was only held by a combination of devastating French artillery fire and reinforcement provided by Joffre and Franchet d'Espérey himself. All the same, it continued to pose a threat to the French offensive right up to the end of the battle.

For Kluck, it was Maunoury's attacks on his right which were the major cause of anxiety. To meet them, he moved his II and IV Corps from behind the Grand Morin to his right, leaving only a thin screen to hold the gap up to Bülow's front.

There was a lingering morning mist over the battlefield in that area between the French Fifth and Sixth Armies where the British were to attack and it was through this they began their advance. The effect can only have been eery. On either side of them battles were raging, but here there was silence and desertion. The chilling, uneasy suspicion of commanders is perfectly understandable, but it was without foundation. They had simply struck in an area which the enemy had denuded to try to meet the threats elsewhere. The British were being offered the chance of slicing through the thin links attaching the First to the Second Army and so bringing total disaster to the invaders. All that was needed was for them to speed up their advance. They did the reverse. Convinced they must be walking into a trap, they slowed down, painfully reconnoitering every mile of the way.

While the British were hesitating, Kluck's II and IV Corps were now making their presence felt. By the 7th, Maunoury's advance seemed to have been held and Galliéni turned to every available means to find reserves to send in.

On the 8th, a fresh division reached Paris, a total of some 6,000 men. They were 40 miles from the front and the railroads choked. It was then that there occurred the most famous legend of this legend-encrusted battle – that of the taxis to the Marne.

The idea was not entirely original. The Germans had already used taxicabs as a means of transporting troops when they made their famous switch from left to right in

Left: German soldiers during the exhausting action which brought the German armies so close to Paris.

Right: Generals Joffre and Foch at the end of 1914.

Below: French troops in a forward trench during the Battle of the Marne.

East Prussia. In any case, Galliéni's headquarters had earlier commandeered 100 taxis for its own use, so it was no sudden brainwave. To this 100 a further 500 were added and, each with five men aboard, they raced, swerved and honked their way to the front, making the journey twice. The romantic view of war, survives on the idea that the tide of battle can be turned by the single, timely and dramatic gesture. Reality is rarely so obliging and the taxis of the Marne are no exception. Though there was no lack of success, courage or élan on the part of either of the attacking armies, what brought the final change of fortune was an act of the German High Command.

With confused reports reaching him, the already unnerved Moltke decided to send a personal representative to take stock. The chosen man was Lieutenant Colonel Hentsch, who when he departed on the 8th had already been given full power to coordinate a retreat if necessary.

Hentsch went first to the Fifth, Fourth and Third Armies, all of whom were holding their own, if without confidence about the final outcome. When he arrived at Bülow's headquarters he found only a deadening gloom. That day the Second Army commander had received reports from air-reconnaissance of columns, that both the British and the French were marching in his direction. What it meant was that the gap between his own and Kluck's forces had at last been discovered and was being penetrated in such strength as threatened to cut them off from one another. In response to this he had ordered retreat.

At the First Army headquarters, Hentsch found similar depression. Its commanders wanted to break off the battle and orders for a retreat had gone out. These, Moltke's representative confirmed, but adding that the direction taken should be northeastward so as to bring the First and Second Armies together. By the 11th, the order to retire had been extended along the entire German line. By the 12th, the Battle of the Marne was over. The German armies were behind the Aisne.

If the British and French could look on themselves as victors, they were only partially so — a fact to be agonizingly reaffirmed by the events of the next four years. The Marne was a victory of lost opportunities such as even the vicissitudes of warfare can rarely exhibit. Some of these have already been discussed, but there was a still more important one. The French line formed roughly the three sides of a box enclosing the Germans. Had sufficient strength been applied at the top edges of the two open sides, at Senlis on the left and Verdun on the right, the Germans could well have faced the direst disaster. Pressure came only from one side and Sarrail, commanding the Third Army, on the right, asked in vain for reinforcement.

He was certainly entitled to it. By his stubborn defense of Verdun, which he had been expected to give up, he had helped to frustrate Moltke's revised battle plan, as well as preventing the movement of forces westward and against Maunoury.

Almost without exception every crucial battle has generated its long and often acrimonious debate. On the one side the defeated commanders declare that had their advice been heeded at this or that point the result would have been different. On the other, those of the victors, each vying to prove himself the instigator of the successful train

of events. It would be amazing if the Marne, whose results were so far-reaching, should have been an exception. Indeed, the battle remains something of an enigma because the arguments of the claimants are so difficult to substantiate or refute and they themselves so numerous.

They include Foch, whose true role we have seen, but the principal of them must, of course, be Joffre and Galliéni, both of whom have their jealous partisans, as if one or other was totally and alone responsible for the victory from inception to enemy retreat. In fact Joffre, in view of his later profession of prescience, showed a surprising reluctance to make the first moves. Certainly he intended to go over to the offensive and was laying plans and making preparations to that end. However he had in mind both a different place and a different time. If, in the realization of his intentions, the French capital should, by chance, fall he was prepared to accept this with equanimity.

It was Galliéni who first saw the necessity of holding the capital and it was his *coup d'oeil* which took in the opportunity Kluck was so imprudent as to offer when he turned his flank and tried to slip past the defenders. Joffre was slower to react, but he had, it should be remembered, been through great disasters and knew a further one could spell the destruction of his country. To his credit it must be recorded that once convinced, he wholeheartedly supported Galliéni and gave the orders for a general offensive. It is possible that had Galliéni been Commander in Chief, as he might so easily have been, the result of the battle would have been more decisive. When the laurels for the Marne victory are handed out, there should, therefore, be not one, but two wreaths.

CHAPTER 9

PÉTAIN AT VERDUN 1916

At Christmas 1915, the German Commander in Chief, Erich von Falkenhayn, wrote his customary annual report to his Supreme War Lord, Kaiser Wilhelm II. The year had not been without achievement. In the west, every attack by the Anglo-French coalition had been held; in the east, there had been the stunning, apparently unending chain of victories which culminated in defeats for the Czar's army along their entire line and would shortly bring the country to the peace table. For the foreseeable future, it could be said that the Russian threat to the German homeland had been removed. If a similar result could be obtained in the west, the war would be finished with Germany the victor. The trouble was that this desirable state of affairs had eluded German arms as consistently as it had those of her enemies. The problem was simple: the amalgam of trenchline, barbed wire and well-placed machine gunners so greatly favored defense as to nullify every assault. Even the old maxim that victory went to the biggest battalions applied no longer. The greater the numbers of the attackers, the greater their slaughter.

What was the answer? Various ideas had been devised by both sides, including the ill-fated Anglo-French Gallipoli attacks intended to outflank the enemy far to his right. Falkenhayn in his report to his Kaiser offered yet another and in its originality it deserves to be seen as one of the most brilliant plans of the entire war. So far, all endeavors had been directed toward the seizure of militarily important objectives. Not only had these been strongly defended, but given the nature of trench warfare even if a withdrawal were forced, it had always proved possible to dig more trenches further back and all that could be gained before the impetus of the attack was lost would be a narrow band of territory.

Falkenhayn proposed that, instead of going for the comparatively flexible trenchline, they go for a fixed objective whose loss the enemy dare not contemplate. In other words, it would not necessarily be a truly strategic objective at all, rather one whose capture would be politically disastrous. What especially distinguishes this notion was Falkenhayn's apparent understanding, an understanding unique among generals of both sides, of the totally changed nature of modern war. The morale of the army was also the morale of the people, and the politicians, as the people's representatives, were vital. If their confidence was shaken, this might spread to the whole of the fighting forces.

However, this quality of originality is one which has largely been attributed to Falkenhayn by the hindsight of later writers, and the real enigma is whether he had really thought this far ahead or whether he simply wanted to obliterate what was in fact a menacing salient. If, in so doing, this also produced depression and perhaps a collapse of morale that would be a bonus, but not one Falkenhayn was so sanguine as to expect.

The salient in question, at Verdun, bulged out between Avocourt in the north and St Mihiel in the south with the town itself at its very heart, a fortress city blocking the Valley of the Meuse, whose origins went back to Gaulish times. However, since 1870, it had become the center of a double ring of some 21 forts which had been dug out of the chalk and reinforced with concrete. Of these, 10 stood

A German 21cm Mörser ready to shell French fortifications at Verdun.

on the west bank of the river and the rest on the east. The latter included those which the world was to know so well: Fort Douaumont, Fort Vaux, Fort Souville, Fort Tavannes. This ring of defenses stood on the north side of the salient while most of its south side was made up by the Plain of Woevre, itself backed by the Meuse Heights.

The reason why this salient poking into the enemy belly had survived was precisely because of the fortresses. Yet the French General Staff had so far failed to appreciate this that they had actually drained them of much of their garrisons and were in the process of taking away their guns as late as January 1916.

The policy rested on the thesis that fortresses, however strong by earlier standards, could not withstand modern artillery. The proof was said to be the experiences of Belgium in August 1914 and of Russia in the spring of 1915. This overlooked the fact that the Belgian forts had actually provided a very important first line of defense seriously dislocating the German timetable and thus facilitating defensive deployment. The Russian forts, on the other hand, were out of date and derelict from lack of funds devoted to their maintenance. The coming battle was to show that the Verdun forts were the most impressive.

Unfortunately for the military-intellectuals, the French people and politicians had not caught up with the latest vagaries of their thought. To them, therefore, the fortresses of Verdun remained as the epitome of their nation's will to resist. Joffre, having not deigned to explain what he was doing and why, found as soon as danger loomed he had to reverse his policy or risk censure and possible dismissal. In the process, countless young men died in 'the Hell of Verdun.'

Whether or not Falkenhayn's plans had the radical intention later attributed to them, they were accepted by his Supreme War Lord with as much enthusiasm as he propounded them. The assault would be entrusted to the Fifth

Army under the German Crown Prince, Wilhelm, known derisively to the British public as 'Little Willy.' Its aim would be to force the French into a bloody defense of the fortresses by attacking along the northern side of the salient.

There was early disagreement between the Commander in Chief and his immediate executive commander, the Crown Prince, who wanted a wide-fronted assault on both sides of Meuse where Falkenhayn himself wanted the initial assault to be limited to the east back only and got his way.

Successful as the attack in Russia in the previous spring had been, it was felt that a number of lessons had been learned from it. In that battle secrecy and, therefore, surprise, had been subordinated to the requirement of overwhelming force. One of the ways in which German intentions had been revealed had been by the digging of advanced trenches opposite the positions to be attacked as the means of reducing the distance which would have to be crossed. In the event, such was the effect of the enormous bombardment, the Russians were in no state to man their trench defenses anyway.

In France, Falkenhayn wanted, if possible, to retain the element of surprise and, with the Russian experience in his mind, decreed that no advance trenches should be dug. The shock troops would rely on the artillery barrage alone to give them cover as they raced across no man's land, in some places as much as half a mile wide.

Falkenhayn also set great store by attacking at the earliest possible moment. The closer to spring and the campaigning season they came, the better they could expect the French to be prepared. Indeed, they might preempt with an attack of their own. The date was set for 13 February, the attackers hoping for the best as far as weather was concerned.

The Germans' desire for total surprise was not realized. French intelligence got wind of the planned attack in late January and passed their information on to GQG (*Grand Quartier Général* or General Headquarters) where it was ignored. Joffre had decided that the next enemy offensive was going to be in Champagne and no intelligence would change his mind.

Thus, the French defensive system at Verdun comprised a group of forts bereft of guns and troops and a single trenchline, backed up by a subsidiary trench which was only partly usable. There was a considerable shortage of men and materiel; wire was incomplete in many places; and very little shell-proof cover had been prepared. As the signs of enemy activity grew more obvious, while still insisting it was part of a bluff, Joffre sent some reinforcements in the form of two *Territorial* (reserve) divisions. Mercifully, the Germans did not get the weather they had hoped for and were forced to postpone until 21 February.

Nonetheless, even with this respite and the addition of two corps, sent at the last moment, there were only three divisions on the right or east bank of the Meuse, with two on the left and a further three to the south of the fortress-area facing east. There were no reserves at hand.

On the 21st, the attack came, heralded by the explosion of a 14-inch shell on the Archbishop's Palace. The barrage, along a 15-mile front lasted the entire day reaching its height at 1600 hours. Three-quarters of an hour later, on a 'dry, cold day,' the shock-troop squads of bombing parties made their dash across no man's land, probing for the weakest points in the enemy defenses. They struck along a four-and-a-half mile front between Bois de Haumont and Herbefois on the east bank of the Meuse.

In fact, as Basil Liddell Hart says, this narrow frontage favored the defenders who, unlike the Czar's peasant soldiers of 1915, did not wilt and turn tail under the bombardment. The initial assault was actually something of a disappointment with two of the attacking divisions back at their start line by nightfall. However, a footing was gained in most French trenches and one regiment of 2000 men suffered losses of no fewer than 1800.

That sector of the line of which Verdun formed part came for administrative purpose under Groupe d'Armées du Centre or Central Army Group (GAC for short), under the command of Langle de Cary. On the night of 21st he was an extremely anxious man. If the German succeeded in breaking through here, in the very center of the French line and successfully turned a flank a great disaster threatened. They did not, of course, appreciate at this point that

Falkenhayn's real intention was to compel them to fight for the forts, rather than to force a breakthrough.

Joffre, at Army headquarters, however, seemed unperturbed. The attack at Verdun was a feint. The real attack was coming in Champagne. In 1914 with German phalanxes apparently invincible in their march on Paris, he horrified the politicians by his declaration that to him, as a strategist, 'Paris was a town like any other.' He did not propose to hazard his forces in its defense. He had similar thoughts about Verdun which was only a minor provincial town and not the national capital. As in 1914, he was once more to find himself outmaneuvered by the politicians whom he loathed and against whose power he had done his best to insure himself. The fall of the Verdun fortresses would inflict an intolerable blow on French morale. Unwillingly, he was forced to recognize that he must fight for the area he had himself done so much to demilitarize or else there would be no war left to fight. This volte face was only to come about after further disasters.

On the 22nd, the Germans widened the front of their attack. Although the dual pressures of artillery and continual infantry assault produced an advance of only about three and half miles, the French trenchlines were crumbling away and many of their second positions falling. Even more worrying, a battalion of North African Tirailleurs holding a crucial sector of the line broke and fled, leaving a gap filled only with difficulty. Langle de Cary then decided much to Joffre's annoyance to evacuate the Plain of Woevre on the south side of the salient though it had not yet been attacked. They now had their back against the Meuse Heights, but Langle de Cary was so pessimistic he doubted whether the east bank of the river could be held at all. While this was happening, the Grand Quartier continued in its conviction that the attacks were a diversion and insisted on holding back the greater part of their reserves. GAC was assisted only in the most niggardly way.

In December 1915 Joffre had been given as Chief of Staff, General de Castelnau, who by chance was a former commander of GAC. He had watched events with singular disquiet throughout, but never more so than on the night of 24 February as the evidence of Langle de Cary's depressed state of mind began to emerge. He took the unprecedented step of actually awakening the Commander in Chief from his slumbers to suggest a visit from the Grand Quartier must be made at once. Having got permission, Castelnau drove through the night to see the GAC commander, arriving to find he had been preceded by a telegram from Joffre ordering that the front north of Verdun be held at all costs. Any order to retreat would result in a court martial for the officer responsible.

Fresh catastrophes were on their way at the very moment that Castelnau was at the front. For it was on that day that Fort Douaumont fell and that by a humiliating incident. The drawbridge had been left down and a patrolling German unit simply walked in. Its garrison, consisting of 23 gunners who manned its single usable turret were taken prisoner as they slept, having let it be said, collapsed from exhaustion. This was the kind of event whose effect on public morale the government most feared.

What emerged from Castelnau's visit was the certainty that Verdun could only be saved if GAC's forces were augmented by reinforcements capable of holding the entire left bank against German attacks. If Falkenhayn's intention was indeed that of bleeding the French, they were now playing into his hands. GQG decided the scale of reinforcement called for nothing less than an army and decided on the Second under the command of Henri-Philippe Pétain, then 60 and a general. He had begun the war as a colonel and even this promotion had come late. Born at Cauchy-la-Tour in the Pas de Calais, he had been successively a lieutenant in the *Chasseurs Alpins* and a captain attached to the general staff of the XVI Corps at Marseilles. In 1902 he served as an instructor at the École Normale du Tir, at Chalons and by 1906 was an assistant instructor at the War School. In 1912 he got his colonelcy and shortly after the beginning of the war was given the temporary rank of brigadier general. He had fought in Belgium and on the Marne.

During his army career he had shown certain gifts, though none outstanding. Indeed he was hardly the sort of officer likely to rise to prominence had it not been for the exigencies of war. By nature he was cautious and dour, characteristics which reflected themselves in a tendency to rely on strategic withdrawal rather than offensive action in the face of a superior enemy.

Why then was he chosen for a task which seemed to call for precisely opposite qualities? The explanations are innumerable, but one is forced to agree with Richard Griffiths that it was probably simply because his Second Army had been taken into the army reserve. Whatever the reason, he was chosen and was summoned from the bed of a Paris brothel to receive his new orders.

After a meeting with Joffre, he drove to the little French town of Souilly which he made his headquarters. At midnight he was in full command. Not long afterward he found himself back in bed, not this time with a mistress, but with double pneumonia. This did not stop him from setting to work with a vigor which would have been astonishing in a healthy 40-year-old. Although an attempt to retake Douaumont, ordered by Castelnau failed, he began establishing that form of defense he has often been credited with inventing: 'the elastic defense.' (In fact, it was the Germans who first used it in September 1915). In essence it consisted in a three-layer line. An 'Advanced Zone' was intended to cushion the first shock. Behind this was a 'Battle Zone' in which the real struggle took place, and behind that a

Above: Officers enjoy themselves at the Cafe de la Paix in 1916. There was some bad feeling that the people back home were enjoying themselves while the soldiers were dying.

'Reserve Zone' into which the defenders could withdraw if necessary. As the advancing Germans began to encounter this system the going became harder and harder. By 28 February their attacks had been halted and, according to their own sources, a deep depression was setting in.

What was more French reinforcement was beginning to arrive. This, however, posed its own problems. All but a single light railroad being in German hands, there remained only the narrow Bar-le-Duc to Verdun road. A brilliant and determined engineer, Major Richard, was given the task of making this single artery sufficient for all needs. He divided the road into six sections each with its own improvised workshop for the repair of vehicles. So that there was no obstacle to movement, any vehicles that broke down were simply manhandled off the road. The stream was unending: one driver in the early days sat at his wheel for 50 hours at a stretch. The statistics of the achievement are incredible. Every day 1700 trucks travelled down the road in each direction, an average of one every 25 seconds. Between 27 February and 6 March, 190,000 men, 23,000 tons of munitions and 2500 tons of military materiel were brought up. Even when a sudden thaw on 28 February caused the trucks to sink to their axle-trees in mud, it was somehow kept open, Richard drafting in men to throw gravel under their wheels. It was little wonder that the 40-mile stretch became known as

La Voie Sacrée, the Sacred Way.

With the road working efficiently, it proved possible to rearm the forts, while artillery, directed by Pétain personally, mounted 'offensives' meant to compensate for deficiencies in infantry. Pétain also saw that, despite difficulties of transportation, the only way men could tolerate the strain of the battle was by speedy replacement. This meant that the Second Army would have to go beyond its own resources of men and therefore brought him into collision with Joffre who could see his entire forces going through the Verdun 'mincer.' All the same, Pétain got his way and in this, many observers, believe lies the ultimate success of the French. The Germans entrusted the entire battle to the Crown Prince's army, replacing casualties as they occurred. It was their morale which dropped first. If the troops were quickly relieved, commanders got more cavalier treatment. At the first signs that the strain was telling they were summarily dismissed.

Pétain's gifts for the defensive were being exercised to the full and, strange as it may seem in the light of the accepted image of Verdun, he was actually extremely sparing of human life. Unfortunately, this was negated by the demands from above for offensive action, such as the attempt to retake Douaumont, which brought heavy French losses, usually without result.

While the French were doing exactly what Falkenhayn was supposed to want them to do, that is pouring men on to the battlefield and throwing away lives in futile attempts to recapture objectives of purely prestige value, the German Commander in Chief found his own ideas so revolutionary that he could not cope with the implications and he began to think in terms of territorial gains. On 4 March the Crown Prince called on his army to try to reach Verdun 'the heart of France.' This time the assault was to be made on both

Below: François Flameng's view of Les Eparges and Woevre, east of Verdun, drawn before 1916. During the fighting at Verdun the German line ran between those points.

Map showing the Verdun area with the following labels:

BRIEULLES
DANNEVOUX
MONTFAUCON
WAVRILLE
VAUDONCOURT
Meuse
VII Res Corps
XVIII Corps
AZANNES
SENON
Fifth Army (Crown Prince)
CONSENVOYE
Bois de Caures
III Corps
VI Res Corps
BRABANT
HAUMONT
SAMOGNEUX
BEAUMONT
MAUCOURT
XV Corps
BETHINCOURT
FORGES
ORNES
ETAIN
Le Mort Homme
CUMIERES
DOUAUMONT
Fort Douaumont 25 Feb
Territory regained by French forces Oct-Dec 1916
Cote 304
CHATTANCOURT
CHARNY
BRAS
FLEURY
VAUX
Fort Vaux
AVOCOURT
VII Corps
XXX Corps
Fort Souville
HERMEVILLE
Orne
MONTZEVILLE
Fort Bois Bourrus
Fort Belleville
Fort St. Michel
Fort Tarannes
EIX
MORANVILLE
Fr Third Army
REGICOURT
THIERVILLE
VERDUN
Fort Belrupt
Fort Moulainville
CHATILLON
Fort Sartelles
Fort Chaume
Fort Regret
BELRUPT
Fort Rozellier
HAUDAINVILLE
Fort Haudainville
HAUDIMONT
FRESNES
Fr Second Army
Fort Landrecourt
Fort Dugny
DUGNY
II Corps
WOEVRE
Meuse
DIEUE
LES ESPARGES
V Corps

Legend:
FRONT LINE, 21 FEBRUARY 1916
" " 24 FEBRUARY
" " 9 APRIL
" " 8 AUGUST
GERMAN ATTACKS
FORTS
WOODS
FRENCH COUNTERATTACK
MILES 0 ———— 5
KILOMETERS 0 ———— 8

© Richard Natkiel, 1980

Left: The destruction at Fort Souville following intensive shelling, 26 July 1916.

Right: The damage inside Fort Douaumont.

Below: The battlefield outside Fort Douaumont. The fort fell to the Germans on 25 February but it was recaptured by the French on 24 September 1916.

sides of the river.

The reason for this was that Pétain's artillery 'offensives' had proved so destructive that the Germans realized they could not advance on the east bank as long as they were threatened by the guns of the west, in particular those of Côte de Poivre and the Mort-Homme ridge.

The bombardment began that very day and continued for two more. On the 6th the infantry began scaling the hillsides. They came up against stubborn resistance from the French employing their 'elastic defense' system. Furthermore, the defending artillery used with matchless skill, succeeded in destroying all the German 17-inch howitzers and blew up an entire gunpark with 450,000 rounds of heavy shell, an achievement which affected the attackers for the whole of the rest of the battle.

By the 8th the two hills were still in French hands. By the 9th the Germans had so far admitted failure as to convert the struggle into one of attrition. This second date has frequently been regarded as that which marked the end of the Germans' hopes, though fighting was to continue until the end of the year.

By the end of March Falkenhayn was plainly losing faith in the result himself and looking for another place to attack. On the other hand the Crown Prince still believed in ultimate victory and continued the effusion of his men's blood.

As the physical battle was turning against the attackers, so was another. No less than their barbarities of 1914, the Battle of Verdun and the dogged French resistance swung world public opinion against the Germans. The French did their utmost to exploit this. Army cameramen braved the thick of the fight to secure dramatic newsreel shots which were shown all over the world. The Russian Czar, who was given a Royal Command Performance, was so moved that he ordered his spent and ill-equipped forces to mount a vast offensive round the area of Lake Narocz in Lithuania. The result was a complete bloody disaster, the Russians losing between 110,000 and 120,000 men. It probably made a greater contribution to the army's disillusionment and so to the revolution than any other single one of the Russian's ill-managed offensives.

There was one other unlooked-for result. This was in the German High Command itself and its origins were also on the Eastern Front. The joint commanders there, Hindenburg and Ludendorff, noting the failure to gain a decision at Verdun, were soon whispering in the ear of their Supreme War Lord whenever he chanced to visit their headquarters. Time, surely for a change at the top and who better to succeed as Commander in Chief than the two men whose every effort had been crowned with triumph, as witnessed by the latest Russian disaster round Lake Narocz.

Early April found the Germans once more on the attack. The chosen sector this time was south of Douaumont in the Bois de la Caillette. For two days, 10–11 April, the shock waves kept coming at the defenders. They produced little in the way of results and the losses were horrific. On the second day a counterattack was ordered, though there were a mere 100 men available for it. The small company retook all that had been gained so painfully. The German commander was relieved, but this could do nothing to cure the underlying malaise: it was not any longer the French who were displaying the symptoms of demoralization.

At the end of April, Pétain's area of command was extended as he took over the leadership of GAC with one of his most aggressive lieutenants, General Robert Nivelle, taking command of the Second Army. The change only reflected the growing renown of Pétain and the feelings of confidence he had inspired in all circles, not least those of government. This was scarcely to the taste of Joffre and

GQG. Not only was Pétain outshining them, he also made no secret of his distrust and dissatisfaction with their conduct of the struggle as a whole and was now in a position to find influential listeners for his complaints.

In these circumstances Joffre's order that Douaumont be retaken seems very like an attempt to regain the limelight. Pétain was strongly opposed, but Nivelle and his own subordinate commander, Charles Mangin, were enthusiastically in favour and the assault was launched on 22 May. It was a failure for which, despite his opposition, Pétain took full responsibility. The effects of this ill-timed offensive actually went further than the immediate disaster. Within days the Germans were themselves attacking once more and succeeded in capturing the objective which had previously eluded them: the Mort-Homme ridge.

Above: the mud and chaos around Fort Douaumont.

Top left: An aerial view of Fort Douaumont before it was pounded to rubble by German shelling.

Top right: General Nivelle was Pétain's subordinate and after Pétain's promotion to command the Group of Armies of the Center, was in command at Verdun.

Left: Inside Fort Douaumont following its capture.

Right: French poilus struggle across the battered landscape which has become featureless from the intensive bombardments.

In the first week of June, the Germans were pressing southeast of Douaumont toward Fort Vaux. A garbled telephone message (the German military telephone system was notorious) had led to a report appearing in the German press that the fort had fallen no less than three months' earlier and, indeed, two of the officers who were supposed to have taken it had actually been awarded the *Pour le Mérite*, the highest military decoration, for their feat. Now it was really in the front line and the struggle which centered upon it was to take its place in French military annals beside the Foreign Legion defensive battle at Camerone in Mexico.

Permanent garrisons had been installed in all the forts and Vaux was held by 300 men under Commandant Raynal. Resistance continued after the Germans had succeeded in surrounding it and even after an attempt at its relief by Nivelle had failed. Adding to the trials of the defenders was the fact that they could not reach the latrines. The attackers declared that they were as much repelled by the stench emanating from the fort as by the hail of fire which greeted them. In the end surrender was forced by the exhaustion of water supplies. On 7 June a small handful of men, unshaven, filthy and ragged emerged from their stinking holes to be greeted by Germans with arms at the present. Raynal himself was received and congratulated by the Crown Prince.

This was a low ebb in French fortunes comparable only with the early days of the offensive and there is no doubt

Left: The central corridor at Fort
Vaux, which fell to the Germans
on 7 June.

Right: French prisoners are
questioned after being taken at
Verdun.

Below: German troops inch their
way forward.

that at this time Pétain was sinking into gloom and was
ready to criticize GQG for lack of support. Actually his stric-
tures were without foundation. In January a meeting of
the Chief of Staff of all the Allies, including Russia, had
taken place at Chantilly, and plans were laid for coordinated offensives on both Eastern and Western Fronts.

The western effort was to be a combined Anglo-French
one on either side of the Somme and originally the French
were to have had the greater share in it, their 40 divisions,
to the British 25 striking along a front of 25 miles. Such
was the drain on the army created by Verdun that Joffre

was forced to serve notice that only 16 of his divisions
would be fit to participate and that their frontage must be
reduced to eight miles. Thus, it could not be said he was
failing to respond to Pétain's appeal for reinforcements.

However, in June came fresh German blows with Forts
Souville and Tavannes which formed part of the inner
ring of defenses, as objective. During the preparatory
bombardment a new form of shell, filled with phosgene
gas, was used. It paralyzed the French artillery and in the
subsequent assault Belleville Height and the village of
Fleury were both taken while the advance was approach-

ing the two forts.

Pétain was forced to beg Joffre to speed up the Somme offensive. The decision no longer rested with the French Commander in Chief, however, since the British were now playing the predominant role. There had already been repeated calls for relief action and Haig had to some extent responded by taking over a greater length of line to allow more troops to be sent to Verdun. On the other hand, he was understandably reluctant to fritter away the forces he was amassing for major action in minor, improvised attacks of doubtful value.

As the fall of Souville and Tavannes would threaten the

entire defensive structure it was necessary to contemplate a large-scale withdrawal including the sacrifice of the town of Verdun itself. Plans were drawn up in the deepest secrecy and even when Joffre somehow found four more divisions to send to the front it seemed unlikely that this would change the course of events.

However, on 26 June, Pétain received a piece of news which welcome as it would have been only a short time before, at this moment presented him with new difficulties. The preparatory artillery bombardment for the Somme offensive had begun Joffre told him, but as a result it was imperative that the east bank of the Meuse was held at all costs. He undertook to take full responsibility himself for what might follow in pursuing this course.

He was not asked to bear it. On the 24th, Falkenhayn's nerves finally gave out and he stopped the flow of ammunition to the Verdun front. On 1 July, the Somme offensive opened and if this disappointed in its result at least it had the effect of giving the men of the French Second Army much needed relief.

There was one final, desperate assault on Souville in early July, but at GAC Pétain had sensed that the crisis was over. The problem he now faced was that of keeping his aggressive-minded lieutenants in check and as early as 11 July, they had launched an attack which produced no tangible results despite considerable losses. No more offensives were to be attempted until adequate strength had been recruited, though planning to this end was begun. Preparations were meticulous. Pétain put himself in charge of overall planning, with Nivelle in charge of detail. To reduce casualties as far as possible, artillery was to be extensively employed and Mangin, an artilleryman in origin, was made responsible for arrangements.

It was late September before the Commander in Chief of the GAC decided that the time to strike was arriving. It was late October before it came. On the 24th, a bombardment of withering ferocity fell upon the German trenches and it culminated in the use of the new 'creeping barrage' by which the guns registered so that their shells exploded just ahead of the advancing infantry forcing the defenders to stay in their dugouts.

The whole operation was handled with the greatest efficiency and though only three and a half divisions were used in the attack, their success was brilliant. Within a short time the forts of Douaumont and Vaux were back in French hands. Mangin's troops who conducted the attack took 6700 prisoners and German losses in killed and wounded were proportionately high.

On 15 December there was yet another French thrust, which, like the last was limited in its objectives. Of the five German divisions in line, four were from the Somme front, and it is indicative of their shortage of reserves that it was necessary to 'rest' troops by sending them to a sector hardly less active than the one they had left. The results were even more spectacular than those of October. Over 11,000 prisoners and 115 guns fell into French hands and one of the features of this phase of the struggle was the success of the air force. French flyers had been playing an increasingly significant part and now had total air supremacy.

It could be said that this was the end of the battle. The cost to both sides had certainly been enormous. The entire strength of the Second Army was 24 divisions, yet no fewer than 65 divisions of the French Army had travelled the Voie Sacrée between March and June. This large number was, of course, a result of Pétain's policy of speedy rotation and in fact the Germans had used only 47 divisions up to August, so their troops spent more time at the front.

French losses amounted to 315,000; German to 281,000. It is instructive that whereas, in the early part of the battle the ratio of French to German casualties was of the order of 8:7, in the final offensives of October and December, it had completely reversed itself to 7:8.

Before one considers the victor of Verdun one must first ask oneself whether it was a victory. This has always been a controversial question, even in France where there is the most enormous, justifiable pride in Verdun. It is a little difficult to understand why. If an attacking force achieves its objectives, that, certainly, is victory. If a defending force

Left: French prisoners from Verdun.

Right: A dead soldier. By May the number of French casualties had risen to 185,000. The final cost of holding Verdun for the French was 275,000 men.

frustrates them and finally compels them to abandon their endeavors, that is one of another kind — a defensive one. When, in addition, the defenders, taking the offensive themselves, recapture all that has been snatched from them who can deny that they have shown their supremacy? Although the words of King Pyrrhus may apply, 'One more such victory and we are undone,' the Germans having failed in their aims, the Second Army achieved a victory.

It was also a victory of another kind. It galvanized the French spirit. Far from being broken by the disasters of the weeks of late February and early March, they were if anything exulted by them. They did not end the conflict with a sense of imminent defeat, but of ultimate decisive triumph. It was the German spirit which began to flag, both here and at the Somme.

If this counts as victory then its laurels certainly belong to Henri Philippe Pétain, a man who perhaps more than any other First World War general, understood the importance of morale. In the last stage of the First World War Pétain began writing his own account of Verdun. In it he describes how he used to watch the young French soldiers marching to the battlefield, confidence displayed in their very tread. Then he describes their condition when they returned, often only in a matter of days: 'Their gazes, impenetrable, seemed fixed on a vision of terror; their walk and their attitudes betrayed the most complete exhaustion; they were weighed down with terrifying memories; they scarcely replied when I cross-examined them, and in their troubled minds the joking voices of old soldiers awoke no echoes.'

Of course in the Battle of Verdun as in every other battle there was, on the victor's side, an element of luck without which the decision would have gone the other way. At least twice, fate smiled on the defenders. The first time was when the Germans were compelled to delay the start of their offensive because of bad weather. The eight-day reprieve enabled the French to reinforce in some degree. The second was when Falkenhayn chose to attack on the east bank only, allowing the guns of the west bank to interdict the assault.

This does not diminish Pétain's achievement. Given command of this crucial, embattled front when everything seemed about to be lost he did all that could possibly have been asked of him. Nor did he complain at the hopelessness of the task as others had done and were to do later. He went to the front, established his command post at the Mairie of Souilly and coolly and methodically set in being his 'elastic defense' against which the enemy threw himself in vain. He instituted his artillery offensives. He improved communications. He introduced the rotation of troops. He rearmed the forts. Most of all he obeyed to the letter Joffre's injunction that there was to be no retreat.

Throughout the struggle, the French, supposedly at their poorest in defense, held back attacks from vastly superior forces. This alone was a fantastic achievement and a tribute to Pétain's ability to inspire confidence in his troops. As a consequence, two weeks after the first assaults had been launched, he was already causing the enemy deep heart-searching and making him recast his aims.

Finally when the moment was right, he swung from defensive to offensive. Even then he curbed his impetuous and overambitious subordinates, Nivelle and Mangin, who wished to make every engagement into a march to Berlin, and persuaded them to limit themselves to lesser, more attainable objectives, a form of tactics which was to be generally adopted by everyone, including the British once it had sunk in that the Somme and Passchendaele were catastrophes.

The scale was, of course, infinitely smaller, but one is to some extent reminded of Wellington and the Peninsular War. Where Pétain differs from Wellington, however, is that the success of Verdun did not prove to be repeatable in bigger terms. When he became Commander in Chief of the French armies in succession to Joffre, then Nivelle — and perhaps this was another result of Verdun — his natural caution and defensiveness took over and not until bestirred by Foch as Allied Generalissimo in 1918, did he bring himself to think offensively.

HAIG AND THE BLACK DAY 1918

Writing his memoirs during the autumn and winter of 1918–19, Erich Ludendorff, a man not given to linguistic flights, coined a phrase. '8 August (1918) was,' he wrote, 'the black day of the German Army in the history of the war.' It is a phrase which historians have employed to exhaustion ever since.

Just exactly what he meant by it is made clear in the chapter which he began with those words. It was the day when the British, Australian, Canadian, American and French forces were released in a massive offensive between Moreuil and Albert, the latter perhaps the most battle-scarred and fought-over town on the entire Western Front. The defenders behaved in ways which, the author describes, he would 'not have thought possible in the German Army.' That is to say, they behaved as troops everywhere behave when they know they are totally, irrevocably overpowered.

In fact, their behavior was astonishingly restrained compared with what had happened in other armies on similar occasions. If their collapse was less complete than might have been expected, it was enough finally to snap their commanders' nerves. Only six days later, on 14 August, the German emperor, Wilhelm II, ordered that peace feelers should be extended. Some 12 weeks after that he was a crownless exile, cooling his heels at a Dutch frontier post, begging his neighbor monarch to grant him sanctuary.

Yet through almost the whole of that year the stars of both the German army and of its commander had seemed to be in the ascendant. Ludendorff was 53 and effectively master not only of the military machine, but also of the government.

Everywhere, except in the West, German arms had been crowned with triumph. Rumania, which had imprudently joined the war on the side of the anti-German coalition, the Entente, the previous year, had been forced to make peace. Negotiations to the same end were currently going on with Russia's post-revolutionary Bolshevik government. The campaign of unrestricted submarine warfare was producing such drastic losses to her shipping that it seemed as if Britain would soon be subjected to a blockade as harsh as that Germany had experienced through the whole war, though it was true one of its results had been to bring America into the war.

The victories in the east, besides giving Germany access to the grain wealth of the Ukraine, also freed the forces engaged there and the not inconsiderable numbers in prisoner-of-war camps, a total, all in all, of perhaps a million men.

By contrast, the position of the Entente had never looked poorer. Besides the Russian defection, there had been mutinies in the previous spring which had infected half the French Army. These had been quelled but its reliability in an emergency was uncertain.

In Britain the carnage of Third Ypres had led to such doubts about the generals' ability to conduct the war that the politicians had intervened. David Lloyd George, the Prime Minister, had insisted that reserves must remain in Britain so that the Commander in Chief was not tempted to throw men into yet another 'Big Push' without prior consultation. This, added to enormous casualties, meant the British Expeditionary Force was so short of men that

Field Marshal von Hindenburg and General Ludendorff in May 1918.

divisions had been reduced from 12 to nine battalions. There were only 26 of these reduced divisions holding a line roughly from Ypres to Soissons. Furthermore, their defensive deployment was in the process of being changed. They were now to be organized 'in depth' on the German pattern, with an advanced line to break the impetus of an attack. There would be a 'Battle Zone' in which the main struggle would take place and a fall-back or 'Reserve Zone.' Unfortunately, these changes were far from complete and in many places the Reserve Zones existed only on maps or as turfs removed from the ground to show where they were to be.

The sole light amid the gloom was America's involvement, but the British and French were making exactly the same complaint the French made about the British in 1914 and 1915, they were being desperately tardy about arriving. Thus, Ludendorff could report to his emperor with some confidence on 13 February 1918, 'the Army in the West is waiting for the opportunity to act.'

Germany needed to act quickly. The trickle of American troops would, if later rather than sooner, become a flood. The augmentation of the Royal Navy by that of the United States was likely to reduce, or even nullify, the effect of the U-Boats. France, notwithstanding her losses at Verdun in 1916, was recovering and making heavy drafts upon her colonial empire. Britain seemed to possess a bottomless pit of reserves from her dominions.

Germany, on the other hand, was fast reaching the end of her manpower resources. If she did not use her temporary advantage to inflict a decisive reverse, she must sue for peace. At best, she could only hope for an indefinite prolongation of the war of attrition, a possibility the German people, including the armed forces, seemed unwilling to resign themselves to.

Ludendorff had divined, rightly, that at present the British were the major adversary. Once their forces had been crippled, France would be a smaller challenge and might well be inclined to make an end. Deep in the British rear lay the three main Channel ports, Dunkirk, Calais and Boulogne, with the shortest sea routes from Britain. Not only were they vital to the movement of men and supplies, their loss would also mean that the BEF would be in the trap most dreaded by British military planners. At the same time, the Americans would only be able to reach France by the much longer routes further down the channel. Ludendorff had deluded himself that the Americans were unwilling allies (though this was far from true, once committed they were determined to see the war out) and that faced by such difficulties they might decide the fight was not worth the candle.

He knew that what he had in mind was incredibly ambitious, requiring the kind of continuous advance against the

Left: Field Marshal Sir Douglas Haig with General Rawlinson, General Byng, General Lawrence and General Birwood at Cambrai on 31 October 1918.

Below: German artillery moves forward during the Battle of Kemmel in April 1918. The final German offensive was launched on 21 March 1918 and was supposed to win the war before American troops arrived on the front in force.

enemy no offensive on either side through the whole course of the war had so far been able to produce. He did not underestimate this in his report to the emperor. 'It will be an immense struggle,' he admitted, 'which will begin at one point, continue at another and take a long time.' On the other hand, though 'it was going to be difficult,' he promised that it would be victorious.

His British opposite number, Field Marshal Sir Douglas Haig, was four years his senior. A Scot, related to the famous whiskey firm of Haig, he was educated at a British public school and then at Brasenose College, Oxford. He was a cavalryman and had seen service in the Sudan, South Africa and India. In August 1914 he had been in command of the I Army Corps which fought at Mons, on the Meuse and at Ypres. By 1918 he had been Commander in Chief of the British forces on the Western Front for nearly three years, having succeeded Sir John French.

He has sometimes been described as an unimaginative soldier, this epithet being applied to him because of his inability to find an answer to the problem of assaulting an enemy whose forces lay behind broad belts of barbed wire which they could protect with machine guns. Since, however, no commander in any of the belligerent armies found a total answer to this problem it is perhaps a little less than fair. His greatest weakness, apart from vanity — a fault of which few commanders are free — was in a certain hidebound orthodoxy. Believing that, whatever temporary aberrations might appear, the cavalry would always come back to their own once the infantry had broken through, he was suspicious of innovation. Nonetheless once he had been won over he was very willing, perhaps overwilling, to use any new weapons which were thrust in his hand. The early weeks of 1918 saw him waiting with his now attenuated forces for the enemy attack he knew must come and about which, despite German attempts at secrecy, increasing information was arriving at his headquarters.

By the third week of March Ludendorff's preparations were complete. Something like 750,000 men opposed a British force about 300,000 strong. In the early hours of the morning of 21 March 2500 guns opened up along a 44-mile front from Croisilles to La Fère using not only high explosive, but also gas shell.

Churchill, who chanced to be present, has described the tumult in his usual graphic prose, 'The enormous explosions of the shells upon our trenches seemed almost to touch each other, with hardly an interval in space or time. Among the bursting shells there rose at intervals, but almost continually, the much larger flames of exploding magazines. The weight and intensity of the bombardment surpassed anything which anyone had ever known. . . . Daylight supervened on the pandemonium and the flame picture pulsated with a pall of smoke from which great fountains of the exploding "dumps" rose mushroom-headed.'

Over that spring and summer such bombardments were to become almost commonplace. Indeed, they were surpassed. On 25 April the bombardment was so concentrated that, though it lasted only two and a half hours, the men were reduced to gibbering imbeciles. The artillery bombard-

ment was, of course, only the precursor. It was followed by the attacking waves of infantry. The Germans had assembled their large forces into two army groups. That of the Crown Prince Rupprecht of Bavaria comprised the Fourth, Sixth, Seventeenth and Second Armies and held a sector from the Belgian coast to just north of St Quentin. The army group commanded by the German Crown Prince was made up of the Eighteenth, Seventh, First and Third Armies and its line stretched from St Quentin to just south of Reims.

Like every other commander on the Western Front, Ludendorff had learned, at great cost to his troops, one important lesson from the war. As Sir Basil Liddell Hart puts it, the depth of advance before the momentum of an offensive was lost equalled roughly half the frontage attacked. In other words, if the attack was made along a front of 16 miles, the attacking troops could expect to penetrate to a depth of eight miles. Not only was it rarely possible to exceed this, but the loss of life involved became more and more disproportionate to any gains made. Thus the battle on which Ludendorff had now launched his men was not to be a single offensive, but rather a whole series each abandoned when the consolidation of enemy resistance made the going too hard. Nevertheless each, in terms of men and materiel involved, was vast enough, even if compared with other big battles of the war.

First in one place, then in another, the armies struck. In the initial assault, on 21 March on the Somme, British casualties exceeded 100,000 and 500 guns were lost while

the Third and Fifth Armies were pulled back by their commanders with as much speed as was possible. Never, in the whole war, was terrain ceded in such large areas as it was during this series of offensives.

Next the battle moved to Flanders with the crucial Channel ports as the objective. At one point the attackers were outside Hazebrouck, whose fall would endanger the entire British communications network.

Further south Arras, Péronne, St Quentin, Amiens and La Fère were all threatened or taken. Beyond them the Chemin des Dames, Soissons, Fismes and Reims were each on the front line and soon to be behind it.

The French, by this time themselves involved, began to fear for Paris. Pétain, their Commander in Chief, convinced that under pressure the British would make for the Channel ports and, in all probability, embark, disposed his forces as a semicircular shield around the capital. By doing so without consultation he made it very difficult for his allies to maintain continuity of line.

By April his fears seemed about to be realized. The Channel ports were no longer in jeopardy, but with the fall of Château-Thierry Paris was only 30 miles away.

Left: Artillery bogged down in mud during the German advance in March 1918.

Below: German storm troops march through what is left of the village of Messines during the Battle of Kemmel.

Far left: French soldiers find cover in a quarry at Ribécourt on the Oise in April 1918.

Already the city had been subjected to shelling from 'Big Bertha,' the enormous cannon which the Germans had emplaced at Laon, 75 miles away.

The sense of urgency and impending disaster which gripped the city was comparable only with those early weeks of the war until the First Battle of the Marne broke the Germans apparently inexorable advance. Everyone who could began leaving for less exposed regions of the country and the Prime Minister, Georges Clemenceau, must if anything have fuelled this panic by his statement to the Chamber of Deputies, 'I fight before Paris, I fight in Paris, I fight behind Paris.'

May, June and July all saw further German gains in one place or another along the line. Nor can it be said that there was a substantial abatement in July. As late as 2 August Ludendorff was still nursing further offensive plans. Yet Ludendorff completely lost his nerve shortly afterward.

To trace the origins of the 'Black Day' one must go back a month to 4 July. It was on that day that an Australian division recaptured Hamel and Vaire Woods, south of the Somme. This was accomplished by infantry and the new weapon – tanks. They worked in close cooperation and with the infantry and were present in large numbers, not

in the 'penny packets' so derided by Liddell-Hart. It was by no means the first time tanks had been so used during the series of actions which began on 21 March. Indeed, as early as 24 April there had actually been, as harbinger of what was to come 22 years later, a purely tank battle with 13 British engaging 13 German ones.

On 18 July the French general Charles Mangin had used the fast Renault tanks to considerable effect on the Marne. The Germans, including Ludendorff, admitted that they were totally taken by surprise and attributed their failure to withstand the attack very largely to this. Rawlinson, who had replaced Gough as commander of the British Fourth Army, was eager to see tanks used as a weapon in a major offensive to replace the artillery barrage which, by this nature, gave the enemy preliminary warning of what he could expect.

Another reason for the sudden reversal of German fortunes was Allied reorganization. On 26 March, largely as a result of appeals by Haig, who saw the need for total coordination, particularly now the large American armies were beginning to arrive, General Ferdinand Foch was made supreme commander of all the Allied forces. For the first time in the war there was unified direction. One imme-

Left: German heavy artillery in action during August 1918.

Right: Men of the 10th Australian Battalion occupy a trench near the Bois de Crepy on 10 August 1918.

Far left: Georges Clemenceau, the French Prime Minister in 1918.

Below: An Australian Field Artillery unit which took part in the Allied offensive on 8 August 1918, at Villers-Brettoneux.

diate result was that reinforcements, drawn from any national army, could be sent to the assistance of any other. This was quickly put to use in Flanders where French troops were sent to support the hard-pressed British units.

Foch was the apostle of '*l'attaque*,' the overzealous application of which in the early days of the war had proved so costly and disastrous. His belief in attack was undiminished, but he had learned much from experience and had acquired a marvellous intuition about battles and the moment to strike.

At a conference on 24 July attended by the British, French and American commanders, the question of a counterstroke was considered and Haig drew attention to Rawlinson's proposal. Foch saw its potential. If the stroke were to be delivered east of Amiens this might free the Paris-Amiens railroad, not actually in German hands but within such easy reach of their guns that it was unusable. This, in turn, would open up the entire lateral communica-

tions network. He therefore ordered the French First Army on the right of the British Fourth to participate and put it under Haig's orders, though he was to use only its two left-hand corps on the first day of the attack. To the south of Amiens, the German line bulged out in a salient which embraced the recently captured towns of Moreuil and Montdidier. Foch accordingly ordered the French Third Army to the south side of this salient to attack in concert with the First. This was not the end of his intervention. The offensive was scheduled to start on 10 August. He had it moved forward two days — one of the very few times such a thing had happened. More often offensives were postponed.

Since the essence of tank attack was surprise, this had to be safeguarded at all costs. Every conceivable subterfuge was adopted. No two conferences were held in the same place. Reconnaissance was carried out in the most elaborate secrecy. Divisional commanders were not even told of the attack until 31 July. The troops themselves were told only

36 hours before.

All troops and other movements were made at night. As guns were slipped into place they registered on target by joining in each day's regular bombardment of the enemy, while other guns, already in place, remained silent. It was hoped the enemy would not realize that the batteries were being increased. In this way 1000 extra guns were brought in. During the course of about a week, up to D-Day, the ranks of Fourth Army were virtually doubled. There

were six fresh infantry divisions, as well as two of cavalry. The tank force numbered 600, of which about 420 were combat ones. The British air force had amassed about 374 planes and the French had an even bigger force, though it was not to be used until the 9th.

Such was the secrecy that not even the War Cabinet in London were told. The story is related of how William Morris Hughes, the Australian premier, who had come to London with the express purpose of demanding that his

Right: The American entry into World War I was decisive. The Germans were demoralized further at the prospect of facing fresh troops.

Below: On the trail of the Hun by W J Aylward.

Left: General Sir Herbert Plumer, the commander of the British Second Army.

Far left: General Hubert Gough was one of those blamed for the German breakthrough in 1918.

Below: An Australian 18-pounder is rushed into action near Merricourt on the Somme in August 1918.

own country's troops should be taken out of line, was in the full flow of rhetoric when the news came that they were at that moment advancing upon the enemy.

The Australian Corps, under General Sir John Monash, certainly one of the war's finest commanders, was on the left of the Canadian Corps who were to strike the main blow. It was also part of the secrecy that, since they were known as shock troops, they were kept out of the line until a few hours before zero hour.

The British contribution included the III Corps and the Cavalry Corps, plus a reserve division. The United States 33rd Division was in reserve to III Corps. The French First Army were providing three corps and were also to use tanks, but as these were all light it was decided that they should be employed for exploitation rather than for the initial shock. Altogether there were 21 divisions, British, French and American, with another four in reserve. They faced 14 German ones of the Eighteenth and Second Armies.

Surprise was total. The 8 August was a morning of dense fog, 'rendered,' as Ludendorff puts it, 'still thicker by artificial means,' and it was out of this that 456 tanks, rumbling forward along a 14-mile frontage, rolled down on the German trenches. The six divisions holding those trenches were barely more than skeletons thanks to casualties in the previous weeks of battle, and averaged about 3000 men each. An attack was so unexpected that none of the usual meticulous German defensive preparations had been made.

Coincident with the tank advance, the guns opened up and the infantry surged forward. South of the Somme were the Canadians, next the Australians and north of the river the III Corps. It was only here that there was any check, partly because the British troops had fewer tanks in order to maximize the strength of the main punch and partly because the trench works and rusting barbed wire of the old Somme battlefields checked the advance. Reserves which had been set in motion as soon as the attack began arrived shortly afterward.

No less effective than the tanks were the armored cars which chased up the roads, penetrating far behind the German lines. One of them even reached a staff headquarters where its members were taken by surprise over breakfast.

One may perhaps take up the account from the German point of view. Ludendorff tells us that six or seven divisions, thought to be battle worthy, were completely broken. Whole bodies of men gave themselves up to a single enemy soldier. Retiring units, encountering those going up the line, shouted abuse at them, 'Blacklegs,' 'You're prolonging the war.' Officers lost their influence. A crack alpine corps broke down into parties of men of all ranks, who, according to its official history 'wandered wildly about, but soon for the most part finding their way to the rear.'

Of course attempts were made to hold the advancing enemy, reserves were hurried from the north, thus depleting the forces of Prince Rupprecht. Troops, rushed in trucks, often had to fight as they scrambled off their tailboards. There was an inevitable confusion of units. The

Bottom right: The wreckage of enemy transports on the road to Abancourt in August 1918.

Below right: The bodies of German soldiers on the road to Chipilly during the Allied advance in August.

Right: A soldier of the London Irish Rifles fires at a sniper during a patrol in Albert, August 1918.

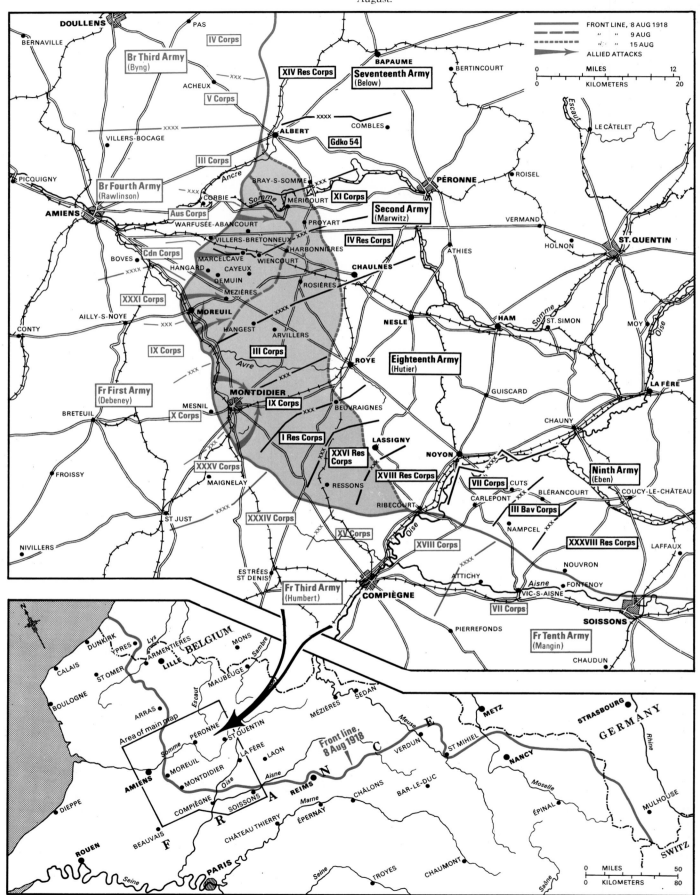

FRONT LINE, 8 AUG 1918
"   "   9 AUG
"   "   15 AUG
ALLIED ATTACKS

MILES 0 ... 12
KILOMETERS 0 ... 20

DOULLENS
BERNAVILLE
PAS
IV Corps
Br Third Army (Byng)
ACHEUX
V Corps
VILLERS-BOCAGE
BAPAUME
BERTINCOURT
XIV Res Corps
Seventeenth Army (Below)
COMBLES
ALBERT
Gdko 54
III Corps
PICQUIGNY
Ancre
BRAY-S-SOMME
PÉRONNE
ROISEL
Br Fourth Army (Rawlinson)
CORBIE
Somme
MÉRICOURT
XI Corps
Second Army (Marwitz)
VERMAND
AMIENS
Aus Corps
WARFUSÉE-ABANCOURT
PROYART
ST. QUENTIN
VILLERS-BRETONNEUX
IV Res Corps
ATHIES
HOLNON
BOVES
Cdn Corps
MARCELCAVE
HARBONNIÈRES
WIENCOURT
CHAULNES
HANGARD
CAYEUX
DEMUIN
ROSIÈRES
XXXI Corps
MÉZIÈRES
HAM
Somme
ST. SIMON
MOY
Oise
MOREUIL
ARVILLERS
NESLE
AILLY-S-NOYE
HANGEST
CONTY
IX Corps
III Corps
ROYE
Avre
Eighteenth Army (Hutier)
GUISCARD
LA FÈRE
Fr First Army (Debeney)
MONTDIDIER
BEUVRAIGNES
CHAUNY
BRETEUIL
MESNIL
IX Corps
X Corps
LASSIGNY
NOYON
Ninth Army (Eben)
I Res Corps
XXVI Res Corps
XXXV Corps
XVIII Res Corps
VII Corps
CUTS
BLÉRANCOURT
COUCY-LE-CHÂTEAU
FROISSY
MAIGNELAY
RESSONS
CARLEPONT
III Bav Corps
RIBECOURT
NAMPCEL
XXXVIII Res Corps
ST JUST
XXXIV Corps
Oise
LAFFAUX
NIVILLERS
XV Corps
XVIII Corps
NOUVRON
ESTRÉES ST DENIS
Aisne
FONTENOY
ATTICHY
VIC-S-AISNE
Fr Third Army (Humbert)
COMPIÈGNE
VII Corps
SOISSONS
PIERREFONDS
Fr Tenth Army (Mangin)
CHAUDUN

N
DUNKIRK
YPRES
Lys
ARMENTIÈRES
MONS
BELGIUM
Sembre
CALAIS
ST OMER
LILLE
BOULOGNE
MAUBEUGE
ARRAS
Escaut
Area of main map
PÉRONNE
ST QUENTIN
MÉZIÈRES
SEDAN
Meuse
METZ
STRASBOURG
GERMANY
DIEPPE
Somme
LA FÈRE
LAON
VERDUN
ST MIHIEL
NANCY
Rhine
AMIENS
MOREUIL
MONTDIDIER
Oise
Aisne
Front line, 8 Aug 1918
NANCY
Moselle
MULHOUSE
COMPIÈGNE
SOISSONS
REIMS
Marne
CHÂLONS
BAR-LE-DUC
ÉPINAL
ROUEN
BEAUVAIS
CHÂTEAU THIERRY
ÉPERNAY
Sabre
SWITZ
Seine
PARIS
TROYES
CHAUMONT
MILES 0 ... 50
KILOMETERS 0 ... 80

There was, naturally enough, jubilation among the Allies. This was caused not so much by the seizure of ground, for the war had seen too much of this kind of victory, as by the evidence of sharply declining enemy morale. During the advance a number of important documents were captured. These plainly showed that the Germans were reaching the end of their manpower reserves.

It was only to be expected that the German press should seek to reduce its significance and turn it into a purely local victory of no strategic value. Such was their success in playing down every Allied victory in the coming weeks that the German people were left stunned and unbelieving when they learned in November that an Armistice was being sought. This accounts for much of the confusion and the sense of betrayal which was felt after the war.

A great deal of hard fighting remained before that day was reached. There is no room for doubt, however, that the 8 August 1918 saw the beginning of the process. It is debatable how much of the credit is due to Haig.

Sir Basil Liddell Hart quotes an exchange of letters between Foch and Haig which started on 12 July, a time when the Germans were still in the ascendant. It proposed that the British offensive then being planned should be in the north round the Lille industrial area, a place of enormous value to the French. The objective was to be the freeing of

drain on precious reserves was enormous.

So far as the attackers were concerned the only major resistance was encountered in villages which could offer some kind of cover for the defenders. The pace of the first day could not be maintained on the 9th. The Germans were enormously reinforced with something like 14 divisions. The day saw the beginning of the attacks by the French Third Army south of the salient, but they were late — the Germans were already withdrawing.

By the 12th, the struggle was largely over, though with sporadic fighting occurring up and down the line. Haig's reaction was to bring it to an end and launch his Third and First Armies southeastward to strike the enemy in flank and reduce pressure on the Fourth and French First Armies. Foch, on the other hand, was demanding that the advance on the present line should continue. After a fierce quarrel Haig agreed to the extent of ordering localized probing attacks, but Rawlinson, the Fourth Army commander, and Sir Arthur Currie, commanding the Canadians, were so strongly opposed to this that they actually countermanded orders, once leaving the French in a bad spot. Foch was forced to concede, as he was always prepared to do once convinced.

The struggle on 8 August — 'The Black Day' — had produced a famous victory. More important still in a war distinguished by its extravagance with human life, it had been extremely economical. British losses during the four days were 22,000, and the French were about 20,000. German losses, never made official, have been estimated at something like double these combined figures, though this is likely to be exaggeration. What is certain is that the British and French together took some 30,000 prisoners.

the Bruay mines and the protection of the communication center of Estaires. This was not at all to Haig's liking, as he made plain in his answer of the 17th. He saw 'no advantage in an advance over a flat marshy region.' His counter-proposal was to push forward the Allied line to the southeast and east of Amiens to safeguard the town and railroad. This would be a combined operation with 'the French attacking south of Moreuil and the British north of the Luce.' He added that to 'realize this object I am preparing plans secretly for an offensive north of the Luce, direction east. . . . In liaison with this project the French forces should, in my opinion, carry out an operation between Moreuil and Montdidier.' This was, of course, the plan which was subsequently followed.

However the fundamental tactical idea, that of using tanks in mass came from Rawlinson after witnessing the success of the attack at Hamel on 4 July. Thus there is some justification for claiming that this was a British concept.

Left: An artillery convoy moves to the front, a watercolor by Georges Scott.

Far left: The result of the German offensive in April 1918, painted by François Flameng.

Below: Large areas of northern France were left totally devastated by four years' fighting.

On the other hand, Haig never pretended to be solely responsible for the strategic plan which brought such a gratifying victory. The claim was made on his behalf by the British press, all too eager to show that a general of their own country could actually win a battle decisively. It was then taken over by historians writing shortly after the war.

However, in one respect Haig's philosophy of the war had been justified. This was in his theory of 'the last reserves.' He had held consistently to the view that in the final analysis the Entente must always win, simply because they had greater human resources. This view has been the subject of bitter parody and in the process his actual opinions have been distorted. He was not saying the war should be conducted without consideration for human life because the Entente possessed a huge reserve of manpower, as some detractors have made it appear. He was simply stating, in times when the entire battlescape looked its blackest, that there were still some grounds for optimism. The enemy simply could not win because the mathematics were against him.

When first enunciated, this doctrine had been based on

the fact that Russia with its seemingly endless numbers was in the struggle. Though they had fallen out, they had been replaced by another, and far more significant ally — the United States.

Through the spring and summer, Ludendorff's troops kept running into groups of American soldiers. At first they were simply small groups, fitted out with a hotch-potch of British and French equipment and hurried into position to aid their allies. Although they were brave, often suicidally so, they were no match for the cunning German veterans. Ludendorff dismisses them with contempt (he never had a good word to say for a brave opponent).

However by June they were becoming a force to be reckoned with. On the first day of that month a machine-gun battalion of the 3rd Division showed its mettle in the defense of Château-Thierry. On 6 June the 2nd Division launched an attack which won a village and part of Belleau Wood. By July, the time of the Second Battle of the Marne which set a limit upon the enemy advance, there were 25 divisions in France. They played a significant part in that battle as they did at Amiens on 8 August. By the second half of the month, their Commander in Chief, Pershing, was able to gather all his units together into one cohesive whole which was to be increasingly important over the coming months.

There was another element in the victory: the tanks which came out of the fog upon the ill-prepared German trenches. At their inception tanks were named, very accurately, 'land-ships.' Haig, like Kitchener, the British War Minister, had been a late convert to the tank. In Haig's case, this was only to be expected from a cavalryman.

In the end, Haig was if anything rather overpersuaded of their value and allowed their use first at the Somme in 1916 and later at Cambrai in ways which were in direct breach of their designers' principles. Colonel Swinton had declared they should be deployed in large numbers, sweeping suddenly down upon an enemy and wiping out his defenses. At the Somme, only 60 out of the 150 machines ordered reached France and for mostly technical reasons only 32 actually got to the start line. Nine accompanied the infantry, nine were overtaken by the foot soldiers, nine broke down and five ended up in battlefield craters. In this way the shock to the enemy which could have come from a massed attack was frittered away. He was warned of the existence of a new weapon and actually given the idea of making it himself. Luckily, the German General Staff was, if anything, even more purblind than the British. Had they not been the story of the March offensive might well have been quite different. However Rawlinson's proposal to employ tanks to spearhead the 8 August attack was not original. He was simply reverting to the principles of Swinton.

The view of a German general, Zwehl, quoted by Liddell Hart, is instructive, 'It was not the genius of Marshal Foch that beat us, but "General Tank."' Another German general corroborates, 'The arrival of tanks on the scene had a most shattering effect on the men. They felt quite powerless against these monsters which crawled along the top of the

Right: General Douglas Haig sits next to Arthur James Balfour, then Foreign Secretary, at the Trianon Palace.

Below left: After the Black Day the Allied troops advanced rapidly. German prisoners are rounded up near Bellicourt, 29 September 1918.

trench enfilading it with continuous machine-gun fire and closely followed by small parties of infantry who threw hand grenades at the survivors.'

The tank was one element in German disintegration and, if important, still only one. There was besides a supreme war weariness, made worse by the hardships imposed on them through the Allied blockade. The German government took no steps either to control the money supply through taxation or to introduce a fair system of rationing. It was *sauve qui peut* on the Home Front with appalling suffering imposed on the poorer sections of the community. This made them a breeding ground for discontent.

If German morale was cracking, so too was that of its Chief of Staff. Ludendorff, in the late summer of 1918, seemed to be in a disturbed state of mind. On the one hand, he failed to apprehend the fundamental truth — that his country had lost the war and its only hope lay in making peace. On the other, at the first signs of disaffection he was thrown into a state of hysteria. One might feel some pity for his condition were it not that of all the generals of all the belligerents he is the one least able to excite such reactions. Even the stolid Hindenburg is redeemed by a certain generosity of spirit which could lead him to commiserate with a beaten foe. All Ludendorff could do in the same circumstances was gloat.

Least of all does one find in Ludendorff's memoirs, as one frequently finds in Hindenburg's praise of the skill of an opposing general. In one sense this is history's loss for nothing is more useful in trying to assess the merits of a commander than what his opponents say about him. Thus, in the case of Foch, Zwehl's dismissal of him forces us to reassess his contribution to the victory.

The resonance of Foch's reputation, one which finally brought him to be Commander in Chief, was based on his supposed success at the First Marne. Liddell Hart has shown how relatively unimportant, indeed hazardous to the enterprise as a whole, his conduct actually was. However one has to recognize that he brought special qualities of character to the office of Generalissimo. Of these, the greatest was his open-mindedness. He could quarrel with a fellow general, as he did with Haig, but finally admit his antagonist had been right. What was more he could continue to work with him thereafter without rancor. It was from this magnanimity that there sprang his ability to understand, as no other French general and few British had, the need to act in strict cooperation with allies. He gave Haig command of French units when this seemed desirable and sent French troops to reinforce the British ones (and was criticized for it) when he saw the necessity.

His other gift was that already referred to — of divining intuitively the state of the war. Thus in July 1918 while others could see only German victories he was declaring 'the edifice is beginning to crack.' He was right, of course. This is not just second sight, like Rommel's famous '*Spitzenfingergefühl*' or feeling at the fingertips. A commanding general has access to an enormous number of sources of information, among them the results of the interrogation of prisoners. These plainly indicated that the German spirit was breaking.

If we lack an appraisal of Foch from the other side of no man's land, we do have one of Haig. Like Ludendorff a change had come over the British Commander in Chief during the last months of the war. The change in him, however, was precisely the opposite one. It made him, as one enemy commentator put it, 'the master of the field.' Foch provided the authority, the aura of confidence, the essential coordinating skill. Haig, in the last weeks of the war, was the strategist.

The question which presents itself is this: would the war not have been a very different thing had Foch been Commander in Chief of all the Allied Forces and Haig his Chief of Staff from its earliest days. Haig may not have been the only figure behind 8 August. He was certainly no Marlborough, Wellington or Moltke. On the other hand, it is hard to see how victory on the 'Black Day' or in the days and weeks which followed could have come about in his absence.

CHAPTER 11

GUDERIAN AT THE MEUSE 1940

Heinz (Schneller) Guderian became interested in tanks and armor during World War I and elaborated his own doctrines thereafter. During the invasion of France, in May 1940, he was able to put to the test, at a pivotal point of the battle of France, his basic *Achtung! Panzer!* doctrine which finally made his reputation.

Guderian was the rather typical descendant of a Junker family from the Warthegau, his father being the first military man of this rather austere and poverty-stricken family. Born on 17 June 1888 to Lieutenant Friedrich and Clara Guderian he was taken all over Germany by his parents who were posted to obscure garrison towns of the Wilhelmian Reich. While the parents were stationed at St Avold (Lorraine) Heinz went to cadet schools, at first at Baden, and then in Berlin. He liked military life and even enjoyed the rigid obedience and discipline, but he also exhibited intellectual independence in argumentation, an exceptional quality in a young Junker bent on a military career.

In 1907 he graduated from the war school at Metz (Lorraine), which he had found rather tedious and was posted to serve in his father's regiment. As a serving officer he showed rather catholic cultural tastes: he enjoyed hunting, was keen on shooting and at the same time was keenly interested in architecture and countryside – paysage. Once again his intellectual independence was discernible in a healthy criticism of his surroundings. It was noted that he found it difficult to make friends, especially with his elders – this was to mark him for the rest of his life. It would always be easier for him to make friends and inspire his juniors. However, with seniors, be they his brother officers or politicians, he somehow failed to communicate.

In 1912 Guderian was engaged to a young girl, Margarete Goerne, whom he married the following year, just before going to the War Academy in Berlin. The marriage brought him in contact with brother officers Wilhelm Keitel, Erich von Manstein and Colonel Count Rüdiger von der Goltz. Both Keitel and von der Goltz were to exercise a profound influence on his life and career. As it happened Guderian never finished his war course; it was dissolved when World War I broke out in August 1914, and Guderian went back to his previous posting (shortly before going to the War Academy he had been on a radio course), the Heavy Radio Station No 3, part of the 5th Cavalry division. He was 26 years of age, specializing in signals, full of patriotism and determined to win the war for Germany. Instead the war proved his *Wanderjahre*: he was to become familiar with tanks and acquire the experience necessary to combat them.

His first combat experience taught him that without good signals bravery was useless. His I Cavalry Corps under General von Richthofen was given the task of cutting off the French Second Army by radio, but the French had broken the German code and were prepared for this drive. When despite this the corps actually performed the task and found itself north of Soissons in the enemy's rear the army superiors were unaware of this brave achievement and it went to waste as the German cavalry unprotected against enemy fire could not press home its advantage. Later Guderian insisted on having radio receivers in each individual tank — the protected cavalry could press home an advantage. While his experience was varied — he fought in initial battles, flew as an observer in aircraft, became a staff officer — he never had to face tanks directly, but heard of the French and British use of the tanks, as a terror weapon at Cambrai, Amiens and elsewhere. When the massed French tanks reversed the German breakthrough on the Marne he was, for the first and last time, involved in the

Guderian was never far from the action during the Blitzkrieg 1940.

mobile defense of the area of the Marne and Vesle. These experiences marked him for life.

Just before the armistice Guderian was sent to Italy on a liaison mission, and had to escape back to Germany to avoid capture. He witnessed the collapse of the German Reich in Munich, but then made his way up to Berlin where he was appointed to the Eastern Frontiers' Protection Service to keep an eye on the Poles and the Russian Bolsheviks. After a spell with the Iron Division in the Baltic lands, he was finally sent to the Inspectorate of Transport Troops in 1922. This in fact meant that he was back to his hobby-horse, mechanized formations. Furthermore he developed his ideas theoretically by writing articles for the *Militär-Wochenblatt*: he came to advocate the *Stosskraft* (dynamic punch) as the latest combat method. In 1914 the new combat methods had been mobile artillery and infantry machine guns — with technological changes the tanks (armor) became the new method. In 1927 Guderian was promoted to major and henceforth devoted all his military efforts to the theory and practice of armored warfare.

As early as 1928 Major Guderian founded a tactical center for tanks and their cooperation with other arms. Then he went to Sweden to drive the tanks in person — by the Treaty of Versailles Germany was not allowed to manufacture armor — and he also made a technical study of them. Back in Germany Colonel Guderian insisted on improvements in communications, not only with the tanks but also with field telephones and teleprinters. In fact he was obsessed with the formation of a new command, a sort of army within an army, incorporating elements from each arm. However both the new political leaders — the Nazis had come to power in 1933 — and the pro-Nazi military leaders, Blomberg and Reichenau, did not care for Guderian's ideas for the moment. It is true that so far Guderian had only produced halfbaked ideas and semidigested inno-

vations. For example, at this stage, he saw the future Panzer divisions as weapons of defense. He knew that the Germans had helped the Russians to develop the tank and he intended to use the German armor in defense of the Eastern Frontiers. However all this changed in 1934 when the new Commander in Chief, General von Fritsch, created the *Panzertruppe* and Colonel Guderian became the Chief of Staff of this new branch.

With Guderian's official appointment the theoretical justification of his belief in technological advance and armor finally came. Already a year before he had Captain Liddell Hart's articles on deep penetration by armored troops translated into German and the following year he published an article on another favorite subject: tactically, tank commanders needed to ride ahead of their squadrons controlling them by means of individual radios. While Guderian's book, *Achtung! Panzer!*, contained all the tactical innovations gained from long practical experience, the basic strategic ideas were in a sense the military response to Hitler's innovatory ideas on the war, and in particular to his concept of the Blitzkrieg, which was the main reason why this book attracted so much attention in 1935 when it came out.

Learning the lesson of the last war Guderian thought that the remedy of the Allied tank failure lay in the fully mechanized (Panzer) division moving into battle at equal speed: medium 'breakthrough' tanks armed with machine guns and cannon up to 75mm; antitank guns and motorized infantry following closely behind; tactical air forces delaying defenders' reserves and airborne troops capturing important enemy points in the rear. Guderian also spelled out his strategic doctrine, 'concentrated blow applied with surprise on the decisive point of the front, to form the arrow head so deep that we need have no worry about the flank.' As he seriously thought of applying these doctrines Guderian was suddenly replaced as Chief of Staff by Friedrich Paulus, and as a result the Panzers were split up as in all the European armies. Guderian's influence with the upper echelons and Hitler seemed to have vanished.

Guderian was most disconcerted by this development and at one stage lost his nerve, walked up to his old colleague Fritsch and blamed him for ignoring his advice on armored forces. Before Guderian could be punished Fritsch and Blomberg were suddenly dismissed by Hitler and he was appointed general officer commanding the XVI Corps and was detailed by Hitler to perform special duties involving the Panzers. Thus on 12 March 1938 Guderian led the Panzers into Austria and subsequently had to remedy the many breakdowns that occurred on that march.

Although an established armor theoretician and practitioner Guderian did not have it all his way. On the contrary top army echelons proved consistently hostile, especially after the adverse reports on the armored troops from Spain. Guderian had to counter criticism of himself in print. He praised strategic speed, 'the tanks had to keep moving despite enemy's defensive fire, thus making it harder for the enemy to build up fresh defensive positions.' In 1938, when military exercises had gone disastrously wrong, Hitler forced General Beck's resignation, and at long last showed a personal predilection for Guderian. Their dialogues became very close and personal but this in turn provoked

Left: Winston Churchill, General Gamelin and General Gort in France on 8 January 1940. This was during the Sitzkrieg when the Allies and Germans faced each other over the Maginot Line.

Right: During the *Drôle de Guerre* phase the soldiers indulged in propaganda exercises. They put up slogans on their lines to demoralize the enemy.

personal hostility from other senior officers. Thus Hitler sent Guderian's XVI Corps to occupy the Sudetenland and subsequently enabled him to visit Britain to do a little more research in British armor progress. Hitler used Guderian to divide top army officers, but though he thus obviously favored him, he made no headway with his obsessive armor warfare because of army obstruction. It took the Polish campaign to vindicate Guderian's theories.

In August 1939 Guderian, who late in 1938 was appointed general in command of Mobile Troops, conducted last-minute exercises with the 3000 tanks at his disposal and came to the conclusion that they could be put to good use in the coming Polish campaign, since the communication systems had been completed and supplies had been improved considerably since the disasters in the previous year. Thus according to him six Panzer divisions and four light divisions, aided by massive air intervention, would achieve a complete defeat of Poland in a few days, a feat that would have taken the remaining 45 German Army formations weeks to accomplish. Although his new XIX Corps exhibited certain shortcomings it in fact accomplished its tasks in a few days with perfection so that when Hitler visited the battlefield Guderian proudly showed him round proving to him the effectiveness of armor. As far as Hitler was concerned Guderian was preaching to the converted and Hitler sent him to Brest-Litovsk, to finish off Poland which gave him the opportunity to put into practice the strategic envelopment, from north to south by massed Panzers. The 4th Panzer Division, badly used, had taken a hammering and the German army had failed to take Warsaw from the south. Because of lack of command the operations had become confused, but ultimately Guderian's personality carried the day. On 13 September the 18th Polish Division surrendered and the following day Brest-Litovsk was taken after hard fighting. Guderian's Corps suffered only four percent casualties in the campaign, accomplished all its tasks and only half of the Panzers needed major overhaul. Although the generals minimized

the effect of the tanks (and Guderian) on the campaign, because of 'enemy impotence,' Hitler became convinced that the new weapon would win him the war. Guderian became his key to the victory and he was to be proved right in the near future.

Hitler was already planning the war in the West and on 28 September 1939 the German General Staff set to work on plans for the invasion of France. Both the Commander in Chief, General Walther von Brauchitsch, and the Chief of Staff, General Franz Halder had very little time to plan the operation, and even less confidence in its success. According to them the Maginot Line had to be outflanked and the *Schwerpunkt* of the whole operation was to be north of the difficult terrain of the Ardennes, in the general direction of Namur. On the right wing, Holland was to be subdued and simultaneously on the left wing, a relatively strong force would be sent through the Ardennes to reach the river Meuse between Givet and Sedan. No one seems to have liked this plan of 'improvisation'; its execution was several times postponed and after 10 January 1940, when a staff officer with the details of the plan made a forced landing in Belgium, it was abandoned.

Long before its abandonment the plan was criticized by Hitler, and Generals von Manstein and von Rundstedt. Hitler suggested that the drive through the Ardennes be enlarged and be aimed as the main attack across the Meuse toward Amiens and the Channel coast, thus cutting off the northern enemy wing. Manstein wanted a complete destruction of the northern enemy wing through a strategic encirclement similar to that of the elder Moltke, which the younger Moltke failed to carry out in 1914, and which Manstein and Rundstedt had successfully achieved in Poland only recently. Rundstedt sent the General Staff his suggestions which were remarkably similar to those of Hitler but with military meat in them (Hitler's were based on intuition and dreams).

Inevitably Guderian was brought in as the foremost Panzer expert. Early in November 1939 General Wilhelm

Keitel called him in for consultations. He explained that the fear of bad weather which would cause the immobility of the armor was the reason for the delay in the war against France, and he wanted to know if a strong tank force could pass through the Ardennes. Guderian was in a doubly strong position: he knew the terrain in 1914 and 1918 and from this knowledge he derived the certainty that Panzer divisions could pass through the difficult area. However after this general consultation Guderian was told that his XIX Corps would lead the drive on Sedan and he immediately considered his two Panzer divisions and one motorized division as insufficient and requested seven divisions as a minimum to ensure success. Needless to say this request was made when the Meuse operation was still subsidiary to that of Namur.

In January 1940 the war games were held by Rundstedt and they demonstrated the feasibility of the main attack against Sedan. However, Guderian's insistence on the armor spearheading the attack through the Ardennes, executing the crossing of the river Meuse and then advancing deep into France, was treated with scorn. General Halder thought Guderian's ideas senseless and Rundstedt supported him in this opinion. Nonetheless Guderian had

the courage to contradict them both and insisted on his suggestions: a surprise drive *en masse* would drive a wedge so deep and wide in the enemy's front so that the Germans would not need to worry about their flanks. Some generals were prepared to support Guderian in his stand and among them Manstein tipped the balance in Guderian's favor. On 17 February 1940 Manstein saw Hitler and described to him Guderian's plan, infecting him with his own enthusiasm. Next day Hitler saw Brauchitsch and Halder, told them that Guderian's plan was in fact his: that was the way France would be attacked and destroyed.

Thus on 24 February 1940 *Plan Gelb* (invasion of France) was revised and became *Plan Sichelschnitt*. Guderian's concept of the breakthrough in the Ardennes became the *Schwerpunkt*; in addition he personally was going to execute this new plan. Army Groups A, B and C under Generals von Rundstedt, Fedor von Bock and Wilhelm von Leeb remained responsible for the previous sectors, but Army Group B's role was relegated to become, in Captain Liddell Hart's words, a matador's cloak, while Army Group A with its Fourth (General Kluge), Twelfth (General List) and Sixteenth Army (General Busch) came to occupy the pivotal point of the *Sichelschnitt*. General Wilhelm List created a special Panzer Group under the command of General Ewald von Kleist, who in turn had under him General Reinhardt's XXXXI Corps, Guderian's XIX Corps and General von Wietersheim's XIV Corps. The center of both the strategic blow and the tactical fighting was Guderian with the 1st, 2nd and 10th Panzer Divisions and the crack motorized SS regiment, Grossdeutschland. The Kleist group, with some 1260 tanks (out of the overall 2800) was given priority of air support, while advancing, and massed bombing while actually crossing the Meuse. This arrangement made heavy artillery redundant, as otherwise the operation would have been greatly retarded; the narrow and bad roads in the Ardennes would have made it impossible for the artillery to follow up closely on the tanks — shell supplies would also be delayed. In March 1940 Hitler finally asked Guderian what he proposed to do

Left: A German soldier keeps a careful eye on activities behind the Maginot Line.

Below left: German troops take cover as shell bursts beside them.

Right: A propaganda picture showing a storm boat being used in an amphibious attack.

Below: Another posed picture demonstrating the usefulness of rubber boats. These boats were used during the crossing of the Meuse.

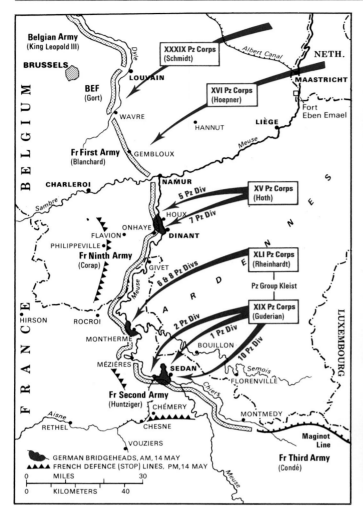

after the crossing of the Meuse, to which Guderian replied that he would have made a dash for Amiens or Paris, or the best optimal option, for the Channel coast. Hitler fell silent refusing to divulge his final objective, that is the destruction of France. However, Guderian's tactical victories implied in all three options would inevitably lead to the final objectives.

In the meantime Guderian concentrated on hard training. Tank crews practiced in the Eifel mountains under their general's personal supervision. There were map reading exercises, but the tanks could not fire their guns, because of shortage of ammunition. Infantry and engineers practiced river crossing, ferrying tanks across and then pontoon building. The river Moselle was almost identical with the river Meuse. During these two months of preparation all the ranks came to know their general personally; he inspired confidence and even enthusiasm. Many survivors speak of his fairness, paternal care and the personal share of the bustling general in the hardships of the exercises and combat itself. While his men had complete faith in his plans, Guderian's superiors, and occasionally even himself, were full of doubts. Guderian had no confidence in Brauchitsch and Halder, for although they were belated converts to his plans, they continued to vacillate. List wanted infantry to lead the assault across the Meuse. Rundstedt, like Hitler, refused to consider deep penetration after the Meuse. Kleist had no experience of armor. Busch never thought that Guderian would get across the Meuse, while Bock, whose

Army Group was relegated to a secondary role, thought the whole operation was crazy, 'You will be creeping along, 10 miles from the Maginot Line for your breakthrough, hoping that the French will watch inactive. You are massing all those tanks on the narrow roads in the Ardennes as if there was no such thing as air force. You then hope to operate as far as the coast with an open flank, some 200 miles long, with the French army letting you through.' However Guderian, after all the planning and practical exercises, became totally confident. He felt sure that if he were allowed to execute his plans, total victory would be his. Between 8 and 10 May 1940 the tanks of XIV, XXXXI and XIX Corps, stretching some 100 miles from head to tail, began to move into their border positions with Sedan as their principal axis of movement. Guderian together with 1st Panzer Division passed through Luxembourg and was ready on the Belgian-French border for the attack through the Ardennes. On the day of assault Kleist suddenly deflected the 10th Panzers from Sedan to Longwy much to Guderian's annoyance. After long arguments the 10th Panzers did not go to Longwy, but changed their axis of advance causing the 1st, 2nd and 6th Panzers to become entangled. Fortunately the French and British Air Forces made no attack on these confused columns and Guderian was able to sort out the confusion. However, everywhere else the schedules were kept and the advance was smooth. Halder described it as 'a very good marching achievement.'

The first battle on the frontier river Semois was decided

before the bulk of XIX Corps arrived in the area. The elements of the 1st Panzers forced the French 5th DLC to retire voluntarily toward the Meuse and at night Guderian took up quarters in the Hotel Panorama at Neufchateau, where he luckily escaped injury after a shell hit the hotel. On the following day Guderian again narrowly escaped injury after an air attack at Bouillon. Still with his 2nd Panzers lagging slightly behind Guderian took advantage of the French withdrawal and pushed his 1st Panzers toward Sedan, where its forward tanks arrived in the evening of 12 May. By nightfall the 10th Panzers also reached the Meuse at Bazeilles, but the 2nd Panzers were nowhere and Guderian flew to Kleist's headquarters for further orders. He found out that General Rommel's 7th Panzers had also reached the Meuse at Dinant and that the Kleist Group would attack generally across the Meuse in the afternoon of the following day. On the flight back Guderian's pilot lost his way and almost landed behind French lines.

On 13 May the Meuse was crossed by infantry and motorcyclists on either side of Sedan. The crossing was preceded by air bombing lasting five hours which demoralized the second-rate French troops facing the Germans. All through the night the Germans were constructing tank ferries and pontoons under Guderian's personal supervision. After the infantry had established bridgeheads on the other side of the Meuse, Guderian paddled across the river to meet his old friend, Colonel Balck, who greeted him with a joke 'about joyriding in a canoe on the Meuse' and the news that the bridgehead, three miles wide and six miles deep, was secure. The first tanks would begin the crossing at dawn. The French had no idea that the crossing would be attempted only one day after the German arrival and on the whole held their fire to save ammunition until the time when the real assault occurred. They were badly let down

Top left: German motorcyclists press into France.

Above left: Crossing the Meuse in Belgium on 13 May 1940.

Top right: A German bicycle squad cross the Meuse-Scheldt canal in Belgium on 13 May 1940.

Above right: The remains of a French column in Sedan wood, 14 May 1940.

Right: The Blitzkrieg left a trail of destruction in its wake as witnessed from this picture of a Channel town.

by the air forces, which because of the lack of communication between the French and the British commands, failed to support them. The Germans had to do some hard fighting to dislodge the French from their bunkers and defensive lines. However, once they had achieved it the French High Command panicked; General Corap, commanding the Ninth Army ordered general retreat.

However, on the French side it was not all panic and confusion, despite the fact that General Lafontaine had under his command only reservists. General Grandsard in command of the Meuse sector, at first thought that nothing was lost, but then he heard reports that German tanks were in the La Marfée area. The reports were incorrect, but they panicked two of his colonels into issuing withdrawal orders. General panic spread and Colonel Pourcelet committed suicide. While Lafontaine withdrew his headquarters to Chémery, Grandsard sent two tank battalions and two infantry regiments to counterattack. Behind them General Joseph Georges, who had dismissed Corap and replaced him by General Henri Giraud, was sending up to Sedan the powerful units of the Third Army and 3rd Motorized Division. Lafontaine ordered a two-pronged attack at dawn, but the units were not ready in time for the counterattack, which started at 0700 hours after the German tanks had already crossed the river in force. To plug the hole on the Meuse, between Dinant and Sedan, the French command ordered its 1st, 2nd and 3rd Armored Divisions into counterattack. However, the 1st Division found itself without fuel and was subsequently annihilated and the 2nd entrained and was cut into two at Hirson. Only the 3rd actually went into battle against Guderian, but from the beginning proved rather static, went into battle piecemeal and was chopped up not even by German armor but mainly by the *Sturmpionierbatallion*. Initially the situation looked rather ugly for the Germans, but after they had put out of action about one half of the French

tanks, the French turned round and withdrew. During the battle Guderian went back to La Chapelle to prepare orders for the following, decisive day of the battle. After the French counterattack the bulk of the French 55th and 71st Divisions had ceased to exist, fleeing in disorder. The contagion of fear and panic also spread to Grandsard's X Corps boding ill for the French armies still resisting the German assault. By the 14th Guderian's bridgehead was 30 miles wide and 15 miles deep and some 85 Blenheim bombers had been shot down over the German pontoons.

On the 14th Guderian's 1st Panzers moved resolutely against Chémery and the 10th Panzers supported by the SS Grossdeutschland pushed onto the high ground of Bois Mont Dieu, both effectively preventing the French XXI Corps from assembling for a counterattack, though hard fighting went on for the rest of the day at the key village of Stonne. Guderian was presented with a gap of 12 miles through which he could drive his 1st Panzers in order to break toward the west and the Channel thus fulfilling his wildest hopes and executing Plan *Sichelschnitt* to perfection. At noon Guderian reported to Rundstedt on the state of the battle, 'in the very middle of the bridge, while air raids were in progress.' Then he issued new orders to his men: they were to cross the Ardennes canal and head west. This was a risky decision for the Second and Ninth French Armies were still counterattacking and far from routed. The southern flank at Stonne was only guarded by the battered Grossdeutschland and 10th Panzers, while there was a

Left and below: Pz Kw 38(t) tanks in action in May 1940. These were Czech tanks which were redesignated to serve with the Wehrmacht. They equipped the 7th and 8th Panzer Divisions.

Bottom: Another French town is engulfed by flames.

Far left: The destruction of the Maginot Line.

Below left: German soldiers pick off snipers in a small town in Lorraine.

possibility of the Allied air forces succeeding in destroying his pontoon bridges. Guderian therefore moved forward his AA batteries to ring off the bridges: they brought down some 112 aircraft and Guderian's armor went on crossing the Meuse. By nightfall Colonel Balck had reached his objective at Singy and the French Generals Flavigny and Huntziger caused confusion by issuing frequent and contra-

dictory orders. The greatest battle that Guderian had to fight on that day was with his superior, General Kleist: the latter insisted that Guderian consolidate his southern line before pushing westward to Rethel. Guderian was both furious and bitter, as he saw himself deprived of the fruit of his tactical victories. In the end Guderian won the argument, for the generals on his flanks, Hoth and Rein-

Above: Panzer Mark IIIs (background) of Guderian's Panzer Corps cross the cornfields of northern France.

Left: General Heinz Guderian leaves a meeting. Note the censor has whited out his briefcase.

Top: French forces surrender to the Germans in Lille, 29 May 1940.

Top left: The retreating British blew up this bridge over the Meuse in Belgium, 22 May 1940.

Opposite top: Guderian plans the advance of his Panzer Group at Langres, June 1940.

hardt, were also on the point of breaking through.

Some French historians speak of 15 May as the day France lost the war, but for Guderian it was a day of miracles: considering the nature of the terrain it seemed miraculous that his Panzers succeeded in breaking through. French counterattacks were constantly delayed for technical reasons and the Germans were able to deploy antitank guns and rush up reinforcements. However, the progress of 1st and 2nd Panzers on that day was not impressive as they had to smash through tough French defense not yet confused and demoralized by superiors' orders. General Brocard was dismissed, but the French armor was nevertheless routed by all three German generals, Guderian, Hoth and Reinhardt. Nonetheless Guderian once again had a verbal battle with Kleist who ordered him in the evening to stop advancing. Some lively shouting on the telephone won Guderian 24 hours of freedom to advance: at dawn he arrived at Bouvellemont where the 14th French Division, after fighting hard all day under the future Maréchal de France de Lattre de Tassigny, had just withdrawn from the combat. There and then Guderian personally briefed the commanders for the operation ahead and instead of resting the troops set out immediately.

On 16 May Guderian thrust forward only with his 1st and 2nd Panzers, leaving the 10th Panzers and SS Grossdeutschland on his southern flank as insurance, and in deference to Kleist. By the end of the day they were 40 miles west, at Dercy on the Serre, while XXXXI Corps reached Guise on the Oise. Guderian then issued orders by radio for the following day: advance was to continue. However, he did not reckon with Kleist whose orders he was disobeying. Kleist's HQ monitored Guderian's orders and issued counterorders. Kleist also asked Guderian to report to him the following morning. When they met, on 17 May 1940, the exchanges were acrimonious as both men were on edge and their nerves were rapidly giving up from excitement. Kleist accused Guderian of deliberately disobeying his orders and Guderian offered his resignation on the spot. When Rundstedt heard of it he immediately ordered Guderian to remain in charge and sent General List to act as mediator. Once again Guderian won the day: his HQ remained on the spot, but his troops were permitted their reconnaissance in force. Guderian just lay his communication wires and dashed away from his HQ to join his advancing tanks. At the end of that day his vanguard passed Montcornet and Laon. However, on that very day Hitler joined Brauchitsch and Halder and the rest of the generals in their fear of French counterattacks from the south: thus the disputes continued, while the Panzers rolled on gathering speed as they went. The very speed of their advance made it impossible for the French armies to react.

Guderian's 2nd Panzers reached St Quentin on the 19th and Abbeville on the following day. At long last the French command ordered counterattacks, but because of the changes in the personnel the offensive was disrupted and never materialized.

On 21 May, near Amiens, British 4th and 7th Tank Regiments and the French 3rd Mechanized Division unexpectedly turned on General Rommel's 7th Panzers and caused a lot of anxiety, but Guderian would have dealt

with the threat as efficiently as he did at Montcornet, when de Gaulle's 4th Armored Division counterattacked. Though Guderian was ultimately halted for four days (24–27 May) which saved Dunkirk for the British Expeditionary Force, his tanks took Boulogne and Calais thus cutting Allied armies into two. On 28 May Hitler put him in command of a special Panzer group which then transferred to the east and which during the three weeks of June 1940 cut across eastern France from Châlons sur Marne to Pontarlier and Belfort. France was defeated and on 22 June signed an armistice.

Even with hindsight it is thought by military experts that the battle of France, had it been fought along the lines of Plan Gelb, would have ended in a deadlock. It can be said without exaggeration that Guderian's tactical and strategic ideas embodied in Plan *Sichelschnitt* virtually decided the issue. The factors of surprise and the psychological blow that the French command and armies suffered after the Germans emerged from an impenetrable area and effected their main breakthrough were decisive for this battle of France. In the execution of his ideas Guderian acted resolutely having absolute faith in them, while his political and military superiors and masters dithered.

At this stage of World War II he was years ahead scientifically and technologically of his friends and foes and his achievements on the battlefield fully vindicated his theories and doctrines. Thus for Guderian the battle of the Meuse, which really meant the battle of France, was a perfect victory, never to be repeated again so clearly by any other warlord.

CHAPTER 12

ROMMEL
AT TOBRUK 1942

Even before the beginning of World War II the great desert wastes of North Africa captured imaginations in Europe and America, and once the Desert War began in earnest, much of the romance formerly engendered by the French Foreign Legion became attached to the new desert warriors. The Afrika Korps, the Long Range Desert Groups, the spies sipping gin in the clubs of Cairo and the British Tommies sweltering in their tanks — all have won a place in modern mythology.

The campaign in North Africa fascinated the British, especially Churchill, to the extent that soon its strategic considerations became almost overshadowed by the psychological need to win a great victory over the Afrika Korps and its legendary leader, General Erwin Rommel, the Desert Fox.

Churchill described Rommel as a master of war. Troops on both sides called him, half-affectionately, 'that bastard Rommel.' During the desert campaign he was to become the best-known and — even in Britain and America — practically the most popular figure in the Middle East owing to his brilliance, his courage, his skill at tactical maneuver (often when gravely outnumbered), and his old-school, chivalrous attitude toward his opponents. To the British soldier he came to be the personification of the bold feint, the surprise attack, the wild chase. 'Rommel is coming!' would race through the troops with the speed and destructive force of a forest fire. In fact, General Sir Claude Auchinleck, Commander in Chief of the British forces in the Middle East, found it necessary to issue an extraordinary Order of the Day, begging his senior officers to 'make every effort to destroy the concept that Rommel is anything more than an ordinary German general,' adding that, 'this matter is of great psychological significance.'

Thus Tobruk, a Libyan port not far from the Egyptian frontier, soon assumed an importance far beyond its already vital strategic position. The British captured it from the Italians early in the desert campaign in January 1941. In the spring of 1941 the stubborn Australians who held out for some nine months against Rommel's siege became the heroes of the Empire. When the fortress fell to the Afrika Korps a year later it was a crushing blow to the British public and Rommel's reputation was not so much made, as confirmed.

Erwin Johannes Eugen Rommel was born in Heidenheim, near Wurtemburg, on 15 November 1891. He was one of five children, and his sister describes him as a gentle, amiable, dreamy child. He was small for his age, and had little interest in books or sports. In his teens, however, he suddenly woke up and developed a keen interest in physical fitness, spending every spare minute on his bicycle or on skis in winter. The dreamer became a hard-headed, practical youth who did well in school and studied aircraft design as a hobby.

In June 1910 Rommel joined the army. Since the family had no military connections (his father and grandfather were both schoolteachers and mathematicians), he joined the 124th Infantry Regiment at Weingarten as an 'officer cadet' — which meant he had to serve in the ranks before going on to a *Kriegsschule* (War Academy). He was promoted to corporal in October, made sergeant by the end of December, and was posted to the Kriegsschule at Danzig in March 1911.

As a young officer he was good at drill and was found to be especially effective in training recruits. Like the young Montgomery he showed an unusual interest in the smallest details of military organization — but unlike Monty, he was never argumentative, usually preferring to listen rather than talk.

During World War I Rommel showed the boldness, independence, and understanding of the value of surprise that he would use to such advantage almost 30 years later. He stood out almost immediately as a perfect fighter. He was cold, cunning, ruthless and untiring — quick to make decisions and incredibly brave. Before the year was out he was transferred to a mountain battalion where his remarkable eye for country and his ability to ignore physical discomfort were invaluable. By the end of the war he had been wounded twice and had been awarded not only the Iron Cross Classes I and II, but also Germany's highest decoration, the *Pour le Mérite* (corresponding to Britain's Victoria Cross or the United States' Medal of Honor).

His comrades from this period invariably describe him as having *Fingerspitzengefühl* — a sixth sense, or intuition in his fingertips. To a man they credit his tactical genius — and remark that he was often disliked by his fellow officers, from whom he expected as much as he did from himself.

At the end of the war Captain Rommel turned to peacetime soldiering, ending up in November 1939 a colonel in command of the Kriegsschule at Wiener Neustadt. He lived a quiet life with his wife and son. He did not smoke, drank little and did not care much about food or parties. He was good with his hands, and could make or fix almost anything.

Like the majority of his fellow countrymen, Rommel considered the Nazis upstarts but saw Hitler as an idealistic patriot who had a chance to pull the country together and oust the Communists. This attitude may have come more easily since he was not part of the snobbish Prussian clique that included so many regular army officers. He was never close to Hitler, which perhaps is one reason why his admiration took so long to wear off, but through their brief encounters in the early days of the war the Führer developed a liking for him. Rommel, for his part, admired Hitler's magnetic personality, his amazing memory for technical detail and his physical courage.

At the outbreak of World War II he was commander of the Nazi headquarters in Poland, but by June 1940 Hitler had given him a fighting command, at the head of the 7th Panzer Division of the 15th Armored Corps (the 'ghost division' that took part in the stunning German advance through France).

On 15 February 1941 Rommel — now a newly appointed lieutenant general — was given command of 'the German troops in Libya.' His mission was to assist the Italians and help prevent a British breakthrough to Tripoli. In two months General Sir Archibald Wavell's small force of three British divisions — 31,000 men, 120 guns and 275 tanks — had chased 10 Italian divisions totalling over 200,000 men across 650 miles. Though Hitler still could not bring himself to take North Africa all that seriously, he had finally realized that if something were not done quickly Britain

itself the Afrika Korps.

For the next two years they would be fighting over a featureless wasteland of gravel and scrub, dotted with ancient wells and the tombs of sheiks. The battlefield stretched almost 400 miles, from Dernia in the east to El Alamein in the west. The only road followed the coast and the only dominant physical feature was a 500-foot escarpment that faced north to the coastal plain, descending from the limestone plateau where the armies maneuvered. With no natural obstacles and few civilians or settled areas except along the coast, the campaign over the vast spaces of the desert bore more resemblance to a naval war than to a traditional European land conflict. Rommel was the first to really appreciate the fact and fully exploit the freedom of movement it gave him.

In addition to their human adversaries, the soldiers fought the desert. Temperatures rose to 120 degrees in summer, with the stones and sand remaining burning hot until late at night. The sand got everywhere, in hair, underclothes and machinery. Insects and sand fleas viciously bit everyone, general and enlisted man alike, and dysentery was rife. The men grew lean and tanned, with skins like leather. The infantry prepared their positions, the sappers laid their mines and the tank crews sweated in their tanks, all in a climate which many would have considered impossible for Europeans to work in at all.

The existence of a common enemy – the desert – may have been one reason why the North African campaign was, by and large, a 'gentleman's war.' Neither the Afrika Korps nor the British abused prisoners. In fact the former, according to Rommel's biographer Desmond Young, treated enemy prisoners of war with, 'almost old-world courtesy.' Compliments were traded sincerely: Rommel often described General Wavell as a 'military genius' and commander of the highest order, while for his part Wavell sent Frau Rommel a copy of his lectures on generalship inscribed 'to the memory of a brave, chivalrous, and skillful opponent.'

Much of the credit on the German side must go to Rommel himself, who had strong feelings about correct behavior and observance of the 'soldier's code' – feelings that were shared by most officers in the regular German Army (with a few notable exceptions). He was also lucky not to have any SS troops attached to his command and to be a continent away from headquarters.

Perhaps because of their early successes in the hostile desert environment most people believed that the Afrika Korps was an elite volunteer unit, specially trained in desert warfare. On the contrary, the men of the Afrika Korps had, for the most part, been plucked out of the Balkans and set down in Tripoli with no preparation at all. They were typical German soldiers – strong, brave, well-trained and disciplined. They were not well-suited to the desert. Many, especially the very young, the very blond and the older veterans of World War I, could not adapt to the new, rigorous conditions as well as did the British colonial troops or even the British regular troops. Few had ever been out of Europe before and they lacked the British experience with foreign lands. It was difficult, for example, to teach the troops not to drink untreated water. German doctors knew less than their British counterparts about

would gain control of the entire southern shore of the Mediterranean, leaving Europe's 'soft underbelly' dangerously exposed.

Rommel built up his army and eventually had at his disposal two Panzer divisions (the 15th and the 21st), an Italian armored division (the Ariete), a German infantry division (the 90th Light), a 'frontier group,' two Italian motorized divisions and four Italian infantry divisions. They formed Panzer Army Africa but the German section called

Left: Rommel and General Gariboldi on a tour of inspection in Tripoli, February 1941. This was shortly after Rommel arrived in North Africa.

Far left: General Sir Archibald Wavell was Commander in Chief, Middle East until July 1941.

Far left below: German tank passes through a Libyan town.

Below: General Rommel and General Cavallero inspect Italian troops in Tripoli.

tropical medicine and German field hospitals were far inferior. Sickness, especially dysentery and jaundice, took a heavy toll in the early days of the campaign.

The Afrika Korps had some definite advantages as well. Though they had less transport, their weapons outclassed those of the British, especially in the beginning, and they understood more about using them effectively. The men had better prospects of leave and received more news from home (even publishing their own paper, *Oase*) — all of which improved morale. In addition, they were a homo-geneous group, in distinct contrast to the Eighth Army, which was a mixture of divisions from every part of the far-flung British Empire.

With all this it was Rommel who, by sheer force of personality, turned the Afrika Korps into a tough, resiliant and self-confident fighting unit. He taught each man to give 150 percent and never to admit defeat, so that even as they were marching down to the prisoner-of-war ships in 1943 they were able to hold their heads high.

Life in the desert was unfamiliar and rigorous, but

Right: Destroyed British tanks, knocked out during Rommel's whirlwind advance through North Africa in 1941.

Right below: Following the fighting at Ras el Madauar on 30 April 1941, the first British prisoners to be taken outside Tobruk are marched into captivity.

Below: Rommel decided to bypass Tobruk and push forward into Egypt but the city remained a thorn in his side. Rommel looks at the front following the failure of the British Battleaxe operation, 19 June 1941.

Rommel found it no problem becoming 'desert-worthy' (a British term that came to be applied to anyone and anything that functioned effectively in the desert). Though no longer young, he was in fine physical condition, with a Spartan streak that made him take pride in the stoic acceptance of discomfort and fatigue. As one of his officers put it later, 'he had the strength of a horse . . . he could wear out men 20 and 30 years younger.'

He did not need much sleep nor did he care much about what he ate. He insisted on getting the same rations as his men, which were not too good (the food that arrived on the supply ships was usually much too heavy for the desert and did not include fresh fruit or vegetables). He was appalled when he discovered that there were three 'classes' of rations in the Italian army — for officers, noncommissioned officers and enlisted men. His reading was limited to newspapers or books about military topics. Though he did have a mild interest in North African history, the story that he was an avid classical scholar and archeologist was fabricated by the German propaganda machine.

Most days he was awake by 0600 hours, and by 0630 hours — always shaved and in uniform — would be off on his daily round of the positions. Sometimes he took to the air, piloting his own Stuka, but more often he drove himself around the desert in his Volkswagen. He would arrive at the most isolated post without warning — and it was an unlucky senior officer who was caught in bed after 0700 hours. Nor were the visits mere formalities. He missed

the rest of the evening.

Rommel was fond of young people and was very definitely a 'front-line type' so it is not surprising that the men of the Afrika Korps idolized him. The younger officers admired him especially, for his concern with their problems, his tactical ideas and his skill in navigating the desert. General von Esebeck has said, 'He had a smile and a joke for everyone who seemed to be doing his job . . . he had a very warm heart and more charm than anyone I have ever known.'

His charm and sensitivity, however, were not always so apparent to senior officers, with whom he could be impatient and, on occasion, brutal. He was in the habit of giving orders directly to subordinates instead of following the chain of command and would fly into a rage if his orders were questioned. He also tended to rely on his own judgment much more than his superiors would have wished – and the fact that he was so often right probably did not make the pill any easier to swallow. His relationship with his nominal Italian superiors was difficult, while he made no secret of the fact that he considered Halder and many others on the German General Staff fools who had no practical experience of war. Even Hitler, who liked him personally, would become angry at his 'veiled' reports to headquarters.

One of his most distressing habits was that of dashing around the battlefield, often taking his Chief of Staff with him and leaving no one with authority at headquarters. While his officers fretted in the rear, Rommel would be racing around the front.

In fact, his system of command was not haphazard. He and his commanders did not lead attacks from the front line out of a love for histrionic heroism, but because of their

nothing, from a badly positioned machine gun to an inadequately camouflaged minefield. At the end of the day he would eat dinner in about 20 minutes, listen to the news on the radio and write his daily letter to his wife. Official papers or letters to the survivors of his World War I battalion (with whom he was a faithful correspondent) took up

knowledge of armor tactics and troop psychology. Time and again a tank crew who had halted their vehicle would hear a loud knocking on the side. Opening the turret they would discover Rommel's aide, Lieutenant Freiherr von Schlippenbach using a crowbar as a knocker, while the General, standing up in his car, shouted, 'On your way! Attacks don't succeed by standing still!' Though he took great personal risks, his uncanny instinct for what the enemy was going to do saved his life on several occasions.

His admirers will point out that this personal command was one of the main reasons why the Germans were consistently ahead in the speed of decision making and velocity of movement. Even his harshest critics will admit that he was extraordinarily brave, wonderful with his men, generous with praise and willing to admit when he was wrong. In sum, he was the best possible commander for the desert war at that time.

Since he was frequently outnumbered Rommel's main strategy was to keep his troops concentrated in the open and to use the vast desert for wide maneuvers to strike at the flanks or rear of the enemy, dividing their forces so that they could be more easily disposed of. The British frequently appeared to be cooperating with this strategy. As he commented to one captured brigadier near the end of Operation Crusader, 'What difference does it make if you have two tanks to my one when you spread them out and let me smash them in detail? You presented me with three brigades in succession.'

His main contribution to tank tactics was the use of a screen of self-propelled antitank guns to shield his Panzers as they advanced, withdrew or refuelled. Although he had several 105mm guns with tapered barrels that were highly effective against tanks, his most valuable weapon was the 80mm gun with high-velocity, armor-piercing shells that could inflict serious damage on both tanks and aircraft. Wherever possible artillery, rather than tanks, was used for defense, to help keep tank losses down. The tanks themselves – 22-ton Mark IIIs with 75mm guns – were superior to Allied tanks. German tank recovery operations, too, were much better organized than those of the British.

Rommel used every ruse he could think of to deceive the enemy or keep them off balance. The first order he issued upon landing in Tripoli was for the construction of dummy tanks. Transport trucks – some with airplane engines mounted on the back – were often used to raise dust and create the illusion of a major army on the move. Captured vehicles were used not only for extra transport, but also to confuse the British.

Another innovation was the formation of 'combat groups.' These were about the size of a battalion and consisted of a tank company, a mixed company of antitank artillery and flak (75mm guns, 50 and 37mm antitank guns and 20mm flak) and a column of armored reconnaissance trucks and radio cars. These groups proved to be the ideal fighting unit for the desert and Rommel had several under his personal command, constantly flinging them into the hottest parts of the battle.

Mines were a major weapon for both sides. Probably in no other theater did they play such an important role. Air support was important too, for it could seriously damage the enemy's supplies and communications lines. However in the desert it had limited effectiveness against front-line

Far right: German soldiers climb aboard a tank during a lull in the fighting, 1941.

Bottom: The main square in Tobruk after it had fallen to the Germans, 21 June 1942.

Below: Rommel observes the fighting from a safe distance during the Crusader Battles at Sidi Azeiz, December 1941.

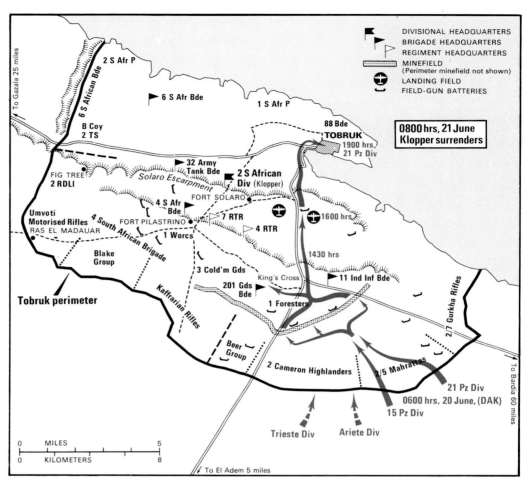

DIVISIONAL HEADQUARTERS
BRIGADE HEADQUARTERS
REGIMENT HEADQUARTERS
MINEFIELD
(Perimeter minefield not shown)
LANDING FIELD
FIELD-GUN BATTERIES

0800 hrs, 21 June
Klopper surrenders

To Gazala 25 miles
2 S Afr P
6 S African Bde
6 S Afr Bde
1 S Afr P
88 Bde
TOBRUK
B Coy
2 TS
1900 hrs,
21 Pz Div
32 Army Tank Bde
2 S African Div (Klopper)
FIG TREE
2 RDLI
Solaro Escarpment
FORT SOLARO
4 S Afr Bde
1600 hrs
Umvoti Motorised Rifles
RAS EL MADAUAR
4 South African Brigade
FORT PILASTRINO
7 RTR
4 RTR
1 Worcs
Blake Group
1430 hrs
3 Cold'm Gds
King's Cross
11 Ind Inf Bde
201 Gds Bde
2/7 Gurkha Rifles
Kaffrarian Rifles
1 Foresters
Tobruk perimeter
Beer Group
2 Cameron Highlanders
2/5 Mahrattas
21 Pz Div
0600 hrs, 20 June, (DAK)
15 Pz Div
To Bardia 60 miles
Trieste Div
Ariete Div

0   MILES   5
0   KILOMETERS   8
To El Adem 5 miles

strength. It could hamper enemy movements but it could not check an advance on the ground.

The real battle in the desert war was for vital supplies, especially gasoline and water. The drivers of the supply trucks on both sides are the, largely unsung, heroes of the campaign. Doggedly they drove their three-ton trucks across burning sands when tires burst like balloons, through freezing nights and through sandstorms, where visibility was reduced to about a yard and lookouts had to lie along the front fender clinging to the radiator, to shout directions to the driver. In the Afrika Korps the motor-cyclist was the jack-of-all-trades, fetching ammunition, directing the artillery and carrying the wounded to dressing stations – all with a vehicle that could not have been more inappropriate for the terrain.

During 1941 and 1942 the campaigns in North Africa were inconclusive. In January 1941 the British took Tobruk, along with the rest of Cyrenaica, from the Italians. On 31 March Rommel, who had been told to submit plans for a counteroffensive by 20 April, sent out a reconnais-sance raid in force. The British front collapsed in front of him and within 12 days, to everyone's surprise, he had recaptured Bardia, only a few miles from the Egyptian frontier.

He had recovered all the territory the Italians had lost except Tobruk, which he had under siege, and which the British were determined to hold at all costs. The British were embarrassed. Their prestige sank to a new low in the United States and the Soviet Union, and their future seemed dark indeed. More important, they found themselves with

a very precarious hold on the eastern end of the Mediterranean, and it became obvious that now they were going to pay for their neglect of tanks and tank warfare between the wars. The British tanks were too slow and their guns too weak to compete with the German equipment.

While Rommel turned his energies to asking Hitler — in vain — for four more Panzer divisions, Churchill had no such reservations. North Africa was no sideshow to the British; in fact, it was to become the focus for the greatest single military effort of the Empire. On 12 May a large convoy survived the risky journey across the Mediterranean and delivered 238 tanks to General Wavell. Wavell, knowing that Rommel was short of supplies and fuel, launched an immediate offensive (Operation Brevity) to relieve Tobruk. However, instead of retreating, Rommel counterattacked and chased the Eighth Army back into Egypt.

With his tanks outnumbered four to one, Rommel had to come up with a way to keep the British off the coast road to Tobruk. He deployed his small forces to create a bottleneck where they could stall the next offensive until he had time to bring up his reserves. Meanwhile, he joined Admiral Raeder in pressing for a decisive offensive against Egypt and Suez — a concept which Hitler did not even begin to understand. The Führer continued to give Barbarossa (the campaign against the Soviet Union) top priority — one of his greatest blunders of the entire war.

The British offensive, Operation Battleaxe, came on 14 June. It was Wavell's last chance if he wanted to hang on to his Middle East command — and it failed. By the end of the day he had lost almost half his tanks in Rommel's trap.

True to form, the Germans counterattacked on 15 June. The result was another stalemate.

Churchill immediately removed Wavell and sent General Sir Claude Auchinleck as his replacement, giving him everything he had denied his predecessor — men, materiel, and time to prepare. By the time Auchinleck launched Operation Crusader on 18 November 1941 he had an unprecedented superiority in men and equipment — almost three times as many planes, twice as many tanks and tens of thousands more men. The British neutralized their numerical superiority by dividing their force up into several attack groups for speed and flexibility — which the wily Rommel simply dealt with one at a time.

The result was another draw. Though Rommel could blunt Crusader with his skillful flanking movements and unexpected counteroffensives, he was not strong enough to defeat it. With no prospect of reinforcements in the near future, he decided to conserve his forces and withdraw to safety in Tripolitania. The siege of Tobruk was abandoned on 7 December and the Axis forces moved back across Cyrenaica, their withdrawal punctuated by a series of sharp, savage battles that took a heavy toll among the (mostly Italian) infantry units.

On paper it looked like a major victory for the British. However they were exhausted and their supply lines were long and weak, while Rommel was sitting at his home base in strong, well-prepared defensive positions. At about the same time the Germans belatedly realized that Malta was the key to the North African supply routes and began immediate steps to neutralize it. In August 1941 some 35

Right: A Messerschmitt Bf 110
prepares for take off. The Bf 110
could not compete with Spitfires
and Hurricanes and was used in
the 'safer' theaters in North Africa,
the Balkans and Russia.

Below: Sand is raised during
shelling.

percent of Rommel's supplies and reinforcements had ended up on the ocean floor. In October about 63 percent had been lost and by November the flow was down to a trickle. However, in January 1942, after a concentrated effort by the Luftwaffe, not a single ton of supplies was lost.

Rommel had no thought of remaining on the defensive a moment longer than necessary, and the arrival of several dozen tanks on 5 January was enough to start him planning the next offensive – over the strenuous objections of his nominal superiors. Near the end of the month the worst weather in many years – alternating sandstorms and heavy rains – offered the perfect cover for a surprise attack. The Panzer Army Africa emerged from its defensive positions in one of Rommel's most brilliant displays of opportunism and agility.

Benghazi, with its large supply of fuel and ammunition, fell on 25 January but by 8 February the Axis impetus was

spent. Rommel was brought to a standstill at a line stretching from the coast just west of Gazala, south to the British-held oasis at Bir Hacheim. As both sides settled down to build up their resources, activity dwindled to a series of harassing operations as the British long-range desert groups and German frontier patrols stabbed at each other's lines.

Thus, in spring 1942, little had changed since the beginning of the year. The Eastern Front was still static after the long winter. In the Pacific the Japanese continued their string of victories – and continued to ignore Hitler's request for a strike against India. In North Africa Axis and Allied forces still faced each other at the Gazala Line, some 50 miles west of Tobruk.

The Gazala Line was a 40-mile-long stretch of minefields from the coast to Bir Hacheim. Minefields alone cannot stop tanks for long, so Auchinleck and Ritchie had studded the line with a series of static field dispositions, called 'boxes.' These strongpoints stood alone, like medieval castles. They were usually circular barbed-wire entanglements about two miles in diameter, enclosing a minefield with listening posts, machinegun nests and gaps covered by artillery. The garrison, about the size of a brigade, was supplied for long-term defense and had two jobs: to guard the minefield so that the enemy could not clear a path through it at leisure and, in the event of a breakthrough, to form pockets of resistance to harass his flanks, rear and communication supply lines. The armored and motorized units ranged around behind the boxes, ready to fall on the enemy forces while they were tied up at the strongpoints.

On paper it was a brilliant defense system, but it made the same mistake as Crusader, splitting up the forces into small groups. Rommel, by consolidating his forces, was able to trample through the defenses in one great sweep before the superior British numbers and newly arrived American Grant tanks began taking their toll.

For his Tobruk offensive Rommel fielded 561 tanks, though the only really effective ones were 280 German medium tanks – the 228 Italian 'sardine tins' and the German light tanks hardly counted. The British Eighth Army had some 850 tanks, including 167 Grants, with 75mm guns that out-shot everything on the battlefield except Rommel's 19 Mark III Specials. Rommel's 50mm, 75mm and 88mm artillery pieces were far superior to the little British two-pounders. In the air the German Luftflotte 2, with 542 aircraft, opposed the 604 planes of the Desert Air Force.

On 26 May Rommel made his move. At the unusual time of 1400 hours, General Crüwell led a magnificently staged feint on the Gazala Line. In the center was the 361st Regiment of the 90th Light Division (hot-headed, hard-fighting ex-Legionaries), on the flanks were Italian infantry divisions and behind them all raced the motorized units, recovery vehicles – anything with wheels – throwing up enough sand for two entire Panzer armies. Behind all this activity, at 2030 hours, Rommel gave the code word 'Venezia,' and five divisions moved due south, navigating by the stars, toward Bir Hacheim.

As dawn broke the Italian Ariete Division (20th) split off to engage the Free French garrison at Bir Hacheim, while the rest of the force – the two Afrika Korps armored divisions (15th and 21st, with the 90th Light and reconnaissance units on the left) – moved up behind the Gazala defenses toward the coast. Soon they encountered the 3rd Indian Motor Brigade Group, newly arrived from Egypt, and within minutes an alarmed report was speeding to Headquarters, 7th Armored Division, 'We have the whole bloody Afrika Korps in front of us!'

By midday on the 27th Rommel had good reason to congratulate himself. His force was making considerable progress against various British armored units, in fulfillment of his first principle of desert warfare: concentrate your own forces and split the enemy's, destroying them at different times. Later, led astray by faulty intelligence, he decided that he had disposed of the British armor and could safely attack the rear of the enemy infantry positions. It was a costly error. His divisional commanders, throwing caution to the winds, flung themselves into the attack without adequate artillery support. Suddenly they were attacked on all sides by British armored units – including the new Grants and American six-pounder guns. By late afternoon the armored units were encircled in the north and the 90th Light immobilized in the east, both cut off from their supplies and transport. Tactical cohesion broke down and the attack foundered in the desert south of Knightsbridge. By nightfall the 15th Panzers were completely out of gasoline and the 21st was only slightly better off.

All through the 28th Rommel worked desperately to get supplies up to his strike force, which was continually being harassed by British tanks and all the British air power in the area. Typically, he never once thought of withdrawal. During the night General Crüwell was ordered to try to break through the main Gazala Line to help, but before the operation could begin his plane was shot down and he was captured. The day was only saved from total failure by the Italians. The Ariete abandoned their attack on Bir Hacheim and made their way up to Bar el Harmat where they set up an effective antitank screen to cover the supply columns. Meanwhile, the Trieste and Pavia divisions had managed to clear preliminary gaps in the minefields from the west, where they were crossed by the Trigh el Abd and Trigh Capuzzo caravan routes. Unfortunately both gaps were covered by the 150th Brigade Group box at Got el Ualeb – of which Rommel was not yet aware.

Rommel spent all day on the 29th snaking supplies to his stranded armor, while the delayed attack on the main Gazala Line failed to break through the South African and 50th Division positions. The tank action that day around Knightsbridge, in blazing heat and blinding dust storms, was some of the fiercest so far in the campaign. At sundown both sides fell back exhausted and Rommel pulled his forces into a close defensive formation between the Sidra and Aslagh ridges – the area that came to be known as 'the Cauldron.'

On 30 May the General decided that his only hope lay in breaching the British minefields from the east, to regain contact with his main supply bases. Thus, on the 31st, he massed his artillery to protect the minefield gaps. Ritchie, thinking the strategic withdrawal was a retreat, cabled Cairo: 'Rommel is on the run!'

Rommel was not retreating but consolidating. For the next two days he tried in vain to knock out the 150th Brigade box. If the British had mounted a decisive armor attack immediately they would have had the Afrika Korps virtually at their mercy. Instead, they softened up the air offensive and launched a series of ill-coordinated jabs at Rommel's antitank guns. Those guns had been set up on the ridges in accordance with Rommel's second principle: everything possible must be done to protect one's own supply lines and to upset, or better still, cut the enemy's. Ariete and 21st Panzer were well dug in and resisted all attacks. By the time a major attack was launched on 5 June, it was too late. Got el Ualeb had been overrun, almost before Ritchie realized what was happening.

Rommel's brilliant improvisation had turned a near-fatal encirclement into a wedge driven deep into British territory – and once again the British had been defeated by their inability to move their forces rapidly, as a unit.

Next, Rommel turned his attention to Bir Hacheim – the southern key to the whole Gazala Line, and a serious

Left: Italian artillery shells the retreating British columns after the fall of Tobruk.

Right: Those who surrendered at Tobruk are taken into captivity, 22 June 1942.

threat to his flanks and rear. That battle lasted a week, and turned out to be the hardest fought so far in the campaign. While it was still going on, the tank commanders of the Afrika Korps in the north — by now used to Rommel's tactics — reacted to the British attack against the Ariete at Got el Ualeb on 5 June with a fierce counteroffensive on 6 June and another on the 7th. Though both these attacks were repelled, the British suffered heavy losses and it was clear that the balance was shifting in Rommel's favor.

Finally, on 10 June, the Free French outpost fell. It was an important event: not only could Rommel's supplies now move freely, but a large number of tanks, artillery pieces and planes were released for the next move — to Tobruk itself.

Immediately he gathered most of his force for the move north. On 13 June Ritchie sent a force to the southwest for an attack on the Axis flank, but Rommel, concentrating his heavy artillery, lured the British forces into an ambush. In the heaviest blow struck so far by either side, about 300 British tanks were lost, compared to about 70 for the Axis. The engagement marked the final turning point in the battle. By 15 June Rommel had cut the coast road just west of Tobruk, and the British were forced back to their third defensive line near El Adem and Rezegh. This stand did not last long; both El Adem and Rezegh were abandoned by 18 June.

In 1941 the Australians had held Tobruk for nine months, until Rommel's withdrawal to the west. That winter the Middle East Command in Cairo had decided that without naval support it would be impossible for the fortress ever to be held again in isolation. London had been informed and had — they thought — agreed, but on 15 June Auchinleck received a telegram from the Prime Minister, 'Leave as many troops in Tobruk as are necessary to hold the place for certain.' At length a compromise was reached. Tobruk was to be 'temporarily' invested while a new strike force was built up near the frontier.

The main part of the garrison was to be formed by the 1st South African Division with General Klopper — a major general of one month's standing — named commander of the stronghold. The port's physical defenses, while not in good shape, were hardly weaker than they had been in April 1941. The barbed wire, tank traps and well-placed gun emplacements were still there. Equipment was, if anything, a bit better. There were two partial medium-artillery regiments and the garrison was strong in field artillery. Although there were no antitank regiments, there were about 70 antitank guns, including 18 six-pounders. In antiaircraft guns — 18 37mms and a number of Bofors — the strength was about the same, and there were about 55 tanks. The strength of the garrison was also about the same — some 35,000 men. There was one important difference — and it was one which Klopper, none too sure of himself or his position, was ill-equipped to deal with. This time the defending troops were exhausted, their morale was low, and the camp was filled with a feeling of insecurity and impermanence.

As Tobruk prepared for battle the South Africans took up their positions along the northern coast and the western and southern perimeter from the sea to the El Adem road. East from there were the 2nd Camerons, 2/5th Mahrattas and 2/7th Gurkhas. Near the Palestrino Ridge in the center were the 201st Brigade headquarters, the 3rd Coldsteam, and the Sherwood Foresters. Meanwhile, the rest of the Eighth Army made their way toward the defenses at the Egyptian frontier.

As usual, Rommel had devised a ruse for capturing Tobruk. Only his infantry approached the western perimeter, while his mobile forces swept on past, to give the impression that he was heading straight for the border as he had done the year before — and sending messages in clear to reinforce the illusion. Just before Bardia he and the 90th Light Division turned back to join the Afrika Korps assault divisions and the XX Italian Motorized Corps, who had been waiting southeast of the city. He was using the same plan he had intended for 23 November 1941.

Rommel's zero hour was 0520 on 20 June. As the first rays of sunlight began to creep over the desert the long black lines of tanks, trucks and infantry slowly started to move forward. From far away there came a faint drone. As it grew louder small black dots appeared on the horizon

which, as they drew nearer, resolved themselves into a wave of Stukas and Ju 88s. Every airworthy Axis plane in North Africa had been pressed into service for the battle. As the heavy artillery began to fire, the planes released their bombs and quickly got out of the way of the next wave, operating a shuttle service between the defense perimeter and El Adem airfield, 10 miles away. They pounded open a gap 600 yards wide. Behind them, under cover of the artillery barrage and half-hidden by smoke and dust, German and Italian sappers raced forward to lift the mines and bridge the tank traps, and tanks and infantry raced through the gaps. As they moved forward they lit red, green and purple flares and the Stukas dropped their bombs just ahead of the advancing, multicolored smoke screen while the other planes and the artillery blasted the enemy's rear with bombs and shells.

The timing of the entire operation was perfect. Panzer Army Africa might well have been on maneuvers. The first shock troops broke into the fortress from the southeast. A second group breached the defenses in the south, along the El Adem road, soon after. As tanks poured into the city they fanned out and headed for the harbor, while parachutists were dropped behind the lines to disorganize the defenses and protect the supply dumps from demolition.

Inside Tobruk the situation was chaotic. General Klopper — his headquarters bombed out, his radio and telephone wrecked and his code book destroyed, lost the last vestige of control. Disconsolately he and his staff watched the Panzers race past their headquarters on their way to capture the fuel dumps in the harbor. Some British troops broke out to the east. Others fought grimly on, while still others, like the South Africans in the west and southwest, hardly realized anything was happening until the 90th Light came up on their rear.

By dawn on 21 June Tobruk was a pile of ruins. The streets were a maze of rubble and in the harbor the masts and funnels of sunken ships rose pathetically from the water. General Klopper gave his compass and staff car to seven young men from the South African 6th Brigade who were determined to escape, saying, 'I wish I was coming with you.'

A few hours later a small party of officers set off in a truck, a little white flag fluttering over the hood, and at 0940 on the Via Balbo Klopper officially turned the city over to Rommel. Soon after, a large white flag was hoisted over 6th Brigade Headquarters by South African native drivers.

The signal to surrender created even more confusion. Some units never got it. Others, like the 3rd Coldstream, decided to ignore it and try to escape. The Cameron Highlanders, along with the remnants of some of the Indian brigades, held out for more than 24 hours — surrendering only after being told that if they did not the Germans would concentrate every piece of artillery in Tobruk on their

Left: British prepare to reenter the city following the German withdrawal after the Battle of El Alamein.

Left center: Tobruk harbor in flames.

Below: Rommel with General Bismarck in the summer of 1942.

Far left bottom: British prisoners waiting to go into camps after the fall of Tobruk.

position. Finally giving in, they marched down to the prisoner of war cage in parade formation, with the pipes skirling 'The March of the Cameron Men.' As they approached every man along the way — prisoner and German sentry alike — snapped to attention.

After two years in British hands Tobruk had fallen in two days — and despite Rommel's anger at the extent of the destruction effected by British demolition squads on vehicle parks and fuel dumps, he still had captured enough to carry him on his drive to Egypt. The fall of Tobruk came as a shattering blow to the British public (as Churchill had known it would), as well as to the Australians and South Africans. General Klopper came in for most of the criticism, but he was not entirely to blame. The decision to invest Tobruk at all had been, in General Bayerlein's phrase, 'a fatal decision.' Though a more experienced general might have made more progress toward pulling the garrison into shape in time, there was also confusion among the British High Command. For example, Auchinleck realized full well that Rommel was almost certain to stick to his original plan and attack from the southeast. When Ritchie flew into Tobruk on 16 June to confer with the defenders, he warned Klopper to pay special attention to the western perimeter.

On the Axis side there was great enthusiasm. Mussolini, hitherto anxious to move slowly and consolidate positions, flew to North Africa with a white horse to ride when he entered Cairo at the head of his troops. Hitler was ecstatic, calling the victory an 'historic turning point' of 'decisive import' for the whole war. On 22 June Rommel received a message from the Führer informing him that at the age of 49 he had just been appointed Germany's youngest Field Marshal. He celebrated that night with canned pineapple and a small glass of whisky, but after dinner he wrote his wife, 'Hitler has made me a Field Marshal. I would much rather he had given me one more division.' Still, he was in unusually good spirits — it was the high point of his career as well as for the North African campaign.

Typically, he saw it not as an end, but as the springboard to a new Egyptian campaign. True to his cardinal rule — never give the enemy a breathing space — he did not celebrate too long. The next day his Order of the Day read, 'Soldiers of the Panzer Army Africa! Now we must utterly destroy the enemy! During the coming days I shall be making great demands upon you once more, so that we may reach our goal.' Gathering the tired but triumphant Afrika Korps together he set off in pursuit of the Eighth Army and that ultimate goal — the Nile.

He would never get there. Hitler, by discontinuing the attack on Malta and refusing to send Rommel adequate supplies, would make defeat in the desert inevitable. Later, the Field Marshal would find himself presiding over another fiasco — the defense of Normandy — and still later would come involvement in the plot against Hitler and, eventually, suicide.

All this was in the future; in June 1942 the Desert Fox was as he is still best remembered — dashing, resourceful and brave, racing across the desert with the tanks of the Afrika Korps, heading for the pyramids of Egypt.

CHAPTER 13

MONTGOMERY AT ALAMEIN 1942

In the last week of June 1942 the entire British position in the Middle East seemed to be on the verge of collapse. The battered Eighth Army was in retreat, racing for the Egyptian frontier before, beside, and sometimes even behind Rommel's victorious *Panzerarmee Afrika*. The loss of some 80,000 men since 26 May and the wreckage of two armored corps testified to the triumph of German professionalism over superior numbers. The victory seemed so complete that Rommel, even though he had orders to halt on the frontier and await the capture of Malta, was given the go ahead to continue the chase and take advantage of the panic and resulting disorganization of the British Army.

On 25 June the Commander in Chief for the Middle East, General Sir Claude Auchinleck, flew to Mersa Matruh from Cairo to take over personal command of the Eighth Army from General Ritchie and begin a planned, more orderly withdrawal to the defenses at El Alamein. By 30 June this had been accomplished and the two armies had begun arranging themselves, with Auchinleck hurriedly organizing the defensive positions for a last-ditch stand while Rommel pulled his exhausted troops together for another major effort. If the seemingly indestructible Afrika Korps could overcome this obstacle there would be little to stand in their way as they drove toward the Nile and the Suez Canal. Less than a hundred miles away, in Alexandria, papers were destroyed and government offices evacuated with a speed the Desert Fox himself would have envied, while the Royal Navy hurriedly sailed out of the harbor.

Waiting in the wings was another actor – a complex, often abrasive character who would soon achieve a reputation in the desert that would equal Rommel's and receive the impressive title, Field Marshal the Viscount Montgomery of Alamein, KG. In July 1942, of course, he was simply Major General Bernard Law Montgomery, a largely unknown career soldier in the Home Army, with more than a month to go before his first appearance on the desert stage.

Montgomery came from a large family (the fourth of nine children) that had no military tradition. Born in 1887 in Moville, County Donegal where his father was an Anglican bishop, he lived in Ireland for two years before the family was moved to Tasmania, Australia. From all accounts, including his own, he had a very unhappy childhood which he describes as 'a series of fierce battles, from which my mother invariably emerged the victor.' He was, again in his own words, 'a dreadful little boy' – a loner, in constant rebellion against an unusually strict, methodical mother, given to rudeness, bullying and hysterics.

In 1902 the Montgomerys returned to England, and young Bernard became a day pupil at St Paul's School where he asked to be enrolled in the 'army class.' This involved no commitment to the military; the 'army class' was simply designed for those more inclined to practical than academic pursuits. His mother's overreaction to the news decided him – from that moment on he was determined to be a soldier. At home he remained surly and morose, but at school he was energetic, good at cricket and other games (where his agility and stamina earned him the nickname 'Monkey'), and a confident, effective leader. He had no hobbies, no intellectual interests, and very little scholastic ability but he did have a driving ambition.

General Bernard Law Montgomery with on his right General Brian Horrocks.

At the age of 19 – in January 1907 – he entered the Royal Military College at Sandhurst. The year began well and after six weeks he was among the few oustanding cadets promoted to lance corporal. The honor gave him an overinflated sense of power and he soon became the ringleader of a gang that persecuted other cadets. Following an unfortunate episode in which Montgomery set fire to someone's shirt tails, resulting in severe burns, he was demoted to gentleman cadet (the lowest possible rank) and was not allowed to graduate with the rest of his class. This reverse finally gave him the impetus he needed to settle down and work and for the first time he began to demonstrate the concentration on and devotion to minutiae, the passion for precise detail, that was to play such a large part in his future professional life.

On leaving Sandhurst he joined the Warwickshires – primarily because they had a good reputation but also because he liked their cap badge. He was anything but a typical officer – he was not good on a horse, had no military connections or social graces, and showed only intermittent respect for authority.

In 1913 back in England after a tour of duty in India, the young officer began for the first time a systematic study of the art of war. However, within three weeks after World War I began he was in France learning about war on a much more practical level, at the first battle of Ypres. Badly wounded in this, his first major action, he was sent back to England, promoted to captain, and awarded the DSO – one of the highest honors the British Army can bestow (usually reserved for majors and above) for sustained gallantry in the face of the enemy. For a subaltern to win it was very rare, and generally meant that he had just missed the Victoria Cross.

During the 1920s and 1930s he began to stand out as a radical crusader for military efficiency – a quality badly needed in the British Army at the time. In 1927 he married Betty Carver, a widow with two sons. The next year their son David was born and Montgomery's life was completely changed. He and his wife were inseparable and she used affection and gentle mockery to soften his rough edges. Her death, only ten years later, 'utterly defeated' him for the first time in his life. From that time he built a wall around himself which was impenetrable and he began living only for his work.

By 1938 he was established in his first command, with the 8th Infantry Division in Palestine, where the flood of Jews fleeing Nazi persecution in Europe had led to a state of emergency among the native Arab inhabitants. This grounding in Palestine was useful. He was noted for his insistence on methodical operations from a firm base (though he was fighting guerrillas), his egotistical view of his troops and their operations, and his ability to make sure that 'his' achievements were known and recognized in the proper quarters. He was also ruthless with the incompetent or inefficient.

When World War II began Montgomery went to France with the British Expeditionary Force, in command of the 3rd Division. With his usual energy he immediately began strenuously training his men, instilling a professional attitude toward war that helped immeasurably in maintaining order during the British retreat to and evacuation from Dunkirk. At this time too he made a great impression on his corps commander, Alan Brooke – which was to be of great value to him in his future career.

After Dunkirk he was stationed in England, climbing the ladder of promotion fairly rapidly, despite his abrasive personality and his unfortunate habit of displaying an insolence that approached insubordination toward officers he did not respect (a group which included General Auchinleck). By 1941 he was a Lieutenant General in charge of the South Eastern Army – England's main anti-invasion force, the post he still had at the beginning of July 1942.

Meanwhile, in Egypt, Auchinleck and the Eighth Army were racing against time to prepare for Rommel's arrival. The area chosen for their final defense was a narrow neck of land, about 40 miles wide, between El Alamein on the Mediterranean and the Qattara Depression in the south. The northern half of the passage was featureless, so much so that almost imperceptible rises in the ground were to assume great tactical significance in the months that followed. South of the long, low rise called the Miteirya Ridge and the rocky, austere Ruweisat Ridge came the smooth swelling of Alam Nayil, with the Alam Halfa ridge some 14 miles to the east. These were followed by a series of abrupt escarpments. Further south the going got even worse as the rocks disappeared in a stretch of eroded channels and soft sand – 'devil's country' that continued to the edge of the great Qattara Depression and the steep cone of Mt Himeimat, which rose to almost 700 feet, dominating the landscape like a great pyramid.

Auchinleck had chosen well. If they were properly disposed, there was no way the defenders' flanks could be turned in that narrow passage, especially since the rough going in the south would drastically hinder the movement of Rommel's armor. It was just as well, for the British had their backs against the wall. With such a small area for maneuver, the loss of even a little territory could be vital in the defense of Egypt. Rommel, on the other hand, could afford to lose much more ground without really damaging his position. Similarly, in the race for supplies and reinforcements, gaining a slight advantage would not materially affect Britain's position. Even if they could push Rommel back to the border and ease some of the pressure on Alexandria, it would not be enough for them to destroy him

altogether. For the Axis, though, even a small increase in men and materiel might be enough to allow Rommel to make a temporary breakthrough, which could be all he needed to clear the way for his dash to the Nile.

So, Auchinleck threw everything he could muster into the breach at El Alamein. It was just enough. Time and again during July Rommel sent his tired troops against the British line only to see them thrown back. The balance of forces was too even to allow either side to launch a successful counteroffensive.

Meanwhile, behind the lines, the race for supplies went on. Britain had by far the longer and more dangerous supply route, but thanks to President Roosevelt a stream of men and equipment – including American tanks and planes – flowed steadily into Alexandria. Rommel, for his part, could obtain little from a country already strained to its limits by the demands of the Russian campaign. In air power, he could not match the Desert Air Force, now supplemented by American forces. Though possession of the port of Tobruk helped ease his supply lines it was attacked almost nightly. His convoys, too, were under constant bombardment by planes and submarines, as German strength was diverted from Malta to the Eastern Front.

Once again, the attitudes of the Axis and Allies toward the North African campaign were to be the deciding factor. Hitler never really appreciated its significance, always putting it second to his Napoleonic dreams of conquest in Europe. Even before Italy entered the war in June 1940, the Mediterranean and the Balkans had played a major role in British strategic planning. After America entered the conflict in December 1941, it was agreed at the Acadia Conference that one of the Allies' primary aims must be to 'tighten the ring' around Germany – and the southern section of that ring ran along the North African coast.

Thus, Auchinleck was under considerable pressure to do more than fight a holding action. Churchill, not the most patient man at the best of times, was having political difficulties at home and was anxious for a spectacular victory to bolster his fading popularity. In addition 'the Auk,' as the Commander in Chief for the entire Middle East, could not give his full attention to affairs in Egypt, and was frequently distracted by developments in Iraq, Syria and elsewhere. In short, his command was too large and complex and there was too much expected of him in too short a time – a fact Churchill subsequently admitted when it was too late.

There were, however, some severe problems within the Eighth Army, and Churchill fastened on them eagerly. There were indications that Auchinleck's command organization was not working harmoniously or effectively, and he seemed reluctant to change it in any way. Furthermore, morale was at a new low. The ups and downs of the desert war had sapped the battle-hardened veterans' confidence in their ability to finally defeat the Afrika Korps. The uneasy stalemate in July, with its frequent orders for attacks that turned out to be futile, accompanied by unfortunate, if necessary, precautionary plans for withdrawal, further weakened morale.

Churchill decided that only a drastic change could restore the Eighth Army's confidence. Early in August he visited Cairo with General Sir Alan Brooke (now Chief of the Imperial General Staff) and after a whirlwind tour of several Eighth Army units the two concluded that 'a new start and vehement action were needed to animate the vast but baffled and somewhat unhinged organization.' The Prime Minister's temper was not improved on this visit by Auchinleck who, lacking Montgomery's facility for public relations, made few provisions for his comfort or entertainment and left him more or less to his own devices.

Churchill's plan embodied changes in both organization and personnel. On the organizational side, Persia and Iraq would retain the name 'Middle East Command' while Libya, Egypt, East Africa, Palestine, and Syria would be detached to form the 'Near East Command.'

Within this new organization, he proposed, Auchinleck would take over the Middle East, with communication lines running back to India. For the Near East he had a new team: General Sir Harold Alexander would be installed as Commander in Chief, headquartered in Cairo; Brigadier 'Strafer' Gott would take over as commander of the Eighth Army (though Gott himself felt that he was too tired and lacking in new ideas to handle the job effectively); and Brooke's protégé, Montgomery, would be named British Task Force Commander for Operation Torch (the Allied landing in North Africa, scheduled for early November). The day after the decision was made, the plane in which Gott was returning to Cairo was shot down and the general killed. Thus two unknown Luftwaffe fighter pilots can take credit for bringing together what was to become the winning British command team. Montgomery was given Gott's job as Commander, Eighth Army, while Lieutenant General Kenneth Anderson took his place with General Eisenhower.

Alexander — considered by many to be the best strategical brain in the Empire — was the ideal man for the labyrinthine politico-military atmosphere of wartime Cairo. He had a wealth of battle experience, having participated in most of the major battles of World War I and in many minor operations from the Baltic to India's Northwest Frontier; from this he had developed not only sound military judgment, but also a real knowledge of British and Indian soldiers that helped him avoid psychological mistakes when dealing with them. Above all, he was an ideal supervisor — good at decentralization and knowing not only how much freedom to give his subordinates, but also when and how to step in if things were going wrong. He was a perfect superior officer for Montgomery — who was a strong, demanding and extraordinarily difficult subordinate.

The new commander of the Eighth Army stood out in

From left to right: Mr Anthony Eden, General Brooke, Air Chief Marshal Tedder, Admiral Cunningham, General Alexander, General Marshall, General Eisenhower and Montgomery with Winston Churchill in the center.

god-like figure. In addition, many of his superiors considered him to be a first-class leader who took no risks and was the epitome of the dedicated professional. This professionalism was one of his best qualities – along with his ability to instill it in the men under his command. He had the great gift of being able to explain the most complicated plan so that it appeared simple and when he had finished a briefing everyone present knew exactly what he had to do – and more importantly why he had to do it and how his job fitted into the overall operation. In fact, Montgomery had more flair than the rather diffident Alexander when it came to talking to men and officers in public. His high-pitched rasping voice could hold an audience spellbound.

In his work, Montgomery was devoted to precision, caution and control. His plans – which usually relied more on overwhelming force than on skillful maneuvering – left nothing to chance. Most of his experience had been gained in the staff room rather than on the battlefield, and where Alexander deplored useless loss of life for humanitarian reasons, Montgomery hated it because it reflected poor organization.

Though he could with justice claim that he never lost a battle, he often made disastrous miscalculations. Some of these stemmed from his naturally unpredictable, eccentric state of mind, but also especially since the death of his wife, the world outside his work had become shadowy and unreal. This narrow view often hampered his decision-making capability. As will be seen later, the battle of Alamein is a good example of a certain victory that was very nearly turned into defeat.

Upon their arrival, the new command team faced three major challenges: to improve the Eighth Army's morale; defeat Rommel's next attack, which intelligence sources predicted for 26 August; and prepare their own offensive (codenamed Operation Lightfoot), which was to be coordinated with Operation Torch.

Montgomery immediately plunged into the work of training and reorganizing the worn-out army, directing all his energies toward making it an efficient, professional and

distinct contrast to his urbane superior officer. He was not a likeable character – still less the sort of 'decent chap' so highly prized by the upper echelons of the British army. His conspicuous lack of some of the more civilized virtues led many of his contemporaries to see him as a vain, opinionated showoff, an overrated general and a thoroughly unpleasant human being; even Churchill once referred to him as 'a little man on the make.'

To many of the men who served in the ranks in the desert or, later, in the Normandy landings he was an almost

Above left: English prisoners of war near El Alamein, 18 June 1942.

Left: A German 88mm antiaircraft gun bombarding British tanks. The 88mm gun ended the Desert supremacy of the Matilda tank.

Above right: German soldiers move forward to El Alamein in July 1942.

Right: Members of the Afrika Korps eat their meager rations during the summer of 1942.

above all tidy organization. His first move was to establish communications with all levels of the army, so that when the time came each of the tens of thousands of soldiers would not only know what to do, but would have the incentive to do it.

He was the first British general consciously to project an 'image' to his public. Every unit was visited, and as the general's cold gaze balefully fell on the idle or incompetent, as the rasping voice explained, questioned, or ordered, he was striving to leave an impression on every man's mind. Even his studied informality of dress and his famous hats (an Australian slouch hat carrying all the 9th Australian Division badges or a black Tank Corps beret) were carefully calculated to make the 'new boy' part of the desert world as soon as possible. Pressmen and photographers appeared in unprecedented numbers and the Army public relations staff began working overtime.

In addition to his famous 'no bellyaching' order, Montgomery stressed in his talks to the troops that there were to be no plans for withdrawal, and that in the future there would be fewer risks, no short cuts and – most important – no more failure. The organizational changes he announced included training the X Corps, under General Lumsden, as a mobile *corps de chasse* to rival the Afrika Korps, and the relocation of Eighth Army Headquarters from the eastern end of Ruweisat Ridge to the seashore at Burg el Arab – for closer cooperation with the Desert Air Force and for greater comfort. This firm line was well received. Some of the men were too disillusioned to accept this unknown general, who had just arrived from England with his 'knees still pink,' at face value, but most were at least content to give him the benefit of the doubt.

The next problem, preparing for Rommel's expected attack in August, was easier to solve. Soon after his arrival Montgomery had several conversations with Auchinleck himself, with General Ramsden (commander of the XXX Corps), and with Brigadier de Guingand (Auchinleck's Brigadier General Staff) – all of whom described their plans for the defense of the British position. In talks with Churchill and Brooke on 19 August, these became Montgomery's plans – but in fact the prelude to Alamein, the battle of Alam Halfa, was eventually to be fought on a plan constructed by General Dorman-Smith (first Auchinleck's principal Operations Officer and then, until Montgomery's arrival, Deputy Chief of Staff) and approved and initiated by Auchinleck, and utilizing fixed defenses which had for the most part been dug before he ever left England.

By the end of August Rommel was still at a disadvantage, despite the reinforcements that had arrived earlier in the month. He still had only 200 German tanks with guns, and was so short of fuel that, though he still planned to sweep around the southern flank, he had to then use a much shorter turning radius than he would have liked. The British, by contrast, had 713 tanks in the forward

area (including 164 heavy Grants). All infantry and artillery divisions now had 6-pounder antitank guns, and the Desert Air Force had achieved complete control of the skies over Alamein. In addition, Rommel was so ill that he could not get out of his truck. He was not only under constant medical supervision but had a replacement ready on the spot at all times. Gause, his Chief of Staff, was also sick — none of which helped morale in Axis army.

Nevertheless, after nightfall on 30 August, Rommel's four veteran divisions — the Afrika Korps and the Italian XX Corps — began to work their way through the British minefields. The barriers were thicker than they had expected, so that as dawn broke on 31 August they were only just emerging instead of being already in position in front of Alam Halfa Ridge.

Almost from the beginning it was obvious to Rommel that the offensive would not succeed. In fact, he considered a withdrawal at 0800 hours on 31 August but he allowed himself to be dissuaded by General Bayerlein. The only doubt in Montgomery's mind was whether Axis troops really were committed to an attack from the south. By 1100 hours it was clear that this was the case, and British forces were rearranged accordingly. At 1300 the Panzer Army stopped to refuel and then, instead of swinging out to the center and eastern end of Alam Halfa in an encircling movement, they turned sharply north and drove straight

for the heavily defended western end — a move dictated by the shortage of fuel and the time they had lost in the minefields.

The attack stalled and that evening Allied aircraft inflicted heavy damage on both equipment and personnel (for the first time US pilots in Liberators, Mitchells, and Kittyhawks were flying alongside the RAF). The next night, 1–2 September, saw more of the same. By dawn — battered and almost completely out of fuel — the Afrika Korps began to edge away to the west.

At this point Montgomery, by launching a counter-offensive across Rommel's communication lines, could probably have encircled all the Axis armor and finished the Battle of Alamein before it ever began. With his usual desire to avoid any risk of failure and his unwillingness to deviate from a set schedule, he declined pressing the issue. By the night of 3–4 September the opportunity had passed and the Panzer Army was executing another of its skillful withdrawals. On 6 September they were out of danger, and in addition had established a bridgehead on the eastern edge of the British minefields, astride Mt Hememat.

In later years Montgomery was to claim credit for devising the defensive tactics that prevented British armor from being destroyed piecemeal as in previous engagements. However, according to General Renton, Gott (ordered before his death to prepare a detailed defense against the

Left: Soldiers of the Afrika Korps advance during the first Battle of El Alamein.

Left below: Members of the Afrika Korps fry some eggs, which they probably made up from captured British rations.

Left bottom: An early morning wash.

Far left below: Grant tanks of the 22nd Armored Brigade advance to the front line in column, south of El Alamein.

German attack) had never intended 'unleashing the armor.' Despite this, and despite the missed opportunity, Alam Halfa was a model defensive battle that reflects great credit on both the planners and the field generals (Montgomery and Gott's replacement, Horrocks) who executed those plans.

Characteristically, Churchill wanted to follow up Alam Halfa with an immediate offensive in mid-September. Equally characteristically, Montgomery refused, determined to do nothing until the odds were overwhelmingly in his favor. Eventually the Prime Minister gave in, but he did so with bad grace. He insisted that the Eighth Army must have a decisive victory over Rommel before Operation Torch was launched on 8 November.

Much has been written about whether or not the battle of Alamein was really necessary. From a strictly military point of view, it probably need not have been fought at all. Rommel would have had to withdraw as soon as Operation Torch began, and would have found himself sandwiched between two forces almost immediately. There were several political reasons for a strong British offensive: to dispel any feeling in the United States that American boys were being killed to salvage British prestige in an obscure colonial backwater; to influence the ambiguous French population in Morocco and Algeria; and to help Churchill quell the atmosphere of distrust that marked his dealings with Stalin.

Thus, throughout September and October, Allied naval and merchant vessels delivered supplies and reinforcements to Egypt. Workshops hummed, supply depots swelled and spread and camps were crowded with pale-skinned soldiers.

Rommel knew that a major attack was imminent, but did not have enough transport to withdraw to a better position at Fuka. All he could do was dig in and make his

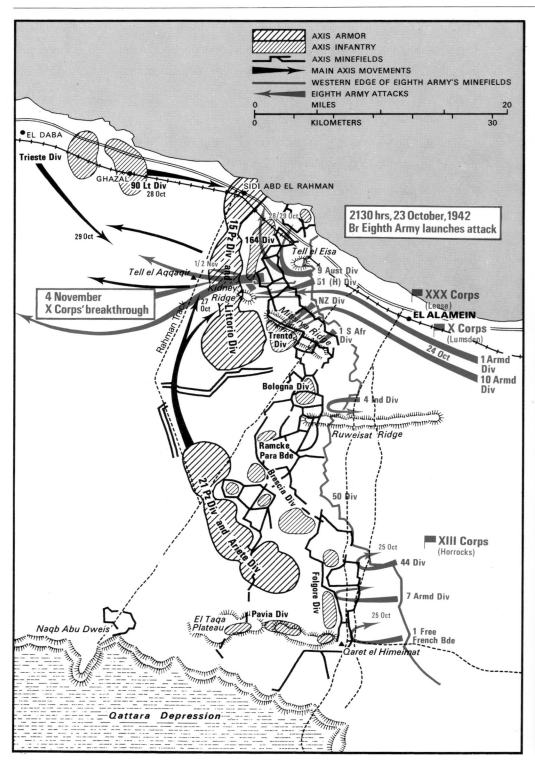

AXIS ARMOR
AXIS INFANTRY
AXIS MINEFIELDS
MAIN AXIS MOVEMENTS
WESTERN EDGE OF EIGHTH ARMY'S MINEFIELDS
EIGHTH ARMY ATTACKS

MILES 0 — 20
KILOMETERS 0 — 30

EL DABA
Trieste Div
GHAZAL
90 Lt Div 28 Oct
SIDI ABD EL RAHMAN
29 Oct

2130 hrs, 23 October, 1942
Br Eighth Army launches attack

15 Pz Div and 28/29 Oct
164 Div
Tell el Eisa
Tell el Aqqaqir 1/2 Nov
Kidney Ridge 9 Aust Div
27 Oct 51 (H) Div
NZ Div
XXX Corps (Leese)
4 November X Corps' breakthrough
EL ALAMEIN
Littorio Div
Miteiriya Ridge 1 S Afr Div
Trento Div X Corps (Lumsden)
24 Oct 1 Armd Div
Bologna Div 10 Armd Div
4 Ind Div
Ruweisat Ridge
Ramcke Para Bde
Brescia Div
50 Div
21 Pz Div and Ariete Div
XIII Corps (Horrocks)
25 Oct 44 Div
Folgore Div 7 Armd Div
25 Oct
El Taqa Plateau Pavia Div
1 Free French Bde
Naqb Abu Dweis
Qaret el Himeimat
Qattara Depression

positions as strong as possible. In the weeks before he went on sick leave his Italian and German sappers laid half a million mines across the entire front, in a honeycomb pattern. Within the cells of the pattern they planted 'Devil's Gardens' — patches of mines and any other booby traps they could evise, such as harmless-looking wooden poles attached to huge charges of explosives. They employed other tricks too — like burying metal rods which would register on metal detectors, to camouflage the gaps in the minefields. One group actually followed a British sweeping squad, laying mines behind them, so that the

reconnaissance group that came later found a nasty surprise waiting. Another group crept into British minefields and removed detonators, then reburied the harmless mines. (Needless to say, the British sappers were engaged in many similar activities.)

German and Italian infantry units were strung out all along this line. The biggest problem to be faced, however, was deployment of the armor. There was no way Rommel could obey his own cardinal principle and consolidate the armor because he did not have enough fuel to then move it effectively. In addition it was too vulnerable to Allied air

attack when it was on the move. Out of necessity, then, he placed his 15th Panzer and Littorio Divisions in the north, where they formed into three mixed groups; 21st Panzer and Ariete did the same in the south. His only reserves, 90th Light and Trieste, guarded the coast at El Daba against an amphibious landing. On 22 September, having done all he could to secure his position, Rommel finally left for Germany to have his swollen liver and constantly inflamed throat seen to.

By the end of September 1942 Montgomery had 195,000 men to Rommel's 54,000 Italians and 50,000 Germans.

A British 25-pounder fires at enemy positions near El Alamein in the early morning, September 1942.

He had 1029 medium tanks to the Germans' 496, 1451 antitank guns as opposed to 800 (of which some 85 were the impressive 88s) and 908 pieces of field and medium artillery against the Panzer Army's 500. All this was in addition to virtually unlimited supplies of ammunition, fuel and other stores.

Most welcome of all the new weapons were 300 Sherman tanks – products of American tank expertise combined with

British battle experience. With their powerful 75mm guns, they were a match for anything the Germans could field. Almost as valuable were the 'Priests' — 105mm self-propelled guns, on Grant tank chasses, with a flashless charge to make them harder to locate. British antitank guns — the 6-pounders — were good weapons, but had one defect. Most of them were carried on Austin trucks, called 'portees,' with high, bulky silhouettes that made them horribly conspicuous against the desert horizon.

For their part, in addition to the Panzer Mark IIIs and IVs, the Axis forces had the most formidable piece of artillery in the desert — the famous 88, with its 21-pound armor-piercing shot that could kill an enemy tank at 3000 yards. One of these guns could hold off an entire tank squadron for some time. The German infantry was armed with the Spandau, described as a 'vicious' machine gun with an intimidating sound like a racing car revving up.

Throughout September and the beginning of October, Montgomery collected more supplies, trained his army and laid his plans with his usual supreme confidence. Starting with Auchinleck's general concept of a breakthrough in the northern sector of the line (previously both sides had always swung first south, then north toward the coast in an attempt to encircle and trap their opponents), he developed his own system of organization and tactics. His would be a methodical, controlled operation; he still did not have enough confidence in the level of training throughout the army to give his forces much independence. In this he was probably justified. During their years in the desert the veterans of the Panzer Army had come to understand each other so well that they fought as a team almost by instinct. The British army, on the other hand, had always been a much more unwieldy, less homogeneous group and could not achieve that instinctive cooperation.

Just two weeks before Operation Lightfoot was scheduled to begin, Montgomery radically changed his plan of attack. Contrary to all the tenets of desert warfare, which held that armor should be destroyed first (at which the infantry would collapse almost of its own accord) he decided to have his armored divisions simply hold off the Axis tanks while the British infantry units destroyed their opponents in the main defensive system.

Four infantry divisions of the XXX Corps were to open two gaps through the German minefields: one in the north toward Kidney Ridge and another in the south over Miteirya Ridge. Right behind them would come the three armored divisions of X Corps, who would pass through the gaps and establish Report Line 'Pierson,' some two miles beyond the infantry's goal, 'Oxalic.' There they would set up defensive positions until the Axis infantry had crumbled, then chase and cut off the remnants of the Panzer Army. If the infantry had not reached Oxalic by dawn of the first day, the tanks would have to fight their own way out to Pierson.

Meanwhile, XIII Corps and 7th Armored Division would mount a diversionary attack in the south to pin down the 21st Panzer/Ariete groups. Other features of the long, elaborate campaign to deceive the Axis about the real focus of the attack included fake radio traffic and troop movements, and the building of a dummy pipeline. Even though General Stumme (who had assumed command on Rommel's departure) incorrectly believed that the attack would take place in the south, he was never in any doubt about the actual timing of the offensive, despite all the British efforts to hint at a late-November date.

As D-Day drew nearer, confidence in the British command — which had been noticeably shaky — began to improve. Their unprecedented superiority in men and materiel, their complete control of the air, and the efficiency of their intelligence operations made the outcome apparently certain.

By contrast the axis forces were in an unhappy position — short of food, water, fuel and supplies. To make matters worse, the German High Command appeared to have no conception of the gravity of the situation. On 23 October — D-Day — a visiting staff officer told Stumme that in OKW's opinion there was no danger of a British offensive in the near future.

But that evening, as the full moon rose, 882 guns opened fire, in Montgomery's words, 'like one battery' along the 38-mile front. The German guns, under orders to conserve ammunition, remained silent. Twenty minutes later the barrage lifted and more than 70,000 British soldiers and 600 tanks moved out toward the 12,000 men

Left: German troops in full retreat following the defeat at El Alamein, November 1942.

Below: Italian infantry during the battle at El Alamein in November 1942.

Below left: A 5.5 Range gun is fired raising a cloud of dust round the gunners' feet at El Alamein in November 1942.

of the Italian Trento and German 164th Infantry divisions.

Most of the carefully laid traps in the Devil's Gardens had been destroyed by the artillery barrage. Getting through the gaps should have been a simple process. However there were too many men and too much equipment packed into the narrow lanes. The infantry, with no room to fight, could not get to the western edge of the minefields. The sappers could not clear lanes for the armor and the tanks piled behind everyone where they attracted a good deal of enemy fire. Soon the area looked to General Carver like a 'badly organized car park at an immense race meeting held in a dust bowl' and into the immobile mass deadly and accurate fire continued to pour from the Axis guns. The cumbersome crowding together of infantry and armor had turned out to be a major blunder.

The problem was accentuated by the gulf of sectional pride and professional suspicion that traditionally separated infantry and armored forces in the British army. The German services achieved close cooperation naturally; their similar training and all their experience were geared toward establishing a harmonious working relationship. As Corelli Barnett has pointed out, asking the British services to do the same was, 'like asking an estranged man and wife to make love.'

All through the 24th the attack continued. The Littorio and 15th Panzer Division had moved in to form a containing line which, by fighting desperately, they managed to hold through the thunder of the artillery and the rain of Allied bombs. The X Corps (Montgomery's '*corps de chasse*'), with orders to fight its way through if necessary, was still 1000 yards from its objective at the western end of the northern gap. At the southern gap, the armor commanders (Gatehouse and Lumsden) were understandably reluctant to send their tanks over the crest of Miteirya Ridge and down through uncharted minefields under heavy German fire. Montgomery, however, insisted that the objectives must be met. One regiment, the Staffordshire Yeomanry, tried to get through — and lost all but 15 of its tanks. Finally, at dawn on the 25th, the British armor had

either to withdraw back behind the ridge or risk being caught without cover within range of the powerful 88s. In the southern sector of the front the diversionary attack had gone astray, leaving the 7th Armored Division and 44th Division stranded in the German minefields. Already the offensive seemed poised on the brink of failure.

Early on the 25th Rommel was recalled from his sick bed and boarded a plane at Wiener Neustadt. That evening he landed at El Daba and immediately set off for the front. The report he received next morning from General Ritter von Thoma was anything but encouraging, despite the fact that his men were still managing to hold the British off. General Stumme — a courageous soldier, but one who lacked Rommel's desert experience, instinct and luck — had been killed in the front line. Most of the Devil's Gardens had been destroyed by artillery fire. The men were becoming demoralized and much equipment had been lost as the result of bombing and strafing by Allied planes, which were flying virtually uncontested missions. The 15th Panzer had only 31 tanks still in working order and no new supplies of fuel or ammunition could be expected.

Rommel's only hope lay not in withdrawal at this point,

but in a swift, violent counterattack. He immediately issued orders to mass all mobile forces in the northern sector. For a time he was afraid to bring 21st Panzer and Ariete up from the south, but it soon became evident that Montgomery was himself moving troops north. Rommel gambled, and on the 27th pulled 21st Panzer and half the artillery up to help with the attack. He also brought the 90th Light and Trieste forward from El Daba to reinforce the coastal positions.

Montgomery, meanwhile, had spent most of 26 and 27 October in seclusion, rearranging his own plans and dispositions to create a force that could break the stalemate. The 7th Armored Division was to be moved north into reserve where it would join the New Zealanders, the 10th Armored Division, and the 9th Armored Brigade. The 1st Armored Division would continue to press westward around Kidney Ridge, and the Australians would begin a move on the coast road.

On 27 October Rommel launched his offensive — but this time the British stood their ground and the attack was beaten off. On the 28th he tried again. The offensive was broken up by Allied air strikes before it ever got off the ground. The Axis forces braced themselves for the British counteroffensive — but nothing happened. Montgomery was still reorganizing his troop dispositions and would not move until the last man was in position.

The stalemate continued for some days. Though the fighting was fierce — often getting down to hand-to-hand combat in the German artillery positions — the British could make no real progress. Still, it was obvious to both Montgomery and Rommel that no amount of skill or courage could change the final outcome in view of the Allies superiority in manpower and materiel.

Montgomery hoped to launch his great breakthrough, Operation Supercharge, on the night of 31 October–1 November; but when the time came, X and XXX Corps were still too disorganized, and the main event had to be postponed for 24 hours. General Morshead and the Australians did mount a preliminary attack on the Axis coastal positions to 'keep Rommel occupied,' and managed to trap part of the 164th Division against the sea for a time, before

they were beaten off by a violent counterattack from the 90th Light and a battle group from the 21st Panzer.

As night fell on 1 November, most of Rommel's men had been in action for nine days without a break. The northern front, which had been gradually pushed back until all his minefields were in British hands, had been broken in many places. The new front was being desperately defended with too few guns, too little armor and the remains of decimated divisions.

That evening Montgomery followed his well-established pattern. There was a three-hour artillery barrage, followed by carpet bombing by the RAF and an infantry attack. At 0100 on 2 November, 400 tanks rushed through the gap (there were 400 more still in reserve) to meet Rommel's 90 German and Italian 'battlewagons.' Supercharge appeared to be going well in the beginning but amazingly, as the last great tank battle of the desert war developed on 2 November around Tell el Aqqaqir, the attack stalled. The Desert Fox was fighting one of the best tank actions of his career. In fact, at one point his counterattacks almost broke through the British salient. Behind the German lines, however, the infantry units had begun slipping away toward Fuka, while von Thoma prepared to continue the delaying action — with less than a third of his men and only 35 tanks remaining to face the reserves which Montgomery now threw into the breach.

All through 3 November the British X and XXX Corps attacked over and over again. It seemed that the Afrika Korps would never break. While the veterans of the German and Italian armored divisions were carrying out their phenomenal defense, held together by loyalty and the force of Rommel's personality, the rest of the Panzer Army was continuing its withdrawal. It was a difficult day for Rommel, but when he saw that, 'the enemy was operating with . . . astonishing hesitancy and caution,' he considered that he had a good chance of pulling through.

On 3 November, however, in response (he thought) to a message he had sent Hitler regarding the withdrawal, Rommel received a personal telegram from the Führer — containing one of his famous 'stand fast . . . victory or death' exhortations. Rommel hesitated, but eventually decided that he, who had always insisted on unqualified obedience from his own men, could not depart from that principle himself. All orders for withdrawal were cancelled. Montgomery had received help from an unexpected source.

At 0800 on 4 November Montgomery threw 200 tanks against the northern sector of the line, held by the 90th Light and the remaining 22 Afrika Korps tanks. Shortly before noon von Thoma was captured. In the center another 100 tanks descended upon the battered, badly equipped survivors of the Italian Tank Corps who, though outflanked, fought until their last tank was destroyed. At 1730 Rommel finally decided to pull his men out against orders. As it happened, it was a decision Hitler was happy to accept; his original message had been the result of a mix up at headquarters, but he had been reluctant to cancel it for fear of losing face.

The battle of El Alamein was over and the chase, which would finally end in Tunisia, had begun. On 5 November an exuberant Montgomery held a press conference to announce his 'complete and absolute victory.' Victory it was, but the fruits of that victory, which Clausewitz has said must be gathered in vigorous pursuit, were not to be forthcoming. Again, Montgomery had a chance to destroy the Panzer Army – but again he chose a cautious, methodical approach that kept the huge British war machine lumbering along one step behind Rommel. Of course, he had laid plans for a pursuit, but adequate provision had not been made for the problem of getting several divisions, under two different corps commands who were not on the best of terms, through the narrow salient. In addition, the men were physically exhausted after 12 days of hard fighting, and were disinclined to snatch the initiative even if Montgomery had ordered it. His staff were mentally fatigued and simply not up to coordinating operations.

By the time Montgomery could mount a full-scale chase, Rommel and the Afrika Korps already had a day's head start. Worn out and pounded down to less than 5000 men, some 11 tanks and about 25 antitank guns, they were still the masters of desert warfare – and proved it as they conducted their skillful, fighting retreat. Bringing up the rear, as usual, were the sappers – blowing up roads, laying mines, and setting booby traps almost under the Eighth Army's noses, eventually leading Radio Cairo to report that, 'the advance of the Eighth Army is meeting little resistance, but is seriously hampered by the . . . engineers.' With no transport, much of the German and Italian infantry had been overrun. Tens of thousands were captured, but others made it back – like the Ramcke Parachute Brigade, which managed to hijack a complete British transport column and join Rommel at Mersa Matruh in high style.

Four times Montgomery sent his pursuing troops on tight turns toward the coast, hoping to trap the Axis forces – only to find that they had already gone. On 7 November the heavens opened, and heavy rains turned the sands around Mersa Matruh into an impassable bog. Later Montgomery was to claim that it was the rain that gave Rommel his chance to escape – apparently forgetting that it fell equally on both armies, and that Rommel had already left before the Eighth Army appeared.

After Mersa Matruh the two forces settled down to a 1500-mile march across the desert, with Rommel in the lead and the Eighth Army slowly and methodically bring-

ing up the rear. The British commanders begged to be allowed to race ahead and force Rommel into a battle where he would surely be destroyed. They pleaded in vain. Montgomery was not only obsessed with order and method; to a large extent he had fallen under Rommel's spell as completely as the men of both armies. The set-piece battle was his area of expertise and the open desert was Rommel's country. Who knew what unexpected things he might do, even with only 10 tanks and no gasoline? It was much safer to stay behind, knowing that approaching from the opposite direction were 80,000 American and 25,000 British soldiers who would close the mighty pincers on the doomed Panzer Army.

The British victory at Alamein, coming as it did after a year of defeat, was greeted by ringing church bells, banner headlines and an outpouring of public adulation for Montgomery that continued, almost unabated, throughout his subsequent career as Field Marshal, leader of the British troops during the Normandy landings and commander of the British forces in Western Europe.

No reasonable person would deny that Montgomery was the victor at Alamein nor that Rommel was defeated. There is no need, on the other hand, to depict him as a savior who appeared out of the blue to singlehandedly snatch a defeated army from the jaws of disaster and disgrace. At the time he assumed command, the Eighth Army was anything but defeated. It had just stopped Rommel in his tracks and paved the way for a major offensive which, as Montgomery himself saw most clearly, they were at the time too weak to begin. As we have seen, many of the innovations at Alamein were based on planning done by Auchinleck and his staff, while many of Montgomery's own plans led to chaos and confusion. Time and again his inability to improvise and his unwillingness to take chances gave Rommel the breathing space he needed to regroup and prolong the action.

The fight that the Panzer Army put up in the face of impossible odds is one of the great actions of World War II – and therein lies the only real basis for naming Montgomery the victor. For despite Britain's overwhelming numerical superiority in men and arms, it required one man with Montgomery's self-confidence, determination and rapport with the ranks to elicit the dogged, unremitting effort that eventually resulted in the British victory.

PATTON
AT THE BULGE 1944

The Battle of the Bulge, often known as the Ardennes Offensive, has been called — by no less an authority than Winston Churchill — the greatest American battle of World War II. Adolf Hitler, reacting to a situation in Germany that had been steadily worsening since the Allied landings in June and the abortive 20 July bomb plot, had developed a bold, imaginative plan to re-capture the initiative in the west. As a result, in one month, from mid-December 1944 to mid-January 1945, the American army lost some 80,000 men and an enormous amount of equipment opposing this last great German offensive of the war.

There were many heroes at the Bulge — most of them unsung — but of them all one name stands out above the rest, that of General George Smith Patton. Flamboyant, swaggering and controversial, he was a man of contrasts: deeply religious and fiercely profane; a hot-tempered, ruthless fighter with a kind heart; a fiery tank commander and a learned military theoretician. With the Battle of the Bulge he hit the apex of his career, amply justifying the German High Command's opinion, expressed in an analysis of Allied generals, that he was 'the most dangerous man on all fronts.'

George Patton probably never considered anything other than a military career. This was perhaps because there was a strong soldiering tradition in his family. In June 1909 Patton graduated from West Point a year behind the rest of his class, owing to his problems with mathematics. During the 1930s he devoured military studies by Erwin Rommel and Heinz Guderian, Basil Liddell Hart and John Fuller, on infantry and armor tactics and strategy. How-ever, in his own reports he had to be very cautious as by this time he had made some highly placed enemies. He felt it necessary to make his points obliquely, while still paying lip service to traditional concepts.

In 1938 General George Catlett Marshall was appointed Deputy Chief of Staff, and fortune finally turned in Patton's favor. Marshall, an astute judge of character, was not put off by Patton's flamboyant eccentricity. Furthermore, he was himself convinced of the importance of armor in the new army and had read enough between the lines of Patton's reports to recognize the latter's preferences.

Patton would never have agreed that 'war is hell.' He saw it as the only area in which he could excel. In 1940, when the German blitzkrieg rumbled through the Low Countries, France and Poland, he was tremendously ex-cited — not only by war itself, but by the thrill of seeing the armor tactics he had been studying for 20 years actually being put into practice. Immediately he began hounding Marshall for a job nearer the action and at the same time tried to obtain a commission in the Canadian army.

In July Marshall came through. The first two American armored divisions were established and Patton was sent to Fort Benning, Georgia to organize one of the brigades, comprising 2nd Armored Division. He was promoted to Brigadier General and he soon took over command of the division, and began to turn it into a polished, dashing outfit. Meanwhile, his own extraordinary personality was doing much to publicize the new service and he was rapidly gaining a reputation as America's top tank expert.

However Patton was never just a tank specialist. His extensive reading and study had given him a broad view of war that enabled him to see parallels that many others missed. He was greatly interested in weapons and tactics used by the other services, and in fact held strong views on almost every aspect of the conduct of war. His insistence on military discipline was well-known — to the extent that an especially snappy salute was often called a 'georgepatton.' Discipline, he believed, saved lives, and as for the outward manifestations, 'If you can't get them to salute when they should salute and wear the clothes you tell them to wear, how are you going to get them to die for their country?' His opinions even extended to the placement of military cemeteries. This interest in minutiae and the ability to fit them in a broader context were traits he shared with another great field commander, Erwin Rommel. The two tank men had other things in common, for instance their habit of roaming the front lines during battles.

Thus beneath the flamboyance, the emotionalism and the 'goddamming,' the discerning eye could spot Patton's dedication to his profession. One of his superiors who recognized that professionalism was Marshall, another was Dwight David Eisenhower. The two were old friends who had served together in the tank corps during World War I. However Ike's support during the next war, often against considerable opposition, was not solely based on friendship. He saw Patton as a 'master of pursuit' — the sort of general, like Napoleon or Grant, who could spur his men on to chase the enemy even when they were dropping from fatigue. For his part, Patton had early predicted that Eisenhower would one day be his superior officer. 'Ike,' he said, 'You will be the Lee of the next war and I will be your Jackson.'

In 1942 Patton was given command of the Western Task Force for Operation Torch, the Allied landings in North Africa. Although the responsibilities of the post gave him less direct contact with the armor than he would have liked, both he and his men profited greatly from their experience in Africa, learning lessons that would be put to good use later in France. Later, at the head of Seventh Army in the dust and heat of Sicily, he demonstrated his talent for pursuit to the full. He also showed his quick temper and after hitting a soldier in public was demoted. Finally, in January 1944, he was called to England and placed in charge of Third Army.

Operations in the European Theater were controlled by Supreme Headquarters, Allied Expeditionary Forces (SHAEF) and the Supreme Commander, General Eisen-hower. Under him were two army groups. The British 21st Army Group, under General Montgomery, consisted of the British Second Army and the Canadian First Army. General Omar Bradley was in charge of the US 12th Army Group — US First Army, commanded by General Hodges, and Patton's Third Army. In the Normandy landings, the First Army would hit the beaches at Cherbourg; the very exist-ence of Third Army would be kept secret until at least D-Day+10, when they would land south of the First Army positions and execute Patton's favorite type of operation — a breakthrough followed by an advance clear across France to Germany.

The choice of commanders shows that Marshall, as much as he appreciated Patton, was not blind to his faults.

Although on the face of it Patton was best qualified to lead the invasion, his impetuousness, Marshall knew, would be a grave handicap in that intricate political/military situation: Patton 'needs a brake to slow him down . . . someone just above him and that is why I am giving the command to Bradley.' That Patton recognized his own limitations is shown by the fact that he accepted the news of his former subordinate's promotion with good humor. He had always considered himself more of a tactician than a stratigist; in any event, he much preferred commanding at the field army level where there was more personal contact with the men.

The Third Army's embarkation from Southampton was delayed; it was 5 July — almost a month after Operation Overlord — before they finally slipped ashore in France and dug in near Néhou, some 15 miles south of Cherbourg. While Patton champed at the bit, the transporting of men, equipment, and supplies took its deliberate course, and it was not until 26 July that Cobra was ready to strike.

Following a mammoth aerial and artillery bombardment by the First Army, the Third Army took off, racing through the gap at St Lô, past Avranches on the 30th, and on through France in an extraordinary sweep that brought them to the outskirts of Paris by 21 August.

The Third Army could easily have entered the French capital, but instead Patton was instructed to bypass the city to the north and south and continue his advance toward Germany — an order that did not sit well with the men of Third Army, who objected strongly to the news that Montgomery, with his British troops and the US First Army, were parading the streets as liberators.

Their tempers were not improved by the sudden realization that their supplies were being drastically reduced. Montgomery, whose forces were some 100 miles behind Patton's, had been arguing for some time that instead of advancing on a broad front, there should be one decisive thrust — in his northern sector. Thus most of the supplies should be diverted north, while the troops in the south fought a holding action. Near the end of August Eisenhower had arrived at a compromise: Montgomery's push into Belgium would be given priority until he had taken Antwerp. Then the offensive would revert back to the 'broad front' advance.

Furious with Eisenhower, whom he felt had sacrificed an early victory to Montgomery's 'insatiable appetite,' Patton ordered his advance units to keep moving until their gas tanks ran dry — and then to get out and walk. Third Army took Verdun, surrounded Metz, and reached the Moselle River (where they linked up with the Franco-American Seventh Army under General Alexander Patch). There — about 30 miles from the great industrial complex in the Saar valley and less than 100 miles from the Rhine — they finally came to a halt at the beginning of September.

The dash across France was a triumph for both Patton and his staff — which was, at his insistence, the fastest, most efficient, and most professional in Europe. Ever since landing in Europe his orders had been to 'advance and keep on advancing' and that is exactly what his army had done — repeatedly bypassing areas where the Germans tried to make a stand, never giving the enemy time to stop and organize a counterattack. But though the attacks were

violent and unrelenting, they were never wild or uncontrolled; staff planning was always at least two operations ahead of the one in progress so that the army advanced with great speed, but never in haste.

The support between Patton and his Third Army staff was mutual and wholehearted, whether the battle was being waged with the Germans, 12th Army Group Headquarters, or even SHAEF. When the gas crisis was nearing its height, Patton received a bitter complaint that Third Army officers, disguised as First Army men, had stolen quite a lot of fuel. With a straight face, he replied, 'I'm very sorry to hear First Army lost some of its gas. But I know none of my officers would masquerade as First Army officers. They wouldn't stoop to that, not even to get gas SHAEF stole from us.' The merit promotions that several young officers in the Quartermasters Corps received shortly thereafter were probably coincidental.

Antwerp was finally taken on 4 September, and Third Army's supplies began arriving again. But in the meantime the German High Command had taken advantage of the 'miraculous' — if to them incomprehensible — respite to reinforce the area. Though Patton was able to force the Moselle in several places, he had to scrap his plans for a lightning offensive and settle down to a hard, set-piece battle.

Nancy, the old capital of Lorraine, fell to Third Army on 15 September. A few days later Patton received word that he was to stop and assume the defensive while Montgomery and the 21st Army Group made another concentrated assault, north of the Ardennes. If there was one thing Patton hated, it was defensive warfare; he considered it not only a waste of lives, but even un-American. His method of defense, therefore, was to maintain a continuous series of local actions with units small enough to escape Group Headquarter's notice. If questioned about his actions he would reply that he was only 'rectifying the line.'

As winter drew near the weather became almost as bitter an enemy as the Germans. The Third Army was receiving minimal supplies, not only of fuel and ammunition but also of other essentials like raincoats, sleeping bags and winter clothing. Despite the efforts of the Quartermaster Corps, who commissioned what they could from local suppliers and improvised the rest, sickness (especially trench foot) began to increase alarmingly. At one point the sickness rate equalled the battle casualty rate.

Patton spent most of October hounding headquarters for permission to attack, while Bradley and Montgomery argued the merits of a single attack in the north against the Ruhr versus a two-pronged assault against both the Ruhr and the Saar Basin. Eventually Bradley won his point, and Patton was called to Group Headquarters to receive the good news that the winter campaign would involve a series of assaults along the entire front with an attack on the Saar as one of the two main efforts.

Little planning was necessary since they had been working on a Saar offensive for over a month, but the precarious supply situation had to be taken into account. There were also some new troop dispositions. For this offensive, the VIII Corps would stay in the Ardennes under the control of the First Army. The Third Army would get the III Corps in its place, along with two divisions presently en route from the States, the 10th Armored and 95th Infantry Divisions. On 3 November Bradley reported that Montgomery had not yet set a starting date for his half of

the attack, but would probably not be ready before 1 December. Patton, however, could go whenever he was ready.

The Third Army attacked on 8 November. Metz was under Patton's control by the 19th (though the last of its forts held out for another month). Mud and slush hindered the offensive and it became more and more difficult to maintain momentum. However, after six weeks of fighting, the Third Army had reached the Siegfried Line along most of its front and in one place had driven a wedge some 30 miles beyond it. Overall, they had advanced approximately 50 miles. Meanwhile, the northern offensive had been stopped in an attempt to take the Roer River dams and in the south the Seventh Army had reached the Rhine and turned to join up with Patton's right flank.

Although the Allies' headlong advance had been slowed down and their initial complacency dispelled to some extent by the Germans' stubborn resistance, it was obvious

Top: A German assault gun is hidden in a farmyard as part of the preparations for the Ardennes Offensive.

Left: The Siegfried Line is breached by the 39th Infantry, 9th Division, 3rd Armored Division on 15 September 1944.

Far left above: Generals Eisenhower and Patton consult a map in North Africa in March 1943.

that they would come out the winners in a battle of attrition. However, as early as September Hitler had conceived an audacious plan. In one last, gigantic offensive, he would strike westward, penetrate between the First and Third Armies to Antwerp (thus depriving the Allies of their main supply port), and mop up the British and Canadian Armies along the Belgian-Dutch border. The attack would be mounted through the Ardennes, where he had made his great breakthrough in 1940. He knew that this area was thinly held by only four infantry divisions. Field Marshal Gerd von Rundstedt was called out of retirement to lead the offensive (code named Operation Autumn Fog), and the task of quietly collecting an attack force began.

In November a few high-ranking German staff and army officers were informed of the plan. The generals were astounded. They pointed out that Germany was much weaker than it had been in 1940 and furthermore was facing a much stronger enemy. Together von Rundstedt, Field Marshal Walther Model (Commander of Army Group B) and the three field commanders — SS General Sepp Dietrich (Sixth SS Panzer Army), General Hasso von Manteuffel (Fifth Panzer Army) and General Erich Brandenburger (Seventh Army) — concocted an alternative plan with more limited objectives. Hitler remained adamant. He was convinced that he could shatter the 'unnatural' Allied union with one blow — and the generals were too intimidated after the purges that had followed the July Plot assassination attempt to put up much of an argument.

On the evening of 12 December several senior German

field commanders were taken to a secret rendezvous at Hitler's bunker near Frankfurt where many of them learned for the first time of the great offensive, scheduled to begin in only four days. Over 250,000 men – 28 divisions, nine of them Panzer units – had been assembled, along with some 2500 tanks and field guns. It was a puny force compared to the resources Rundstedt had controlled on the same front in 1940, but the Germans were not without some advantages.

Although the men themselves were not as fit or well-trained as the previous strike force, their ranks were stiffened by the addition of several fanatical, well-equipped SS units. In addition, the Germans still had superiority in some important weapons. The Mark V Panthers and Mark VI Tigers, for example, had thicker armor and wider tracks than the new Shermans and were armed with the more powerful 88mm guns, compared to the Sherman's 76mm.

There were also some desperate problems. Allied bombing raids had left the German transportation network a shambles and there was a grave fuel shortage. Hitler was counting on his last, most important weapon to help overcome these difficulties – the advantage of surprise. Despite the lesson of 1940, Allied planners persisted in viewing the Ardennes, with its heavily wooded hills cut by deep valleys where tank advances could easily be stopped, as unsuitable for a large-scale armor action. From the German point of view, however, the forest was a natural blind for concentrations of ground forces, while the drier high ground gave the tanks more maneuverability.

The night of 15 December was cold and a thick mist covered the German troops as they moved into position along their 70-mile front between Monschau and Echternach. Bad weather had been forecast for the next several days and was essential to the success of the operation, to protect the German supply columns from Allied air strikes. On the 16th they moved out. In the north, the main strike

Left: German troops captured by the 82nd Airborne Division are lined up on a road near Hierlot, Belgium.

Far left above: A Marder III with a 75mm gun in December 1944.

Far left: A knocked out American half-track near Malmédy, 16 December 1944.

Top: A German soldier examines an American half-track, which has been converted into an ambulance.

Top left: Soldiers of the Adolf Hitler Division on the road between Malmedy and St Vith, 17 December 1944.

force, Dietrich's Sixth Panzer Army had orders to advance northwest, past Liège to Antwerp. Manteuffel and the Fifth Panzer Army, in the center, were to push through the Ardennes and then turn north to the Meuse and, eventually, to Antwerp.

The offensive came as a complete surprise to the Allies — and unnecessarily so. Third Army advance units had been keeping a faithful eye on the German troop dispositions ahead of them. As early as 4 December Patton's G-2 (Intelligence) staff had sent a Special Estimate to Group Headquarters that concluded, 'A large build up of troops and supplies is clearly in progress opposite the southern (Ardennes) flank of First Army.' Allied planners were blinded by overconfidence. Montgomery and Bradley were both convinced that Hitler lacked the resources for a major offensive, and that he would be forced to throw everything

he had into stopping their drive toward the Rhine. An unofficial rebuff was the only reply to the Third Army's communique, and on 15 December Montgomery issued his own assessment of the situation, 'The enemy is at present fighting a defensive campaign . . . he cannot stage major offensive operations. . . . He has not the transport or

A German tank passes some of the many US prisoners taken during the first days of the Ardennes Offensive, 17 December 1944.

the petrol . . . nor could his tanks compete with ours in a mobile battle.'

He could not have been further from the mark. In the north, where Sepp Dietrich (a good fighter, but one with little understanding of armored warfare) was in command of the attack, the Allied line buckled, but held at several key points. In the center, however, the brilliant tank commander Manteuffel led the Fifth Panzer Army in a rapid advance that recalled the days of the first blitzkrieg. Despite the terrible weather, the delays in bringing up fuel supplies and occasional pockets of American resistance, it was more than 48 hours before his headlong rush was stopped, not far from Bastogne, by the US 10th Armored Division's desperate delaying action.

Manteuffel was assisted by the confusion in the American ranks sown by Operation Greif. This consisted of about 40 jeep loads of English-speaking German commandos in American uniforms, led by the famed Otto Skorzeny, whose mission was to infiltrate the American lines and turn signposts, cut communication wires, kill messengers and do anything else they could think of to create confusion. One captured commando started the rumor that a unit was on its way to kill Eisenhower with the result that the Supreme Commander was virtually immobilized in his Paris headquarters for some days, to his extreme annoyance. In the attempt to weed out the infiltrators, soldiers in the battle area were frequently required to identify themselves by correctly answering questions like 'Who won the World Series in 1938?' Bradley had to spend some time convincing a zealous sentry that Springfield, not Chicago, is the capital of Illinois.

As the German attack began on 16 December the Third Army was preparing for its own offensive, an elaborate assault on the Siegfried Line. On the 17th, when they finally got a clear idea of what was happening, several staff members were optimistic, thinking that the action to the north would take some of the pressure off their own operation. Patton, however, was less sanguine. Realizing that he would probably have to postpone his own plans, he set the staff working to prepare three contingency plans, in case they were asked to move north and help.

This proved to be a wise precaution. The next day he was ordered to Luxembourg to confer with Bradley, and by the 19 December he was attending a meeting of senior American commanders called by Eisenhower at Verdun. Ike decided that the First and Ninth Armies, in the north, should be temporarily placed under Montgomery's command — a plan that greatly displeased Bradley. All were agreed that the Allies' first move should be a counterattack by the Third Army from the south, against the underbelly of the Bulge. There were only two problems. First, how far north could General Jacob Devers stretch his 6th Army

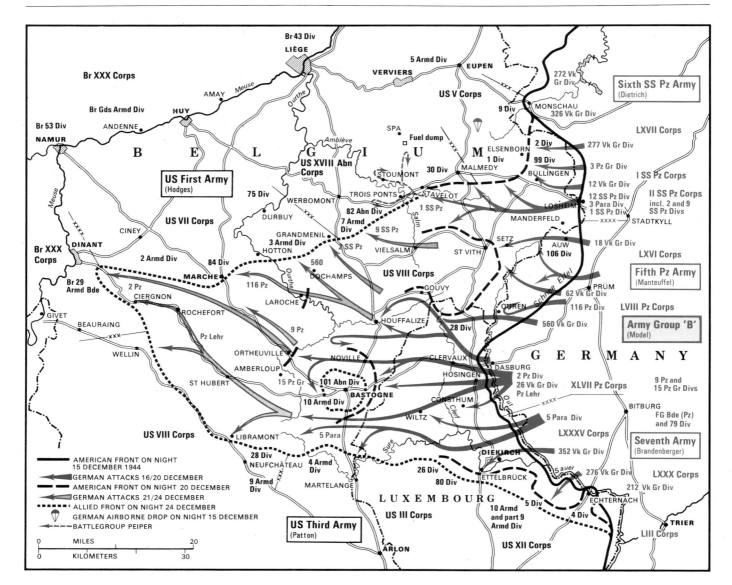

AMERICAN FRONT ON NIGHT 15 DECEMBER 1944
GERMAN ATTACKS 16/20 DECEMBER
AMERICAN FRONT ON NIGHT 20 DECEMBER
GERMAN ATTACKS 21/24 DECEMBER
ALLIED FRONT ON NIGHT 24 DECEMBER
GERMAN AIRBORNE DROP ON NIGHT 15 DECEMBER
BATTLEGROUP PEIPER

Group to cover Patton's territory (eventually he moved up to a point between Saarlautern and Saarbrücken), and second, how to time the attack. As far as Patton was concerned, he could begin immediately and be in position to attack by 21 December. This was no idle boast. His staff had completed their plans and were simply waiting for him to telephone a code word, telling them which of the three to put into operation.

It was eventually decided that Middleton's VIII Corps would be returned to the Third Army. Most of it was already engaged in and around Bastogne, but the remainder would be launched on the left against the nose of the German salient, toward St Hubert. The XII Corps would attack on the right, toward Echternach. In the center would be Milliken's III Corps — 80th and 26th Infantry and the 4th Armored Divisions — with their main objective the liberation of Bastogne (by now surrounded and under siege). This was a calculated risk, for it meant leaving the 50-mile north-south front, including the Saarlautern salient, in the hands of just XX Corps (two infantry and one armored division) and a Cavalry Group. Though too weak to withstand an attack in force, they were in a position to fight a strong delaying action — as well as prevent a surprise attack through that sector.

General 'Nuts' McAuliffe, commander of 101st Airborne Division.

Members of the 101st Airborne Division leave Bastogne to relieve a neighboring village, which has been under attack for 10 days, 29 December 1944.

Returning to his own headquarters, Patton found his staff under a considerable amount of tension, which he characteristically dispelled as soon as possible. Standing in front of the room, he glared into the tense faces before him and after a few moments announced impressively, 'This will get the bastards out of their holes so we can kill all of them. Now go to work!'

And work they did, racing against time to shift a three-corps, north-south battle line to a four-corps line running east-west in the Ardennes and north-south in the Saar. Hundreds of units had to be moved quickly and efficiently. At Patton's orders they disregarded all blackout regulations and drove north with headlights blazing and throttles wide open. Thousands of miles of telephone wire had to be laid and an entire communications network, capable of remaining operational despite both cold and enemy attack, had to be established. A new supply system had to be set up, and thousands of tons of supplies shifted for distribution or storage in new depots and dumps. Hospitals had to be organized and stocked, maps and orders distributed, and hundreds of other details taken care of. During those busy days Patton worked as hard as anyone else, but neither then nor later would he claim credit for the gigantic effort. His staff, he said, had performed a miracle. (He was willing, however, to take credit for being the unofficial 'ray of sunshine and backslapper' for both his men and his superiors.) At 0600 on 22 December, right on schedule, III Corps attacked.

Meanwhile, the 'Battered Bastards of the Bastion of Bastogne' – the 101st Airborne Division along with remnants of the 10th Armored Division, the 705th Destroyer Tank Battalion and several other units, all under the command of General McAuliffe – had been grimly hanging on under the German barrage. Although it was a tiny town with a population of about 3500, Bastogne was the junction of seven major paved roads, and the key to the defense of both the Ardennes and the Meuse beyond. Theoretically the Germans could have bypassed it on secondary roads, but these were in such bad condition that during bad weather they were practically impassable for military vehicles.

On 18 December, when Manteuffel's spearheads had advanced to within 15 miles of Bastogne, there were few American troops in Bastogne. However, by that evening, when his main force arrived, the town was already occupied by the 101st Airborne. Starting the previous evening from Reims, over 100 miles away, the 'Screaming Eagles' had beaten the Germans in a decisive race. Eventually, and with some difficulty, Manteuffel got his divisions around Bastogne, but he was forced to leave a substantial force behind to deal with the junction. In addition, the delay had given the Americans time to bring up more reserves, and Manteuffel was getting no support from Dietrich's army (which had been cornered by the US First Army near Stavelot on 19 December). Although an advance guard would eventually reach the Meuse at Dinant on the 24th, they would soon be cut off and forced to withdraw.

In effect, the German advance was stalled at Bastogne where, as an anonymous GI put it, 'They've got us surrounded, the poor bastards.' On the 22nd General Heinrich von Lüttwitz, commander of the German XLVII Armored Corps, sent a message to McAuliffe demanding he surrender the city. McAuliffe's reply, 'Nuts!' is probably unequalled in the annals of military history, both for its brevity and the amount of confusion it caused in the enemy camp before it was deciphered.

The III Corps began their attack first, in the worst possible weather conditions. They were hampered by fog, a howling blizzard and below-freezing temperatures. Their winter equipment still had not arrived. They had no snow-camouflage material, parkas, hoods or fur gloves. Blankets were cut up for shoe liners, and yards of white cloth purchased from the French were run up into camouflage suits by the Quartermaster Corps. Along with their improvised equipment the men carried a prayer for good weather,

Left: The 101st Airborne Division held out in Bastogne from the 19 to 26 December when the 4th Armored Division relieved the 'Battered Bastards of the 101st.'

Right: German prisoners of war are marched into captivity.

Right center: Soldiers of a Volks Grenadier Regiment are taken behind the lines. Some 14,279 prisoners were taken during the first days of the counterattack by Allied troops.

Right bottom: More German prisoners.

Far right: Members of the 101st Airborne who held out in Bastogne.

composed some days before (for the Saar offensive) by Chaplain James H O'Neill.

On the east of III Corp's front, the 80th Division was attacking toward Ettelbruck and the Sûre River, while in the center the 26th Infantry Division was aiming for Wiltz. The terrain these two divisions were moving through was some of the worst in the entire Ardennes. On the left was 4th Armored Division which had orders to 'drive like hell' straight for Bastogne. The division was divided into three units, or combat commands. Combat Command A (CCA) went north along the Arlon-Bastogne road. Combat Command B (CCB) took secondary roads through the hilly, wooded countryside from Hubay-la-Neuve (about 5 miles west). Their orders called for them to gradually move closer to each other until they were advancing on parallel paths about two miles apart. Combat Command Reserve was held back; they were rarely committed to action.

The two 4th Armored units made good progress at first, but Combat Command A was slowed down by an American minefield originally laid to stop the German Seventh Army. In the afternoon they ran into a German rifle company guarding the demolished bridges at Martelange, and were held up until about 0300 on the 23rd. Combat Command B met little resistance on the back roads, and by noon had advanced nearly 12 miles – to within 7 miles of the center of Bastogne. They too, however, were stopped cold until the small hours of the morning, this time by a German company holding a blown bridge at Burnon.

The next morning, 23 December, was crisp and cold with clear, sunny skies. As the air filled with the sound of Allied aircraft, a jubilant Patton shouted at one of his staff officers, 'Hot damn! That O'Neill sure did some potent praying! Get him up here, I want to pin a medal on him!' For the next five days pilots of the XIX Tactical Air Command (TAC) flew until they were out of ammunition, touched down to

reload and refuel, and took off again. As darkness fell a Night Fighter (P-61) Squadron took over.

Patton called Milliken and ordered the 4th Armored to 'stop piddling around' and get on toward Bastogne, by-passing centers of resistance and mopping them up later. Combat Command A followed orders. At midday they tried to go around the village of Warnach. However, the Germans put up a desperate defense of the area, and Combat Command A was not able to either take Warnach or bypass it for 24 hours. Thus by noon on 24 December they were still several miles behind the point Combat Command B had reached 48 hours earlier.

Combat Command B fared little better. At noon on the 23rd they took Chaumont, which had been under heavy aerial assault all morning, but the sun thawed the open ground, and the Shermans became thoroughly bogged down. When the Germans realized what was happening they moved an assault-gun brigade onto the hills above the village, knocked out every tank they could see, and retook the town, inflicting heavy losses on the Americans. By the end of the day Combat Command B had managed to struggle only as far as Hompre – a distance of just over five miles. They were still three miles from the Bastogne perimeter.

Patton fought the Battle of the Bulge in his own fashion – tearing around the front lines in his open jeep, which was specially equipped with a loud horn and blazoned with huge red stars. One of his cardinal rules was that 'Commanders (and their staffs) should be seen physically by as many individuals of their command as possible, certainly by all combat soldiers.' Patton, in his worn field overcoat, visited battle lines and unit headquarters, showing up at every critical point. No one saw him moving backward. He always flew to the rear, to save time and avoid the possibility of lowering morale.

On the day before Christmas the Bastogne defenders received the message from Patton, 'Christmas Eve present coming up. Hold on.' It was not to be. Though the attack had picked up somewhat once the planes were in the air, the tanks and infantry still moved slowly over the winding, icy mountain roads, battling fiercely for every foot of ground. This did not conform to Patton's theory of warfare, and he called Eisenhower several times to apologize for the delay.

Despite its initial success the offensive, in von Rundstedt's and Model's view, was a failure by 22 December. The field commander, Manteuffel, came to the same conclusion on Christmas Eve. He knew he could not hold the narrow salient much longer. On 25 December he pulled troops from the eastward advance and turned them toward Third Army. In the east, in VIII Corps' sector, he launched power-

ful Panzer attacks against St Hubert, Rochefort, and Hotton; another series of heavy assaults hit Bastogne itself. In addition, in the west, General Brandenburger initiated several violent battles in the Saarlautern region.

Combat Command A and Combat Command B had each been strengthened with a battalion of infantry, and even Combat Command Reserve had been thrown into the fighting, on a line parallelling the Neufchateau-Bastogne road. Again, they could not make a decisive breakthrough. Patton's temper on Christmas night was precarious, and it was not improved when Bradley relayed Montgomery's suggestion that Patton fall back to the Moselle and 'regroup' (as Montgomery himself was doing). Patton's reaction was that it was a 'disgusting' idea.

The 26 December was an important day in the battle. Manteuffel finally received official permission to establish defensive positions, though Hitler was still convinced that Germany could ultimately take Bastogne, drive through on the Northern Front and capture Antwerp. That same day, Bastogne was relieved. In the afternoon, advance elements of the 4th Armored Division's Combat Command B made contact with the 101st Airborne Division's outposts on the perimeter, and that evening five tanks from Combat Command Reserve entered the city itself. In the afternoon Combat Command Reserve's two battalion commanders, Lieutenant Colonels Creighton Abrams and George Jacques had decided to ignore the battle plan for the day and drive straight through Assenois to Bastogne. While most of their forces engaged the Germans at Assenois, the five tanks plus an infantry half-track slipped away from the battle and into the beleaguered town. The first exchange between Captain William Dwight and General McAuliffe was typical of the American command style that Germans, and other Europeans, found so difficult to understand. 'How are you General?' asked Dwight. 'Gee, I'm mighty glad to see you,' replied McAuliffe. The latter's first message to Patton was typical of the Bastogne defenders' attitude throughout the conflict, 'Losses Light. Morale high. Await-

Right: A scene from the Ardennes
Offensive: German soldiers go
through the possessions of dead
American soldiers. The US boots
have been removed as they were
much better than standard
German issue.

Far right: Troops of the 82nd
Airborne Division march German
prisoners to the rear on 29
January 1945.

Below right: The outskirts of
Bastogne.

Below: US trucks bring supplies
into Bastogne after it was relieved,
22 January 1945.

ing orders to continue the counteroffensive.'

Bastogne, of course, was still an island in hostile territory, though it now had a thin, tenuous lifeline to Third Army – whose first priority had to be widening and protecting it. The Germans, who rallied with their usual speed and efficiency. launched two heavy attacks on the narrow corridor that night. Two convoys loaded with food, ammunition, medicine and other supplies were ready to go on the 27th, but the journey was still too hazardous. Instead, Patton, after much haranguing of headquarters, arranged for an airlift to drop as many supplies as possible. Though SHAEF air mission to evacuate the wounded had to be scrubbed, XIX TAC improvised their own lift. A small squad of L-5s, with P-47 fighter support, dropped several teams of medics into the town. By the evening of the 27th the corridor, though still dangerous, was passable. The siege of Bastogne was over; the battle was just beginning.

On 28 December the weather closed in again and the Germans began their next attack on Bastogne in a heavy snowstorm. In addition to the reinforcements from the west, Manteuffel had been able to call on German units in the northern sector, where Montgomery was still refitting his four armies. Within a few days, at the height of the battle, Third Army would be facing elements of 12 German divisions in the Bastogne area alone. Seven other divisions would be thrown against the newly-reinforced VIII Corps in the west and XII Corps in the east.

The German attacks were held off on all fronts. This success was due in large part both to XIX TAC, which managed to keep its planes up almost all day despite the weather, and to the artillery units. The new proximity fuze had given an added boost to the artillery, which proved to be indispensable in the days ahead. Still Hitler would not consider a German withdrawal. Despite Manteuffel's pleas that he be allowed to pull out while he still had a chance to save his army and despite Guderian's warning that the Eastern Front was a 'house of cards,' the Fuhrer – in a harangue lasting several hours – ordered the western offensive resumed.

For the next eight days, from 28 December to 5 January, it was attack and counterattack, as Germans and Americans engaged in a series of murderous battles. On the 28th Patton launched a three-corps offensive toward St Vith, Houffalize and Echternach. The Germans quickly countered with their own series of attacks. On 30 December Manteuffel threw five divisions against the Bastogne corridor but the weather cleared for a few hours and they were beaten off with help from XIX TAC. On 1 January the Luftwaffe launched a well-coordinated attack on Allied airfields. Although they inflicted some damage, they also lost about 300 of their own planes. That same day there were seven attacks on Bastogne, and a force of eight German divisions moved against the Saar. None succeeded.

On 3 January there was another heavy attack against Bastogne, by two German corps (nine divisions). While all German eyes were on Third Army, First Army was allowed to finally make its move in the north. By the 5th it was obvious, even to Hitler, that the great Ardennes Offensive had failed. There were still some weeks of hard fighting left, but from then on the outcome was never in doubt. The Germans executed a skillful withdrawal. Operating behind

Right: A US soldier examines the slogan written on a wall in Trois Vierges, north of Clervaux. It reads 'Behind the last battle of this war lies our victory.'

Far right: Mark V Panthers were the best German tanks of the war and took part in the Ardennes Offensive.

Below: German prisoners en route to the rear following their capture by troops of the 102nd Division.

infantry attacks, armored groups moved at night to avoid the worst of the air assaults and fought from prepared positions during the day. The German experience in Russia proved to be of great value; Hitler's forces were much better than the Americans at travelling through the deep snow and over the icy roads. The withdrawal was accelerated on 12 January when the Russians launched their mammoth winter offensive.

Finally, on 16 January, First and Third Armies linked up at Houffalize. It was an especially satisfying meeting for Patton, since First Army was represented on this historic occasion by a task force from his old unit, 2nd Armored Division. Now there was only mopping up to be done. On 23 January St Vith was retaken. The Battle of the Bulge was over.

Hitler had gambled and lost — and in losing, had not only sealed the fate of his armies in the west, but had also made

defeat in the east inevitable. Overall, Germany lost more than 120,000 men (killed, captured and wounded) during the Battle of the Bulge, along with some 800 tanks and guns, 1600 planes and 6000 vehicles. American losses included about 80,000 men and over 700 tanks and guns. The Allies could replace their men and equipment, while Hitler had staked everything on this last throw.

There were three basic reasons for the failure of the German offensive. First, Hitler grossly overestimated his own capabilities vis-à-vis the Allies. Not only did he discount the effect of superior Allied numbers of men and equipment, but he also had little regard for the American soldiers' fighting ability. He was misled, as were many in the German High Command, by the Americans' generally casual approach to military discipline and appearance. The second reason for the failure was the Allies' overwhelming superiority in the air — and the five days of clear weather

(23–27 December) that enabled it to be used to such effect. Last, but not least, was Third Army's unexpectedly quick assault on the southern flank of the Bulge. The Germans never dreamed that Patton could turn and strike so hard in such a short time.

This chapter has looked at just one aspect of one engagement in Europe, but it is one which Patton described as 'one of the greatest exploits of military history . . . executed under the most difficult and trying conditions and against tremendous odds . . . and also without any help from anyone.' He was not praising himself, but was speaking, quite sincerely, about his army. Characteristically, he always maintained that he had had 'Damned little to do' with the victory. Speaking of himself in the third person, he emphatically told the press, 'All he did was to give the orders. It was the staff of his headquarters and the troops in the line that performed this matchless feat.' During his

last staff briefing in Europe he told his men, 'There is probably no commander in Europe who did less work than I did. You did it all.'

Patton was not indulging in a show of false modesty, he believed every word he said. What he did not say, however, was that in addition to giving the orders, it had been he who turned Third Army into an efficient, well-knit team capable of carrying them out, and who kept that team functioning at peak performance under extremely difficult conditions. Though his reputation was made as a tank commander, his army's success at the Bulge depended far less on spectacular tank battles than on the superb coordination of armored, infantry, artillery, air and support forces. The arguments, the political gaffes, the eccentricities, aside, Dwight D Eisenhower has summed him up best, 'George Patton was, if nothing else, a man who knew his business.'

CHAPTER 15

ZHUKOV
AT BERLIN 1945

oviet military leaders during World War II were in a more difficult position than their Western counterparts in so far as the formulation of strategy was concerned, because their warlord, Josef Stalin, insisted on formulating it himself, giving them an even smaller leverage in executing it than Western generals had. It is true that in the case of Marshal Zhukov he was always consulted and much of his advice was accepted by Stalin. However, Zhukov could never be sure of anything. Like Hitler, Stalin insisted on controlling even tactical maneuvers of the various armies under Zhukov's command. The result was that all the strategic planning had to be done through Stalin and the credit for it had to go to him. Much of the tactical execution went to Stalin who not only controlled reserves, but also frequently altered the lines of attack by groups of armies (fronts), armies, corps and even armored units. Thus only within these narrow limits could Soviet generals establish their reputations and claim credit for the battles of Stalingrad, Leningrad, Kiev or Berlin.

Together with Marshal Vasilievsky, Zhukov was in a particularly privileged military situation, for from the very beginning of the war he was Stalin's most intimate adviser on strategy and his personal 'executor' on the spot, at the battles. Thus, for example, Zhukov had to inspect the battle ground at Stalingrad long before Stalin finally decided on the strategy of that battle. Zhukov reported his findings to his master and advanced his suggestions. Together with other commanders, who were only allowed to comment on their partial tasks, Zhukov often forced on Stalin his military decisions. However equally often Stalin, who had additional sources of information (the NKVD channel, for example) and his own finger on the strategic reserves button, made his decisions contrary to the military advice proffered. However, Zhukov had an additional privilege; that of commanding the groups of armies — especially toward the end of the war — in the execution of Stalin's

strategies which he had helped to formulate. Naturally in the latter capacity Zhukov had a greater scope for initiative and decision making, but even then his powers were limited by Stalin's frequent checks and especially, famous telephone calls, which usually resulted in changes of tactical plans and operations. Thus, with all these reservations in mind, we can now consider Zhukov's role in the Battle for Berlin.

Stalin earmarked his favorite general, Zhukov, for the final Battle of Berlin in October 1944. Zhukov himself says that this was exactly one day after he had persuaded Stalin to halt the Warsaw offensive, whose objectives were unclear to him, and which was draining Soviet military resources quite disproportionately to its significance. Zhukov had only just returned from Bulgaria and was inspecting the Warsaw front: the 1st Byelorussian and 1st Ukrainian Fronts came to a halt on the line Praga–Warsaw–Vistula–Jaslo. Stalin therefore began to plan subsequent offensives, the most crucial of which would be the one from the Vistula to the Oder carried out by the 1st Byelorussian Front. This would bring this group of armies to the

Top: Members of the Hitler Youth visit Hitler at his headquarters during March 1945. At the end of the war these youths were called upon to fight and defend Germany.

Left: Josef Stalin and President Roosevelt at the Yalta Conference in February 1945. At this conference the future of Europe was discussed.

Berlin strategic zone, to be ultimately responsible for the final operation of the war, the taking of the Nazi capital, Berlin. This was Stalin's idea in October 1944, when he appointed Zhukov Commander in Chief of the 1st Byelorussian Front, one of the most brilliant staff officers, Colonel General M S Malinin as his Chief of Staff and Lieutenant General K F Telegin, one of the most experienced political officers, as members of the War Council.

The thrust toward Berlin was to be prepared most thoroughly by Zhukov and Malinin during November 1944. Although Germany still had some 7,500,000 men engaged in the war, Hitler was draining the Eastern Front of its best units in preparation for a counteroffensive in the west, the Battle of the Bulge. Zhukov's strategic planning included a feint thrust in the south toward Vienna in order to draw German reserves from his front, which was approved by Stalin and GHQ (Stavka). However, once again Stalin forced on Zhukov several strategic limitations. First of all, Stalin neglected to reduce East Prussia and in particular Königsberg (Kaliningrad) during the summer of 1944, and though he had now charged General Ivan D Chernyakhovsky and his 3rd Byelorussian Front with this task, he still persisted in ignoring the danger to the offensive operations of the 1st Byelorussian Front, about to be launched. Next Stalin charged General Konstantin Rokossovsky and his 2nd Byelorussian Front to strike in the general Northwest axis, from north of Warsaw, and coordinate this advance with Zhukov's, running along the axis south of Warsaw to Poznan (Posen) and in the general direction of the Oder and Berlin. Even to a politician, Stalin, this joint, coordinated operation, must have appeared as two separate missions to be treated on their own, but once again Stalin insisted on coordination so that he could more easily control the actual operations. Finally, Stalin also arrogated the time factor: although the preparations for the offensive were completed early in December 1944 he refused to give it the go ahead, because he thought heavy mud and poor visibility would make it impossible to exploit Soviet armor and mechanized superiority and set the tentative D-Day for 15–20 January 1945. Zhukov and Malinin had therefore unexpectedly more time for the planning and rehearsing of the offensive, which for the first time in the war was taking place on unfamiliar territory. All supply routes were in Poland and the military intelligence seemed unfamiliar coming from the secret service and aerial reconnaissance rather than from guerrilla sources. In addition the Red Army soldiers badly needed political explanations and education, in order, as Zhukov put it, to prevent irresponsible acts by Soviet soldiers in the liberated territory. Thus Zhukov was able to hold staff maneuvers in which he was able to consider logistical support (only two comparatively small bridgeheads), and he was able to resolve more or less the problem of cooperation with Rokossovsky's armies by staging maneuvers on the army scale. In a sudden change of direction Stalin launched both armies against Germany on 12 January 1945, apparently in response to Churchill's requests.

For this last but one push forward the combined Soviet forces were staggering: some 164 infantry divisions; 32,143 guns and mortars (7:1 superiority); 6460 tanks and self-propelled guns (5:1); 4772 aircraft (17:1); and

2,200,000 men (5:1). On Zhukov's front there was one infantry division for each 1.3 mile. Artillery barrage on a depth of 2 miles lasted $2\frac{1}{2}$ hours, and after two days of hard fighting Zhukov's forces forced a wedge of over seven miles. After two or three days tank armies were used in support of the infantry and they swiftly developed the offensive pursuing the Germans through breakthroughs up to 62 miles every day. After six days of the offensive General Guderian had to contend with Soviet penetrations of up to 100 miles on a 310-mile wide front. Cities began to fall – Warsaw was finally liberated – or were besieged and all the objectives were reached ahead of the schedule by 25 January 1945. By this time Stalin wanted to stop Zhukov's offensive to wait for the final defeat of the Germans in

Right: General Koniev and Bokov in 1943. Koniev was in command of the 1st Ukrainian Front in 1945.

Left below, center and bottom: On 12 November 1944 a parade was held for members of the Volkssturm in Berlin. The Volkssturm units were made up of older men and younger boys who were to defend the fatherland during the last battle. In the bottom picture Goebbels takes the salute at a march past.

East Prussia, but Zhukov objected: instead he asked for one more army to strengthen his right flank and then ordered a forced advance into the fortified Meseritz zone. For he knew that once the line Bydgoszcz-Poznan was reached by his and Rokossovsky's armies, all future operations, which also meant the ultimate one against Berlin, would be decided on the spot.

However, Stalin's telephone calls slowed down Zhukov, especially since Stalin refused to send Zhukov reinforcements which he needed if he was to reach Küstrin on the Oder river, on the direct line to Berlin. As Rokossovsky's offensive north of Bydgoszcz got stuck Zhukov's right flank was now in real danger and though the Oder was reached on 3 February Zhukov was obliged to halt, even though he had issued a directive to his armies and staffs to get ready to resume the final offensive on Berlin by 9 February: after all Berlin was only some 50 miles from Küstrin, which his armies had by then reached. Perhaps rather optimistically Zhukov ordered that Berlin be taken seven days later. This was indeed strategy on the spot, but rather different from what Stalin had in mind.

Instead of letting him dash for Berlin he obliged Zhukov to remain on the defensive and wait for the other commanders, Rokossovsky and Koniev to complete their tasks. Above all he was apparently worried about Zhukov's right flank: Soviet intelligence indicated that the Germans had the Third Panzer Army in Western Pomerania and it could either reinforce the East Pomeranian forces or defend Berlin. In either case it presented a risk to Zhukov's further operations. Although Marshal Chuikov, who was Zhukov's subordinate, claimed subsequently that the 2nd Byelorussian Front could have dealt with any threat along the Berlin axis, Zhukov came to agree with Stalin. On 1 March 1945 assisted by Zhukov's troops Rokossovsky's 2nd Byelorussian Front had finally broken through and reached the Baltic Sea by cutting the German armies into two. However, the attacking Russians still had their supply bases on the Vistula and this fact also forced Zhukov to

slow down. Marshal Koniev in the south reached Breslau with his 1st Ukrainian Front, but failed to cross the river Neisse as envisaged. Delays were caused on this front by the erratic supplies of ammunition, by the slowness of aviation to afford him cover, and above all by the lagging of the pontoon bridge units. Still by the end of March 1945 all these tasks were accomplished and Stalin summoned Zhukov and Koniev to Moscow to discuss the final phase of the war and let them into a few political secrets.

It must have been an unpleasant surprise for Zhukov to find on arrival at the Stavka in Moscow his chief rival, Marshal Koniev (who had been exclusively a field commander) in on the final operation. He probably suspected that his warlord might trick him of the final war trophy, Berlin, or make him divide the laurels with that 'simple' Koniev. Indeed Stalin was already engaged trying to outmaneuver the Allies. Before the generals arrived he had received a telegram about an Allied attack on Berlin under Field Marshal Montgomery, so Stalin now decided to give the operation top priority. However, fooling the Allies was child's play compared to tricking his own generals. General Eisenhower, with his straightforward naivety, informed Stalin, against Churchill's advice, that his main thrust in Germany would aim at Erfurt, Leipzig and Dresden and not at Berlin as the British had desired. Stalin in turn told Eisenhower that Soviet main effort would also aim at Dresden and would be launched in the second half of May. After misleading Eisenhower he immediately turned round to his generals and urged them to take Berlin before the Western Allies. Thus Stalin trusted neither his Allies nor his generals: now he played one general against another as he had previously done with the American and British allies.

In Moscow both Zhukov and Koniev wanted to take Berlin now that they had been given the green light. With his tough soldier mentality, Zhukov wanted a frontal attack by his 1st Byelorussian Front while the 'politician' Koniev favored a coordinated attack. Other general staff leaders

Left: One of the last pictures taken of Hitler in March 1945 at his bunker headquarters in Berlin.

Left below: A Volksturm member demonstrates the use of a Panzerfaust antitank weapon.

Right: Some of the men and boys who defended Berlin in May 1945.

Right below: Russians shell the Reichstag and Reichschancellery on 24 April 1945.

proposed pincer movements. Stalin had made up his mind long before and was just playing games with his generals. After all he had approved Zhukov's frontal proposal on 27 January 1945 and two days later Koniev's coordinated drive; the latter was a thrust from Silesia and river Neisse toward the river Elbe with the right flank of the 1st Ukrainian Front helping the 1st Byelorussians cut the German Army Group Center to ribbons, south of Berlin. However, both these proposals considered the reduction of the capital to be the business of the 1st Byelorussian Front, that is Marshal Zhukov. This would have been the climax of the Marshal's war career and might in a political sense overshadow even the real achievement of the Generalissimo. Cleverly Stalin pretended to consider all the proposals anew

and rather predictably approving the Zhukov-Koniev plans making them dependent on the demarcation line between the two army groups: the line ended at Lubben on the Spree river and Stalin added rather explicitly 'Whoever gets there first, will also take Berlin.' Thus he not only usurped all the prestige from the overall success of the operation, but made sure that he would control it throughout its stages up to the final point at Lubben, when he could decide which of the competing generals would be allotted the final trophy. As was usual with Stalin, once he had defined and safeguarded his own interests, he gave a completely free hand to Zhukov and Koniev to prepare and execute the operation on the spot.

As always Zhukov took the coming battle of Berlin most seriously. Immediately after his return from Moscow he invited all the commanders to his HQ and the conference lasted two days (5 to 7 April 1945). Commanders of support services also took part in these staff conferences and all the command aspects of the offensive were thoroughly thrashed out. Between 8 and 14 April Zhukov ordered more detailed war games to be conducted by the lower commanders. To Zhukov's mind the taking of Berlin whose defenses were well prepared and covered some 350 square miles was a formidable task and he ordered six aerial surveys by Soviet reconnaissance aircraft. Above all he was worried by the overextended lines of communications which made supplies difficult to move into operational zones. After all the 1st Ukrainian Front also had its supply problems, but it was the supplies of Rokossovsky's 2nd Byelorussian Front (which was to launch its offensive five days' later) which complicated Zhukov's supply situation. Up to the last moment Zhukov could not be sure that his supply requirements would be met in time.

Since 'never before in the experience of warfare had (Zhukov) been called upon to capture a city as large and as heavily fortified' he had his engineers construct an exact model of the city and its suburbs in planning the final assault: every street, square, alley, building, canal, bridge and the underground network were taken into consideration. Detailed assault maps compiled from aerial surveys, captured documents and interrogation reports of POWs, were supplied to all commands down to company level. Zhukov's next consideration was how his offensive would make the maximum impact on the enemy and he decided to launch it during the two hours before dawn. In order to avoid confusion among his own troops 140 powerful searchlights were to be used to illuminate the German positions. Thus Zhukov personally took care of every detail of the operation even allowing his politruks [political commissars] to take care of his men's morale, which they did without unduly disturbing the general, who was completely absorbed in military matters.

When Zhukov began to plan the Berlin operation, following his great victories between the Vistula and Oder rivers in January–March 1945, he knew practically nothing of Allied military plans. Stalin must have let him know that General Bradley had turned down the operation against Berlin on account of heavy casualties. However he did not add Bradley's rider 'that after all that the Americans would have to hand over Berlin to the Russians because of the zonal agreement.' Still Zhukov never cared much about casualties and this time he was prepared for any sacrifice, since Stalin himself urged the final operation for political reasons. In March 1945 Zhukov had to face new factors on the German side as soon as he had finished his own regrouping in his zone of operations between Schwedt and Gross Gastroze and particularly in the Küstrin sector, where

he had a beachhead on the other side of the Oder river. These new factors would make the Berlin operation even more costly. After the February breakthrough the German High Command reorganized its Berlin defenses into three zones: the first, main defense line and two other, rear ones. The first line ran along the left banks of the rivers Oder and Neisse and consisted of three fortified networks protected by minefields, in the depth of three to six miles. The second

defense line was some 10–14 miles from the first defense line and was connected by means of strongpoints, up to the depth of two miles. The third line was seven to 14 miles from the second one, but was unfinished when Zhukov launched his attack.

On paper Zhukov, who had detailed intelligence reports, had to face a formidable enemy: for the defense of his capital Hitler concentrated round and in Berlin Third Panzer Army and Ninth Field Army from the Army Group Vistula and Fourth Panzer Army and Seventeenth Field Army from the Army Group Center. Thus only in regular troops Hitler, who assumed the defense of the capital himself, had some 48 infantry divisions, four Panzer divisions and 10 motorized divisions. In addition Berlin contained a garrison of some 200,000 men, of whom, however, 200 battalions were Volkssturm formations. Youths of 16–17 years of age were specially trained in Panzerfaust handling and were to perform miracles in the defense of the capital against the Russian tanks. Hitler also concentrated some 72 percent of his Luftwaffe in Berlin. Altogether Hitler amassed some 1 million men, 10,400 guns and mine-throwers, 1500 tanks and self-propelled guns, more than 3,000,000 antitank rockets (Panzerfaust) and 3300 air-craft. Thus he had one division per nine kilometers of the front; 17.3 guns per kilometer of the front; and in the Küstrin sector, where a frontal attack was rightly expected, he had one division per three kilometers and 66 guns and 17 tanks per kilometer of the front. (A greater density was achieved only at Kursk in 1943, where the Russians deployed 105 guns per kilometer in the expected sector of

Soviet sharpshooters pick off snipers in the streets of Berlin. The fighting for the inner city was bitter and lasted for five days.

attack.) In addition Hitler put in charge of the Küstrin defenses General Heinrici, an acknowledged Wehrmacht expert on defense fighting: once again Heinrici succeeded in withdrawing from the first line of defense under cover of darkness, so that the Russian artillery barrage hit empty trenches. While the German High Command was a little disorganized, because of Hitler's interference, German soldiers were in a desperate mood, ready to defend the capital to the last shot, largely because of the barbaric behavior of the Red Army in East Prussia.

To bring this impressive edifice of Hitler's power crashing down Zhukov was given a free hand in the actual offensive planning and execution and Zhukov could also alter the deployment of the Stavka's strategic reserves. Stalin, nevertheless, sent another of his watchdogs to check on his tough general in the shape of General Vasiliy Sokolovsky, whom he appointed as 1st deputy commander. In fact Sokolovsky, who was a finer soldier than Zhukov himself, strengthened Zhukov's forces rather than restricting him and Stalin never forgave him, giving him no appointment of distinction while alive.

Zhukov's front was narrowed from some 200 to 120 miles and for his frontal assault it was in fact some 28 miles. In this narrow zone Zhukov concentrated a group of shock armies consisting of Forty-seventh Army, Third and Fifth Shock Armies, Eighth Guard Army, Third Army and First and Second Tank Guard Armies. His artillery had some 250 guns and minethrowers per kilometer of the assault sector and some 1000 rocket guns (Katyushas) which were mobile. From the air Zhukov was supported by Colonel General S Rudenko's 16th Air Fleet and the Polish Air Corps. With the support from the 2nd Byelorussian front (Marshal Rokossovsky) and 1st Ukrainian front (Marshal Koniev) Zhukov had an overall control over some 2,500,000 men, 41,600 guns, 6250 tanks and 7500 aircraft.

According to Zhukov's plans, approved previously by Stalin, this formidable force was to be launched on 16 April 1945 to annihilate frontally the opponent in the Küstrin sector, break through toward Berlin which at first would be bypassed from the northwest and southeast and then stormed. The tank armies were made responsible for the exploitation of the breakthroughs: on the second day the Second Tank Guard Army would fight its way in the northwesterly direction while the First Tank Guard Army would do the same in the southeasterly. As a result the Fifth Shock, Eighth Guard and Sixty-ninth Armies would take the city on the sixth day of the operation. Zhukov was to achieve this straightforward success provided that Koniev's Twenty-eighth Army and Third Tank Guard Army succeeded in breaking through into Berlin from the south and Rokossovsky's armies covered his right flank, for they were not intended to launch their offensive attacks until two days later.

It was only on 15 April 1945 that Stalin informed the Supreme Allied Command of the Berlin operation; up to then he maintained that the Russian main thrust would be in the direction of Leipzig. Then at 0300 hours on 16 April Zhukov's barrage was started, but Heinrici had anticipated

it and evacuated the first line at 2000 hours the previous evening. During the murky dawn, before 0500 hours, the barrage was shifted farther on, and the four armies (Third and Fifth Shock, Eighth Guard and Sixty-ninth) pulled out of their positions to attack the empty German line in the light of some 143 search lights (zenith projectors). However, the terrain was extremely difficult for tracked vehicles and the German Flak and antitank guns were hitting Russian tank formations with alarming accuracy. Men on both sides exhibited incredible heroism and in many sectors Russian attackers were beaten back. At 1430 hours, after some nine hours of fighting, Zhukov's armies hardly dented the German lines, advancing only some four miles, without achieving a breakthrough. Zhukov suitably depressed telephoned Stalin, who after listening to him quietly, just told him that Koniev's armies had already crossed the Neisse river. However, Zhukov was given permission to react to the circumstances of the battle as he deemed necessary.

Zhukov immediately changed his tactics ordering the 16th Air Fleet to make massive air strikes: enemy gun positions were to be blasted to smithereens; Soviet artillery was ordered to pulverize German strongpoints on the heights and tank armies were thrown into combat even before the important Seelow Heights had been reduced. Tough fighting continued – in the sector of the Third Shock Army the German 309th Infantry Division lost 60 percent of its men – and the Russians only succeeded in destroying the first line of German defense. That night Zhukov once

Soviet troops in the heart of Berlin. Thousands of civilians died in the intense house-to-house fighting.

Right below: Su 122 assault guns
move into Berlin April 1945.

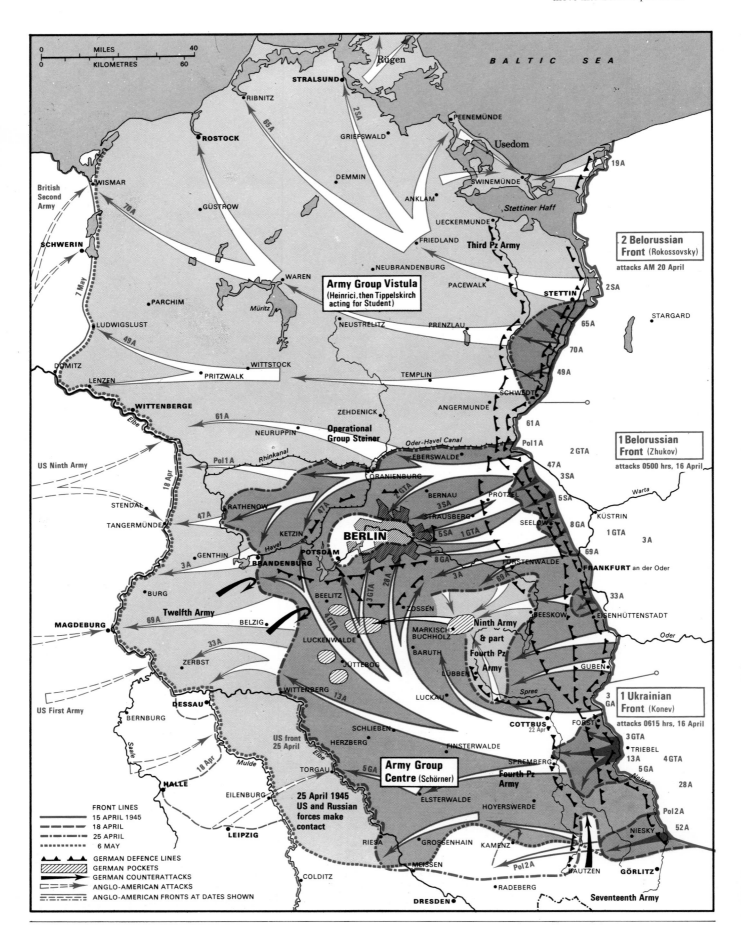

MILES 40
KILOMETRES 60

BALTIC SEA

Rügen

STRALSUND

RIBNITZ

PEENEMÜNDE

GRIEFSWALD

Usedom

ROSTOCK

DEMMIN

ANKLAM

SWINEMÜNDE

19A

British
Second
Army

WISMAR

GÜSTROW

UECKERMÜNDE

Stettiner Haff

2 Belorussian
Front (Rokossovsky)

attacks AM 20 April

SCHWERIN

WAREN

FRIEDLAND

Third Pz Army

NEUBRANDENBURG

PACEWALK

STETTIN

2SA

STARGARD

PARCHIM

Müritz

NEUSTRELITZ

PRENZLAU

65A

70A

49A

LUDWIGSLUST

DÖMITZ

LENZEN

PRITZWALK

WITTSTOCK

Army Group Vistula
(Heinrici, then Tippelskirch
acting for Student)

TEMPLIN

SCHWEDT

ANGERMÜNDE

WITTENBERGE

Elbe

NEURUPPIN

ZEHDENICK

61A

61A

Operational
Group Steiner

Oder-Havel Canal

Pol1A

2 GTA

1 Belorussian
Front (Zhukov)

attacks 0500 hrs, 16 April

US Ninth Army

Pol1A

Rhinkanal

EBERSWALDE

47A

3SA

Warta

STENDAL

RATHENOW

ORANIENBURG

BERNAU

PRÖTZEL

5SA

KÜSTRIN

TANGERMÜNDE

47A

KETZIN

BERLIN

STRAUSBERG

SEELOW

8GA

1GTA

3A

3A

GENTHIN

Havel

POTSDAM

5SA

1GTA

69A

BRANDENBURG

8GA

FÜRSTENWALDE

FRANKFURT an der Oder

BURG

Twelfth Army

BEELITZ

28A

3A

69A

33A

69A

BELZIG

3GTA

ZOSSEN

Ninth Army

BEESKOW

EISENHÜTTENSTADT

MAGDEBURG

33A

LUCKENWALDE

MARKISCH-
BUCHHOLZ

& part
Fourth Pz
Army

Oder

ZERBST

JÜTTEBOG

BARUTH

LÜBBEN

GUBEN

WITTENBERG

13A

LUCKAU

Spree

1 Ukrainian
Front (Konev)

attacks 0615 hrs, 16 April

DESSAU

US First Army

BERNBURG

COTTBUS
22 Apr

FORST

3 GA

3GTA

SCHLIEBEN

FINSTERWALDE

TRIEBEL

Saale

HERZBERG

SPREMBERG

13A

4GTA

5GA

18 Apr

Mulde

TORGAU

5GA

Fourth Pz
Army

28A

HALLE

EILENBURG

Elbe

25 April 1945
US and Russian
forces make
contact

Army Group
Centre (Schörner)

ELSTERWALDE

HOYERSWERDA

Neisse

Pol2A

LEIPZIG

US front
25 April

NIESKY

52A

FRONT LINES

15 APRIL 1945

18 APRIL

25 APRIL

6 MAY

GERMAN DEFENCE LINES

GERMAN POCKETS

GERMAN COUNTERATTACKS

ANGLO-AMERICAN ATTACKS

ANGLO-AMERICAN FRONTS AT DATES SHOWN

RIESA

GROSSENHAIN

KAMENZ

MEISSEN

Pol2A

BAUTZEN

GÖRLITZ

COLDITZ

RADEBERG

Seventeenth Army

DRESDEN

again telephoned Stalin who excitedly reprimanded him for throwing in the tanks prematurely, altering their axis of attack to a sector (Seelow Heights) where defenses were too strong. Zhukov defended himself with ability by pointing out that he was annihilating the enemy and drawing to his sector Berlin reinforcements. Stalin, however, ended the conversation on a sour note: if Zhukov got stuck, Rokossovsky would take Berlin from the north and Koniev from the south. Still Zhukov's progress on the 17th would decide the issue.

On the following day, albeit the Seelow Heights were taken, the Germans did manage to bring up reinforcements, especially antiaircraft artillery from Berlin, and their stubborn resistance slowed down the advance of Zhukov's armies. Zhukov personally inspected his armies before they lurched forward and used all sorts of gimmicks to boost up Soviet morale. Thus fighter pilots covering and supporting the Eighth Guard Army, threw down on the advancing guards, four parachutes with copies of the keys to Berlin captured by the Russian armies in the Seven Years' War, together with an unhelpful exhortation: 'Forward to Victory — here are the keys to the city.' However, the three reserve divisions and 50 tanks which the enemy threw into the battle slowed down the attacking Russians considerably. On this decisive day of the 17th Zhukov's armies failed to annihilate the second enemy line and Stalin told Koniev in the evening to direct his tanks to Berlin from the south and gave him a free hand to take it.

Zhukov later claimed that he had asked Stalin to shorten his attacking front even further concentrating thus the most formidable firing power to blot out the enemy, but Stalin refused. As it was Zhukov acted feverishly and in

despair, as he saw his competitor, the dullard Koniev, taking from him the final prize. He ordered night raids on enemy strongpoints by the 18th Air Fleet. At 1030 hours the Eighth Guards and First Tank Guards, went forward after an artillery barrage lasting half an hour. Colonel General V I Kazakov, one of the greatest artillerymen, personally directed fire and even he had to acknowledge defeat, when the two attacking armies were brought to a standstill. General M Ye Katukov's First Tank Guards performed all sorts of miracles with their armor, while Chuikov's Eighth Guards pressed on despite their losses. However, this was all a crawl reducing plans to nought and upsetting time schedules. Only early in the morning of the 18th did the right flank of the attacking armies (Forty-seventh and Third Shock Armies) break through the second line of defense: Zhukov's armies were two days behind the schedule. Stalin had to issue new directives to the supporting armies: on the 17th Koniev was ordered to advance on Berlin from the south; a day later Rokossovsky was ordered to drive in the southwestern direction instead of northwest. Thus during the operation itself the offensive was transformed into a tripartite one, instead of the envisaged frontal attack with subsidiary attacks on the right and left flanks.

Zhukov continued his inspections to resolve the problem of non-advancement. He realized that the Russians did not have a full picture of German defenses and above all failed to take into account the water obstacles which the Germans skillfully exploited against the advancing tanks. He also noticed that his armies wasted a lot of artillery fire and the air forces failed to penetrate enemy defenses and annihilate them. Thus on the 19th Zhukov decided on new tactical changes. He considered that his right flank, consisting of the

Forty-seventh and Third Shock armies, managed a great enough break to continue its drive in northerly and north-westerly directions in order to take Berlin from the western side. Colonel General P Belov's Sixty-first Army also changed the direction of its advance along the Hohenzollern canal, while the First Polish Army, Forty-seventh Army and Third and Fifth Shock Armies were ordered to strike in a southwesterly direction instead of westerly. Orders went out to continue the offensive night and day until modified objectives were achieved. Although this concentric attack did drive the Germans from their second defense line they swiftly reorganized and dug in in the unfinished third line. The Ninth German Army took the overall command in this sector, and was reinforced with three infantry divisions and the 11th motorized division, SS Nordland, withdrawn from the northern sector.

Thus on the 18th, when the Third Shock Army advanced to Bazlow where the SS reinforcements were concentrated, they were halted. Both Shock Armies waited until the night and then, after an artillery barrage, they fought all night in close hand-to-hand fighting to overcome this strongpoint. However further advance was checked by increased German air activity: in groups of 30 the Luftwaffe attacked Russian ground formation and kept them pinned down in this sector all day long. However, the right flank once again achieved a breakthrough of some 14 kilometers in the third line and Zhukov poured his tanks into this breach. This was taking his armies further away from Berlin rather than into the city. It was only on the 20th that the Forty-seventh and Third Shock Armies finally broke through the third line and Colonel General S Bogdanov's Second Tank Guards wheeled through to reach Ladeburg–Zepernik pass north of Berlin. The guns of the Third Shock Army closed on Berlin sufficiently to fire two symbolic shots at the city. However, it was only a day later that the Russian armies reached the northern fortifications of the city and the northern ring of the *autobahnen* and began systematic shelling. On that same day the Fifth Shock Army and XII Tank Guard Corps pushed their way into north eastern suburbs achieving the original objective of the whole 1st Byelorussian Front after considerable delay.

The toughest fighting had to be undertaken by the Eighth Tank Guards and First Tank Guards; even as late as 19 and 20 April both armies were bogged down on the third line of defense. Although their superiority was threefold the fanatical resistance of the 23rd Motorized Division, SS Niederland together with five Volkssturm battalions and a special antitank brigade, was able to halt these two elite tank armies. In the sector Fürstenwalde the Germans counterattacked several times; only on 21 April did the units of these two Russian armies reach the outer defense of Berlin at Petershagen and Erckner, while they were supposed to storm the inner city next day.

By 20 April Zhukov more or less acknowledged the failure of his frontal attack on Berlin and decided to wheel round the city and by joining with Koniev's and Rokossovsky's forces cut the capital off, encircle it and storm it afterward. Although Koniev's Third and Fourth Tank Guards were three times as distant from the city as Zhukov's tanks, they reached the outer defenses of the city, after some tough fighting, on the 19th. In this drive to northwest, far

from their original objective, the hardest nut to crack was the Zossen defense strongpoint, which among other things was GHQ Wehrmacht-Heer. On the approaches to Zossen at Barut the Third Tank Guards had to fight hard and once in the defense perimeter it took them two full days to reduce the strongpoint. The terrain was not ideal for tank maneuvers and when at night on the 22nd the strongpoint was taken the Germans blew up and flooded the underground GHQ. However, the Fourth Tank Guards commanded by Colonel General V Gordov, succeeded even better, for after their breakthrough at Kottbus, they drove on toward Potsdam, on the western side of Berlin, ripped a hole in the German defense round Zossen, forcing the German Ninth Army to give up its stand to avoid outflanking. On 21 April Koniev's tanks crossed the outer belt road, but further progress was slowed down by tough resistance especially round the Teltow Canal. It was only on 25 April that Koniev's forces finally made contact with Zhukov's and Berlin was encircled. Even before Lelyushenko's tanks made contact with Perkhorovich's soldiers their respective supporting air forces were strafing and bombing each other. On that day Stalin ruled that the city be divided between Koniev's and Zhukov's fronts along a line running 150 meters west of the Reichstag building. Also on that day Koniev's soldiers reached the river Elbe near Torgau and met for the first time American troops.

Zhukov was however completely absorbed in directing

the fighting. It became clear that Hitler had ordered General Steiner's Army Group and General Wenck's regrouped Twelfth Army to counterattack and relieve the city. While the former's attack never really got off the ground, the latter was joined by a hastily assembled group isolated in the south of the city, consisting of one motorized, three infantry and a tank division, and they managed to break through between Koniev's Twenty-eighth and Third Guards. Once again there was some tough fighting around the town Barut, but within a day the Third Tank Guards were let loose on these improvised and desperate formations and they broke them up. This was the final masterstroke that Hitler conjured up and sent Keitel to implement. Afterward the Germans acknowledged that they were surrounded in the capital, but rather than surrendering they were willing to go through the *Götterdammerung* while Hitler was alive and even after. Although their situation was hopeless they continued to form Volkssturm battalions and at the time that the capital was encircled they had some 200 battalions in position. Apart from the Volkssturm they had some 80,000 infantry in the city from the various armies defeated on the approaches. Hitler and the Nazis could feel quite safe from the security point of view, for there were some 32,000 police troops. Throughout Berlin at strategic crossroads, squares and parks the defenders had built block houses and pillboxes, tanks were dug in and mobility was assured by the underground railroad network. However, there was

also the population to feed and no antiaircraft cover: the Russians now concentrated the 16th and 18th Air Fleets on heavy bombing. Thus in the night of 26 April 563 aircraft of the 18th Air Fleet dropped on Berlin some 569 tons of bombs.

Despite all this Zhukov found it quite a task to reduce and take the city fiercely defended by some 300,000 fanatical soldiers under the leadership of General Weidling. However, once the breakout efforts by the Germans had been contained, both Zhukov and Koniev, after Stalin had altered still further the respective front boundaries on the 29th, threw themselves onto the hapless city with great gusto. According to Stalin's ruling Zhukov's troops would take the central sector of Berlin and in compensation Koniev's armies were permitted to divert to Prague, which had risen against the Germans at this time, and liberate that capital. Zhukov ordered the Third Shock, Forty-seventh Armies and two tanks armies, First and Second Guards, to storm the inner city from the north. Zhukov said that he had thrown in the two tank armies, because they had no other operational tasks and their impact would further destroy the morale of the enemy. It seems, however, that the employment of the tank armies in storming a city was the result of political pressure. Stalin had been worried by the lack of progress and by delays, and was once again suspicious of his allies. He thought they just might make a dash for Berlin from Torgau, despite an agreement with him. He therefore ordered Zhukov to engage in this costly way of taking a city. The tanks undoubtedly demoralized the German defenders, but at a cost: in the street fighting they could not maneuver and were sitting ducks for the youthful defenders with their Panzerfausts. In all parts of the city hand-to-hand fighting continued from house to house, block to block, street to street.

The central sector of the city (9th district) was particularly well prepared for the defense. The river Spree formed a natural defense in the north, while in the south the Tiergarten and other parks and boulevards were transformed into fortresses. The Reichstag building in the very center, at which the two Russian armies were both aiming, was a veritable Festung which reinforced walls, railings all round, and windows covered with cement plates. All the streets leading to the Reichstag were barred with barricades and fierce fighting continued from one to another. The Third Shock Army forced its way into Moabit and reached the Spree and were poised near the Moltke bridge to fight its way to the Reichstag. The Fifth Shock Army and the Eighth Guards with the First Tank Guards were making their way into the center from the east and southeast. Koniev's Third Tank Guards and Twenty-eighth Army were making their way through the southern defenses. The Second Tank Guards, after taking Charlottenburg, began to make their way into the center from the northwest. In the west the Forty-seventh and Thirteenth Armies together with Fourth Tank Guards were protecting the rear of the storming armies.

To demonstrate how fierce the fighting was we must realize that on 29 April, when units of the Thirteenth Army stormed and captured the Moltke bridge they were some

By 1 May 1945 Soviet troops had reached the Reichstag. These troops are carrying a flag which they were to fix on top of the building.

300 meters from the Reichstag, which Stalin, for his own reasons, singled out as the ultimate objective. Still it took the Russians three days before they could hoist their flag over the ruined building. In the night of 1 May the Third Army command ordered Colonel Zinchenko with his 756th Regiment to get the flag on the building for symbolic reasons, even though the Reichstag block was still largely in German hands. Sergeants Yegorov and Kantariya, under heavy fire, scaled the ruins, and the Russian generals and soldiers in the center could see their flag flying. The Russians give enemy casualties in that sector (2500 dead and wounded, 2604 captured; 28 guns destroyed, 59 captured together with 15 tanks), but not theirs, for their symbolic gesture must have cost Zhukov and Stalin heaps of casualties.

Even after the Soviet flag had been unfurled on the Reichstag fierce fighting continued. The Russians had no idea what was happening in Hitler's bunker under the Reich Chancellery and on 1 May they thought the Germans would surrender, when at 1500 hours General Krebs, then Chief of Staff, appeared on the line of the Eighth Guards and asked for parley with the Soviet command. Zhukov refused to see him, but he sent his first deputy, General Sokolovsky to assist General Chuikov in negotiations. They all had Stalin's order to accept unconditional surrender only. Krebs, who was the last military attaché in Moscow before the war, told Chuikov that Hitler had committed suicide and handed over power to Admiral Karl Dönitz and Dr Josef Goebbels. The latter was interested in an armistice, but the Russians told Krebs that it would have to be unconditional surrender: if the garrison surrendered Chuikov guaranteed them their lives, personal belongings to the soldiers and small arms to the officers. At 1800 hours Krebs brought back a message from Goebbels that Russian conditions were unacceptable and the fighting was resumed.

Immediately after artillery barrages the Russians went on with street fighting; it was only now, 1 May, that the Third Shock Army effected junction with the Eighth Guards south of the Reichstag. Only early next day did the Second Tank Guards finally meet Eighth Guards and First Tank Guards in the area of the Tiergarten. Although they had met, not all Berlin was under their control; in fact counterattacks continued in all areas, even in the Tiergarten, where the Russians overran General Weidling's command post. Still the hapless general decided to throw the sponge in and sent Colonel von Duwing to negotiate surrender. For the German command it was all over at 0630 hours when they all trudged into Russian captivity; in the afternoon, at 1500 hours, the rest of the garrison, some 70,000 men, gave up. It was only now, on 2 May 1945, five days before the general surrender, that Zhukov could report to Stalin that he had successfully completed his mission and that the Nazi capital was in his power. At this stage Zhukov had no idea at what cost this had been accomplished. Only in 1963 were we given details: the Berlin operation was described as one of the biggest of the war. Some 3,500,000 men fought in it equipped with 50,000 guns, 8000 tanks, and over 9000

Above: The scene on the Kurfürstendamm, a street that was famous for its theaters and clubs.

Left: The final assault on the Reichstag.

Far left: Soviet troops charge down to the Reichstag.

aircraft. The Russians captured 480,000 Germans, 1500 tanks, 4500 aircraft and 10,917 guns. They lost 304,887 dead, 2156 tanks, 1220 guns and 527 aircraft. We can immediately see General Omar Bradley's point: overall Allied casualties for the whole of 1945 operations were 260,000 men.

Immediately after the unconditional surrender, 8 May 1945, Marshal Georgy Zhukov held a press conference at which he briefed the world about his part on the surrender of the German Reich. He emphasized that the German armies were smashed on the river Oder and that the Berlin operation was really a mopping up. He did not mention any other generals who might have helped him with Berlin and generally seemed to play down 'this crowning achievement of World War II.' Still in 1965 when he finally published his memories of this battle, he acknowledged that the Berlin battle was his greatest. It is true that he had planned and executed the three decisive battles in Russia, at Moscow, Stalingrad and Kursk, but in all these battles his participation was overshadowed by that of his warlord, Stalin. It

Right: Jodl and von Friedeburg sign the instrument of surrender with Allied representatives.

Left bottom: Marshal Zhukov signs the four-power pact on the surrender of Germany.

Far left above: Soviet troops congregate in front of the Brandenburg Gate on 2 May 1945.

Far left center: Soviet troops raise the victory flag over the Brandenburg Gate.

Left: The view from the top of the Reichstag as a Soviet soldier raises the Red Flag on 2 May 1945.

was only at this last battle that the warlord singled Zhukov out for the 'crowning glory,' very much against the opposition of the other generals. Stalin not only accepted Zhukov's strategic plans, but gave him a completely free hand in tactical matters. Thus he foisted Zhukov on the 1st Byelorussian Front (and transferred Zhukov's rare friend, Rokossovsky, to command the 2nd Byelorrusian Front), and then determined, by means of demarcation lines that the 1st Byelorussian Front and not the 1st Ukrainian Front take the capital.

However, he also revised the time schedules and demarcation lines to suit his personal and political purposes. It is thought that he delayed Zhukov's dash for Berlin in February 1945 for political purposes – he had the Yalta conference to attend and Allied political pressures to consider. In the final phases of the operation he changed the demarcation lines between Zhukov's and Koniev's armies several times, probably not to give any of his generals total credit. So apart from these two basic considerations Zhukov, for the first time in the war, acted not as a representative of the Stavka, or coordinator, or 'Spasitel' (Savior) as in the Battle of Moscow, but on his own with control over strategic reserves, changing both strategic and tactical plans as he went, with the Stavka (and Stalin) approving them subsequently.

Thus the battle of Berlin had all the characteristics of Zhukov's past battles. Each time he amassed and commanded huge armies; millions of men were involved on both sides; meticulous staff work, preparations and rehearsals were conducted before each battle; each time he expended artillery and air power lavishly; each time armor was massively deployed and used; each time infantry did the actual fighting on a gigantic scale inflicting irreplaceable losses on the enemy; each time his own casualties were enormous. The series of Zhukov's battles was started in 1939 against the Japanese Kwantung armies and even then it was described by the British expert, Professor John

Erickson, as brilliant but costly. This was a characteristic Zhukov had in common with Stalin and may be the key to their wartime close friendship: both believed military aims can only be achieved by sacrificing life on a massive scale.

In addition Zhukov's personal characteristics contributed to this 'crowning victory' of his military career. He was not a popular figure; he never put a human touch to his orders; he had no friends only subordinates and one superior. He had a will of iron, tempestuous temper and almost supernatural determination, all of which characteristics were those of his principal opponent, Hitler. While Hitler could let loose these terrible traits on the German nation and the Wehrmacht to a certain extent, Zhukov let them loose liberally. He spoke rough, acted rough: his terrible threats – obey orders or face the firing squad – were invariably backed up. By means of these unattractive qualities he achieved success everywhere, and though he was a lonely figure, he was respected even by his rough fellow soldiers. The Berlin battle reflected both his character and military genius, and Stalin recognized it by sharing the podium with him at the victory parade in Moscow in June 1945. In Berlin Zhukov proved to the world that he was the master of the twentieth-century mass warfare and had no equal either at home or abroad. However, in strictly military terms he cannot be ranked highly. All the strategic decisions concerning his battle were taken by a committee whose chairman was Stalin. Even his tactical decisions were controlled by the committee, and above all Stalin, who rectified some of the costly errors: the initial frontal assault was a failure, and the line of the armor attack had to be changed several times to achieve success. In the battle for the city Zhukov also made many costly tactical mistakes, which undoubtedly overshadowed the overall achievement. One judgment that can safely be made about his generalship was that he was infinitely ruthless about his men in order to achieve overall success: however, such generals of World War I, were termed butchers of their men, not great generals.

CHAPTER 16

MACARTHUR AT INCHON 1950

The amphibious landing at Inchon, one of the most successful operations of its kind in military history, took place in mid-September 1950, early in the Korean War. Men from 17 member countries of the United Nations fought in that remote country. Yet no great novels or poetry no major cinematic efforts (with the possible exception of the comedy, *M*A*S*H*) have emerged to make the war real to those who stayed at home.

This nonwar – this 'police action' – claimed some 100,000 American casualties alone in three years; at one point US forces numbered 750,000 (the largest American force ever fielded under one command). Quite apart from the numbers of men committed, the technological advances and the war's strategic implications in the Far East, its unique place in military history – as the bridge between traditional conflicts fought to achieve total victory and the more recent 'limited' wars – must mark it as a major military event.

By the same token Inchon – the latest and probably the last great amphibious operation – has often been dismissed as a wild gamble that succeeded only by virtue of incredible luck. In fact, it was much more. Admiral Halsey called this dramatic switch from defense to offense 'the most masterly and audacious strategic course in all history.' David Rees, in *Korea: The Limited War*, has called it, 'a 20th-Century Cannae, ever to be studied.' It stands as a monument to military professionalism – in the formulation of complex plans under great pressure, in the execution of those plans against tremendous odds, but most of all to the strategic genius and iron will of one man – General Douglas MacArthur.

Against all advice and despite all apparent disadvantages, MacArthur decided to divide his Korean forces, using part to hold the besieged Pusan Perimeter and the rest to mount an intricate amphibious attack on the port of Inchon, far behind enemy lines. It was a daring plan that turned out to be the most brilliant stroke of a brilliant, if controversial, career. It also foreshadowed – indeed, its very success contributed to – the downfall of the commander, while the war of which it was a part was to end in stalemate.

Douglas MacArthur was intelligent, courageous, professional, impressive and even attractive – but not loveable. At least a partial explanation lies in the fact that MacArthur himself was, as William Manchester notes in *American Caesar*, 'a great, thundering paradox of a man.' For virtually every strength in his character one must cite a corresponding weakness and for every military triumph a mistake, often caused by overconfidence or an inability to look at facts squarely and objectively.

MacArthur's background is well known. He came from a family with a strong military tradition. His father, Arthur MacArthur, was a popular hero during the Spanish-American War. Following in his father's footsteps he went to West Point Academy and graduated first of his class in June 1902.

MacArthur saw his first action in World War I in February 1918 and at the same time collected the first of his many combat decorations – the American Silver Star and the French Croix de Guerre – for leading a raiding party through the barbed wire. During World War I he was to receive two Distinguished Service Crosses (second only to

Above: MacArthur meets Emperor Hirohito after the Japanese surrender. Hirohito is supposed to have admired MacArthur.

Top: President Harry Truman with whom MacArthur was to have such a difficult relationship.

the Medal of Honor), another Croix de Guerre, six more Silver Stars, and an appointment as a Commander of the Legion of Honor. In this first intensive, practical experience of war he proved that he could operate bravely and effectively in combat. He also turned out to be a fine commander. His men were proud to serve under him, and after the war General Pershing wrote that he was 'the best leader of troops we have.'

In June 1919 he took over as Superintendent of West Point – at the age of 39, one of the youngest men ever to hold that position. A three-year tour of duty in the Philippines (1922–25) strengthened his commitment to that island country. His wholehearted support of Philippine independence made him many friends among the Filipinos – and a corresponding number of enemies among those who preferred the colonial system.

In November 1930 MacArthur returned from another posting to the Philippines to Washington as Chief of Staff with the rank of general. On his retirement from that post in 1935, MacArthur set off once more for the Philippines (this time as Military Adviser to the Commonwealth). This last sojourn in the Philippines proved unfortunate in several respects. It marked the final disappearance of the tough field commander beloved by the doughboys of the Rainbow Division and the emergence of the pompous, theatrical mini-dictator who caused World War II GIs so much amusement.

In July 1941, when the US, Britain, and Holland froze Japanese assets in their respective areas, MacArthur was named commander of the US Forces in the Far East. In the four months between his appointment and the Japanese attacks on Pearl Harbor and the Philippines in December, his relations with Roosevelt were strained – especially as the President resolved to put European affairs first. Even in the days just before the invasion MacArthur remained optimistic about his ability to repulse – or at least delay – a Japanese attack on the Philippines. His staff may have kept from him the knowledge of just how badly equipped his troops were. In addition, he seems to have convinced himself with his own rhetoric and believed that he was invincible.

MacArthur's role in World War II is well documented: the attack on the Philippines that destroyed most of his air force on the ground; the subsequent, badly organized retreat; and the comeback, in the brilliantly conceived New Guinea campaign and retaking of the islands. He soon became a hero to the American public. In part this adulation was based on real achievements. However, the general consistently issued communiques that were greatly exaggerated and gave all the credit to the commander and little or none to the men who did the actual staffwork or fighting. Ostensibly this was an attempt to build up morale at home. However, the practice continued long after there was any need for morale building, and the troops were less than impressed. The cynical, antiauthoritarian GIs did not appreciate 'Dugout Doug's' gold braid and turgid rhetoric. The more accessible Eisenhower ('Ike'), with his friendly smile and relaxed, unassuming manner, was more their style.

Throughout the war MacArthur demonstrated a complete inability to see his theater of operations as part of an overall tactical plan. He bitterly resented his low priority

vis à vis Europe, and conducted a running battle with the President, the State Department, and the Pentagon – all of whom were credited by his staff with being in league with the British imperialists or, worse still, the Communists. Even before the end of the war this complaint had begun to creep into his news releases – and of course it was eagerly seized upon by disaffected Republicans at home.

MacArthur's appointment as viceroy of postwar Japan was one very much to his taste, and he turned out to be a compassionate and generous dictator. Surprisingly, the political reactionary introduced a wide range of liberal policies including civil rights for women, the organization of labor unions and a series of land reforms that completely restructured Japan's feudalistic rural society. Despite his model administration and many achievements, MacArthur on the whole had a bad press at home, mainly the result of his high-handed treatment of correspondents and his public quarrels with those who offended him in any way. The Republicans still hoped to field him as a serious presidential candidate, but his refusal to leave Japan spoiled their hopes. His reason for remaining was that he was absolutely convinced that if he 'returned for only a few weeks, the word would spread through the Pacific that the US had abandoned the Far East.' Though he was nominated at the 1948 Republican convention, he suffered a decisive defeat.

In 1950 the American defense establishment was in a sorry state. Wholesale demobilization after the war had severely reduced its numbers, and the recent unification of the Armed Forces had divided the remainder into hostile cliques, each striving to ensure the survival of its service. The two units under strongest attack were the Navy's air arm and the Marine Corps. In October of the previous year General Omar Bradley had informed the Armed Services Commission that 'large-scale amphibious operations . . . will never occur again.' In the next months of in-fighting a series of savage cutbacks reduced the number of Navy amphibious vessels from 362 (in 1947) to 91. Plans were also afoot to transform the Marines into a 'lightly armed,' largely ceremonial force.

In the Far East MacArthur, now 70, had long since passed retirement age. His military bearing as he paced vigorously through his Tokyo headquarters, his penetrating gaze and his unassailable self-confidence made observers forget his age. The forces under his command consisted of the Eighth Army (four undermanned and undertrained divisions commanded by General Walton H 'Johnny' Walker) and US Naval Forces Far East (one light cruiser and four destroyers). The Seventh Fleet, commanded by Vice Admiral Arthur D Struble in the Philippines, included one carrier, one cruiser, a squadron of destroyers and a division of submarines, and was not under MacArthur's control. As the only senior American commander who still believed in the value of amphibious operations, he requested, and in April 1950 received, a small Marine mobile training unit of 67 officers and men. They were augmented the following month by a tiny, naval amphibious force from San Diego headed by an enthusiastic, capable veteran, Rear Admiral James H Doyle.

Korea had also been occupied after World War II. The Soviet Union controlled the northern half, above the 38th Parallel, while the US occupied the south. MacArthur saw the danger clearly, but he had cried 'wolf' too often and his warnings about the 'Communist menace' went unheeded – as were CIA reports which as early as March 1950 described mammoth build ups north of the 38th Parallel and predicted an invasion for June. In Washington the re-unification of Korea was viewed as a political problem for the United Nations – not as a military or strategic one for the United States.

By June 1950 the only American personnel left in the Republic of Korea were a handful of diplomats and the 500-man Korean Military Advisory Group, which stayed to train and advise the Republic of Korea (ROK) Army. This South Korean army was seen as little more than a police force and was purposely kept underequipped to inhibit 75-year-old President Syngman Rhee's bloodthirsty plans for reunification. The 65,000 combat troops were equipped with cast off, American small arms and old Japanese Mausers, along with a few short-range M3 105mm howitzers and 2.36-inch bazookas. They had no medium artillery or combat aircraft.

In North Korea it was a different story. Beginning with a cadre of some 3000 North Korean veterans who had seen action in China and the Soviet Union during World War II, the Soviet Union had been laboring since 1945 to build up a tough, well-equipped, well-trained army. By 1950 the *In Min Gun*, or North Korean Peoples Army (NKPA), numbered 120,000 men equipped with almost 100 Yak and Sturmovik fighter planes, 150 T-34 tanks, and a vast array of modern artillery and infantry weapons.

At 0400 hours on 25 June eight NKPA divisions exploded over the 38th Parallel. The summer monsoons had just begun. Rain was falling heavily as the big guns opened up along the 40-mile front and the fighters headed toward the South Korean capital of Seoul, less than 50 miles away. The ROK Army – what was left of it – fell back in confusion before the disciplined NKPA columns. Within 72 hours Seoul had fallen and the South Koreans were in full retreat toward the south.

At MacArthur's headquarters the news of the invasion was soon followed by a message from Washington notifying

Far left above: Street fighting in southern Korea on 21 July 1950.

Left: Americans receive first aid during the first few chaotic months of the fighting in South Korea.

MacArthur watches the Inchon invasion from aboard the USS *Rochester* with Vice-Admiral Doyle and General Harris.

him that he would be in command of any US military operations. Immediately he flew out to Korea to assess the situation. He stayed eight hours and spent only 20 minutes watching a battle for the bridges over the Han outside Seoul, but it was enough to convince him that on their own the South Koreans were doomed. Hurrying back to Tokyo, he informed the Pentagon that he needed US ground combat troops and he needed them immediately. Truman received his message at 0300 on 30 June. The Security Council of the United Nations had already met in hasty session to censure North Korea and, in another, had voted to render South Korea 'every assistance.' At 0500 hours, therefore, Truman authorized MacArthur to send one regiment into combat. By lunchtime he had met with his war cabinet and approved not only the expansion of the expeditionary force, but the calling up of American reservists. The US was at war again, though no one had asked Congress for so much as an opinion.

One of MacArthur's first acts on receiving his orders from Washington had been to send a GHQ Advance Command Group to Korea under Major General John H Church, to collect supplies, information about the best places to use American troops and so on. On 1 July the first combat forces — 406 officers and men of the 24th Infantry Division — landed at Pusan, in the southeastern corner of the peninsula. On 7 July MacArthur was officially named Supreme Commander Allied Powers (SCAP) by the United Nations.

Most observers assumed that the tide would turn as soon as the Americans landed in Korea, but it was not to be. Less than 20 percent of the men in 24th and 25th Infantry Divisions had seen action in World War II and even the veterans were out of shape and yearning for their soft jobs back in Japan. They were outnumbered 20 to one, and their 10-year-old bazookas had no effect on the powerful T-34s. As US and ROK forces were driven relentlessly back toward the coast at Pusan, defeatism began to creep into the high command.

However, MacArthur remained confident. Tirelessly he paced and thought, studied maps and talked to himself. As the American defense finally stiffened and held in the small corner around Pusan, a plan began to take shape. Obviously, the NKPA supply lines were dangerously attenuated after their rapid advance. He planned to land in the rear, cut the supply lines and — in one great enveloping action — cut the army off and destroy it. Geographically Inchon, a Yellow Sea port just a few miles west of Seoul, was the perfect spot. The more he thought about it the more he liked it, and the doubts of his staff regarding its obvious unsuitability for an amphibious landing only reinforced his notion that it would be the perfect spot for a surprise attack.

On 3 July the Joint Chiefs of Staff received a message from MacArthur requesting the immediate dispatch of a Marine regimental combat team. JCS had not yet decided whether Marines would be sent to Korea at all but on 4 July, while they were discussing the matter, MacArthur was already outlining his plan to army, navy and air force staff in Tokyo.

The first scenario, code named Operation Bluehearts, called for landing the 1st Cavalry Division at Inchon on 22 July. It was too ambitious. Eighteen days were not enough to plan any amphibious operation, and the plan was finally scrapped when 1st Cavalry was needed to help defend Pusan. The basic idea was still alive. On 9 July MacArthur received a visit from Lieutenant General Lemuel C Shepherd, the new commander of Fleet Marine Force, Pacific in Hawaii. On his arrival Shepherd heard that General Collins (the Army Chief of Staff) had already told MacArthur that no Marines were available. Shepherd's response — an assurance that he could get 1st Marine Division together in three weeks and in Korea in another three — resulted in the first of a series of messages from MacArthur to the Pentagon. Under the bombardment, JCS finally gave way. On 20 July MacArthur was told that he could have his Marine division plus air support in November or December. Back shot another message from SCAP, 'Essential Marine Division arrive by 10 September 1950. There can be no demand for its potential use elsewhere that can equal the urgency of the immediate battle use contemplated for it.' This was enough to elicit a prompt request for more information, and the resulting outline of MacArthur's plan (now code named Operation Chromite) did little to ease minds in the Pentagon. On 25 July they gave in — but almost immediately began to have second thoughts.

One of MacArthur's major problems was an acute shortage of trained men. Truman was understandably reluctant to call back seasoned World War II veterans who were just settling down in civilian life, and raw recruits could be of only limited value against the North Koreans. The crack troops of 1st Marine Division gave MacArthur the edge he needed — men already familiar with the host of specialized techniques required for an amphibious operation who, as the long lines of reservists outside Marine offices showed, were anxious to employ them.

Doyle and his staff of Amphibious Group I (usually called Phib-1) were already at work on preliminary planning for the landing when news came that the Marine division had been approved. They continued work on the naval side of the operation while responsibility for the main coordination and planning was shifted to the Joint Strategic Plans and Operations Group (JSPOG) — MacArthur's 'staff within a staff' which contained representatives from all four services.

A Navy F4U-4 from the USS *Philippine Sea* flies over supply ships in the Inchon anchorage.

Though the planners considered other landing places, such as Kunsan, it was obvious that MacArthur was set on Inchon. Traditionally, landings take place on lonely beaches, not on city waterfronts that are strongly held by enemy forces. Add Inchon's other disadvantages and you have the oft-quoted remark by Lieutenant Commander Arlie G Capps: 'We drew up a list of every natural and geographic handicap – and Inchon had them all.' Or as Doyle's communications officer put it, 'Make up a list of amphibious "don'ts" and you have an exact description of the Inchon operation.'

In 1950 Inchon – the ancient gateway to the Korean capital of Seoul – was an ugly, dreary industrial port about the size of Jersey City. Its sheltered harbor is always ice-free, but that was its only asset. For the rest, it was a planner's nightmare. The only viable approach was through Flying Fish Channel – a tortuous and difficult passage studded with rocks, shoals and islands, where currents ran up to eight knots and which, moreover, was easily mined. It was so narrow in places that one disabled ship would block all others behind it. Once through the channel the landing force would be in a cul-de-sac, facing high seawalls (higher than the standard ladders) and piers, and overlooked by the hills on both the mainland and on the mile-long island of Wolmi Do, which dominated the harbor. The tidal range of 32 feet was second only to the Bay of Fundy, but the only time tides would be high enough to support a landing operation would be on 15 September and again on 11 October. Then, during short periods in the morning and evening, landing craft could get in, but between tides they would be stranded on huge mud flats. On 15 September high tide would be reached at 0659 hours, only 45 minutes after sunrise. This meant that the landing force would have to negotiate Flying Fish Channel in the dark. In the evening the tide rose at 1919 hours – 37 minutes after sunset. On top of everything else, September was the middle of the typhoon season.

August came. In the small rectangle around Pusan 'Johnny' Walker worked frantically to avoid being overrun. In Tokyo JSPOG and the Navy planners continued to collect information for Chromite. Navy intelligence men scoured the country for anyone who knew about Inchon – tides and currents, the trafficability of the beaches and mud flats, the position and heights of the seawalls and so on. MacArthur was assaulted on all sides by demands from Walker for every available man, by worried messages from JCS, by the fears of the naval planners, the arguments of the Marines and even the concern of his own staff. Through it all he remained supremely confident and aloof, blandly refusing to contemplate the impossibility of the operation. As far as he was concerned, the benefits to be derived from retaking Seoul and the nearby Kimpo airfield, cutting the North Korean communication lines and beginning a two-front war outweighed every disadvantage his superiors or staff could present.

His attitude may have been infuriating, but he did ensure that plans for Chromite continued to move forward instead of getting bogged down by the sheer impossibility of the action. By this time almost everyone in GHQ had a hand in some part of the planning, much of which was carried out by word of mouth. Not surprisingly, news of the operation leaked out and soon every correspondent in Tokyo was aware of Operation Common Knowledge. All the planners could do was try to direct their attention to Kunsan and other sites. This task was made easier by the very disadvantages they were trying to conquer at Inchon.

On 20 August two members of JCS, General Collins and Admiral Sherman, came to Tokyo. Ostensibly they came to discuss MacArthur's plans but in reality they were there to dissuade him. The general realized that their conference, scheduled for 23 August, would be vital and that his most important job would be to convince Sherman, the highly respected Chief of Naval Operations, that Chromite was feasible.

After the introductions, the conferees heard nine officers from Phib-I give presentations on various technical details, summed up after 80 minutes by Doyle, 'Inchon is not impossible.' Then MacArthur took the stage, exerting every ounce of his reason, courage and histrionic ability to swing the chiefs. He pointed out that most of the top NKPA forces were concentrated around the Pusan Perimeter, and demonstrated the ineffectiveness of a short envelopment from Kunsan. He recalled Wolfe's surprise landing at Quebec in 1759 and countered the Navy's objections to Inchon by insisting that he had more confidence in them than they had in themselves. Finally, after an impressive 30-minute performance, he reached his closing lines, 'I realize,' he admitted, 'that Inchon is a 5000 to one gamble, but I am used to taking such odds . . .' Here he stared impressively into the faces before him and his voice dropped to a near whisper, 'We shall land at Inchon and I shall crush them.' After almost a full minute Sherman broke the silence, 'Thank you. A great voice in a great cause.' Completely won over, the shaken audience filed quietly out of the room.

Of course, once the spell had worn off, not everyone remained convinced of Chromite's viability. Collins complained that he had been 'mesmerized' and Doyle remarked, 'If MacArthur had gone on the stage you would never have heard of John Barrymore.' However, in the absence of any concrete alternative, and in view of MacArthur's obvious determination to carry on with the operation regardless, JCS finally gave him an unenthusiastic go-ahead on 28 August, only 17 days before the operation began.

The planners from 1st Marine Division, headed by Major General O P Smith, had arrived in Tokyo on 22 August and

in their preliminary briefings had found that much of their spadework had already been done for them. Now they settled down to work out the details of the landing while the Navy concentrated on finding the ships to transport them. The usual assault vessels (attack transports, cargo ships and landing ship, tanks) could not negotiate Flying Fish Channel at night. The first force would be sent in on high-speed APDs (rebuilt destroyer escorts) and an LSD (Landing Ship Dock).

The Marines' task would be to take Inchon and establish a beachhead, seize Kimpo airport, cross the Han River and take Seoul, and occupy blocking positions to the north, northeast and east of the capital. They would have support from General Harris's 1st Marine Air Wing (temporarily named TAC X Corps), operating first from carriers and then from Kimpo, and from 7th Infantry, who would land behind them and advance along their southern (right) flank. Smith would be in charge of operations during the landing. Thereafter General Almond (MacArthur's Chief of Staff, who had been placed in charge of 'X Corps') would be in command.

Amphibious operations are enormously complex and usually take about six months to plan. The fact that Chromite was completed at all in the time available was an amazing feat. In addition, the Marine team had to work out many details normally handled by regimental commanders. The 5th Marine Regiment was still fighting with the Eighth Army in Pusan (so well that Walker claimed he would not be responsible for holding the front if they were pulled out), and the 1st Marine Regiment had not yet reached Japan. As Smith said later, more time would have allowed them to make the operation more precise but it would not have changed its essential character. As it was, the plans for the Inchon landing eventually came to 185 pounds of Secret and Top Secret documents.

As D-Day approached, the already hectic pace became frantic. There was no time for the usual rehearsals, and certainly none for mistakes. On 2 September, in one of dozens of maneuvers and operations, Navy Lieutenant Eugene Clark landed with a small team on a little island about 11 miles southwest of Inchon. Their mission, which they carried out with the help of friendly villagers (many of whom later paid with their lives for their assistance), was to collect first-hand, current information on the height of the

Landing craft head for the beach
at Inchon, 15 September 1950.

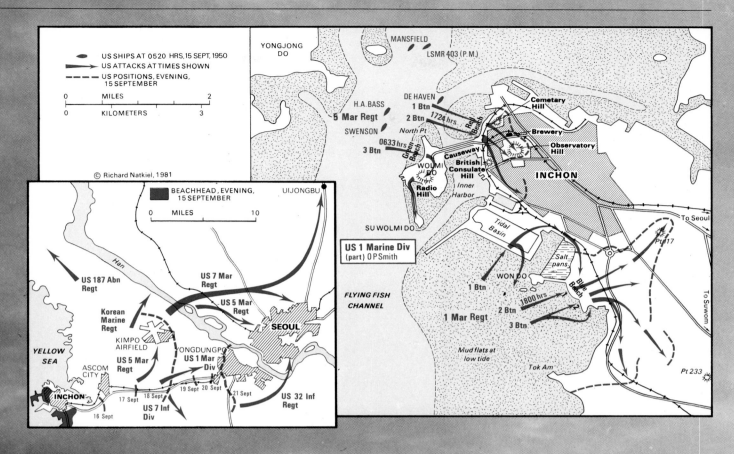

Map legend (top map):
- US SHIPS AT 0520 HRS, 15 SEPT, 1950
- US ATTACKS AT TIMES SHOWN
- US POSITIONS, EVENING, 15 SEPTEMBER

MILES 0 — 2
KILOMETERS 0 — 3

© Richard Natkiel, 1981

YONGJONG DO

MANSFIELD

LSMR 403 (P.M.)

DE HAVEN
1 Btn
2 Btn — 1724 hrs
H.A. BASS
5 Mar Regt
SWENSON
North Pt
0633 hrs
3 Btn

Cemetary Hill
Brewery
Observatory Hill

Green Beach
WOLMI DO
Radio Hill
Causeway
British Consulate Hill
Inner Harbor

INCHON

To Seoul

SU WOLMI DO

US 1 Marine Div
(part) O P Smith

FLYING FISH CHANNEL

Tidal Basin
Salt pans

WON DO
1 Btn

Blue Beach
1800 hrs

1 Mar Regt
2 Btn
3 Btn

To Suwom

Mud flats at low tide
Tok Am

Pt 17

Pt 233

Inset map (lower left):

BEACHHEAD, EVENING, 15 SEPTEMBER
MILES 0 — 10

UIJONGBU

Han
US 187 Abn Regt

US 7 Mar Regt

Korean Marine Regt

US 5 Mar Regt

KIMPO AIRFIELD

SEOUL

YELLOW SEA

ASCOM CITY

YONGDUNGPO
US 1 Mar Div

US 5 Mar Regt

INCHON
17 Sept
18 Sept
19 Sept 20 Sept
21 Sept
16 Sept
US 7 Inf Div

US 32 Inf Regt

seawalls, numbers and positions of gun emplacements, and a host of other vital details. On 10 September Clark reconnoitered a lighthouse on the tiny island of Palmi Do, five miles north of his base, and found to his surprise that it was still in working order. Thus, on the 14th, he and his men were able to make their way to the islands and help guide the landing force to the rendezvous.

On the 3rd, as if fate was determined to make everything as difficult as possible, Typhoon Jane struck, with 100-knot winds and 40-foot waves. The resulting damage took 24 hours to repair. At the same time the Marines from the West Coast were being transshipped to their transports and LSTs; they would take part in the second, evening attack. The process was complicated by the need to transfer every Marine under the age of 18 (some 500 men) to non-

combatant status, to conform with a last-minute Navy decision.

On the evening of 5 September the men of 1st and 5th Battalions, 5th Marines were fighting in the pouring rain to hold down a section of the line at the perimeter. At the stroke of midnight they were silently replaced by infantrymen, marched four miles to the rear, and trucked back over the muddy roads to Pusan. They only knew they were being shipped out to fight another battle — but most had no idea what kind or where it would be.

That same day MacArthur received a request from Washington for details of his plan, which he had been consistently refusing to discuss with JCS. True to form, on this occasion he replied simply that his plans were unchanged. The Chiefs, understandably anxious for more

Arthur would be the first US commander of a unified force to take the field tactically. Doyle was not happy about participating in this historic occasion. Space was at a premium on his flagship and he did not have enough room for even one general – much less the crowd accompanying MacArthur. Doyle, MacArthur and Smith were the only ones with cabins to themselves. From there down, generals doubled up, you had to be a captain or a colonel to get a bunk, and no one with less than two stripes could even commandeer a cot. The flagship sailed out into the teeth of Typhoon Kezia. MacArthur was seasick during the storm, but by dinnertime he had recovered his form and during the meal delivered a long monologue – which some found 'illuminating' and others 'pompous.'

For some time, while the assault force ships had been performing the intricate maneuvers designed to get them all to the right places at the right times without getting in each other's way, softening-up operations had been going on at Inchon. Between the 5th and the 10th air operations had been stepped up on both coasts of the peninsula (to keep the North Koreans guessing), and British warships had begun shelling Inchon from the sea on the 7th. On Wednesday 7 September two American and two British cruisers entered the harbor along with a number of destroyers. As they had hoped, the defenders on Wolmi Do opened fire, exposing their positions; the ships began their own bombardment while from four carriers over the horizon a swarm of heavily armed ADs (single-engine

information, demanded an estimate as to Chromite's 'feasibility and chance of success.' This message, which he received on 7 September, chilled MacArthur to the core. Defiantly – and still without revealing any concrete details – he shot back a reply: 'There is no question in my mind as to the feasibility of the operation and I regard its chance of success as excellent.' The Chiefs knew when they were beaten; there was nothing left but to send a last, grudging cable, 'We approve your plan and the President has been so informed.'

On 10 September meteorologists spotted Typhoon Kezia, heading straight for Japan. They estimated that the storm would hit Kobe on the 12th. That was the day the assault force was due to sail, and there was no way Admiral Doyle could get them out a day early. Somehow he managed and his flagship, *Mt McKinley*, left the harbor at 1030 on the 11th.

The same day MacArthur flew in his official plane, *Scap*, from Tokyo to Kyushu. Behind him, in *Bataan*, came the press. The care and feeding of 86 news correspondents from the United States, Britain, France, Scandinavia and Taiwan was an important part of the operation. Press arrangements had been made on a mammoth scale, and there was practically one Public Information Officer available for every two reporters.

On the 12th, MacArthur and his entourage (five more generals, their aides, and several of his favorite correspondents) rendezvoused with Doyle at Kyushu. Mac-

Left: The day after the landings: the view across the harbor at Inchon.

Far left above: The scaling ladders are ready for the assault of the Inchon sea wall.

Far left center: Marines scale the walls at Inchon in what was one of the fastest amphibious operations ever.

Far left below: MacArthur goes ashore at Inchon on 16 September.

Below: A wounded Marine is evacuated from Seoul on 28 September 1950.

Bottom: An American jeep convoy stops by knocked out North Korean tanks during the advance from Inchon on 21 September 1950.

Douglas Skyraiders) blasted every position they could spot. By the night of 14 September, as 261 ships from seven nations prepared to negotiate Flying Fish channel, Wolmi Do was a pile of rubble or, as one Marine pilot reported, 'one worthless piece of real estate.'

The first 18 ships rendezvoused at the entrance to the channel just after midnight. First in line were three destroyers, followed by the advance assault group carrying 5th Marines (*Diachenko, Wantuck, H A Bass,* and *Fort Marion*). They were followed by three more destroyers, *Mt McKinley* with her load of top brass, and finally the four British and American cruisers. About two hours later, as they passed Palmi Do, a light flashed out; above them Lieutenant Clark, wrapped in a blanket, sat watching the ships slip by. His secret had been so well kept that neither MacArthur nor General Smith had even known of his role in the operation. Both assumed the light had been left burning by the North Koreans, by mistake.

As the ships picked their way through the darkness, MacArthur paced the deck, unable to sleep. He had faced many amphibious landings in his time, but never one like this, at such short notice and under such abnormal conditions. If anything happened to upset the all-important coordination and split-second timing the operation would fail – and he would certainly be held solely responsible. The light from Palmi Do seemed to be a good omen and finally, at about 0230, he managed to fall asleep.

When *Mt McKinley* dropped anchor at 0508 hours, Corsairs from five carriers were already blasting Wolmi Do. At 0520 hours the deliberate, traditional landing procedure began. Captain Sears of the Green Beach control vessel, *Fort Marion*, signalled to Doyle that all was ready, and the Admiral sent the old signal up the yardarm of the flagship, 'Land the Landing Party.' The Marines immediately began embarking, filing past the chaplain one by one to receive his blessings as the landing craft slipped into the water and the first streaks of dawn began to brighten the eastern sky. At 0540 hours the ships began their bombardment. First the destroyers, a few seconds later the cruisers, and finally a

number of rocket ships began shelling the island.

While the terrific bombardment was going on, the men of G and H Companies of the 3rd Battalion, 5th Marines circled in their LCVPs, less than a mile off Wolmi Do. The 5th Regiment of 1st Marine Division was traditionally considered the elite unit of the corps. It had seen action in France, Nicaragua, Guadalcanal, Okinawa and the Pusan Perimeter. Even these tough, experienced veterans had a hollow feeling in their stomachs as they anxiously watched the red and white flag fluttering on *Fort Marion*'s yardarm. The flag was flying 'at the dip'. At 0615 hours it rose and the small ships got into line, then it was let down and the coxswains steered at top speed for Green Beach — a rocky, 200-yard strip on the northwest shore of the island. Almost all fire from the ships stopped, and at 0628 hours, through the sudden silence, roared 38 Corsairs, making one final run. Finally, at 0633 hours, the first ships hit the beach, bow ramps went down, and the green-clad, crouching men burst out with a shout.

As they waded in the Corsairs laid down a protective curtain of machine-gun fire 50 yards ahead of them. Once on the beach, they immediately turned toward their first objectives, which had been drummed into them all the way up the Yellow Sea from Pusan. G Company pivoted sharply to the right and up the northern slope of 351-foot Radio Hill, while two platoons moved along the western shore to seal off the causeway leading to So Wolmi Do (Little Wolmi Do). H Company, meanwhile, marched straight ahead to the opposite shore to secure the base of the causeway leading to the mainland. At 0646 hours the reserves (I Company) hit the beach with 10 tanks of the 1st Tank Battalion, and moved out to mop up North Point.

At 0650 hours 3rd Battalion's commander, Colonel Robert Taplett, landed. Within five minutes an American flag was flying on the summit of Radio Hill. By 0800 hours scattered resistance at North Point and in the industrial complex on the southern end of the island had been mopped up. The only area not cleared was So Wolmi Do, where the NKPA was still putting up a stubborn defense.

MacArthur, who had been awakened by the noise of the bombardment, had been watching the operation from the bow of *Mt McKinley*. He was sitting in the Admiral's chair, wearing his old Bataan cap and a leather jacket, peering through binoculars and paying no attention to his staff as they jockeyed for position with the horde of press photographers. One of his first questions was about casualties. When he was told that about 12 Marines had been killed and perhaps 15–20 wounded, he remarked, 'More people than that get killed in traffic every day.' Then he turned to Doyle with the remark which was, if a bit pompous, still music to the ears of the men whose service had been threatened with extinction only a few short months before, 'Say to the Fleet, the Navy and the Marines have never shone more brightly than this morning.' With a broad grin, he invited his colleagues to join him for coffee. As it was being served he dispatched a jubilant message to JCS. He had been vindicated. His self-confidence increased still further when he heard that the landing force had found the beginnings of intensive fortifications being built on the island. If the attack had been delayed until October he would have been attacking a well-armed fortress. As it was, without the heavy preliminary bombardment his men could easily have been pinned down on Green Beach, just as the landing parties were at Anzio.

At 1115 hours So Wolmi Do was finally subdued. The defenders had been able to prevent anyone crossing the causeway from the larger island, but had been wiped out from the air by eight Corsairs armed with 20mm cannons

As they approached to within 1000 yards of Red Beach on the mainland, Shepherd reminded the general that the area had not yet been taken – and that he made a tempting target. At that Struble turned the boat around and the party returned to the command ship for lunch.

The North Koreans had shared the American planners' opinion of Inchon as a landing spot for a major amphibious assault, and now all the commander of the meager forces there could do was wait for the inevitable. In Seoul, the news of the invasion threw the 16,000-man NKPA garrison (many of whom were administrators or political types) into a state of panic. Reinforcements were hurredly called up and 'volunteers' collected from among the 2,000,000 civilian inhabitants.

The trickiest part of the operation – simultaneous twilight landings on two mainland 'beaches' – was still to come. At around noon the transports carrying the Red Beach assault force, 1st and 2nd Battalions of 5th Marines, arrived in the harbor while relays of Navy Skyraiders and Marine Corsairs bombed everything within 25 miles of the port and the ships began their preliminary bombardment. At 1430 hours the ships started the final bombardment which was to last three hours. Fifteen minutes later, Doyle again sent up the traditional signal and the Marines began filing into their landing craft.

At 1704 hours it began to rain. The first assault group was about a mile offshore, and all eyes were fixed on the flag waving above *H A Bass*, the Red Beach control vessel. At 1722 hours it went up; two minutes later it was down and the first wave was speeding toward the beach, disappearing

and napalm. As the tide receded the Marines solidified their positions while the 17 wounded men were transported back to the ships across a specially made pier. No Americans had, in fact, been killed; they had taken 136 prisoners and found 108 enemy bodies. It was assumed that another 150 lay buried in the shattered gun emplacements.

Admiral Struble invited MacArthur to come for a ride in his barge, to take a closer look at the beaches. The offer was accepted eagerly, and SCAP, along with Generals Shepherd, Almond, Whitney, Wright and Fox – and a crowd of newsmen – climbed aboard. As the boat stood off Green Beach, MacArthur struck a Napoleonic pose in the bow.

Top: Members of the 1st Marine Division mop up Communist positions on 18 September 1950.

Below: Navy Corpsmen evacuate a wounded man out of a helicopter to hospital.

into a thick cloud of smoke from the burning city. As in the morning, the naval fire stopped when the lead boats reached the halfway point, and the planes moved in. The covering fire from Wolmi Do, however, continued. Their position was such that they could keep on firing until the last minute without endangering their own men.

Behind the dense smoke screen the first boats crunched into the seawall; men scrambled up shaky, improvised scaling ladders and tumbled over, just as infantrymen had gone 'over the top' in World War I. Unexpectedly, there was one woman landing on Red Beach as well — Marguerite Higgins, correspondent for the *New York Herald Tribune*, the first woman reporter to advance with front-line troops since Martha Gellhorn covered the Spanish Civil War. At 1733 hours a Marine pilot flashed word back to *Mt McKinley*, 'Scaling ladders are in place and Marines are over the wall.'

On the left, Company A soon captured Cemetery Hill bypassing, for the time being, another vital objective — the gigantic Asaki Brewery. (1st Battalion had been promised an equally gigantic bender if the brewery and its contents were captured intact). By 1845 hours Company E, on the right, had easily taken British Consulate Hill and the lower slopes of Observatory Hill, while in the rear the rest of the 2nd Battalion had control of the area around the Nippon Flour Mill, at the foot of the Wolmi Do causeway.

At this point the Marines had a most unwelcome opportunity to gain first-hand experience of their own Navy's shooting ability. A number of LSTs were scheduled to land at 1830 — within an hour of the initial landing — and unload their supplies (including heavy equipment, fuel, ammunition and napalm) while the Marines secured the beachhead. This would have been a dangerous maneuver at the best of times, especially since the LSTs were expected to pull up side by side on the narrow beach like cars in a parking lot, in rapidly failing light. At Inchon, with rusty,

decrepit, unmaneuverable hulks (all the Navy could produce) manned by inexperienced officers and crews, it was daring. Accompanying the cargo vessels were eight other LSTs, whose job was to distract enemy fire. As they lumbered in, guns blazing, the NKPA fired back, exposing their position. The trigger-happy LSTs, carried away by the battle, returned the fire, and the Marines on the beach were caught in a murderous crossfire while 2nd Platoon, preferring enemy machine guns to the Navy's high-intensity fire, was blasted off Cemetery Hill.

Red Beach finally became quiet at 1900 hours. One Marine had been killed and 23 wounded by US bullets — more casualties than the battalion suffered at the hands of the North Koreans. Despite some confusion on its southern slope Observatory Hill had been cleared and defensive positions had been consolidated before midnight. The port of Inchon was in American hands. Only one objective had not been taken. The brewery caught fire that night and burned to the ground before the very eyes of the thirsty Marines.

While the 5th Marines were fighting at Red Beach for Inchon proper, the 1st Marines were landing on Blue Beach, four miles to the southeast, on the other side of the port. This regiment too had a long and valiant history, from the Boxer Rebellion through Okinawa. It had been disbanded, then hastily pulled together again in only 10 days and rushed to Japan from San Diego. Most of the men were World War II veterans, led by a fearless old warhorse, Colonel Lewis B 'Chesty' Puller — an iron-disciplined, archetypal Marine. Some said that a kind heart beat beneath his crusty exterior, but few among the staff officers, support units or lazier members of his squad, whom he terrorized

Marine tanks oversee operations as Communist troops are marched to a stockade, 26 September 1950.

ruthlessly, would ever believe it.

The approach to Blue Beach (actually, Blue Beaches 1 and 2) had to be made over two-and-one-half miles of mud flats, which have (according to Robert Heinl in *Victory at High Tide*) 'the crust and underlying gooey consistency of solidifying chocolate fudge – but the smell is different.' The run took 45 minutes in 172 tiny amphibious tractors (amtracs), and was carried out in an impenetrable fog composed of smoke from burning Inchon, mixed with rain. Going in blind and unrehearsed, with only four guide boats instead of the 32 they should have had, it is little wonder that the landing soon disintegrated into utter chaos. Amtracs milled around helplessly, landing on the wrong beaches, searching for their commanders, cursing each other, the Navy and life in general. Despite the confusion, the first three assault groups, which had landed on time, managed to secure the beachhead by themselves and by 0130 on 16 September, Puller was able to report that the beach was taken.

The commanders on *Mt McKinley* went to bed happy. Despite the inevitable confusion and disorganization, Doyle had put 13,000 Marines, with their weapons and heavy equipment, ashore on an 'impossible' coast. Both regiments had achieved their objectives on schedule at a cost of only 21 dead, one missing and 174 wounded. Before he turned in, General Smith noted in his journal, 'D-Day has gone about as planned.'

At daybreak on the 16th the Marines moved inland behind their tanks, leaving a detachment of Korean Marines to mop up the town of Inchon. 1st and 5th Marines were due to link up at the '0–3' line, where the Inchon peninsula narrows to a bottleneck. By evening the beachhead had been secured. The 5th Marines held the forward line overlooking Ascom City (a former US supply depot) and Puller, who had had to secure his southern flank, was about

a mile behind. At 1800 hours General Smith took his leave of MacArthur, who told him, 'Be sure to take good care of yourself and capture Kimpo as soon as you can.' Twenty-four hours from the beginning of the operation he took command of land operations, establishing his headquarters on Yellow Beach (Blue Beach 2).

MacArthur was up early the next morning. He set out for shore with his fellow generals and a gang of reporters. As Smith was greeting him the tension that had been building up during the previous weeks broke and, quickly excusing himself, MacArthur doubled over and threw up. A moment later he was himself again, inspecting enemy corpses and listening to Smith's briefing. The group set out toward the front line to find Colonel Puller. Suddenly, to Smith's dismay, MacArthur headed off to see some T-34s that the 5th Marines had just 'killed'. As he approached, a young lieutenant dashed up and tried to stop him. 'General, you can't come up here!' he panted, 'We've just knocked out six Red tanks over the top of this hill!' 'That was the proper thing to do,' MacArthur calmly replied, and continued his climb to the top of the hill. There he surveyed the smoking hulks, disdainfully ignoring the rattle of small arms fire around him. Then, having made his point, he descended and agreed to let Smith take him back. That evening he was able to send another triumphant cable to Washington. Kimpo airfield had been captured.

While the 1st and 5th Marines moved on toward the Han – the next formidable obstacle – Smith continued to work in Inchon, to get the harbor functioning again. There was much to be done. Hospitals had to be established for the wounded and burials arranged for the dead. Supplies were unloaded while the engineers repaired track and got the

The UN forces undertook another amphibious operation on 26 October 1950 against Wanson.

trains running. The civilian population had to be cared for; food distributed, housing found for the homeless, the sanitation system repaired. By the 19th Smith was able to reestablish civilian government in the city and move his headquarters up to Kimpo.

On Wednesday, 20 September, MacArthur came ashore for another tour of the command posts, reaching the front lines in time to watch the 5th Marines cross the Han after two abortive attempts. After a congratulatory word with their commander, Lieutenant Colonel Raymond Murray, and a visit to General Barr (commanding the 7th Infantry on the right flank, near Yongdongpo), he returned to *Mt McKinley*. He was getting anxious about General Walker and the Eighth Army, who were not surging northward according to plan. Instead, they were advancing at a painfully slow rate, against stiff resistance, and he began thinking about setting up another landing, at Kunsan. The North Koreans were bringing up reinforcements at Seoul as well, and it was starting to look as if the capital would become another Leningrad.

The next day General Almond went ashore and assumed command of the X Corps. He would have a difficult time during the next six days. The NKPA commander, General Wol, knew he had no hope of dislodging the American forces – but he could and did use his men wisely to fight a strong delaying action. Although his dual position was not unprecedented, Almond occasionally found himself torn between his responsibilities as corps commander and his duty to MacArthur as the latter's Chief of Staff. His troubles were not eased by the traditional antipathy that existed between the army and Marines, which made his relationship with Smith uneasy and often acrimonious.

By Saturday the 23rd, however, MacArthur decided that Seoul was within reach and ordered his entourage to pack for the return trip to Tokyo. The next morning he received an invitation from Admiral Struble to visit the *Missouri*. The general accepted gladly and as he stood on the quarterdeck for the first time since he had accepted the Japanese surrender there in 1945, his eyes filled with tears. Turning to Struble, he said feelingly, 'You have given me the happiest moment of my life.'

That afternoon MacArthur and his aides boarded *Scap* at Kimpo, the general's mind filled with memories of his triumph at Tokyo Bay and his vindication at Inchon harbor, along with prospects of an early victory in Seoul. As soon as the plane had taken off he leaned back, took a contented puff on his pipe, and asked for a briefing.

Events moved quickly during the next week. On Monday 25 September Puller and his Marines finally took Yongdungpo. On Tuesday the Eighth Army, who had finally broken out of their encirclement and raced north, made the first, tentative contact with the Seoul attackers. By Wednesday the NKPA was on the run and on Thursday Seoul officially fell to the X Corps. On Friday, 29 September, MacArthur, this time accompanied by his wife, landed again at Kimpo and drove into the city. At an impressive ceremony in what was left of the Capitol Building, he handed the reins of government back to Syngman Rhee. Operation Chromite had officially ended. MacArthur's 71,300 troops had defeated 30–40,000 North Koreans at a cost of just 536 dead, 2550 wounded and 65 missing in action.

In the short term, the victory at Inchon bred further triumphs. By 1 October, only 15 days after the landing, the NKPA had fled in disarray across the 38th Parallel. MacArthur's northward sweep to the Yalu River (ordered by Washington based on his opinion that the Chinese would not enter the war) soon had all of North Korea under UN control; on 28 October the general confidently predicted that American boys would be home by Christmas.

In fact, of course, it was only the beginning. A month

Left: MacArthur inspects territories along the Yalu River in Korea, 24 November 1950.

Right: In 1957 MacArthur's statue was unveiled at Inchon: a fitting tribute to a victor.

Right top: After bitter fighting at Chosin Reservoir, Marines of the 7th Regiment leave the shores of Hungnam.

later, on the day after Thanksgiving, 300,000 Chinese Communists stormed the Yalu, sending the UN forces reeling southward once more. On 4 January 1951 Seoul changed hands again. The bitter conflict would not end in equally bitter compromise until July 1953. This unsatisfactory conclusion is probably one factor that has overshadowed the brilliance of the operation at Inchon.

For MacArthur personally, the success of Chromite had much the same effect. Coming at the age of 70, after more than half a century devoted to his country's military affairs, it was his finest hour. In the weeks immediately following

the action he was truly invincible. As Ridgway put it, if he had suggested, 'that one battalion walk on water, there might have been someone ready to give it a try.'

The old soldier was already beginning to fade away. Not in large things (in his misreading of the Chinese reaction he was, for the most part, misled by faulty intelligence), but in smaller matters. For example, he would not take the elementary precaution of issuing winter clothing to the troops in the fall of 1950, just in case the campaign took a bit longer than he expected. More importantly, the success — as usual — went to his head, and he lost all perspective on his situation. An army man of the old school, he was unable at his age to make the switch from the concept of total victory to that of limited war. He began ignoring direct orders from JCS. When his insubordination extended to the President, his Commander in Chief, the situation became intolerable and he was recalled, to end his days as a private citizen.

No one man can ever be credited with the success or failure of a military operation. In the case of Inchon, the Herculean efforts of the men who devised the 'impossible' plan and the professionalism of the troops — especially the Marines — who executed it were ultimately responsible for the victory.

In this case we can give one man credit for the fact that the action took place at all. It is probably safe to say that, although envelopment is a standard tactic in any general's arsenal, only MacArthur had the imagination to conceive this one in the face of such adverse conditions. More important is the fact that, having hit upon the plan, only a man of MacArthur's overwhelming self-confidence and high-handedness could have stuck to his guns so stubbornly and — in effect — forced the entire US military establishment to undertake it against their will. History has not yet passed its verdict on Douglas MacArthur. But at Inchon, where he precluded certain defeat in Korea, he was — in Heinl's words — 'as he would wish to be remembered: bold, judicious, assured and unwavering.'

PICTURE CREDITS

THE AUTHORS

Peter Young was one of the most
decorated soldiers in World War II,
and became the youngest brigadier
in the British army at the age of 25.
On his retirement from military
service he became head of the War
Studies Department of RMA
Sandhurst. He has written and
edited over 20 books on the subject
of military history.
Introduction

Richard Holmes was educated at
Cambridge, Northern Illinois and
Reading Universities. He is currently
senior lecturer in war studies at
Sandhurst and has written a number
of books on military topics.
Chapters 1 and 2

Ward Rutherford was brought up in
Jersey. He is a professional journalist
who has written over 10 books on
recent history. Among the many
titles he has worked on are *Hitler's
Propaganda Machine* and *Blitzkrieg
1940*
Chapters 6, 7, 8, 9 and 10

Douglas Welsh is one of America's
most highly decorated Veterans of
the Vietnam War. A contributor to
The Russian War Machine and *Great
Battles of the World on Land, Sea and
Air*, he has also written a book on
the American Civil War.
Chapters 3, 4 and 5

Patrick Jennings was born in
Chicago in 1937. Educated in both
the United States and Great Britain,
he is a journalist living in London.
Among his works is *A Pictorial
History of World War II*, a subject in
which he has specialized.
Chapters 11 and 15

Judith Steeh was educated at the
University of Michigan and then
became a journalist. She contributed
to the *Olympiad 1936* and co-
authored the *Directors' Guide to the
USA*. She has also written *The Rise
and Fall of Adolf Hitler*.
Chapters 12, 13, 14 and 16

INDEX

Abrams, Lieutenant Colonel C, 217
Acadia Conference, 194
Achtung! Panzer!, 166
Ajaccio, 11
Alam Halfa, Battle of, 197, 199
Ridge, 194, 198
Albert, 149
Alexander I, Czar, 13, 14, 22, 28
Alexander, General Sir H, 195, 196
Alexandria, 193, 194
Almond, General, 244, 249, 252
Alsace, 121
America, 25, 28, 149, 177, 194
Confederate States of, 45, 58
Union States of, 45
United States of, *see* USA
Amiens, 153, 155, 160, 162, 165,
167, 170, 175
Anderson, General R H, 48, 49
Antwerp, 123, 208, 210, 212, 217
Ardennes, canal, 172
Forest, 97, 167, 168, 170, 208,
209, 210, 211, 215, 216
Offensive *see* Bulge, Battle of the
Armistice, 13, 22, 67, 159, 166,
175, 234
Army,
Allied,
Franco-American Seventh,
208
1805, 11, 14, 22, 23
1815, 28, 34, 41, 42
Battalion,
Hanoverian Militia, 28, 37
King's German Legion, 28
Peninsula, 28
52nd, 41
Brigade,
Maitland's Guards, 41
Ponsonby's Union, 37
Somerset's Household, 37
Vandaleur's Light Cavalry,
40
Vivian's Light Cavalry, 40
Cavalry,
British, 28
Hanoverian, 28
Uxbridge's, 29
Chief of Staff, 144
Corps,
Brunswick, 28
Bülow's, 36, 37, 40
Lord Hill's II, 28, 29
Thielmann's, 36
Ziethen's, 40
Division,
General Perponcher's Dutch-
Belgian, 30
Picton's 1st, 28, 34, 37
Sir Alten's 3rd, 28
Force, 154, 163, 187, 221
Franco-British, 123
Anglo-French, 131
Headquarters, General, 130
Reserve, Duke's, 28
Supreme Headquarters, Allied
Expeditionary Force, 207,
208, 219
Australian, 177, 189
Corps, 157
Division, 154
9th, 197
Force, 149
Austrian, 11, 14
Austro-Russian, 13
British, 122, 126, 127, 165, 184,
193, 196, 203, 210
Eighth, 179, 185, 187, 189,
191, 193–197, 199, 202,
205
Home, 193
Fifth, 153
First, 159
Fourth, 154, 155, 156, 159
Home, 193
Second, 207
South Eastern, 194
Third, 153, 159
Brigade, South African 6th, 190
Corps,
Armored, 193
BEF's 11, 122, 125
Cavalry, 157
Tank, 197
I, 151
III, 157
X, 197, 202, 203, 204
XIII, 202
XXX, 197, 202, 204
Division, 144, 149, 177
Artillery, 198

Cavalry, 156
Infantry, 156, 198, 202
New Zealanders, 204
Reserve, 157
South African, 188
1st Armored, 204
1st South African, 189
3rd, 194
7th Armored, 188, 202,
203, 204
8th Infantry, 194
10th Armored, 204
44th, 203
50th, 188
Force, 119, 127, 149, 151,
177, 189, 203, 205
Expeditionary Force, 119, 122,
127, 130, 149–150, 175,
194
General Staff, 162
Group, 21st, 207, 208
Headquarters, 130, 151
High Command, 191
Regiment,
Sherwood Forester, 189
Staffordshire Yeomanry, 203
Warwickshire, 193
2/7th Ghurka, 189
2nd Cameron Highlander,
189, 190–191
3rd Coldstream, 189, 190
4th Tank, 175
7th Tank, 175
Canadian, 207, 210
First, 207
Corps, 157, 159
Force, 149
of Châlons, 89, 91, 94
Confederate, 45, 46, 47, 48, 49,
50, 51, 54, 55, 57, 58,
60, 61, 74, 80
Brigade, 60
Cavalry, 49, 51, 54, 74
Fitzhugh Lee's, 51
Stuart's, 74
Corps, 50
Signal, 54
III, 74
Division, 49, 50
Anderson's, 48, 49, 50, 52
Early's, 54
McLaw's, 49, 50
Rodes', 51
Engineers, 63
Force, 48, 54, 55, 57, 58, 59,
60, 62, 73, 76, 78, 80
Headquarters, 71
Infantry, 52–53
French, 11, 25, 35, 36, 41, 87,
88, 97, 102, 105, 121,
122, 127, 130, 146, 149,
165, 170
Fifth, 122, 125, 130, 131,
132
First, 121, 155, 157, 159
Fourth, 122, 130
Ninth, 130, 131, 172
Second, 121, 138, 139,
141, 146, 147, 165, 172
Sixth, 122, 125, 128, 130,
131, 132
Third, 122, 130, 133, 155,
159, 172
Artillery, 11
Battalion, Tank, 172
Brigade,
cuirassier, 37, 100
Hussar, 37
Cavalry, 13, 17, 20, 34, 97,
122
Lancers, 101
Reserve, 11, 12
Central Army Group, 137–138,
141, 146
Chasseurs Alpins, 138
Corps, 11, 20, 137, 157
Augereau's VII, 11
Bernadotte's I, 16, 17, 20,
22
Cavalry, 36
Davout's III, 16, 22
Douay's VII, 92, 93
Ducrot's I, 92, 93, 100
d'Erlon's, 36–37, 41
Failly's V, 92, 93
Gransard's X, 172
Lobau's VI, 36
Lebrun's XII, 92, 94, 95
Marmont's, 13
Reille's, 36, 41
Soult's IV, 11, 16
II, 101
V, 16
VI, 100

XVI, 138
XXI, 172
Division, 144, 146
Caffarelli's Infantry, 16
Cavalry, 13, 16
d'Erlon's, 22
Friant's, 22
Infantry, 11
Legrand's, of the IV Corps,
16
Oudinot's Grenadier, 16, 20
Suchet's Infantry, 16
Territorial, 137
1st Armored, 172
2nd Armored, 172
3rd Armored, 172
3rd Mechanized, 175
3rd Motorized, 172
4th Armored, 175
14th, 175
55th, 172
71st, 172
'elastic defense' system, 138,
141, 147
force, 92, 119, 149, 160
Foreign Legion, 143, 177
General Staff, 88, 93, 103, 135
Grande Armée, 11–12, 13, 22
Headquarters, General, 137,
138, 142, 144
Imperial, 12
High Command, 172
Imperial Guard, 11, 12, 13, 16,
20, 21, 36, 40, 41, 42
Le Système D, 88–89
National Guard, 89
Regiment, Infantry, 172
Unit, Zouaves, 94–95, 100
Troops, 89, 155, 163, 171
German, 105, 149, 177, 178,
225, 236
Crown Prince Wilhelm's, 139
Eighth, 105, 107, 109, 110,
117, 119
Eighteenth, 152, 157
Fifth, 126, 133, 135–136
Fifth Panzer, 210, 212, 213
First, 109, 110, 119, 122,
126, 130, 132, 133, 152
Fourth, 126, 133, 152, 168
Fourth Panzer, 228
Ninth Field, 228, 232
Second, 110, 125, 126, 130,
132, 133, 152, 157
Seventh, 126, 152, 210
Seventeenth, 152
Seventeenth Field, 228
Sixth, 126, 152, 210
Sixteenth, 168
Sixth SS Panzer, 210, 211
Third, 153, 152
Third Panzer, 225, 228
Twelfth, 168, 233
Wehrmacht, 228, 232
Afrika Korps, 45, 177,
178–179, 182, 184, 187,
188, 189, 191, 193, 194,
197, 198, 204, 205
assault division, 189
Battalion, Volkssturm, 228,
232, 233
'combat group,' 183
Commandos, 213
Corps,
Francois' I, 107, 115, 117
Guard Reserve, 119
Kluck's II, 131, 132
Kluck's IV, 131, 132
Scholtz' XX, 107, 108, 115,
116
1st Cavalry, 165
15th Armored, 177
XI, 119
XIV, 168, 170
XVI, 166, 167
XVII, 107, 115, 117
XIX, 167, 168, 170, 171
XXXXI, 168, 170, 175
XLVII Armored, 215
Division, 146, 157, 159, 211,
219, 228, 232
Infantry, 232, 233
Iron, 166
Light, 167
Motorized, 168, 228, 233
Panzer, 166, 167, 168, 185,
228
Tank, 233
1st Panzer, 168, 170, 171,
172, 175
2nd Panzer, 168, 170, 171,
175
4th Panzer, 167
5th Cavalry, 165

6th Panzer, 170
7th Panzer, 171, 175, 177
10th Panzer, 168, 170, 171, 172, 175
11th Motorized, 232
15th Panzer, 178, 188, 201, 203
21st Panzer, 178, 188, 201, 202, 204
23rd Motorized, 232
90th Light, 178, 187–188, 189, 190, 201, 204, 205
164th Infantry, 203, 204
309th Infantry, 229
Eastern Frontiers Protection Service, 166
Force, 105, 152
East Pomeranian, 225
Formation, 167
General Staff, 121, 133, 162, 167, 182
Group,
  A, 168
  B, 168, 170, 210
  C, 168
  Panzer, 168
  Steiner's, 233
High Command, 105, 106, 108, 133, 141, 202, 207, 208, 220, 227, 228
Inspectorate of Transport Troops, 166
Panzertruppe, 166
Regiment
  SS Grossdeutschland, 168, 172, 175
  124th Infantry, 177
  361st, 187
  Sturmpionierbatallion, 172
Troops, 122, 139, 157, 162, 194
  Mobile, 167
  Police, 233
  SS, 178
Indian,
  Brigade, 190
  3rd Motor Brigade Group, 188
  Division,
    2/5th Mahratta, 189
Italian, 177, 181, 184
  Corps,
    Tank, 205
    XX Motorized, 189, 198
  Division,
    Ariete, 178, 187, 188, 189, 201, 202, 204
    Infantry, 178, 187
    Littorio, 201, 203
    Motorized, 178
    Pavia, 188
    Trento, 203
    Trieste, 188, 201, 204
Japanese, 187
  Kwantung, 237
Metropolitan, 121
of the Meuse, 93
of the Moselle, 89
North Korean Peoples, 241, 242, 246, 247, 248, 249, 250, 252
  Division, 241
  Force, 244
of Northern Virginia, 45, 46, 47, 49, 50, 55, 71, 73, 78, 80, 84, 85
of the North German Confederation, 94
Panzer, Afrika, 178, 187, 190, 191, 193, 198, 201, 202, 204, 205
of Paris, 89
  Garde Mobile, 89, 91
Polish,
  First, 232
  Division,
    18th, 167
of the Potomac, 46, 47, 48, 55, 71, 72, 73, 74, 75, 78–79, 80, 81, 84, 85
Prussian, 25, 28, 29, 87, 88, 93, 102, 103
  I, 94
  II, 93, 94
  III, 89, 93
  Corps,
    Blumenthal's XI, 93, 94
    Guard, 97
    von der Tann's Bavarian I, 93, 94
    I, 29
    II Bavarian, 100, 101
    V, 94, 97
    XI, 97

Division,
  Cavalry, 93
  General Staff, 88, 93
  Regiment, 100
Republic of Korea, 241, 242
  Force, 242
of the Rhine, 89
Russian, 11, 13, 28, 105, 107, 125, 223, 225, 231
  Red, 228
  Vilna, 105, 106, 107, 115, 116
  Warsaw, 106, 108, 115, 119
  Eighth Guard, 228, 229, 231, 233, 234
  Eighth Tank Guard, 232
  Fifth Shock, 228, 229, 232, 233
  First Tank Guard, 228, 229, 231, 232, 233, 234
  Forty-seventh, 228, 231, 232, 233
  Fourth Guard, 232, 233
  Second Tank Guard, 228, 229, 232, 233, 234
  Sixty-first, 232
  Sixty-ninth, 229
  Third, 228, 234
  Third Guard, 233
  Third Shock, 228, 229, 231, 232, 233, 234
  Third Tank Guard, 229, 232, 233
  Thirteenth, 233
  Twenty-eighth, 229, 233
  Corps,
    Artamanov's I, 115, 116, 117
    VII, 115, 223
    XII Tank Guard, 232
    Cossack, 106, 119, 123
  Division, infantry, 224
  Force, 224
  Front,
    Warsaw, 223
    1st Byelorussian, 223, 224, 225, 226, 232, 237
    1st Ukranian, 223, 225, 226, 229, 237
    2nd Byelorussian, 224, 225, 226, 229, 237
    3rd Byelorussian, 224
  Headquarters,
    First army, 115
    General (Stavka), 224
  Guard, 22
    Cuirassier, 21
    Imperial, 20, 21
  Regiment,
    Azov, 23
    Galicia, 23
    Narva, 23
    Preobrazhensky, 21
    Semenovsky, 21
    756th, 234
  Troops, 123, 227
  Zhukov's, 233
Union, 45, 46, 47, 50, 52, 54, 55, 57, 64, 67, 71, 74, 75, 78, 79, 80, 84, 85
  21st Illinois Volunteer, 57
  Brigade, 48
  Cavalry, 46, 52–53, 73
    Buford's, 74
  Corps, 48, 49, 51, 80, 85
    Engineers, 63–64, 74
    McClernand's XIII, 60, 62
    McPherson's XVII, 62
    Sherman's XV, 58, 60, 62, 63
    I, 74, 75
    II, 74
    III, 74, 75, 76, 78
    XI, 48, 75, 80
  Division, 48, 80
    1st, 74
  Force, 53, 54, 55, 58, 59, 62, 63, 68, 71, 75, 78, 80, 84
  Regiment, 48, 54
  Troops, 51, 54, 55, 66–67, 68, 74, 75, 78
United States, 207
  Eighth, 241, 244, 252
  First, 207, 208, 209, 210, 212, 213, 214, 215, 219, 220
  Ninth, 213
  Seventh, 207, 209
  Third, 207, 208–209, 210, 212, 213, 214, 217, 219, 220, 221
  Battalion,

3rd, 248
705th Destroyer Tank, 215
Combat Command,
  A, 216, 217
  B, 216, 217
  Marine Regimental, 242
  Reserve, 216, 217
Company,
  A, 250
  E, 250
  G, 248
  H, 248
  I, 248
Corps, 215
  Quartermasters, 208, 209, 215
  III, 209, 214, 215, 216
  VIII, 209, 214, 217, 219
  X, 252
  XII, 214, 219
  XX, 214
Division, 162, 241
  Rainbow, 240
  1st Cavalry, 242
  2nd, 162
  2nd Armored, 207, 220
  3rd, 162
  4th Armored, 214, 216, 217
  7th Infantry, 244, 252
  10th Armored, 209, 213, 215
  26th Infantry, 214, 216
  33rd, 157
  80th Infantry, 214, 216
  95th Infantry, 209
  24th Infantry, 242
  25th Infantry, 242
  101st Airborne, 215, 217
Force, 194, 239, 252
  Far East, 240
  Korean Expeditionary, 242
  Western Task, 207
GHQ Advance Command Group, 242
Group,
  Cavalry, 214
  Headquarters, 208
  6th, 213–214
  12th, 207, 208
  Regiment, 242
  Troops, 215, 240, 242, 253
Arras, 153
Aslagh ridge, 188
Auchinleck, General Sir C, 177, 185, 187, 189, 191, 193, 194, 195, 197, 202, 205
Augereau, Marshal PFC, 11
Austerlitz, Battle of, 10–23

Bagration, Prince, 20, 21, 22
Balan, 100, 101
Balck, Colonel, 171, 174
Bardia, 184, 189
Barr, General, 252
Barut, 232, 233
Bastogne, 213, 214, 215, 216, 217, 219
Bavaria, 11
Bayerlein, General, 191, 198
Bazaire, Marshal, 89, 91, 92
Bazeilles, 92, 93, 94, 95, 100, 171
Beaumont, 91, 92
Beck, General, 166
Benghazi, 187
Berlin, 106, 121, 125, 147, 224, 225, 226, 228, 229, 231, 232, 233, 234
Battle of, 222–237
Cadet School, 165
  War Academy, 165
Bernadotte, Marshal JBJ, 11, 13, 16, 17, 21, 22
Berthier, Marshal LA, 11, 12
Big Black River, 60, 61
Bir Hacheim, 187, 188
Bismarck, OEL von, 87, 88, 92, 97, 102
Black Day, 148–163
Bläswitz, 20, 21
Blitzkrieg, 166, 207, 213
Blücher, Marshal, 28, 29, 30, 34, 35, 36, 41, 42
Bock, General F von, 168, 170
Bogdanov, General S, 232
Bois de Boulogne, 126
Bois de la Garenne, 100, 101
Bonaparte, N, 10–23, 25, 28, 29, 34, 35, 36, 37, 40, 41, 42, 93, 207
Boulogne, 11, 23, 150, 175
Bradley, General O, 207–208, 209, 212, 213, 217, 227, 236, 240

Brandenburger, General E, 210, 217
Brauchitsch, General W von, 167, 168, 170, 175
Breslau, 225
Brest-Litovsk, 167
British Consulate Hill, 250
Brooke, General Sir A, 194, 195, 197
Bruay mines, 160
Brünn, 11, 13, 14, 16
Buford, Brigadier General J, 74
Bulge, Battle of the, 206–221, 224
Bülow, 125, 126, 131, 133
Burnside, General, 46, 47, 48
Busch, General, 168, 170
Büxhowden, General, 11, 17, 22

Cairo, 177, 188, 189, 191, 193, 194, 195
  Radio, 205
Calais, 150, 175
Cambrai, 162, 165
Camerone, Battle of, 143
Campaign,
  Barbarossa, 185
  Egyptian, 191
  Italian, 11
  North Africa, 178, 184, 191, 194
  Polish, 167
  Russian, 194
Capps, Lieutenant Commander AG, 243
de Castelnau, General, 137–138
Catherine's Furnace, 51
Cemetery Hill, 75, 250
Châlons, 91, 138
Chambersburg Pike, 74
Chambley, 153
Champion's Hill, 60
Chancellorsville, 49, 53, 54, 55, 84
  Battle of, 44–55, 68
Chantilly, 144
Charleroi, 25, 28, 29, 42, 122
Charles, Archduke, 11, 13
Charlottenburg, 233
Chateau-Thierry, 153, 162
Chaumont, 216
Chémery, 172
Chemin des Dames, 153
Cherbourg, 207, 208
Cheryakhovsky, General ID, 224
Chuikov, Marshal, 225, 231, 234
Churchill, Sir W, 151, 177, 185, 191, 194–5, 196, 197, 199, 207, 224, 225
Clark, Navy Lieutenant E, 244–5, 247
Clausewitz, 205
Clemenceau, G, 154
Coalition, Third, 11, 22
Colborne, Sir J, 41
Collins, General, 242, 244
Compiègne, 126, 130, 131
Constantine, Grand Duke, 21, 22
Côte de Poivre, 141
Croisilles, 151
Crüwell, General, 187, 188
Cyrenaica, 184, 185

Daigny, 94
Davout, LN, 13, 16, 20
Dernia, 178
Desert, Air Force, 187, 194, 197, 198
  War, 177, 194
Devers, General J, 213
'Devil's Gardens,' 200, 203
Diachenko, 247
Dien Bien Phu, 66
Dietrich, SS General S, 210, 211, 213, 215
Dinant, 171, 172, 215
Doctorov, Lieutenant General, 17, 20, 22
Donchéry, 94, 95, 97
Dönitz, Admiral C, 234
Dorman-Smith, General, 197
Douay, 92, 93, 100
Doubleday, Major General A, 75
Doyle, Rear Admiral JH, 241, 242, 243, 244, 246, 247, 248, 249, 251
Ducrot, 92, 93, 95, 97, 100, 101
Dunkirk, 150, 175, 194

Echternach, 214, 219
Ecole Normale du Tir, 138
Eisenhower, General DD, 195, 207, 208, 213, 217, 221, 225, 240
El Adem Airfield, 190
El Alamein, Battle of, 192–205
Elba, 28
Elbe river, 226, 232

Brandenburger, General E, 210, 217

Emmitsburg, 74, 78
Entente, the, 149, 161
Erckner, 232
Esebeck, General von, 182
d'Esperey, F, 130, 131

Failly, 92
Falkenhayn, E von, 135, 136, 137, 138, 139, 141, 146, 147
Farrugut, Admiral DG, 58
Ferdinand, Archduke, 11, 12, 13
Ferdinand-Maximilian, 87
Ferdinand VII, King, 25
Flanders, 153, 155
Flavigny, General, 174
Floing, 92, 97, 101
Flying Fish Channel, 243, 244, 247
Foch, General F, 130, 131, 133, 147, 154–55, 159, 162, 163
Fort,
  Douaumont, 135, 138, 139, 141, 142, 143, 146
  Belgian, 135
  Russian, 135
  Souville, 135, 144, 145, 146
  Tavannes, 135, 144, 145
  Vaux, 135, 143, 146
Fox, General, 249
France, Invasion of, 164–75, 177
Franco-British War, 11
François, von H, 107, 115, 116
Frederick, 73, 80
Frederick-Charles, Prince, 93
Frederick, Prince, 28
Frederick, Prince of Scheswig-Holstein, 97
Fredericksburg, Battle of, 46, 47
Frederick-Wilhelm, Crown Prince, 89
French,
  Air Force, 146, 156, 170, 172
  Chamber of Deputies, 88, 154
  Plan 17, 121
  Press, 88, 246
  Resistance, 141
  Revolution, 11, 23
  War Ministry, 124
French, Commander Sir J, 122, 126, 127, 130, 131, 151
Frenois, 97
Fritsch, General von, 166
Froeschwiller, 89
Frogenau, 119
Frossard, General, 89
Fuka, 199, 204
Fürstenwalde sector, 232

Gallieni, JS, 102, 121, 124, 125, 126, 127, 128, 130, 132, 133
Gallifet, General de, 101
Gallipoli, 135
Gaulle, General C de, 175
Gause, 198
Gazala Line, 187–188
Gellhorn, M, 250
Georges, General G, 172
Germany, 23, 28, 87, 89, 105, 121, 149, 165, 166, 194, 207, 208, 217, 220, 224
  Press, 143, 159
  Reich, 166
  North German Confederation, 87, 94, 115
  Reich Chancellery, 234
  Reichstag, 232, 233, 234
Ghent, 29
Giraud, General H, 172
Gneisenau, General Graf von, 30, 34
Goldbach, Valley of, 14, 16, 17, 21
Got el Ualeb, 188, 189
Gott, Brigadier, 195, 198, 199
Gransard, General, 172
Grant, General US, 56–69, 207
Griffiths, R, 138
Guadalcanal, Battle of, 247
Guderian, C, 165
  HS, 164–175, 207, 219, 224
Guingand, Brigadier de, 197
Gumbinnen, 107, 108, 109, 110

HA Bass, 247, 249
Haig, Field Marshal Sir D, 145, 148–163
Halder, General F, 167, 168, 170, 175, 182
Halsey, Admiral, 239
Hancock, Major General WS, 75, 84
Hanover, 22, 110
Han River, 244, 251, 252
Hawaii, 242

Hazebrouck, 153
Heidenheim, 177
Heinl, R, 251, 253
Heinrici, 228, 229
Heth, Major General H, 74, 75
Hill, Lieutenant General APH, 54, 74
Himeimat, Mount, 194, 198
Hindenburg, Field Marshal P von, 104–119, 141, 163
Hitler, A, 105, 166, 167, 168, 170, 175, 177, 182, 185, 187, 191, 194, 204, 205, 207, 210–212, 217, 219, 220, 223, 224, 228, 233, 234, 237
Hodges, General, 207
Hoffmann, General M, 107, 109, 110, 119
Hohenzollern, canal, 232
  Dynasty, 88, 106
  Sigmaringen, 88
Hooker, Major General J, 46, 47, 48, 49, 50, 51, 52, 54, 55, 71–73, 84
Horrocks, 199
Hoth, General, 174, 175
Hötzendorf, C von, 105
Houffalize, 219, 220
Hougoumont, 35, 37, 40
Howard, Major General OO, 51, 75

Iglau, 13, 16
Inchon, amphibious operation, 238–253
India, 25, 34, 151, 187, 193, 195
Isabella, Queen of Spain, 88

Jackson, 60, 61, 68, 69
Jackson, General TJ, 45, 46, 47, 48, 49, 50, 51, 52, 53, 54, 55, 68, 207
Japan, 240, 242, 244, 246
Joffre, General JJC, 120–133, 135, 137, 138, 139, 141–142, 144–146, 147
John, Archduke, 11, 13
John, Prince of Liechtenstein, 17, 20, 22
Johnston, General JE, 60, 61, 62, 67, 69

Kaliningrad see Königsberg
Kazakov, Colonel General VI, 231
Keitel, General W, 165, 167–68, 233
Kidney Ridge, 202
Kienmayer, Lieutenant General, 17, 20, 22, 23
Kiev, Battle of, 223
Kimpo airfield, 243, 244, 251, 252
Kitchener, Lord, 127, 128, 162
Kleist, General E von, 168, 170, 171, 174, 175
Klopper, General, 189, 190, 191
Kluck, General A von, 122, 123, 125, 126, 127, 128, 130, 131, 133
Kluge, General, 168
Knightsbridge, 188
Koblenz, 108, 109, 110
Kollowrath, General, 17, 20, 21
Koniev, 225, 226, 231, 232, 233, 237
Konigsberg, 108, 111, 224
Kunsan, 243, 244, 252
Kursk, Battle of, 236
Küstrin, 225
  sector, 227, 228, 229
Kutuzov, General MI, 11, 12, 13, 14, 17

La Fère, 151, 153
Lafontaine, General, 172
La Haie Sainte, 35, 37, 40, 42
Langeron, General A, 17, 20
Lannes, J, 13, 16, 20, 21
Lanrezac, 122, 125, 127, 130
Last Cartridges, Battle of the, 101
Leboeuf, 89
Lebrun, 92, 94, 95, 101
Lee, General RE, 44–55, 57, 62, 71, 74, 75, 78, 80, 81, 85, 207
Leeb, General W von, 168
Leipzig, 225, 229
Lelyushenko, 232
Leningrad, Battle of, 223, 252
Libya, 177, 195
Liddell Hart, Sir B, 131, 137, 152, 154, 159, 162, 163, 166, 168, 207
Ligny, 34, 36

Lincoln, A, 45, 46, 47, 55, 57, 58, 68, 71, 72
  Emancipation Proclamation, 68
List, General, 168, 170, 175
Lithuania, 105, 141
Little Round Top, 76–78, 85
Lloyd George, D, 149
Lockett, Colonel HS, 63
Lorraine, 121, 122, 208
Louis XVIII, King, 25, 28
Luce, River, 160
Ludendorff, Major General E, 109, 110, 116, 119, 141, 149–150, 151, 152, 154, 157, 162, 163
Lumsden, General, 197, 203
Lüttwitz, General H von, 215
Luxembourg, 121, 170, 213

MacArthur, A, 239
MacArthur, General D, 238–253
Mackensen, A von, 107, 115
Mack, General, 11, 12, 13
MacMahon, Marshal, 91, 92, 93, 95
Maginot Line, 167, 170
Malta, 185, 191, 193, 194
Mangin, C, 142, 146, 147, 154
Manstein, General.E von, 165, 167, 168
Manteuffel, General H von, 210, 212, 213, 215, 217, 219
Marburg, 13
Marmont, AFLV, 13, 25
Marne, Battle of the, 119, 120–133, 138, 154, 165–166
  First Battle of the, 154, 163
  Second Battle of the, 162
Marshall, General GC, 207–208
Martelange, 216
Maryland, 71, 73, 74, 80
Mason-Dixon Line, 74
Masurian Lakes, 105, 109, 115
Maunoury, 128, 131, 132, 133
McAuliffe, General, 215, 217
Meade, Major General GG, 68, 70–85
Medal of Honor, 177, 240
Mediterranean, 178, 185, 194
Mersa Matruh, 193, 205
Metz, 87, 89, 91, 92, 208, 209
  war school, 165
Meuse, Battle of, 164–175
  Heights, 135, 137
  River, 92, 93, 94, 100, 102, 130, 136, 137, 146, 151, 167, 168, 170, 171, 174, 212, 215
Mexican War, 45, 46, 57, 67
Mexico, 87, 143
Mézières, 94, 100
Milliken, 216
Miloradovitch, General, 17, 20, 21
Mississippi River, 57, 58, 60, 62, 67, 68, 69
Missouri, 252
Miteirya Ridge, 194, 202, 203
Model, Field Marshal W, 210, 217
Moltke, Baron HCB von, 86–103, 167
Moltke, H von, 106, 108, 109, 110, 115, 119, 121, 123, 125, 128, 133, 163, 167
Monash, General Sir J, 157
Mons, Battle of, 122, 151
Montdidier, 155, 160
Montgomery, D, 193
  E, 193, 196
  Field Marshal BL, 67, 117, 192–205, 207, 208, 209, 212, 213, 217, 219, 225
Mont Saint Jean, 29, 30, 34, 35, 42
Moravia, 11, 23
Moravia, Mountains of, 14, 17
Moreuil, 149, 155, 160
Mort-Homme Ridge, 141, 142
Moscow, 225, 226, 234
  Battle of, 236, 237
  victory parade, 237
Moselle River, 121, 170, 208, 217
Mt McKinley, 246, 247, 248, 250, 251, 252
Mükden railroad station, 106, 109
Murat, J, 13, 16, 17, 21
Murray, Lieutenant Colonel R, 252
Mussolini, B, 191

Namur, 29, 30, 167, 168
Napoleon III, Emperor CL, 87, 88, 89, 91, 100, 101–102
Narocz Lake, 141
Near East Command, 195
Neidenburg, 115, 116
Neisse river, 225, 226, 227, 229

Netherlands, 28, 29
Neufchateau, 171, 217
New Orleans, 58, 68
Ney, Marshal M, 34, 37, 40
Nicholaievich, Grand Duke N, 105, 106, 119
Nicholas, Czar, 105, 108, 119, 123, 135, 137, 141
Nile river, 191, 193, 194
Nivelle, General R, 141, 142, 143, 146, 147
Nivelles, 29, 30, 34
Normandy landings, 191, 196, 205, 207

Oase, 179
Observatory Hill, 250
Oder river, 223, 224, 225, 227, 236
Okinawa, Battle of, 247, 250
Olly Farm, 92, 94
Olmütz, 11, 13, 14, 16
O'Neill, Chaplain JH, 216
Operation,
  Autumn Fog, 210
  Battleaxe, 185
  Bluehearts, 242
  Brevity, 185
  Chromite, 242, 243, 244, 246, 252, 253
  Crusader, 183, 185, 187
  Greif, 213
  Lightfoot, 196, 202
Orange, Prince of, 28, 29
Oudinot, 16, 20, 21

Palestine, 194, 195
Palestrino Ridge, 189
Palmi Do, 245, 247
Parallel, 38th, 241, 252
Passchendaele, Battle of, 147
Patch, General A, 208
Patton, General GS, 206–221
Pemberton, Lieutenant General JC, 60, 61, 67, 68
Peninsula War, 25, 35, 36, 42, 147
Pennsylvania, 62, 71, 73, 74, 80, 84
Perkhorovich, 232
Pershing, General, 162, 240
Pétain, Marshal HP, 134–147, 153
Philippines, 240, 241
Plan Gelb see France, invasion of
Plan Sichelschnitt see France, Invasion of
Poland, 105, 107, 167, 177, 207, 224
Polish Air Corps, 229
Pomerania, Western, 225
Posen see Poznan
Pourcelet, Colonel, 172
Poznan, 224, 225
Prague, 233
Pratzen, plateau, 14, 16, 17, 20, 21, 22
  village, 14, 20
Prittwitz, General M von, 107, 108–109, 110, 119
Przbyswski, General, 17, 20, 22
Puller, Colonel LB, 250, 251, 252
Puntowitz, 16
Pusan, 242, 245, 248
  Perimeter, 239, 244, 247

Quatre Bras, 29, 30, 34, 35

Raeder, Admiral, 185
Random House Dictionary, 85
Rappahannock River, 48, 54, 55
Rawlinson, 154, 155, 159, 160, 162
Reims, 152, 153, 215
Reinhardt, General, 168, 174–175
Rennenkampf, PK, 106, 107, 108, 109, 110, 111, 115, 116, 117, 119
Reynolds, Major General JF, 71–73, 74, 75, 84, 85
Rhee, S, 241, 252
Rhine, River, 87, 88, 121, 208, 209, 212
Richmond, 45, 48, 49
Richthofen, General von, 165
Ritchie, General, 187, 188, 189, 191, 193
Rokossovsky, General K, 224, 225, 226, 229, 231, 232, 237
Rommel, Field Marshal EJE, 45, 163, 171, 175, 176–191, 193, 194, 196, 197, 198, 199, 200–201, 202, 203–204, 205, 207
Roon, von, 87, 88, 97

Roosevelt, T, 194, 240
Rudenko, Colonel General S, 229
Rundstedt, Field Marshal G von, 167, 168, 170, 172, 175, 210, 211, 217
Rupprecht, Crown Prince of Bavaria, 152, 157
Russia, 11, 13, 28, 105, 121, 135, 136, 144, 149, 162, 184, 185, 220, 241
Russian,
  Air Force, 231, 232
  Revolution, 141
  Stavka, 225, 228, 237
  16th Air Fleet, 229, 233
  18th Air Fleet, 231, 233
Russo-Japanese War, 103, 106
Ruweisat Ridge, 194, 197

Saarbrücken, 89, 214
Saarlautern, 214, 217
Saar, offensive, 216
  Valley, 208, 209, 215, 219
Saint-Hilaire, 16, 20, 22
Samsonov, A, 106, 108, 109, 110, 111, 115, 116, 117, 119
Santon Hill, 14, 16
Savary, General AJMR, 14, 16
Schlieffen, General Count A von, 105, 106, 107, 119, 121, 123, 126
  Plan, 121
Schlippenbach, Lieutenant Freiherr von, 183
Scholtz, 107, 115
Sedan, Battle of, 86–103, 123
Sedgwick, Major General J, 48, 49, 54, 55
Seelow Heights, 229, 231
Seminary Ridge, 75
Senlis, 128, 133
Seoul, 241, 242, 243, 244, 252, 253
Shepherd, Lieutenant General LC, 242, 249
Sherman, General WT, 57, 58, 59, 62, 63
Sickles, Major General DE, 75, 76, 78, 85
Sidra ridge, 188
Siege of Paris, 102, 127
  of Vicksburg, 62–68
Silver Star, 239, 240
Slocum, General H, 80
Smith-Dorrien, 122, 125
Smith, Major General OP, 244, 246, 247, 251, 252
Snyder's Bluff, 58
Soissons, 149, 153, 165
Sokolovsky, General V, 228, 234
Sombreffe, 29, 30
Somme, 122, 144, 145, 146, 147, 154, 157
  Battle of the, 147, 152, 162
Souilly, 138, 147
Soult, Marshal NJ de Dieu, 11, 13, 16, 17, 20, 22, 25, 37
Stalingrad, Battle of, 223, 236
Stalin, J, 199, 223, 224, 225–226, 227, 228, 229, 231, 233, 234, 236–237
Stallüpönen, 107, 109, 115
Staré Vinohrady, 14, 20
St Gond, 130, 131
St Hubert, 214, 217
St Petersburg, 105, 108
St Quentin, 152, 153, 175
Strasbourg, 87, 89
Struble, Vice Admiral AD, 241, 249, 252
Stuart, Major General JEB, 47, 48, 54
Stumme, General, 202, 203
St Vith, 219, 220
Submarine warfare, 149, 194
Sudetenland, 167

Talavera, Battle of, 25
Tannenberg, Battle of, 104–119
Tabb, von der, 93, 94, 97
Taplett, Colonel R, 248
Telegin, Lieutenant General KF, 224
Telnitz, 16, 20
Thoma, General R von, 203, 204, 205
Tiergarten, 233, 234
Tobruk, Battle of, 176–191, 194
Torgau, 232, 233
Treaty, Franco-Russian 1892, 105, 121
Tripoli, 177, 178, 183
Truman, H, 242, 246, 253

Union Navy, 58, 60, 64, 69
United Nations, 239, 241, 253
USA, 57, 162, 184, 199
  Air Force, 240, 242
    P-61 Squadron, 216
    XIX Tactical Air Command, 216, 219
  Joint Chiefs of Staff, 242, 243, 244, 245, 248, 253
  Joint Strategic Plans and Operations Group, 242–243
Usdau, 115
US Navy, 149, 242, 243, 244, 245, 248, 250, 251
  Air Arm, 240
  Amphibious Group 1, 242, 244
  Amphibious vessels, 241
  Battalion, 1st, 245, 249, 259
  Fleet Marine Force, Pacific, 242
  Forces, Far East, 241
  Marines, 245, 247, 248, 249, 250, 251, 252, 253
  Marine Corps, 240–241, 243
  Seventh Fleet, 241, 248
  1st Marine Air Wing, 244
  1st Marine Division, 242, 244
  1st Marine Regiment, 244, 247, 250, 251
  2nd Battalion, 249, 250
  3rd Battalion, 247
  5th Battalion, 249
  5th Marine Regiment, 244, 245, 247, 249, 251, 252
  5th Regiment, 247
USSR see Russia
Uxbridge, Earl of, 28, 29
Uxbridge, Lord, 35–36, 37

Vandamme, 16, 17, 20, 21, 22
Vasilievsky, Marshal, 223
Verdun, 133, 135, 146, 208, 213
  Battle of, 134–147, 149
Versailles, Treaty of, 166
Vicksburg, 57, 58, 60, 61, 62, 66, 67, 68
  Battle of, 56–69
Victoria Cross, 177, 193
Vienna, Congress of, 28
Vimiero, Battle of, 25
Vistula river, 106, 108, 223, 225, 227
Vitoria, Battle of, 25

Walker, General WH, 241, 243, 244, 252
Wantuck, 247
Warnach, 216
Warsaw, 167, 223, 224
  Offensive, 223
Washington, 58, 71, 74, 240, 241, 242, 245, 251, 252
Waterloo, Battle of, 25–45
Wavell, General Sir A, 177, 178, 185
Wavre, 34, 36
Weidling, General, 233, 234
Wellesley, A see Wellington, Duke of
Wellington, Duke of, 25–45, 147, 163
West Point Military Academy, 45, 46, 57, 62, 63, 71, 207, 239, 240
Weyrother, Major General, 14, 17, 20, 22
Wilhelm, Crown Prince, 136, 139, 141, 143, 152
Wilhelm II, Kaiser, 88, 101, 102, 106, 107, 110, 127, 135, 141, 149
Wimpffen, 92, 95–97, 100, 101, 102
Woevre, Plain of, 135, 137
Wolmi Do, 243, 246, 247, 250

Yalta conference, 237
Yalu River, 252, 253
Yazoo River, 58
Yegorov, Sergeant, 234
Yellow Sea, 242, 248
Yongdungpo, 252
Ypres, 149, 151
  First Battle of, 193

Zhilinski, General YG, 105, 110, 111
Zhukov, Marshal G, 222–237
Zinchenko, Colonel, 234
Zokolnitz, 16, 20
Zurlan hill, 16, 17, 20
Zwehl, General, 162, 163